Second Revised and Enlarged Edition

Fundamentals of Cartography

Available text-books by the same author:
— Research Methodology: A Handbook
— Regional Planning: Concepts, Techniques, Policies and Case Studies
— Geography of Health: A Treatise on Geography of Life and Death in India

About the Author

Prof. R. P. Misra (*b.* 1930) is a well known geographer, urban and regional planner, and Gandhian scholar. After his MA in geography from Allahabad University (1954), he worked in the Lok Sabha Secretariat as Assistant (1954-57); National Atlas Organization as Field Officer (1957-60); and Lecturer in Geography, BHU (1960-61). He did his PhD at the University of Maryland, College Park, USA under the Fulbright programme of the USEFI in 1964. On return from the USA, Prof. Misra headed the Post-graduate Department of Geography as Reader (1961-68) and Professor (1968-75). He established the Institute of Development Studies in Mysore University in 1972 and remained its Director until 1979. During this period, he was consultant at UNRISD, Geneva, and UNDP, Teheran, Iran. Then he proceeded to Nagoya, Japan as Vice Director of UN Centre for Regional Development (1979-84). He returned to India as Vice Chancellor of Allahabad University (1984-87). On completion of his three-year term as VC, he was invited as Professor of Geography, and Director, Center for Gandhian Studies and Peace Research, Delhi University (1988-95). Since 1996, he is engaged in the promotion of Gandhian Studies and Peace Research, Environmental Planning and Sustainable Development.

Chapters on the newer nuances of cartography have been contributed by R.B. Singh, Associate Professor of Geography, Delhi University and Vice-President, IGU; Brijesh Misra, Director, Information Technology, Enbridge Gas Distribution, Toronto; and Anupam Pandey, Assistant Professor of Geography, Allahabad University.

Fundamentals of Cartography

Second Revised and Enlarged Edition

R.P. Misra

With Contributions from

R.B. Singh, Brijesh Misra and **Anupam Pandey**

CONCEPT PUBLISHING COMPANY PVT. LTD.

NEW DELHI-110059

© R.P. Mishra

All rights reserved. No part of this work may be reproduced, stored, adapted, or transmitted in any form or by any means, electronic, mechanical, photocopying, micro-filming recording or otherwise, or translated in any language, without the prior written permission of the copyright owner and the publisher. The book is sold subject to the condition that it shall not, by way of trade or otherwise, be lent, resold, hired out, or otherwise circulated without the prior publisher's written consent in any form of binding or cover other than that in which it is published.

The views and opinions expressed in this book are author(s) own and the facts reported by him/her have been verified to the extent possible, and the publishers are not in any way liable for the same.

ISBN-13 : 978-93-5125-014-2 (HB)
ISBN-13 : 978-93-5125-015-9 (PB)

First Published 1969
First Revised and Enlarged Edition 1989
Second Revised and Enlarged Edition 2013

Published and Printed by

Concept Publishing Company Pvt. Ltd.
A/15-16, Commercial Block, Mohan Garden
New Delhi-110059 (India)
T: 25351460, 25351794, *F*: 091-11-25357109
E: publishing@conceptpub.com, W: www.conceptpub.com

Editorial Office: H-13, Bali Nagar, New Delhi-110 015, India.

Cataloging in Publication Data--*Courtesy:* D.K. Agencies (P) Ltd. <docinfo@dkagencies.com>

Misra, R. P. (Rameshwar Prasad), 1930-
Fundamentals of cartography / R.P. Misra ; with contributions from R.B. Singh, Brijesh Misra and Anupam Pandey. — 2nd rev. and enl. ed.
p. cm.
Includes bibliographical references (p.) and index.
ISBN : 9789351250142 (HB)
ISBN : 9789351250159 (PB)

1. Cartography. 2. Cartography—Data processing. I. Title.

DDC 526 23

Preface

This book was first published over 40 years ago in 1969 by *Prasaranga* (Extension Unit), University of Mysore. At that time I was associated with Mysore University as Professor and Chairman, Department of Post-graduate Studies and Research in Geography. My Mysore days (1965-79) were highly rewarding academically, professionally and personally. If I have been able to do something that added to the existing knowledge in the field of geography, development studies and Gandhian philosophy, it is because of the opportunity I had to serve the University of Mysore.

After a three year (1954-57) stint as Assistant in Lok Sabha Secretariat, New Delhi where I learnt how the Parliament of India functions and how to draft reports on the working of various Ministries of the Government of India, my geography background pulled me back to it and I joined the National Atlas Organization [now National Atlas and Thematic Mapping Organization (NATMO)], Ministry of Scientific and Cultural Affairs (1957-61) as a Field Officer, at its Dehradun office. My stay in Dehradun enabled me to secure MA in economics from the local DAV College. And then in 1961, I proceeded to the University of Maryland, College Park, USA for doctoral research. The United States Educational Foundation in India (USEFI) then headed by Dr. Olive I. Reddick offered me a Travel Grant, and the Department of Geography, University of Maryland, then chaired by Late Professor William van Royen offered a Graduate Assistantship to pursue my research. I am so grateful to Dr. Reddick and Prof. van Royen for the help they gave at that critical juncture of my life. But for them, I would have never been writing and updating this book today.

In view of my rural background, I decided to take up the study of House Amish of Lancaster County, Pennsylvania. The House Amish are German migrants; they live even today as their forefathers lived in nineteenth century; even their farming had remained traditional and antique even though green and sustainable. They did not use chemical fertilizers, nor did they use tractors and other agricultural machines. Yet per acre yield in their farms was higher than that is mechanized farms in 1960s, India was at the initial stages of Green Revolution, a revolution, which proved to be Red for all those regions which were rain fed and all those people who did not have access to land. It did, however, make the country self-sufficient in food. I being the son of a farmer, had interest in agriculture and rural development. It occurred to me that Indian farmers could also produce more if they followed the Amish approach to farming. So I visited Lancaster County, stayed there with an American friend for a week and saw closely how the House Amish lived and pursued their main occupation: agriculture.

The House Amish were the most traditional of the farmers in the USA. There were other groups such as Amish, and Mennonites who also followed traditional life style and agricultural practices of their German forefathers. House Amish were, however, the most traditional among them. They were called House Amish because they worshipped at home. They did not go to Church. They dressed the way their ancestors dressed in 18th and 19th centuries; they had horse *buggies* for transport and plowed the land with the help of horses. The other groups were more liberal. Among the Mennonites, there was a group which used cars but painted the bumpers black. The House Amish had their own schools and they did not pay taxes to the government. To sum up, the House Amish pursued not only sustainable agriculture but also sustainable life style. I am told that they continue to follow those practices even today.

I completed my Ph.D. in 1964 under the joint auspices of the departments of Geography and Agricultural Economics, University of Maryland at College Park. The topic of my research was Agricultural Development in a Traditional Society: Case Study of House Amish, Lancaster County, Pennsylvania. After completing the work, I was offered a teaching assignment in a neighbouring college but I decided to return to India in 1965. Many of the colleagues who went with me under the Fulbright Grant stayed back, married, and settled in USA. I had my wife (Rani Misra) and four children back home. My heart prevailed over my mind and I decided to set sail and return home. I stayed in Mumbai for 15 days with my father, who was then working with the Bombay Grain Dealer's Association. I saw an advertisement by Mysore University for the post of Professor of Geography; I decided to apply for it. After an interview, I was offered the post of Reader and Head of the Department of Post-graduate Studies and Research in Geography.

At Mysore, I developed the Department of Post-graduate Studies and Research in Geography at the Manas Gangotri Campus of the University and within seven years made it the mother department of all Universities in South India. In 1968, I launched a Centre for Urban and Regional Studies, which by 1971 became the Institute of Development Studies (IDS), independent of Geography Department. My interest in agriculture development and cartography, continued. And this book is the testimony to my continuing interest in Cartography. A year earlier, in 1968, my first book Diffusion of Agricultural Innovations a byproduct of a research project of the same topic was recognized as the best research publication of Mysore University (1965-68). These two books gave me a good footing in Development Studies and Cartography.

Cartography of the days I learnt in 1960s has changed a lot. In fact the change is so drastic that we cannot find an old time professional cartographer who would sit with black ink, pen, scale, set-squares, and lettering set to make maps. He has now been replaced by a computer cartographer. I am product of an era when even most senior and world renowned cartographers like Prof. Arthur H. Robinson, and E. Raisz used to draft and draw maps manually. Cartography then was not only a science of converting spherical earth flat but also an art of creating maps that were not only informative and educative but also artistic.

Today, satellite imageries are available on internet at any scale and for the whole world. There are programmes to manipulate base maps to create new maps: general or specialized. The days of cartographers sitting on drawing tables to conceive, design and draw maps manually are gone. Cartographers, however, are still needed because map making is a scientific exercise; it is not the job of a layman. There are certain principles to be followed and unless the person designing a map is conversant with those principles the maps he creates may end up in telling lies, rather than truths. It is not just a matter of computer science. He has to be an ICT man who knows the science and art of cartography and can translate his knowledge into maps with the help of computers. He does not do any work manually except operating the computer.

This book consists of 27 chapters divided into four parts. Chapter 1—Introduction tells the scope of the book including the definition of cartography.

Part I: *Understanding Cartography* has eight chapters (2-8); covering nature and scope of cartography, history of cartography, problems related to representing spherical earth on a flat surface, (chapters 2-4), and four basic ingredients of maps—scale, direction, coordinates and projections (chapters 5-8).

Part II: *Sources of 'Cartographic Information'* is broken into six chapters (9-14). Topographic surveys and maps produced therefrom are discussed in chapters 9-11; aerial photo and their interpretation in chapter 12 and collection and processing of statistical data in chapters 14.

Part III: *Map Design and Construction* covers aspects like map design and layout (chapter 15); mapping the terrain, weather and climate data, and socio-economic data are discussed in chapters 15-19. Thematic and Complex mapping is the subject matter of chapter 20; and maps for children, visually challenged, and neo-literates are discussed in chapter 21. Chapter 22 covers problems related to map reproduction, cataloguing and storing.

Part IV: *Automation in Cartography* consists of five chapters (23-27) and explains the nuances of computer cartography, remote sensing and satellite imageries, GIS, LIS, GPS (chapters 24-25). International cooperation in cartography is the subject matter of chapter 26. The last chapter (27) points to emerging trends in cartography and role of maps in planning, development, business, transport and communication.

The book covers all those aspects of cartography which were important in 1960s as well as those which have acquired importance in the first decade of 21st century. Thus book presents a coherent picture of cartography from what was essentially an applied field of making maps to the present stage where theory and automation reign supreme. The cartography as it was practiced up to 1960s provides the basic knowledge, which one must have to create and read maps; it also helps understand the evolution of cartography as a science of visual communication. One who does not know scales and projections cannot understand maps. One who does not know land and air surveying, which were and still are sources of map information cannot understand remote sensing. Similarly, one who does not know the methods of analyzing primary and secondary data cannot make maps, which are informative , analytical and asthetic at the same time.

I am confident that the book would prove useful to the students and teachers of earth sciences in general and geography and cartography in particular and help the reader better appreciate maps so often used in teaching, research, transport, communication, business, and industry.

Allahabad
January 1, 2012

R.P. Misra

Acknowledgements

Fundamentals of Cartography is the product of my sincere efforts to help students of geography, geology, economics, architecture, and town planning to present their research reports better and more illustratively. The first edition of the book was published in 1969, and the second revised and enlarged edition in 1989. The third revised and enlarged edition has been brought out to incorporate latest developments in the field such as Remote Sensing, GIS, GPS, and computer mapping. A map can tell far more about the location and distribution of a phenomenon than many pages of text.

The source material for the book has come from the students I had the opportunity to teach for the last sixty years. But my interest in cartography and the urge to write this book came from Prof. S.P. Chatterjee, Director, NAO (Calcutta) who offered me an opportunity in 1957 to be a part of the team he gathered to make maps for the first National Atlas of India. I consider Professor Chatterjee the greatest geographer, India has produced. He was humane, sincere and highly devoted to mission he commanded i.e. to produce National Atlas of India. I owe a lot Prof. Chatterjee and take this opportunity to express my gratitude to him. Among others who prepared me to take up this work are: Prof. William van Royen, Chairman, Department of Geography and my Ph.D. supervisor (1961-64), University of Maryland, College Park, USA; Prof. Phillip W. Foster, Department of Agricultural Economics; and Mr. Joseph W. Wiedel, Professor, Department of Geography, University of Maryland, College Park, Md. USA. Whatever, is written in the book was learnt at College Park and experimented at Mysore. Dr. A. Ramesh, Professor of Geography, University of Madras; Professors M.S. Masood, P.D. Mahadev and D.C. Jayashankar, at the Department of Geography, University of Mysore went through the first draft of the 1969 edition of the book and helped me improve the manuscript. I am so grateful to them all. I wish could thank them personally to lighten the debt I owe to them but they are no more in this world. I pray God to bless their souls with peace.

The third edition of the book would have not been possible but for the contributions by Dr. R.B. Singh, Professor, Department of Geography, DSE, Delhi

University; Sri Brijesh Misra, Senior Manager, Business Applications, Enbridge Gas Distribution, Canada; and Dr. Anupam Pandey, Assistant Professor of Geography, Allahabad University. In fact, most of the chapters of Part IV of this book dealing with Automation in Cartography have been prepared by them. I appreciate their contribution and thank them for updating the book.

I know this is the last edition of this book. My academic agenda for the next few years is far too crowded to enable me to bring out another edition of the book. Much of my time is now devoted to Gandhian studies and peace science. I am, however, confident that some geographers and cartographers would pick up the threads, and write new and better books on the subject in years to come. People come and go but academic disciplines never die. The depth and breadth of cartography as a theoretical, analytical and applied science is expanding fast. Cartography has moved far beyond the limited circle of making maps manually or mechanically. The era of virtual maps has arrived. The popularity of cartography and its various facets and products would increase as they become readily available through internet, hand-held computers and cell phones. This book is a sort of bridge between manual and digital cartography and would continue to help the future map designers serve humanity with better and more accurate maps.

Revision of the book has been an arduous task. Eyes get tired, as do the hands and mind. I wish I were 20 years younger. But I know that some wishes cannot be turned into reality. There are number shortcomings in the book; they are left untouched not because they can be ignored but because they need more work and time than what I can spare. I have depended on my assistants to brows through the internet and update the text. It is possible that some ideas or even sentences have remained unacknowledged. The illustrations included in the book are the same which were there in the first edition of the book. They ought to have been replaced by new and more relevant ones. Further, some of the material relevant in 1969 is no longer relevant today. I did not remove them for they formed a bridge between what cartography is today and what it was 50 years back.

I am confident that cartography would soon acquire a new status among the academic disciplines in India and new books with greater depth and broader coverage would come out. I would be happy indeed if these books are far better in content, style, depth and breadth than the *Fundamentals of Cartography*, the book you, dear reader, have in your hand.

Allahabad
January 1, 2012

R.P. Misra

Contents

PART III
MAP DESIGN AND CONSTRUCTION

PART IV
IMPACT OF INFORMATION TECHNOLOGY ON CARTOGRAPHY

List of Figures

List of Tables

1
Introduction

While planning for the new geography building at the University of Mysore, I asked the engineer to set aside space for a cartographic laboratory measuring 12 m x 8 m. "I can provide you with only one room of this size; may I suggest, sir, that you use it for mapping and fix the catastrophe (sic.) laboratory in a smaller room", was his reply. I was surprised and shocked at such a characterization of cartography, but the message was quite clear to me. Maps are certainly more popular than cartography, the science and art of making maps. One should not, then, be hard put to understand why this book on cartography begins with a discussion on maps.

WHAT IS MAP?

A map is, usually considered to be a drawing to scale of the whole or a part of the surface of the earth on a plane surface; it is a manually or mechanically created visual model of selected elements of the whole or part of the earth surface. The subject matter of maps is not limited to earth surface; one can have maps of the moon and other celestial bodies as well. We already have maps depicting the surface of the moon and the days are not far off when we would have as detailed maps of the moon or other planets as we have for the earth.

A map gives a picture of one or several phenomena of the earth's surface but such a picture is not the same as a photograph. Photographs show all visible details of the area photographed irrespective of their relevance to the purpose for which they have been taken. The photographer has no control over the selection of the objects within the focal range of the camera. A photograph, no matter from which level it is taken, shows details in their visible shapes and sizes.

Moreover, a photograph shows only those objects which are physically present and visible from the place a photograph is clicked. It cannot, for example, represent a theoretical surface devoid of visible undulations such as income per capita or population distribution. A map, being the mental and manual creation of man, gives all those details which its creator wants to give. Instead of showing the details in there true or visible shape and size, it uses symbols which may or may not have similarities with the shape and size of the objects mapped. Further, it can show facts which are not physically present but are mental images of the reality. It is clear from the land use map of Palni (Fig. 1.1) that many details one may find in a photograph of the city taken from the airplane and satellite have been removed, suppressed, or depicted by symbols. This is the unique characteristic of a map. Map is more concerned with the location and spread (distribution) of earth features both natural and man made than with truly depicting whatsoever is tangible or visible. For example, it may show boundaries which may not be marked on the ground. It may show a pattern derived from the statistical analysis of data but which cannot be seen on the ground. A correct definition of map must, therefore, incorporate all these characteristics. We should, therefore, redefine a map as a symbolic drawing to scale of the visible as well as conceived locational and distributional patterns of the whole or a part of the earth, the sky, or any other heavenly body.

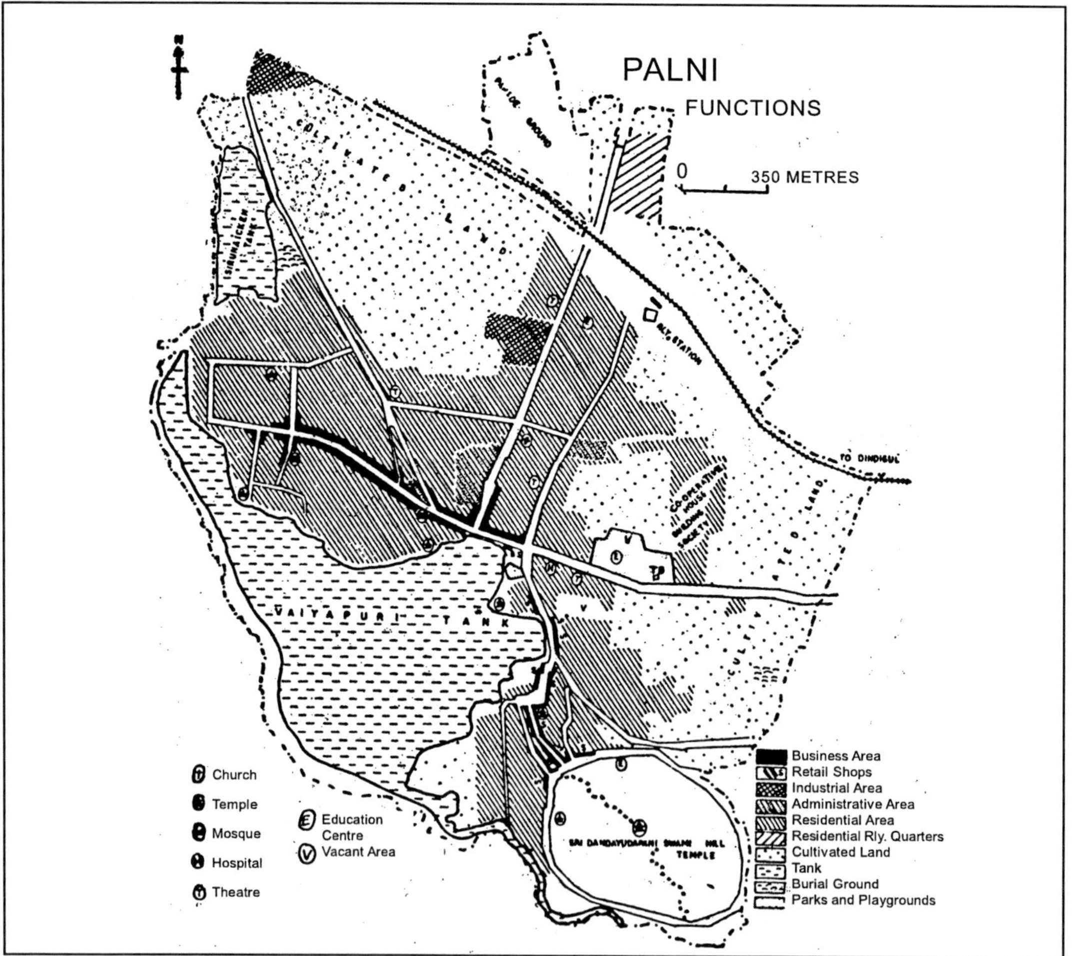

Fig. 1.1: Land Use Map of Palni Town

We associate maps with the earth and not with other heavenly bodies mainly because we were not really concerned with the mapping of the latter until very recently and also because most of the maps, we see deal with the earth and the earth alone. In fact the history of cartography shows that the scope of the term map has been earlier restricted to the portrayal of the earth's land surface alone. In such maps water bodies were shown only incidentally to maintain the desired relationships among the land masses. The maps which gave prominence to water features of the earth's surface were called charts. Charts showed the bathymetric details, water currents, and other details needed by navigators. Such charts are still produced and used. With the introduction of air transportation, a new series of charts, showing details useful for air navigation, have come into being. These charts are often topographical maps with navigational aids superimposed on them.

Certain maps are meant to give very minute details about very small areas like villages and towns. These maps are often called cadastral maps or plans. For example, the city plans show the streets, railroads, water features, prominent buildings, and residential and other types of land uses. The chief purpose of a plan is to show boundaries of property ownerships and the layout of the land, roads, and parks. The cadastral maps are meant to portray in detail the land ownership boundaries in rural areas. Each village in India has a cadastral map that shows the plot numbers and boundaries.

The distinction between maps, charts, and plans is not as clear-cut now as it used to be earlier. For all practical purposes charts and plans are also maps, which portray specialized information on larger scales. It is for this reason that the term map is used in this book to include plans, sketches, diagrams, cartograms, graphs, three dimensional models and globes. A diagram is a drawing designed to show inter-relationships by means of lines, columns, dots, and other symbols. Some of the diagrams are highly simplified and show only the essential elements of the data. Many diagrams are designed to give a combined effect of graphical and pictorial media for an orderly and logical visualization of the relationships between facts and ideas. Diagrams showing interrelations, growth, development, typology, and organization are quite common.

A graph is also a cartographic product. It provides a visual representation of the numerical data by lines. A table of figures may contain a wealth of valuable information but it may mean nothing to a reader who is not conversant with statistical methods. When presented in a graphical form, the data contained in the table become easy to understand. A graph reveals the interrelationships such as trends, and variations from the normal, and therefore, is inherently more interesting. At times diagrams and graphs are superimposed on maps to show the spatial distribution of the data represented by them. Such maps are called cartograms.

A rough drawing of a landscape without scale and co-ordinates and without much concern for details is called a sketch. A sketch is often made of objects or areas for which pictures and maps are not available or cannot be made available readily. Sketches can be drawn hurriedly and can be very valuable graphic aids for researchers. Sketch maps of unsurveyed areas are of extremely useful. For explorers and socio-economic geography surveyors they are indispensable.

Although relief models and globes are also considered to be cartographic products, they differ from maps as they are not done on a plane surface. They, nevertheless, involve the same cartographic principles and techniques in their design and construction. The simulation of the real earth and its surface features in its true shape and details is the objective behind the construction of the three-dimensional models and globes. A globe represents the earth in its true shape. A three-dimensional model represents the whole or a part of the earth's surface showing the vertical relief as well.

TYPES OF MAPS

As each map is unique in its design, content and construction it is a type by itself. On the basis of certain common features, maps can be classified into several types. The following are some of these types:

Types by Relief Representation

On the basis of the amount of topographic details given, maps can be classified into two types:

1. Hypsometric maps, and
2. Planimetric maps

The hypsometric maps are those, which show the relief and the terrain in detail and often at the cost of other details; the large scale topographical maps produced by Survey of India fall in this category. As against these, the planimetric maps lay more emphasis on other details and limit the relief portrayal to a few spot heights here and there. Most of thematic maps representing the cultural features of the landscape fall in this category.

Types by Scale

Taking the scale as the criterion, maps can be classified as:

1. Small scale,
2. Medium scale, and
3. Large scale.

The classification appears to be the easiest one. The terms large scale and small scale are, however, so undefined that there is no universally accepted standard for classifying maps according to scale. What one considers to be large, may appear to be small or medium for others. The same person may consider a map to be large scale for one purpose but small scale for another purpose. As a result, specialized groups of map users set up their own standards for classification.

For the purpose of this book, maps having scales of 1:63,360 or more are classified as large scale maps; those falling between 1:63,360 and 1:1,000,000 as medium scale maps, and those having scales below 1:1,000,000 are treated as small scale maps. The million sheets of the Survey of India and the National Atlas of India are considered to be medium scale maps.

Types by Information

Some maps are general in nature; they give a variety of information ranging from administrative boundaries to roads, and land cover. Some others are thematic nature and give only one type of information. There are still others, which are special purpose maps meant to be used by groups having special perceptual problems such as children, visually challenged, etc. Maps can thus be classified into the following types as well:

1. General purpose maps,
2. Thematic maps, and
3. Special purpose maps.

Among the general maps are included topographical sheets, and wall maps designed to convey information people in general need. Maps dealing with specific themes such as geology, climate, crops, population density, etc., are called thematic maps. The special purpose maps are those which are constructed for a group of people having special reading or perceptual problems. Thus, a map for the visually challenged person will fall in this category. Similarly, the maps for the children and neo-literates are also special purpose maps.

Types by Military Use

There are certain maps which are drawn specifically for the use by the armed forces. From a soldier's point of view, maps can be classified as:

1. General maps,
2. Strategic maps,
3. Tactical maps, and
4. Photomaps.

General maps depict only the broad topographic features and are usually used by the high command for general planning purposes. A map on a scale of 1:1,000,000 or higher is considered to be a general map. Map on a scale ranging from 1:1,000,000 to 1:500,000 are usually classified as strategic maps. These maps are used for the general planning of more concentrated military action. Maps on scale on 1:500,000 or lower are called tactical maps. A tactical map serves as a guide to small units like battalions and patrol units prior to and during the movement near the front line. These maps give the relief and planimetric data and are used in tactical planning by smaller combat units. At times maps having scales of 1:250,000 to 1:500,000 are called strategic-tactical maps. These are show transportation and communication details superimposed on relief and planimetric details. They are used mainly for logistic planning and operations involving infantry and armoured corps.

A photomap is an air photograph with strategic and tactical data superimposed on it. It is not a map in the true sense of the term; it is rather a map substitute. Because of its wide use in the theatre of operations, it has been discussed here. A photomap may constitute just one photograph or it may be a mosaic composed of several of them. The scales of the photomaps range from 1:5,000 to 1:60,000. As the photomaps show the details against their photographic image, they are easily comprehensible to army personnel. There is, therefore, a strong tendency among the troops to use these maps.

The satellite imageries also fall in the category of photomaps. The advantage of satellite imageries is that that they give the picture of the earth surface at a desired interval. Thus the changes occurring in the location and distribution of a phenomenon can also be seen.

Types by Visibility (Real and Virtual Maps)

The concept of real and virtual maps has been well developed by Harold Moellering in his book "Analytical Cartography: Past, Present and Future". He derives the concept from the answers to the following two questions:

1. Is map a directly viewable cartographic image?
2. Is map a permanent tangible reality?

The answers to these questions divide maps into two categories:

1. Real Maps, and
2. Virtual Maps

Real Maps: Any cartographic product, which can be viewed directly and has a tangible permanent reality in the form of hard copy produced manually, mechanically, or electronically. The maps that we use in our day-to-day life are real maps.

Virtual Maps: These maps are of three types.

(i) *Type I*: It is a cartographic image, which can be viewed directly but only as a transient reality as is the case with a Cathode Ray Tube (CRT) map or its latest version the Liquid Crystal Display (LSD) map. It may also be called a temporary map.
(ii) *Type II*: It is a cartographic image of permanent tangible reality but it cannot be viewed directly. It consists of hard copy media, but cannot be viewed without further processing.
(iii) *Type III*: It does not have the characteristics of either Types I or II but can be converted into a real map as readily as the other two types of virtual maps. Computer based information in this format is easily manipulated.

USES OF MAPS

Maps have become so much a part of our life that we cannot dispense with them. We often need them to find the location of a place and distribution of a natural or man made phenomenon. We certainly need them as visual aids in our class rooms. They have already proved their worth as indispensable tools for the synthesis and analysis of statistical data. Urban and regional planners use maps to unravel various planning and development problems and to show the changes in the location and distribution of various phenomena. By giving a visual and integrated picture of the disparities in the development, they influence national policy decisions and international cooperation. Certainty a well-done map is worth several pages of a book.

Maps are used for following purposes:

Location of Places

The most universal use of maps is for locating places and things and finding the shortest and best routes to reach there. Geographic location of an object is determined in relation to the location of other surrounding features. Geographic coordinates too are used to locate a place or object. As mobility of people increases, the use of maps to locate places also increases. In Europe and the USA one cannot find a motorist without a pack of road maps. The use of such guide maps is secularly increasing in other countries of the world too.

Education and Research

Maps are useful and at times indispensable tools for teachers and students of all those disciplines, which have something to do with the location and distribution of natural or cultural features on the earth's surface. One is only too well acquainted with wall maps as classroom visual aids. Many books contain maps as textual illustrations. Good maps save thousands of words and have the capacity to crystallize facts and figures in a fashion that makes them comprehendible and clear. Students will be able to understand more readily if the teachers use maps and diagrams in the class room.

The usefulness of maps as research tools is now well recognized. Maps highlight the problems and show the changes in a fashion that is easy to comprehend. Use of statistical and quantitative techniques in processing of data to be represented on maps has further increased their utility as research tools. The net result of this development is the increasing number of thematic maps.

Planning and Development

Maps are of considerable use in planning and development of a community, region, or nation. While planning for a community such as a village or a city, the jurisdiction of the planning authority must be shown clearly; and past and present, land use must be mapped. The existing property ownership boundaries must be marked accurately.

The density of population in different parts of the community, the educational and recreational facilities and the shopping centers must be planned on the basis of the needs of different areas. Industrial and residential locations should be planned by taking into account the transportation, sanitation, education, recreation and other facilities. Each of these can be presented cartographically to enable people to visualize their inter-relationships. There is no better way of doing this.

Maps are equally useful in regional and national planning. Regional plans often cover areas for beyond the confines of the service areas of the villages, cities and metropolitan centers. The development potentialities can be better gauged by a clear understanding of the location and distribution patterns of the national and human resources and the existing social overhead capital. No other medium can do it better than a map. In the same way maps are useful in showing the progress made in different parts of a region or a nation. The usefulness of maps in this respect is well indicated by the fact that the United Nations Organization has taken up the preparation of an international series of land use maps and almost all important countries of the world have taken up the preparation of National Atlases.

Military Operations

Not many of us realize that wars and conflicts among the nations have been the greatest promoters of map production. No war, big or small, can be executed efficiently without the help of good maps. During the Second World War the Army Map Service of the U.S.A. alone produced about 40,000 different maps and distributed a total of over 500 million copies of them. A single movement of troops requires several different types of maps; and when the operation involves a combined effort from all the branches of the armed forces, the variety of maps needed further increases. The land forces need one type of map, while the air forces another type and the naval forces a third type.

More often than not, military operations have to be carried out in areas, which have not been visited by the armed forces before. These forces have to know the location and other details of topography such as mountains, hills, mounds, rivers, canals, ditches and wells to name a few. Verbal description does not help much. Maps are indispensable for armed forces to win a battle. Many operations have failed mainly because of the lack of good maps. One of the reasons why India could not defend itself when China attacked Arunachal Pradesh in 1962 was the lack of accurate maps.

Other Uses

In recent years the use of maps for propaganda and advertising purposes has increased to a great extent. Maps are also being increasingly used by aerospace science. The use of mapping techniques to represent the surface of moon and other planets and to show the movement of satellites and rockets are examples of new uses of cartographic products.

Maps are certainly of great use to us. We should, therefore, know as much about maps as we can. But knowing maps involves knowing about cartography – the science and art of making maps.

CONCEPT MAPPING

Concept mapping is a non-cartographic technique of representing knowledge in graphs. Knowledge is the product of a series impacts from varied sources. It can also said to be a network of information, concepts, and ideas. If these ideas and concepts are graphed to indicate the sources and the nature of impacts they have on the knowledge, it would produce a graph of networks. Networks consist of nodes (points/ vertices) and links (arcs/edges). Nodes represent concepts and links represent the relations between concepts.

Concept mapping can be done:

1. to systematize the ideas;
2. to help people think and see the reality in its true interconnections;
3. to generate ideas (brain storming, etc.);
4. to simplify complex structure (long texts, hypermedia, large web sites, etc.);
5. to communicate otherwise complex ideas; and
6. to generate understanding or diagnose misunderstanding.

Although concept mapping has been used in one form or the other from times immemorial, it was developed as a systematic technique by Prof. Joseph D. Novak of Cornell University in the 1960s. He concluded that *"Meaningful learning involves the assimilation of new concepts and propositions into existing cognitive structures"*.

A related technique known as Mind Mapping was invented (and copyrighted) by Tony Buzan in the UK. He describes mind maps as: *"a mind map consists of a central word or concept, around the central word you draw the 5 to 10 main ideas that relate to that word. You then take each of those child words and again draw the 5 to 10 main ideas that relate to each of those words."* The two techniques differ only marginally. While the mind map has only one main concept, the concept map may have several. Thus a mind map can be represented as a tree the concept map may have a network representation.

Now a number of concept and mind mapping softwares are available in the market. A number of softwares are now available for concept/mind mapping. Some of them are listed below: Inspiration Software, Inc., CMap 2.0 for Macintosh fetch by gopher, Mindmap by Michael Jetter, SMART Ideas by SMART Technologies, Decision Explorer (formerly called Graphics COPE) by Banxia Software, and Activity Map by Time/ system Int.

REFERENCES

Maps and Their Uses

Andrews, J. H. (1990), 'Map and Language: A Metaphor Extended,' *Cartographica*, vol. 27, no. 1 pp. 1-19.

Bann, Stephen (1988), 'The Truth in Mapping', *Word and Image*, vol. 4, no. 2, 498- 509.

Bertin, Jacques (1983), *Semiology of Graphics: Diagrams, Networks, Maps*, Madison: University of Wisconsin Press.

Curry, Michael R. (2005), 'Toward a Geography of World without Maps: Lessons from Ptolemy and Postal Codes,' *Annals of the Association of American Geographers*, vol. 95, no. 3. pp. 680-91.

Ehrenberg, Ralph E. (2006), *Mapping the World: An Illustrated History of Cartography*, Washington, D.C.: National Geographic.

Harley, J. B. (1987), 'The Map and the Development of the History of Cartography,' in *The History of Cartography* 1, pp. 1-42

Head, C. Grant. (1984), 'The Map as Natural Language: A Paradigm for Understanding,' *Cartographica* 21, no. 1. pp. 1-36.

Huggan, Graham (1989), 'Decolonizing the Map: Post-Colonialism, Post-Structuralism and the Cartographic Connection,' *Ariel* vol. 20, no. 4, pp. 115-31.

Krygier, John and Denis Wood (2005), *Making Maps: A Visual Guide to Map Design for GIS*, New York: Guilford.

MacEachren, Alan M. (1994), *Some Truth with Maps: A Primer on Symbolization and Design*, Washington, D.C.: Association of American Geographers.

Moellering, Harold (2000), Analytical Cartography: An Introduction', *CaGIS*, Vol. 3.

Moellering, Harold (2000), 'Analytical Cartography: Past, Present and Future', *CaGIS*, Vol. 27, No. 4.

Neocleous, Mark (2003), 'Off the Map: On Violence and Cartography', *European Journal of Social Theory*, vol. 6, no. 4, pp. 409-25.

Skelton, R. A. (1972), *Maps: A Historical Survey of their Study and Collecting*, Chicago: University of Chicago Press.

Vasiliev, I., S. Freundschuh, D. M. Mark, G. D. Theisen, and J. McAvoy, (1990), 'What is a Map?', *The Cartographic Journal*, vol. 27, pp. 119-23.

Printing Technologies

Amiran, D.H.K. and A.P. Schick, ed. (1961), *Geographical Conversion Tables*, Zurich: Aschmann & Scheller, 315 p. (for International Geographical Union).

Brunner, Felix (1962), *A Handbook of Graphic Reproduction Processes*, Teufen, Switzerland: Arthur Niggli, p. 379.

Gascoigne, Bamber (1986), *How to Identify Prints: A Complete Guide to Manual and Mechanical Processes from Woodcut to Ink Jet*, London: Thames and Hudson, 1986.

Ivins, William Mills (1943), *How Prints Look: Photographs with a Commentary*, New York: Metropolitan Museum of Art, p. 164.

Leif, Irving P. (1978), *An International Sourcebook of Paper History*, London: Dawson, p. 160. Units of Measurement.

Pearson, Karen Louise Severud (1978), *Lithographic Maps in Nineteenth Century Geographical Journals*, a Ph. D. Thesis, University of Wisconsin, Madison.

Reasonover, John Roy (1946), *Land Measures: French, Spanish and English Land Measures of the United States and Canada, With Units, English Equivalents, Reduction and Conversion Factors*, Houston: Priv. print., p. 124.

Saff, Donald and Deli Sacilotto (1978), *Printmaking: History and Process*, New York: Holt, Rinehart & Winston, p. 436.

"The Analysis of Paper and Ink in Early Maps," in *Library trends* 36, no. 1 (Summer 1987): 85-107.

Woodward, David and Hermon Dunlap Smith (1977), *The All-American Map: Wax Engraving and Its Influence on Cartography,* Chicago: University of Chicago Press, p. 168.

Woodward, David (1975), *Five Centuries of Map Printing, The Kenneth Nebenzahl, Jr., Lectures in the History of Cartography at the Newberry Library,* Chicago: University of Chicago Press, p. 177.

Prime Meridians

Boggs, Samuel Whittemore and Dorothy Cornwell Lewis, "Map Projections," in *The Classification and Cataloging of Maps and Atlases*, p. 81-91, New York: Special Libraries Association, 1945.

Chamberlin, Wellman (1947), *The Round Earth on Flat Paper: Map Projections Used by Cartographers,* Washington, D.C.: National Geographic Society, p. 126.

Haag, Heinrich (1913), *Die Geschichte Des Nullmeridians,* Leipzig: Otto Wigand, p. 111.

Keuning, Johannes (1955), "The History of geographical map projections until 1600," in *Imago mundi* 12 1-24.

Snyder, John Parr (1993), *Flattening the Earth: Two Thousand Years of Map Projections,* Chicago: University of Chicago Press. p. 365.

Snyder, John Parr and Harry Steward (1988), *Bibliography of Map Projections,* U.S. Geological Survey Bulletin, 1856, Washington, D.C.: G.P.O. 110.

Concept Mapping

Buzan, T. (1995), *The Mind Map book. (2 ed.),* London, UK: BBC Books.

Johnson, D. H., Beissner, K., & Yacci, M.A. (1993), *Structural knowledge: Techniques for conveying, assessing, and acquiring structural knowledge,* Hillsdale, NJ: Lawrence Erlbaum Associates.

Lawson, M. J. (1994), Concept Mapping. In T. Husén & T. N. Postlethwaite (Eds.), *The international Encyclopedia of education* (2nd ed., Vol. 2, pp. 1026-1031), Oxford: Elsevier Science.

Novak, J.D. (1991), Clarify with concept maps: A tool for students and teachers alike. *The Science Teacher, 58*(7), 45-49.

Novak, J. D. (1993). How do we learn our lesson? : Taking students through the process. *The Science Teacher, 60*(3), 50-55.

Part I

Understanding Cartography

2
Nature and Scope of Cartography

Cartography (in Greek chartis = map and graphein = write) is the study and practice of making maps of the whole or part of the earth. It combines science, aesthetics, and technique, and is built on the premise that reality can be modeled in ways that communicate spatial information effectively. Cartography is thus a science and art of designing, constructing and producing maps. It includes almost every operation from original field-work to collect data to final printing and marketing of maps. To some cartography signify only the mechanics of preparing maps. Actual drawing is, however, only a part of the total scientific, technical and artistic efforts needed to bring out a map. The processes of designing a map to suit the heterogeneous needs and fancies of the users demand knowledge and skills which range from field work to computer programming.

Map construction has been the main concern of cartography since 2500 B.C. when Babylonian cuneiform clay tablets were produced. Ptolemy's book *Geographia (150 A.D)* helped spark European exploration of the world from about 1500 to 1700, and a new spurt in the production of maps based on the data acquired from these explorations. This tradition of cartography as map production continues to this day. It includes even those maps which have been designed and produced with the help of computers.

A good map can be produced only by a judicious blending and proper coordination of knowledge of earth sciences, communication sciences, computer programming and aesthetics. With increasing use of computers in preparing maps, the role of draftsmen and the instruments which were used to manually draw maps has come down rather drastically. A cartographer need not know how to draft a map manually; but he must know what a map is and how to design it with the help of computers.

It does not, however, mean that traditional approaches to cartography all the way from data gathering, designing, drawing and producing maps have become totally redundant. We still need people who can survey an area with a plane table or prismatic campus, who can design and draft rough maps and who can do fair drawing especially when development projects have to be executed. We need maps for constructing buildings, railways, roads, etc. and in all these cases land surveys have to be done and maps often called plans with exact measurements have to be prepared. This book, therefore, includes all those nuances of cartography which are now treated as traditional and also those which are modern.

SCIENTIFIC BASES OF CARTOGRAPHY

Cartography is a science that has its own individuality. The earth we inhabit is highly complex. It is marked with a great variety not only in its physical configuration and human activities but also changes through time and space. Cartography devises ways and means to bring order and system, generality and simplicity, refinement and legibility and ease of use and comprehension to an almost incomprehensible range of complex details of the earth (both land and sea) and other celestial bodies through a medium, called map. It is thus a science of communication too with artistic leanings. It can be compared with architecture in some respects and may be characterized as a technical science too. But it is something more than that for the subject matter of cartography ranges from a small piece of land to the earth as a whole. Further, it is not only the physical earth that falls in its domain of concerns; the man and his activities in relation to the earth he inhabits is its main concern.

Cartography is closely aligned to earth sciences. Its subject matter is the representation of the earth surface or the surface of any other planetary body. It seeks to represent the huge and spherical earth on a small paper as realistically as possible. To do this a cartographer must have the knowledge of the earth surface either by observation or by study. He must know well the location of various objects and their spatial distribution. Cartographic representations also involve the choice of what is important and rejection or suppression of what is unimportant. Thus a cartographer has to generalize data. It involves prior knowledge of the elements to be generalized and training in reasoning and synthesis. Cartography is, therefore, a geographic undertaking and a cartographer is a graphic geographer or geographic illustrator.

Cartography can also be defined as an auxiliary science that acts as a bridge between techniques, art and earth sciences. As a map is a generalized picture of the earth surface, the cartographer has the difficult task of generalizing the complex details of this surface. To do this successfully, he has to have the background of not only geography but also of other disciplines in which need maps. Cartography is, therefore, a science which entails cooperative efforts of specialists in a variety of fields. The geodesist and the topographic surveyor give the size and shape of the earth and the location of its surface features; the economist, sociologist, geologist, botanist, etc., gives the subject matter to be generalized. The cartographer classifies and generalizes these details and converts them into maps.

This description of cartography may give an impression that various cartographic processes can be carried out independently by different disciplines. This is not so. The surveyor not only surveys but also draws rough sketches. His field drawings are then used to compile maps on smaller scales. The number of original drawings made depend on the colour scheme of the map and the printing process to be used. The financial resources also limit the freedom of action on the part of a cartographer. Even if one uses satellite imageries as base maps, the cartographer has to use it to design a map that meets the demands of his client. It is because of this interdependence of various cartographic processes that most official mapping agencies in the world combine all these operations from original surveys, to designing, drawing, printing and marketing of maps. A writer can be easily separated from the printer of his book, but a cartographer cannot be separated from his map printer. The two have to work together.

The above discussion, points to the conclusion that cartography is neither an experimental science like physics or chemistry nor a social science like economics or sociology. It is a hybrid scientific discipline which employs scientific methods and logic to produce which are accurate and beautiful and at the same time convey intended message to the map reader.

ARTISTIC LEANINGS OF CARTOGRAPHY

Since the aim of cartography is to improve the graphic representation of the earth, it cannot avoid being artistic. A map not only portrays details visually in accordance with certain scientific principles but also in a way that is pictorial and aesthetic. The study of cartography is, therefore, partly a study of map graphics. The cartographic methods of representation and exposition follow the same principles and laws which underlie other types of graphics. And since art is the highest form of graphics, a good map cannot be unimpressive aesthetically.

To what extent cartography is an art is a controversial question. There are cartographers without much artistic skills; there are also cartographers who are artists first. There can, however, be no cartographer who does not have a sense of beauty, proportion and order. And anyone who has these senses can aspire to become a cartographer.

The objectives of cartography are, however, not the same as that of art. Cartography does not aspire to produce the greatest work of art. We have yet to know a map which has been given a place in art galleries as a masterpiece of art. Art and artists require complete freedom of expression; they, often, utilize this freedom to create a piece of work which may be incomprehensible or even meaningless to the common man. Unlike an artist, a cartographer functions under severe limitations set by topographical and statistical details, symbols and colour standards. Moreover, a cartographer can never afford to create a piece of work that is incomprehensible and meaningless to its users. A piece of art is valued for its aesthetic beauty and sensibility whereas a map is valued for its mundane utilitarian values.

But if one defines beauty not only in terms of visual expression and artistic sensibility but also in terms of a sense of proportion, harmony of form and colour and simplicity, a well-done map is always a piece of art. A good map requires a high degree of legibility; it requires expression by methods emphasizing the pertinent and suppressing the less important or irrelevant; and it requires a well balanced harmonic interaction between the individual elements of a map. It is in this sense, and in this sense only that cartography can be said to be an art.

CARTOGRAPHY AS A SCIENCE OF COMMUNICATION

Arthur Robinson (1952) founded the communication school of cartography, with his dissertation which laid emphasis on cartography as science communicating spatial information graphically to the reader via a map. His work triggered spatial cognition research in cartography during the 1960s and 1970s. *Nature of Maps* (1976) authored by Robinson and Petchenik was a part of this blooming of cartography as a communication science.

According to this school, the ultimate purpose of cartography is to communicate facts as they appear on the earth surface as clearly and forcefully as possible. It is distinct from other forms of graphics in the sense that it specializes in communicating facts and ideas about the earth and earth alone. Its role in making maps for the other planets and their satellites is still very limited. Many cartographers prepare maps to satisfy their own intellectual curiosities. Like all creative activities maps give inner satisfaction to its creator. But such maps are very few. Most maps are made to communicate facts and ideas in which people are interested. And hence the success of a map is determined by the effectiveness with which it is able to convey the message. Viewed thus map is a medium of communication and cartography a communication science. We can thus say that one of the main purposes of cartography is to communicate the changing realities of the earth surface.

With the introduction of artificial satellites, it is now possible to get minute details of whole or part of the earth surface. Yet people continue to be ignorant about much of world partly because the satellite imageries are not accessible to all but mainly because unless these imageries are converted into maps, they fail to satisfy the information needs of the people. With the spread literacy and increase in mobility of the people the demand for maps has therefore increased. Cartography has therefore to pay increasing attention to effective portrayal of the facts and ideas about the earth than hitherto done.

A communication system consists of five elements:

1. Source
2. Destination
3. Channel
4. Transmitter
5. Receiver

Source is origin point of a message; and destination is the end point. Between the source and the destination is a channel through which the message passes from source to destination. It may be an individual, telegraph, telephone, cell phone or internet. To allow the information to pass through a given channel it must be encoded in a language, which can be transmitted through the channel. We may call the encoder the *transmitter*. And at the other end there must be a *receiver* to reconvert or decode the message into its original form before it reaches the destination. These five elements, i.e., source, transmitter, channel, receiver, and destination constitute a communication system (Fig. 2.1).

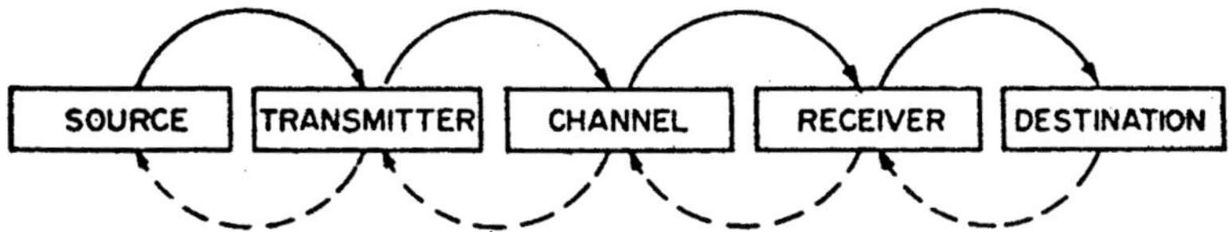

Fig. 2.1: Elements of a Communication System

Broadly speaking, a communication system functions something like this. The transmitter gets a message from the source. He encodes this message into a language that can be fed into the channel selected for the transmission of the message (in this case the map). At the other end, the receiver reads this message and decodes it into a common language.

The likelihood of a message reaching the destination in the form it is intended to reach depends upon a variety of factors. The first is the nature of the message itself. A clearly formulated message coming from an authentic source will have a better chance of reaching the destination than the one which is ambiguous and has emanated from an unreliable source. The efficiency with which the transmitter encodes the message is also important. The noise or disturbance in the channel can distort the message, and finally the attitude of the receiver and his ability to decode the message also matter.

Applying to cartography, the components of the system will be something like this:

1. *Source*: All natural and social sciences concerned with the study of earth and its surface features.
2. *Message*: Ideas and facts about the earth and its surface features; the solar system, its planets and their satellites, and the universe.
3. *Transmitter*: Cartographer who converts these ideas and facts into words, drawings and symbols.
4. *Signals*: The drawings, words, and symbols and their mutual arrangement.
5. *Channel*: Maps and other cartographic products.
6. *Noise*: Poor design or drawing, cluttering of the symbols, irrelevant facts to the detriment of the relevant ones, poor printing, etc.
7. *Received signals*: Symbols, etc., as perceived and understood by the map user.
8. *Destination*: Map users the world over.

A cartographer has to remain sensitive to the reactions of the users of his maps and has to keep them as updated as possible. He has to so develop his maps that they assist the people in clearly perceiving the message intended to be conveyed by them.

Our preceptor sensory mechanisms are always in contact with the real world of things and events. The eyes, the nostrils, the tongues, the ears, the skin, and the nerve endings respond to a variety of stimuli generated by temperature, pressure, odor, sound, vision, and taste. They are the means of perception. When we perceive something, we translate the impression made on our senses by the stimuli into awareness of objects or events. We construct our world of things and events as perceived by our sensory organs and the physical objects, as we know them through sight, sound, taste, smell and touch.

Understanding about the phenonomena we come across results from coordinated perception which is the outcome of multiple impressions recorded through sensory mechanisms. Lack of any sensory receptor eliminates the possibility of complete reception. We cannot acquire a perfect knowledge of the earth and its surface features unless we are in a position to see it with our eyes too. Reading the books and hearing from others give only a partial picture. When we see it with our own eyes, we get experiences and associative feelings which are closer to the reality.

Observation of the whole earth with all its minute details being impossible, man from the very beginning of his civilization, developed the means of portraying it cartographically. The earliest maps were only the generalizations of the reality and were made on earthen tablets. They showed the features of the earth like mountains and rivers to which people were most familiar. Although satellite imageries now enable us to look at the earth more realistically, we still need generalizations to see patterns and orders. And thus was born the idea of map—a graphic model of the earth or a portion thereof. It gives a more realistic picture of the earth than any verbal description can give, and can be fruitfully used to make even satellite imageries far more meaningful.

Maps stimulate thinking and understanding about the earth and other celestial bodies; they also help in attitude formation. Attitude is a mental and neural state of readiness, developed through experience with persons, things and events. It has a directive or dynamic influence upon our response to all those objects and situations with which we come in contact. Our attitude towards other parts of the world depends upon this state of readiness developed through experience with the real world. It can be changed by carefully planned learning situations. Maps are one of the important means to change the learning situations and thus to create a better understanding of the physical and cultural contents of the world. Cartography, the science and art of making maps, is, therefore, a science of human communication too.

EUROCENTRISM IN CARTOGRAPHY

Since cartography developed in Europe first, one sees Eurocentrism in the way we look at the world. Why is it that Europe is always shown in the centre of the world map? Why is Mercator projection conventionally used for world maps? While the 0° meridian pass through UK? Why should we call West Asia as Middle East or South-East Asia as Far East? Why should Europe be called a continent? Answer to all these questions lie in Eurocentrisim.

A continent is a large landmass separated by oceans or very narrow land bridges. Given this definition, North America, South America, Africa, and Australia are undisputably continents. Similarly, the landmass of Eurasia is a continent. But this landmass is divided into separate continents of Asia and Europe. It has no basis in physical geography. Countries like Russia are partly in Asia and partly in Europe.

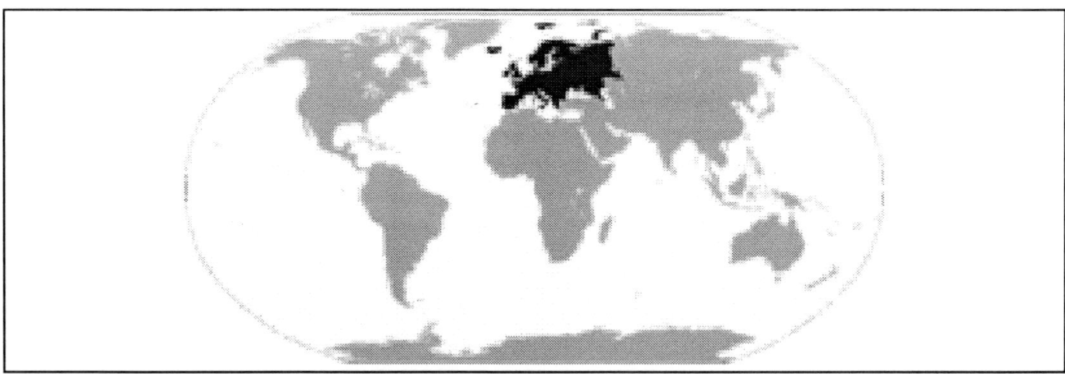

Fig. 2.2: World map Showing Europe Vertically Centered

There is no geographic rationale for breaking Eurasia into two continents except that Europeans did not want to be part of a landmass that is so different and in their view inferior culturally and historically.

One may also ask: Why Mercator projection is so often used for world map? There is political implication of it. The Mercator projection distorts areas further from the equator, making Europe and North America appear disproportionately larger compared to similar sized areas closer to the equator, such as Africa, Central America and Australia. Alaska, for example, is presented as being similar or even slightly larger in size than Brazil, although Brazil's area is actually almost 5 times that of Alaska.

Since 1851, the prime (zero) meridian passes through Greenwich, London. During the Age of Exploration, several other prime meridians were in use. It is true that the current prime meridian places the International Date Line in the Pacific, inconveniencing the smallest number of people, but this alone was not the reason for its choice. The terms like 'Middle East' to describe an area slightly east of Europe, and the 'Far East' pointing to an area eastward and away from Europe. Alternatively, the Western World includes not only Central and Western Europe, but the former European colonies of North America, Australia and New Zealand. All these terms have their origin in Eurocentrism.

IMPACT OF ICT ON CARTOGRAPHY

During the last few decades revolutionary changes have taken place in cartographic processes. Remote Sensing, with the help of artificial satellites equipped with powerful cameras and attendant manipulative techniques enable us to get real picture of the earth surface showing not only the location, colour and pattern of the objects focused on but also elevation and slope. Once these imageries become available, they can be used straight as a picture or converted into maps using computer programmes.

Information technology is now so advanced that maps can be drawn without any of those facilities, which were indispensable earlier. Satellite imageries of the whole world are now available on internet. One can have the image of a house in the city or village at any scale one wishes; it is also to see the whole world, a country, a region, a district, a city or village at different scales. These imageries can then be used to prepare maps both general and thematic to suit ones needs. There is no need to organize large-scale topographic surveys. Maps can be sent by internet to any place within no time. Printing technology too has undergone a dramatic change. The days of making zinc plates for printing of maps are gone. It is now possible to prepare colour maps at home and print them at home.

Thus maps are more readily available than ever before. But the maps made by those who are not conversant with the principles governing cartography can be misleading and convey a very wrong message. That is why there is a need to have book like this: a book that elaborates the fundamentals principles of cartography and helps readers evaluate the maps they come across from to time.

BRANCHES OF CARTOGRAPHY

Because of its wide scope and distinguishable—though not separable—functions, cartography can be divided into three branches:

1. Theoretical,
2. Analytical, and
3. Applied.

A number of other sub-branches such as aerospace cartography, journalistic cartography, geographical cartography, thematic cartography, statistical cartography, and scientific cartography have come to light in recent years. Such sub-divisions do indicate an increasing trend of specialization. They also indicate an increasing interest in cartography and its products. The basic principles and the techniques underlying all these sub-divisions are, however, the same and differences among them are more superficial than real. It is for this reason that all these aspects of cartography are not discussed here as separate branches.

Theoretical cartography deals with the conceptual and theoretical aspects of cartography, such as the critical testing and further development of map design and content, graphical methods of representation, ways of formulation and the establishment of operational standards, and principles of map editing. A theoretical cartographer tries to keep in touch with the map users to know their future requirements and their reactions to the available maps. He tries to devise and design maps which meet the requirements of the users and at the same time function as effective media of visual communication.

Professor Waldo Tobler in the 1960s further refined the approach to cartography and made it more analytical, conceptual, and mathematical. Tobler in his dissertation in 1961 initiated the concepts of map transformations of geographic space. He further refined this view in 1979 Tobler later extended his 1961 conceptualization of cartographic transformations to explain the traditional cartographic map creation and reading process, and extended it further by stating that cartographic transformations "... (Tobler, 1979) occur in map interpolation, filtering, and generalization..." Further advances were made in the transformational approach to analytical cartography by Nyerges (1980), Clarke (1990) and Chrisman (1999).

While these thoughts were stirring the minds of geographers and cartographers, the advent of digital processes such as CRT display images, digital terrain models, and spatial databases forced traditional cartography to change its old definition of was thrust into a conceptual turmoil with the development of new cartographic products, many of which were. These products did not fit into the then conventional definition of hard copy cartography. Morrison (1974) in his article in the The American Cartographer, called for an expanded and extended definition of what constitutes a map. Moellering followed him and initiated research on the problem highlighted by Morrison.

When Tobler (1966) brought the concepts of analytical cartography to the fore, he talked about the half-life of theory (15 to 20 years) *versus* the half-life of technique (3-5 years) and suggested that better insights into cartography could be developed if one laid emphasis on its conceptual and theoretical aspects. Analytical cartography thus focuses on quantitative and mathematical relationships inherent to it. Being quantitative, analytical cartography promotes replication and makes the mother discipline a normative science.

Further, analytical cartography aims at extending the sampling and other theorems, to develop new concepts like Moellering's (1977b) concept of real and virtual maps. Several other authors like Muller (1991) explicitly recognized these needs, Moellering (1991c; 1994) also recognized that the need to develop spatial data standards to fil the wide conceptual gaps in cartography. It is now recognized that the computer may not be able to empirically or heuristically solve certain classes of numerical problems. Saalfeld (2000) has discussed this problem and suggests that the NP of complete or NP-hard problems may not be solved by merely adding more computer horsepower.

Applied cartography includes such functions as the layout and final drawing of maps. It includes knowledge of computer programmes to prepare maps. Applied cartography functions within the framework of certain rules, standards, and guidelines set by theoretical cartography. It is also concerned with the development of improved methods of preparing maps manually or with the help of computers. Instrumentation and automation, therefore, fall within the purview of applied cartography. Topographical and socio-geographic surveys also are considered as aspects of applied cartography.

The major aspect of applied cartography is the technique of actual drawing of maps whether manually or with the help of computers. It may and should include the application of cartographic techniques to other disciplines and problems within its purview. An applied cartographer takes interest in researches leading to the application of cartographic methods in the solution of many of the problems that our society faces today. He helps the planners, the extension agents and teachers in communicating their ideas and views to their clients.

REFERENCES

Albers, S. (1987), "The Mapping Impulse in Dutch Art,' in *Art and Cartography: Six Historical Essays,* Chicago: University of Chicago Press.

Anonymous (1990), *Understanding GIS: The Arc/Info Method,* Redlands: Environmental Systems Research Institute, Inc.

Barber, Peter and Christopher Board (1993), *Tales from the Map Room: Fact and Fiction about Maps and Their Makers,* London: BBC Books, p. 192.

Berggren, J. Lennart and Jones, Alexander (2000), *Ptolemy's Geography: An Annotated Translation of the Theoretical Chapters,* Princeton University Press.

Chrisman, N.R. (1997), *Exploring Geographic Information systems,* New York: John Wiley, See also his, 'What Does GIS Mean?' in *Transactions in GIS,* 3 (2), 1999, pp. 175-86.

Clarke, Keith (1995), *Analytical and Computer Cartography,* Prentice-Hall, Englewood Cliff, NJ, (2nd edition).

Clarke, Keith and Coud, Hohn G. (1990), 'On the Origin of Analytical Cartography', in *Cartography and Geographic Information Science.*

Clarke, Keith, 'Analytical Cartography Revisited', *Proceedings XII Annual Surveying Teachers Conference, Surveying the Future,* Madison, W1: pp. 287-302.

Eschner, S. (1998), *Computer Geoscience as Art,* Bakersfield: Bakersfield Museum of Art.

Greenhood, David (1964), *Mapping,* Chicago: University of Chicago Press, p. 289.

Holmes, Nigel (1991), *Pictorial Maps,* New York: Watson-Guptill, p. 191.

Keates, J. (1984), 'The Cartographic Art,' *Cartographica* 21: 1, pp. 37-43.

Krygier, J. (19895), 'Cartography as an art and a science? *The Cartographic Journal* 32: 6, pp. 3 10.

Miller, J. (1983), *'States of Mind,'* New York: Pantheon.

Moellerings, Harold (2000), 'The Scope and Conceptual Content of Analytical Cartography', in *Cartography and Geographic Information Science,* July.

Monmonier, Mark S. and George A. Schnell (1988), *Map Appreciation,* Englewood Cliffs, N.J.: Prentice-Hall, p. 431.

Morrison, Joel L. (1974), 'Changing Philosophical-Technical Aspects of Thematic Cartography', *American Cartographer,* , pp. 5-14.

Muller, J.C. (Ed) (1991), *Advances in Cartography,* London: Elsevier Applied Science.

Muller, J.C. (Ed) (1991), 'Cognitive cartography: A new heart for a lost soul', in *Advances in Cartography,* International Cartography Association, pp. 201-221.

Nyerges, T. (1991), 'Analytical Map Use', *Cartography and Geographic Information Systems,* 18(1), pp. 11-22.

Nyerges, T. (1991), 'Geographic Information Abstractions: Conceptual Clarity for Geographic Modelling' *Environment and Planning,* 32, pp. 1483-99.

Raisz, Erwin (1948), *General Cartography,* 2nd ed. New York: McGraw-Hill, p. 354.

Raisz, Erwin (1962), *Principles of Cartography,* New York: McGraw-Hill, p. 315.

Rees, R. (1980) 'Historical Links between Cartography and Art.' *The Geographical Review* 70: 1. pp. 61-78.

Robinson, A. (1989), 'Cartography as an Art' in: D. Rhind and D. Clark, Keith and Coud, Hohn G. *Cartography Past, Present and Future: A Festschrift for F.J. Ormeling,* London: Elsevier Applied Science Publishers.

Robinson, A. and Petchenik (1976), *Nature of Maps: Essays toward Understanding Maps and Mapping,* Chicago: Chicago University Press.

Robinson, Arthur Howard (1995), *Elements of Cartography,* 6th ed. New York: John Wiley and Sons, p. 674 .

Southworth, Michael and Susan Southworth (1982), *Maps, a Visual Survey and Design Guide,* Boston: Little, Brown, p. 223.
Tobler, Waldo (2000), 'The Development of analytical Cartography: A Personal Note', in *Cartography and Geographic Information Science*, 27 (3), pp. 189-94
Wood, Denis and John Fels (1992), *The Power of Maps,* New York: Guilford Press, p. 248.
Woodward, D., 'Introduction' in *Art and Cartography: Six Historical Essays,* Chicago: University of Chicago Press.

3
History of Cartography

History of cartography begins with the emergence Homo sapiens on planet earth. They moved from place to place in search of pastures anew and created mental maps of the areas they passed through. Later, when the evolutionary journey of the Homo sapiens advanced and human beings emerged on the scene, movements from place increased and newer methods were found to translate the mental maps into cave paintings. During those prehistoric times, man kept this knowledge in his mind for it was too little to need recording. Nor was there a readily available technique of recording. Even today our so-called primitive people carry mental maps of the areas in which they live or hunt. The first map was thus a mental map.

The first attempt at writing began with cave paintings. One finds animals and humans along with rivers, trees and mountains painted on the rocks. As time passed, the horizons of the world expanded beyond the neighbourhood, community and hunting areas. Gradually, the whole earth became the stage on which man started playing the drama of life. Mental maps thus proved inadequate. New techniques and means were developed to keep the records of his experiences for posterity. Cartography is one of these many techniques. The history of cartography is largely the study of the increase in the accuracy and effectiveness with which tangible as well as intangible contents of the earth's surface were observed, surveyed and portrayed on maps. Maps and all that they have meant to mankind is, therefore, as vast and old a field as the history of art or literature. And the hundreds of thousands of maps which have come down to us have within them an invaluable record for those who are interested in exploring man's past.

As the subject matter of cartography has changed through time so also have its functions. In the past it was more concerned with the measurement and the representation of the shape, size and other broad details of the earth's surface. Today these broad details are well-known to us. We have at our disposal minute details of the contents of the earth's surface. Our problem is how to analyze these complex details and present them in a form comprehensible to map users.

Like science and technology, cartography appears to be of Western origin. It is, however, only partially true. In the East too cartography was developed even in ancient times. But there was no continuity and hence no progressive change. In the West, the tradition of making charts and maps had continued uninterruptedly ever since the days of Greek civilization which had frequent interactions with the civilizations of Southeast and West Asia.

On the basis of available historical records, cartography developed in four distinctive stages. These are:

1. Ancient period (up to 400 A.D.),
2. Medieval period (400 to 1500 A.D.),
3. Early modern period (1500 to 1900 A.D.), and
4. Recent period (1900 A.D. to date).

These stages should not be treated as discrete. There have been periods of retrogression and stagnation, interrupted by periods of rapid development. Further, the theoretical knowledge was not closely followed by its application. Thus many of the processes cut across these stages.

ANCIENT TIMES

This period may be studied under the following heads:

1. Primitive cartography,
2. Greek cartography,
3. Roman cartography, and
4. Asian cartography.

Primitive Cartography

A visit to any good ethnographic museum proves beyond doubt that graphics is not a science of recent origin. Even the most primitive people expressed their ideas and experiences graphically. Primitive people like the Eskimos of the Arctic, the Bedouins of the Arabian Desert, the Polynesians of the Pacific Islands, and the Banjaras of India had a remarkable ability to draw sketches of the areas they ventured to pas through. Such maps drawn on a piece of skin, wood, bone or terracotta indicated the relative position and distances of localities known to them. As these ancients improved their cultural heritage, they started producing better maps—maps which were used as tools to solve problems of day-to-day life. Egyptians used geometrical methods for land measurements and for establishing land ownership lines after each flood in the Nile. These lines were shown on cadastral plans. Many of such plans included specifications for the construction of temples, palaces, canals and roads. Such maps were produced not only by the Egyptians, but also by the Babylonians, the Indians, the Chinese, the Aztecs and the Incas.

Although the mapping of this nature showed a definite improvement through time, its coverage was limited to the part of the earth known to the people. Thus each one had its own notion of the size and shape of the earth. To them the earth was not only flat but also infinite in the sense that beyond their own territorial limits lay unknowns. The first were the Indian who gave serious thought to the shape and size of the earth. In fact, they speculated about the universe, calculated the movement of the stars and planets including the earth. They were followed by Babylonians who gave some thought to the shape and the size of the earth as a whole. They believed the earth to be flat and circular surrounded by sea and heavens.

The knowledge about the size and shape of the earth speculated in ancient times was not converted into maps. It did, however, help their successors in translating their information into sketches and maps.

Greek Cartography

Being close to Babylonians, the Ionians were the first Hellenic people to take interest in scientific thinking. Pressing socio-economic conditions also forced them to pay greater attention to the world beyond their own sphere of influence. They developed trade relations with far lying areas. They prepared itinerary maps showing the stages along the routes leading to such trading areas as the Mediterranean coast and to places like Susa, the then capital of Persia. By 600 B.C. they established maritime settlements from east coast of Spain to the far reaches of Black Sea. They established a number of 'city states'. These city states maintained close relations with the Greek cities.

By 600 B.C. the Greeks had developed Miletus as a centre for geographical studies and cosmological speculations. One of the early products of Miletus was Thales who is considered to be the founder of natural philosophy. Anaximander, who prepared the first map of the world as known then and Hecataeus, who wrote the first book of geography, were also the products of Miletus. Hecataeus believed the earth to be a circular plane, surrounded by continuous belt of ocean with Greece in the centre. (Fig. 3.1).

Another general principle that governed Greek thinking was symmetry in nature. In Anaximander's maps one finds a balance in the features shown north and south of the axis. A generation later, Herodotus, who was also a great traveler and who knew the circumnavigation of Africa by the Phoenicians and the voyage of Scylax down the Indus, further improved the world map and showed Caspian Sea as an inland sea. He did not represent the earth to be circular. He also did not show the northern ocean. (Fig. 3.2).

The military and sailing expeditions of the Greeks provided new information to cartographers. Carthaginians sailed along the Atlantic; Hanno sailed along the West African Coast; Himilco explored the Spanish and Gaulish coasts, the Greek mercenaries organized campaigns against Persia and Alexander invaded India. Greeks learnt from India that the earth was not a flat disc but a sphere. They thus abandoned the flat earth theories of Pythagoras and Plato.

At Alexandria, Eratosthenes calculated the circumference of the earth being 24,662 miles only a few hundred miles short of the correct measurement. Unfortunately, it

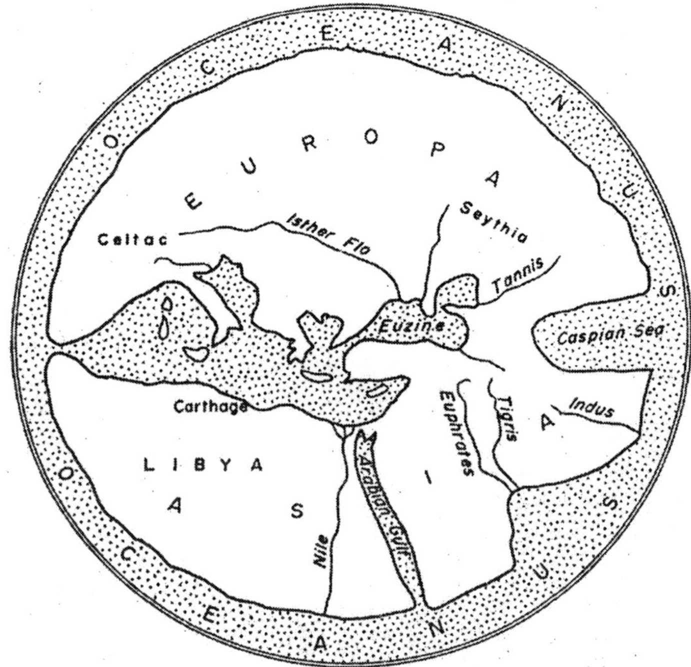

Fig. 3.1: World Map of Hecataeus

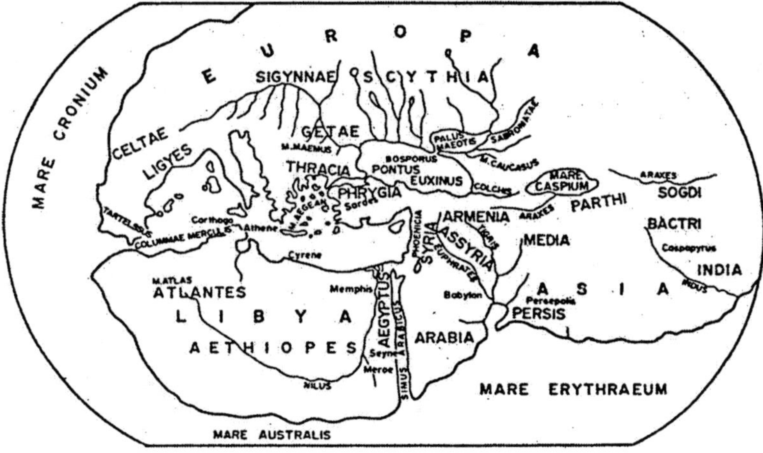

Fig. 3.2: World Map of Anaximander

was not accepted by his successors. Eratosthenes tackled the problem of representing the spherical earth on a plane surface by extending two parallels eastwards, one passing through Gibraltar and Caspian Sea and the other through Egypt and South India. He also established the zero meridian approximately following River Nile but starting from the mouth of River Don (Fig. 3.3). Later on, the ground work prepared by him was improved by Hipparchus who compiled a table of latitudes. He proposed a scheme of 360° of latitudes and longitudes but it could not be given effect to for want of appropriate apparatus and instrument, which could give supporting astronomical data for the purpose. Nevertheless, he laid the foundation for scientific cartography upon which Ptolemy built his concepts further.

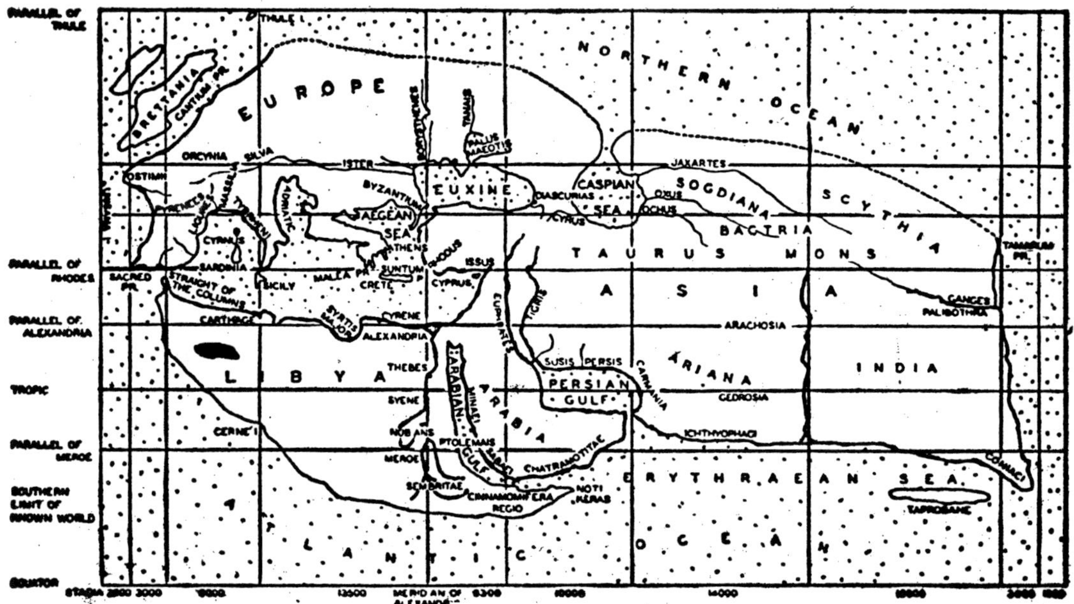

Fig. 3.3: World Map of Eratosthenes

Between Hipparchus and Ptolemy came Posidonius who led the Greek thinkers to pause and consolidate the progress already made. It is known as Posidonian period. The compilations of this period were used by the Romans in constructing their maps. The Romans in turn transmitted this handbook of information to Latin countries of Europe, resulting in the stagnation of cartographic progress until the Renaissance. Even Ptolemy based much of his work on these compilations. He rejected Eratosthenes' calculations of the circumference of the earth and preferred Posidonius' estimates. His estimates were used as late as 15th century A.D. Strabo's Geography in 17 books is one of the several works of this period.

The development of geography at Alexandria reached its pinnacles during the second century A.D. It culminated in the contributions of Claudius Ptolemy. Ptolemy released his 'Geography' in 8 volumes. The first of these volumes deals with the principles of mathematical cartography and methods of representing a spherical surface on a plane surface. The other volumes give the location of places or features of geographical interest. There are twenty-six regional maps and one world map attached to his book. In addition to these, sixty-seven maps of smaller areas are also given. One finds several inconsistencies between the text and the maps. It has, therefore, been doubted whether the maps were done by Ptolemy or someone else. As no manuscript older than twelfth century A.D. is traceable, this doubt does not appear to be baseless. It is now widely believed that the maps were drawn after thirteenth century.

Ptolemy miscalculated the circumference of the earth to the extent that his one degree of latitude was equivalent to 90.4 km in contrast to Eratosthenes' 99.6 km. Thus when converting distances into degrees, he got greatly exaggerated figures and the east-west extension of his world became too long. His concept of Asia was erroneous as he overlooked Peninsular India and estimated the size of Ceylon several times larger than its true size. He also showed Indian Ocean as a land-locked water body. He committed similar mistakes in the representation of South-east Asia and North Africa (Fig. 3.4). All these mistakes were carried forward by later geographers and cartographers and had to do much with the slow rate of growth of cartography in the centuries that followed.

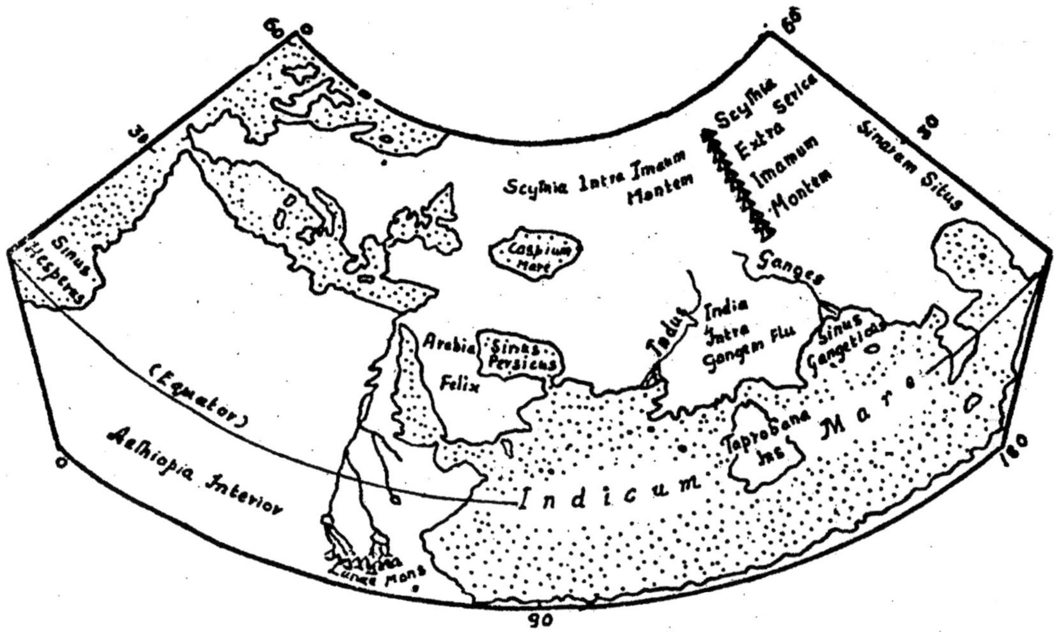

Fig. 3.4: World Map of Ptolemy

Roman Cartography

Unlike the Greeks, Romans did not take much interest in Scientific Cartography and were unconcerned with the achievements of the Greeks in this field. For a long time, they were concerned with the expansion and consolidation of their empire and hence only those maps which assisted them in this endeavour were favoured by them. Thus a map for them was a practical tool to be used in the travels of their officials and campaigns of their armed forces. Their maps showed the road networks and the battle fields; and many of them were nothing more than geographical renderings of written records. The 'Peutinger Table' is an example of this. It was prepared during the 3rd century A.D. For all practical purposes, it is a road map of Roman Empire. The roads are shown by straight lines and the distances in numbers. As true directions are neglected, the shape and relative positions of various features are considerably distorted. It includes Europe, Africa and Asia and contains 534 illustrations, 311 for Europe, 62 for Africa and 161 for Asia.

Another contribution of the Romans was the famous Orbis Terrarum of 'Survey of the World' prepared by Marcus Vispasanius Agrippa in 12 B.C. This map was displayed in Rome for the information of citizens and appears to have contained much greater details than the Table. This is confirmed by Pliny in his 'Natural History'. The shape of the map was circular (Fig. 3.5).

Fig. 3.5: The Orbis Terrarum of the Romans

Indian Cartography: Ancient Times to 800 AD

It is not yet proved that map making was practiced in ancient India. It is, however, now well-known that the knowledge about the universe and the earth was quite advanced in India all the way from Vedic. Even the *Rig Veda* has given detailed knowledge of the astronomical facts and cosmological revelations. During the Siddhantic period these expressions were crystallized into Siddhantas or laws. The post-Siddhantic period gave rise to classical treatises by great astronomers like Arya

Bhatt (AD 476-550), Varahamihir (AD 505-587), Bhaskar (1114-1185) and others. In addition to these, there were works on positive sciences like the *Bhouthika Sutras* which gave astronomical as well as terrestrial details.

Many of the revelations in the ancient times point to the conclusion that some of the so-called discoveries by the West were known to Indians thousands of years earlier. Arya Bhatt discovered the relative movement of the earth and the sun, a thousand years before Copernicus (1473-1543). Even earlier, the *Aitareya Brahmana* said, "the sun never sets or rises; when people think that the sun is setting, it only becomes invisible and thus makes the night below and day to what is on the other side." The *Bhouthika Sutras* stated that the sun was fixed relative to the earth and that the earth was spherical like an egg. Saunaka calculated the earth's distance from the sun which surprisingly approximates to modern estimates. Bhaskara calculated the circumference of the earth to be 24,385 miles (39,967 km), diameter 7,905 miles (12,722 km), and the surface area 396,325,850 sq miles (10,26,48,395 sq. km). He also estimated the atmosphere to extend to a height of over 60 miles (97 km).

Knowledge about the earth appears to be quite extensive in ancient India. It appears that the whole of Eurasia and parts of Africa were frequently visited by our ancients. They divided the earth into seven Dwipas (Fig. 3.6).

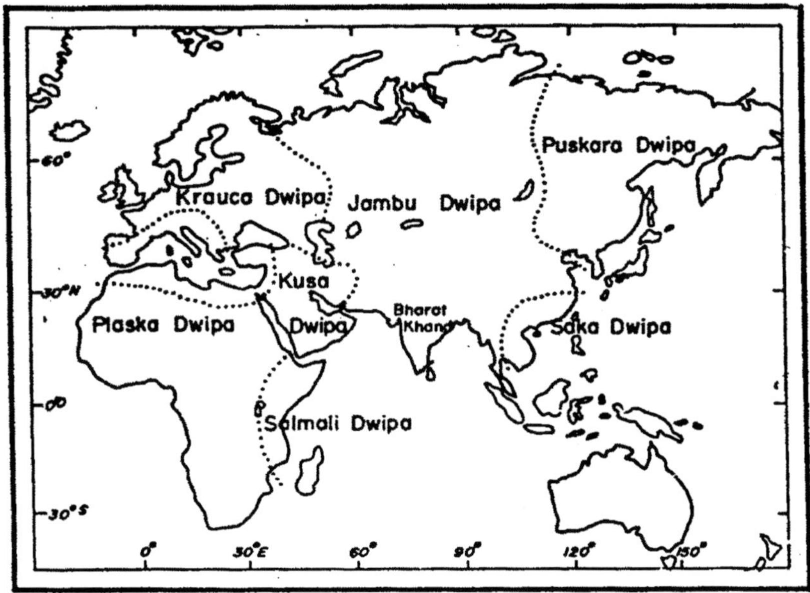

Fig. 3.6: Dwipas as known to Indians before Mahabharata Times (After S.M. Ali)

The earth was undoubtedly considered to be spherical and surrounded by water. The southern waters were considered to be salty and northern ones to be milky. This was the position during the Mahabharata times (5000-6000 BC). But prior to Mahabharata times, oceans were considered to be salty. A diagrammatic representation of the earth as conceived at the time of Mahabharata is given in Fig. 3.7. One interesting feature of this conception is that India was not put in the centre of the earth. The Meru or Pamir Mountain was considered to be the centre.

A study of the maps showing the location of various Dwipas (sub-continents) and the names of the rivers and mountains given in ancient literature prove beyond doubt that Indians in ancient times had frequently visited Europe, Russia, and other parts of the old world. According to Greek scholar, there is hardly a place name in Greece which has not been derived from Indian names. But none of those discoveries appear to have been recorded cartographically. It is possible that even if such records were there, they did not survive the ravages of time. The fact, however, remains that Indians knew more about the earth and its various physical and human contents than the Greeks or Romans had before first century BC.

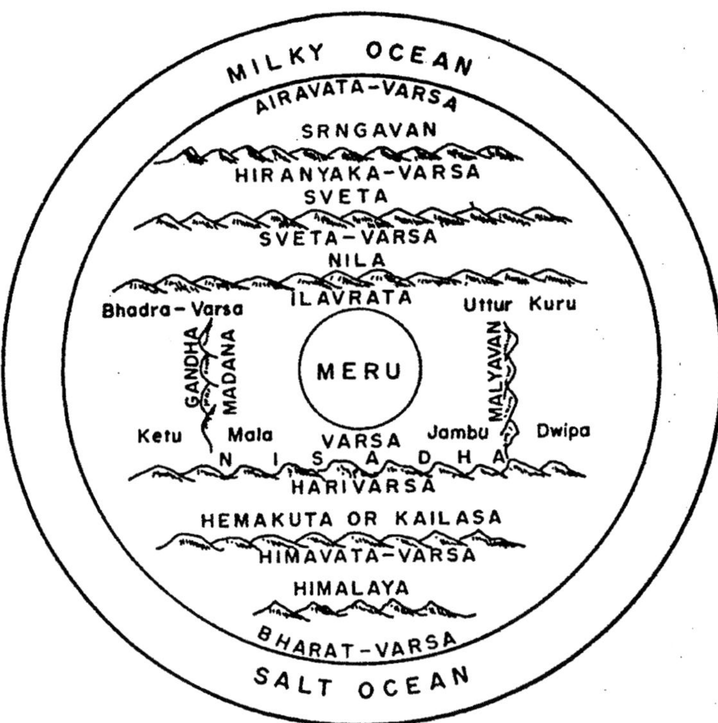

Fig. 3.7: World as Conceived by Indians during Mahabharata Times (After S.M. Ali)

South-East Asian Cartography

To what extent had knowledge of the earth and map making had developed in other Asian countries during this period is not known. It is believed that maps of one sort or the other were prevalent in China, India, Japan and Korea. Considering, the large scale of migration of people of Asia from one region to another, it is doubtful that the routes followed and countries visited were not known. There is every possibility that the intervening seas and oceans were also well charted by the people of China and India.

MEDIEVAL TIMES

Early Medieval Times (AD 400 to 1200)

During the early medieval period there was no major breakthrough in geographical knowledge in Europe. Maps and the past records were used without verification and hence more errors crept in them. Cartography became a copy work of existing maps. Most of the maps of this period are called T-O maps. These maps had east at the top. O represented the circular world whereas the horizontal and the vertical strokes of T represented Don-Nile meridian and the axis of the Mediterranean respectively. Some of the maps of this period were rectangular but with same details as in T-O maps. This innovation might have been brought in to accommodate the Christian view of a four-cornered world.

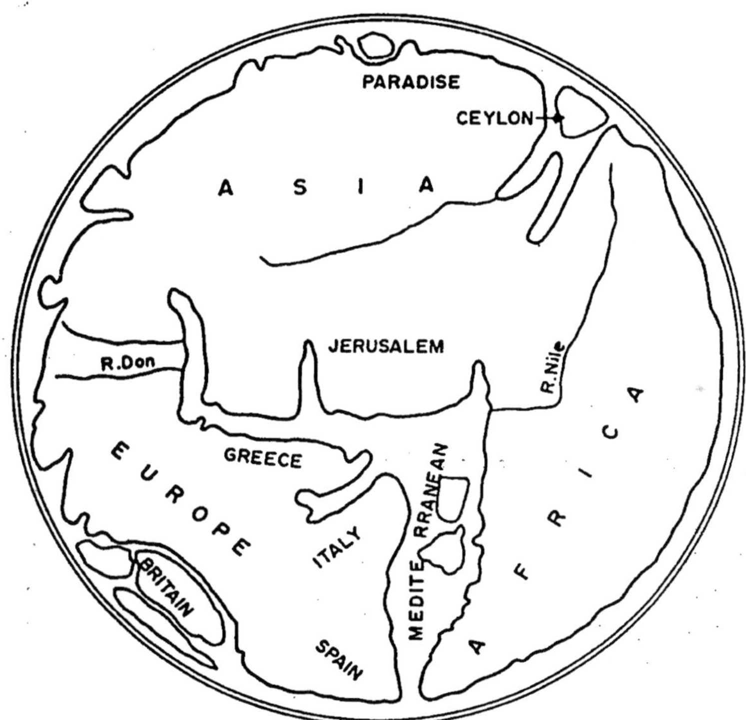

Fig. 3.8: Sketch of a Section of Portolan Chart

The map makers of this period had learnt from the Greeks and Indians that the earth was round but took as a flat disc. *Mappa Mundi,* the main circular map of the world of this period was a modified form of the world map of Marcus Vipsanius *Agrippa* (c. 63 BC-12 BC) a Roman statesman and general. And the modifications were made at the instance of Christian theologians, so that Jerusalem was shown in the centre of the map and the area of Palestine was considerably enlarged to accommodate minute details. The empty space was filled with neatly represented themes from the contemporary history.

The above facts prove beyond doubt that Ptolemy's geography exerted almost no influence on the cartography of early medieval Europe. It, however, did have some influence on the Byzantine and Arab cultures. The text of 'Geography' by Ptolemy was translated into Arabic in the 9th century, and it maps were well-known to the Arab scholars like Masudi. In the 12th century Idrisi compiled a world map using Ptolemy's map as one of the sources. His other sources were the Arabic sailing records and his own knowledge of the world. Similar advances were made in the Far Eastern world too. Chinese had developed mariner's compass, and notable advances were made in India in the calculation of latitudes and longitudes. During this period India's contacts with South-East Asian countries increased further and existing knowledge about the Bay of Bengal and South-East Asia improved. To what extent this knowledge was used in mapping is not yet clear.

Late Medieval Period: 1200 to 1500 AD

By the end of 13th century, use of mariner's compass and construction of navigational charts had become common. Italians were the leading cartographers in this respect. The charts made by them were called Portolan Charts. Nearly twenty of these charts have survived to give us a glimpse of the state of cartography in that period. These charts were made on the skin or parchment measuring 91 × 46 cm to 142 × 76 cm. The names of the coastlines barring the harbours were shown in black; islands, rivers, deltas, rocks and shoals were shown in red. The details of the land features were few and the design thereof was decorative. They were drawn to scale but the units of measurement were not given. Most of these charts show coastline of the Mediterranean and Black seas and part of the Atlantic coastline of Europe with fair amount of accuracy; but representation of the coasts, beyond these limits is not accurate. Even the Baltic coast has been shown crudely.

These charts do not have lines of latitudes and longitudes; nor do they were based on the notion of the earth being a sphere. Instead, they had lines showing approximate compass directions and radiating from two main and several subsidiary centres. The area covered by these charts being small, this deficiency did not lead to any major location error (Fig. 3.9).

These charts were improved after more detailed surveys. This is clear from the fact that the charts prepared subsequently were more accurate. For example, in Carte Pisane (late thirteenth century), Britain is represented in a very crude form but in Perrinus Vesconte's outline of 1327, the representation of the coastline of Southern England is considerably improved. By the beginning of 14th century the Catalonians, mainly Majorcans superseded the North Italians as leading cartographers. The Majorcans revived the classical interest in Asia. This revival is embodied in the Catalan world maps published in the Catalan Atlas in about 1375 A.D. (Fig. 3.10)

Catalan Atlas was based on many sources, the most important among them, were:

1. The circular world maps of the classical and medieval times;

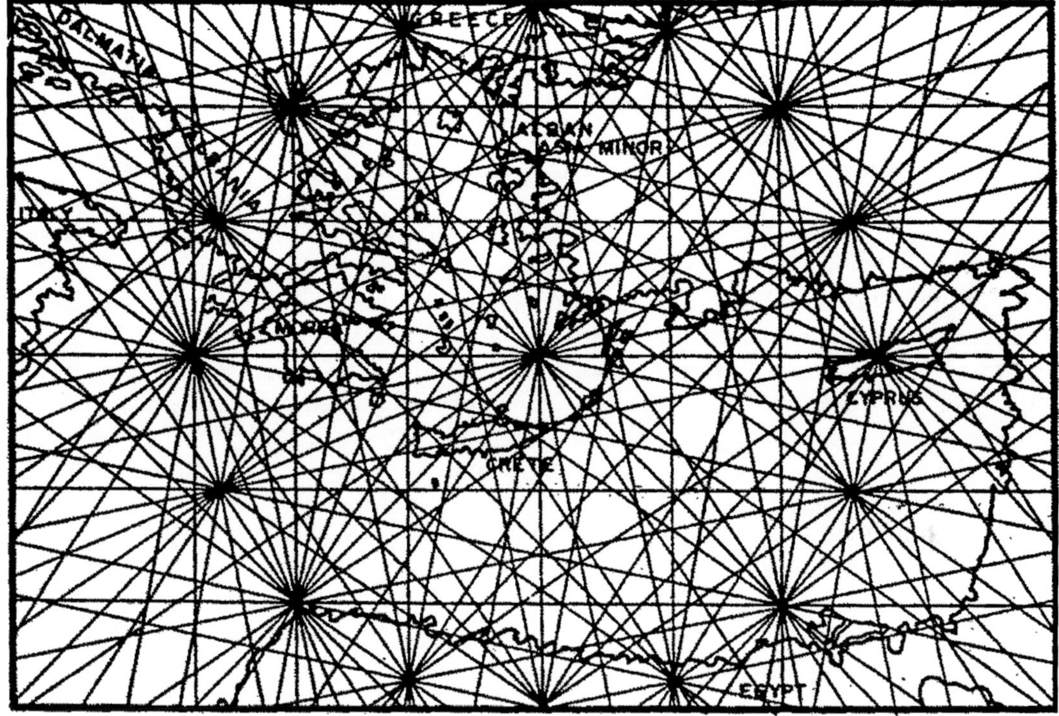

Fig. 3.9: A Section of the Portolan Chart

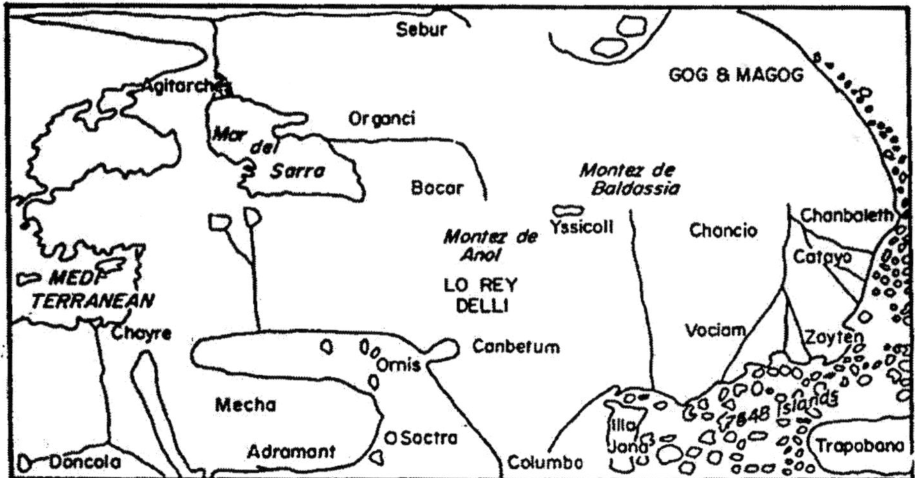

Fig. 3.10: Catalan World Map

2. Portolan charts showing the outlines of the Mediterranean Sea, Black Sea and European coasts;
3. Narratives of the 13th and 14th century travelers such as Marco Polo, and
4. Narratives and charts prepared by Arabs and available at Barcelona.

It was for the first time that the whole of Asia had been represented in one map, although very inaccurately when compared to the representation of Europe. The lakes, the rivers and the mountains together with important places were shown in accordance with the narratives of Marco Polo. Indian Peninsula was emphasized and innumerable islands were shown in the Bay of Bengal and along the coasts of South-east and East Asia.

The importance of these maps lies not in how accurately they represent the world but in the hard work that went in using the available data to present a picture of the earth, which was quite different from the accepted one. The cartographic approach was scientific in the sense that only those details were given for which solid evidence was there. This is evident from the fact that they did not show northern and southern regions of the world, nor did they show the southern part of African continent.

The subsequent cartographic work in Europe was influenced by Ptolemy's *Geography* as its manuscript was readily available there during the first quarter of the 15th century. Moreover, the narratives of Nicolo Conti's travels had also become available after 1447. On the basis of these and other sources an elliptical world map was prepared in Genoa in 1457 wherein the outline of Asia was Ptolemaic but the details appear to have been drawn from Conti. One of the characteristics of this map is the presentation of the South Asian islands in great details. Unlike the earlier maps, it does not show the Indian Ocean as an inland sea.

Another work of great cartographic significance was the world map of Fra Mauro, a monk of Murano near Venice. He started from the usual practice of having Jerusalem in the centre of the map, but did not follow Ptolemy in every respect. He, however, he gave a distorted picture of many of the bays, peninsulas and islands in Asia and Africa.

Martin Behaim, a native of Nuremburg prepared a 50 cm in diameter globe in the closing decades of 15th century wherein were shown equator, the two tropics and the Arctic and Antarctic circles. The equator was divided into 360°; the 80° meridian to the west of Lisbon was also shown. But Behaim accepted Ptolemy's longitudinal measurement of the old world (177°) and add 57° more to accommodate the eastern China. This extended the longitudinal extent of the old world from 131° to 234°. In order to keep the distance of the globe limited to 360° the distance between European and Chinese shores was reduced from 229° to 126° longitude. Another map of importance was Henricus Martellus' world map of 1489. This was also based on Ptolemy's map (Fig. 3.11).

Because of increasing use of maps, Ptolemy's *Geography* containing his world map was multiple was reprinted several times. The first printed edition of the book, without maps, appeared in 1447; two years later Bologna edition came out with maps. Again in 1478 came out a Rome edition. The maps of this edition were engraved on copper. Soon, other editions also came out so that by 1482 four editions with maps were in the market, three from Italy and one from Germany. Ptolemy's influence on cartography reached it pinnacle with the publication of *Geography* at Strasbourg in 1513. The maps of this edition were wood-cut.

The publication of Ptolemy's *Geography* had two important implications for cartography: his ideas became more popular and new methods of printing heralding

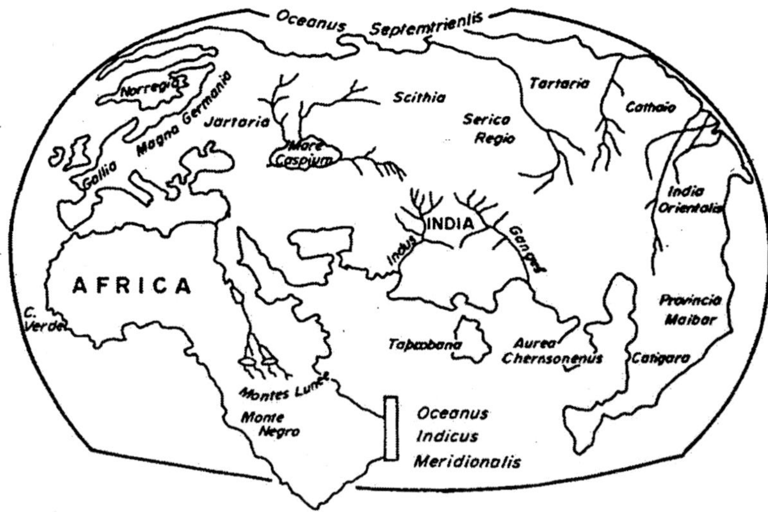

Fig. 3.11: World Map of Henricus Martellus

the multiplication of maps by inexpensive processes. Ptolemy's *Geography* also enabled the people to judge the accuracy of Ptolemy's maps against the data supplied by the discoveries of Diaz, Columbus, Vasco-da-Gama, Cabral, Alfonso-d'Albuquerque and Magellan. These voyages help find the latitudes with the help of the polestar and the mid-day sun.

MODERN PERIOD (AD 1500 TO 1900)

The great discoveries during 15th and 16th centuries contributed immensely to the revival of interest in cartography. During this short period a number of charts were prepared to help the sea farers but only a few of them survive to this day. Henricus Martellus' world map made in 1489 is an important contribution of this decade. Albert Cantino, a Portuguese, prepared Cantino chart around 1500. Another map of significance is the King-Hamy navigational chart of 1502. All these maps had considerably improved the outlines of Asia, and Africa. It can, rather, be said that Cantino chart had entirely disproved many of Ptolemy's concepts and speculated the existence of Pacific Ocean. Some regional charts were also published during the first decade of the 16th century.

Pedro and Jorge Reinel were the two official cartographers in Portuguese court. Pedro made a chart of the Indian Ocean in 1518, and Jorge Reinel made a world map for Magellan's use. Since Maluccas, the chief source of spices lay near the Spanish-Portuguese demarcation line in the eastern hemisphere it had a stimulating effect on the study of cosmology and cartography of the region. In Eastern hemisphere this line of demarcation was 46°37′W while in Eastern hemisphere it was 133°23′E. Maluccas are located at 127°30′E. The Portuguese part was west of 133°23′E longitude.

The earliest among the Portuguese charts of this area surviving from about 1510 to this show that areas west of Malaya are depicted quite accurately but those east of it are latitudinally misplaced. Further improvements were made in Ribeiro's chart published in 1529 which included the whole circuit of the globe between the polar circles with the East Indian archipelago appearing in both the western and eastern margins. But even in this map the eastward extension of Asia continued to be exaggerated by about 20°. He placed Maluccas 7.5° within the Spanish sphere.

The Portuguese chart makers were famous all over the Western world. Leading cartographers were Pedro Reinel and his son Jorge Reinel, Lope Homem and his son Diogo Ribeiro, Fernao Vaz Dourado and Luis Teixeira. Most of the maps and charts were done at Casada India at Lisbon. Spain had its Casa de la Contracacion de las Indians at Seville for the same purpose. Among the important cartographers of this centre were Nuno Garcia de Toreno, Ribeiro and Alonso de Santa Cruz.

The other centres of mapping in Europe in this period were located in Rhineland, Netherlands, Germany, Switzerland and France. By 1500 the real centre of cartographic activity had shifted from Italy to the Iberian Peninsula and Netherlands. Martin Waldseemuller who worked at St. Die in Vosges was the first to use the name America in his world map prepared in 1507 (Fig. 3.12). The Netherlands school of cartography was headed by Gemma Frisius of Louvain (1508-55). Gerhad Mercator was one of his pupils.

By 1525 the German and Netherlands geographers and cartographers had devised geometrical methods of survey with more precise instruments for observation of angles.

Fig. 3.12: World Map of Martin Waldseemuller

The famous geographer and cartographer Sebastian Munster, while at Heidelberg University wanted his friends to survey an area within a radius of 9 to 12 km of the city. He suggested the use of compass for this purpose. The method of elementary triangulation was first described by Gemma Frisius in his *Cosmographia* published in 1533. He suggested the fixing of positions of the objects by intersecting rays. Gradually the foundation for plane tabling was laid by Philip Apian and these methods were in full use by 1570.

Beginning of 16th century saw increased cartographic activity in Europe. As exploratory expeditions increased and methods of surveying improved, the demand for topographical maps by travelers, statesmen, merchants and antiquarians increased by leaps and bounds. Numerous maps of varying sizes and coverage flowed from the presses. The Waldseemuller, Munster, and Laafreri (Italioa) atlases were designed to meet this demand. But it was left to the Flemish cartographers, Ortelius and Mercator to serve in a practical way the public demand for an up-to-date and comprehensive atlas.

Gerhard Mercator was born at Rupelmonde in Flanders in 1512. He was a pupil of Gemma at the University of Louvain. His principal achievements were his globe (15410; his celebrated world map (1569); his large map of Europe (1554), his edition of Ptolemy's *Geography* (1578), and his atlas (1594) published posthumously (Fig. 3.13). Mercator drew the meridians and parallels as straight lines to compensate for the increase in distances between the parallels proportionately. This could not be done on maps having pole ward converging meridians. Thus the area was distorted in his map but directions and shapes were maintained. With the above built in characteristics, Mercator's world map a new conception of the world—a conception which was different from that of Ptolemy.

He showed three major land-masses the one including Asia, Europe and Africa, the other, North and South Americans (erroneously called India Nova) and the third, the continent of Australia. The last of the three landmasses was the result of the Greek conception of the existence of another continent as a counterpart of the inhabited world. The existence of Terra Del Fuego, reported by Magellan strengthened this belief.

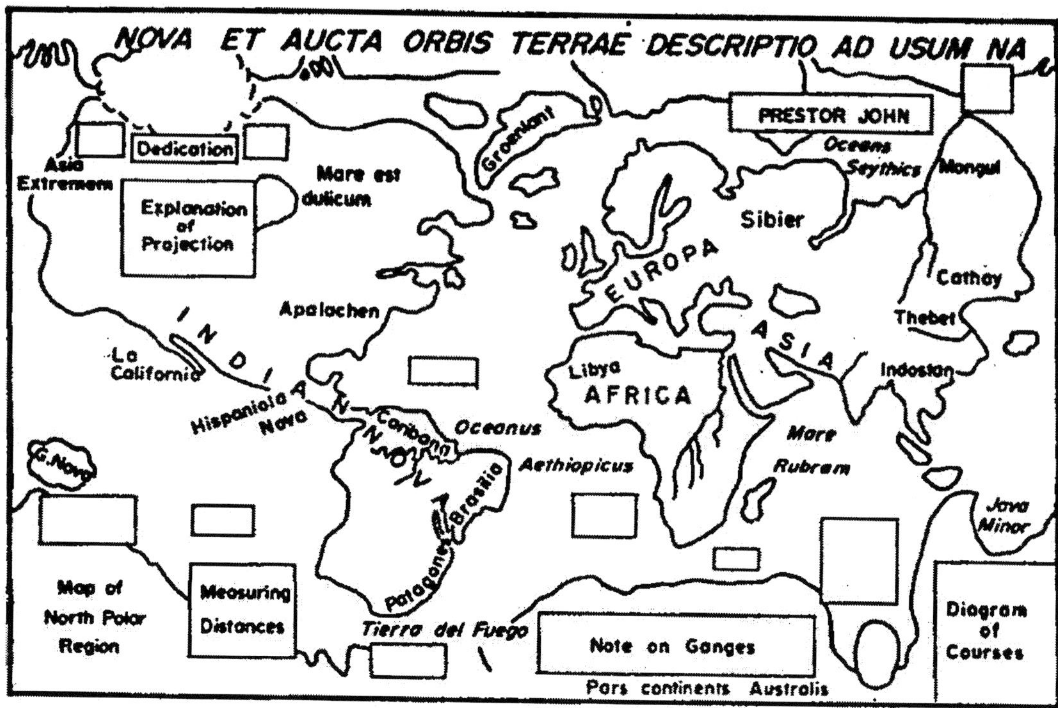

Fig. 3.13: World Map of Gerhard Mercator

More accurate information about Australia, North America, China and other parts of the world were gathered during 16th and 17th centuries. In the meantime, with the development of telescope, logarithmic tables, pendulum clock and level survey and other observational methods were improved. In 1671 Paris observatory was established and latitudes and longitudes of an ever increasing number of places were recorded under the auspices of the Royal French Academy of Sciences. The Academy also took up the preparation of detailed maps of France.

A remarkable contribution to cartography was made by the four generations of the Cassini family. Jean Dominique used the movement of Jupiter's satellite to calculate the longitude of a place. It was a definite improvement over the one based on lunar eclipse. The Cassini, Jacques prescribed complete triangulation survey of France based on Paris meridian. His son, Cassini de Thury almost completed this work in 1784. On completion, the Cassini map comprised of 182 sheets (88 × 55.5 cm). Delisle published a general map of world between 1724 and 1745. In these maps the longitudinal extent of North America was too large. The regional maps of D'Anville had the same deficiencies as the source of information was the same. Cassinis were followed by a succession of competent French cartographers. Important among them are Guillaume Delisle (1675-1726), J.B. Bourguignon (1697-1782) and Jean Baptiste Bourguignon D'Anville (1697-1782).

D'Anville's work was carried on by his son-in-law, Phillippe Buache. He developed better ways of representing relief by symbols to represent mountains and hills and by shades to represent escarpments and valleys. Hans Konrad Gyger (1599-1674), a Swiss cartographer, attempted to show the land surface as seen from above and thus he produced remarkable plastic effect by shading the lower portions, and leaving the higher ones un-shaded. During 18th century, J.G. Lehman developed the principles of hachuring. The technique of showing relief by contours also became in vague in 18th century. The land contours were first used by Milet de Murcan in 1749 and the submarine contours by Phillippe Buache in 1737.

The earliest British map showing spot heights was Physico-chorographical chart of Kent prepared in 1743 by Christopher Pack. A further step in representing the height took place when colours of different shades were used to distinguish successive contours by Adolf Stieler (1775-1836) in his Hand atlas formally titled 'Hand-Atlas über alle Theile der Erde und über das Weltgebäude' and published in 1820. The advances made in the field of mathematics and astronomy during the 18th century also helped improve survey and measuring instruments. The quadrant was improved by John Hadley (1682-1744) added reflecting mirrors and Vernier scale to it in 1730s. Chronometer was developed by John Harrison in 1772. Theodolite (in its essentials) was also developed during this period. It was first employed in Triangulation Survey of England and France in 1787, and later in the Ordnance Surveys of Britain and India. The British were also active in America and India. James Rennell, the first Surveyor General of India, (1742-1830) published his map of Hindustan in 1779. The first edition of his *Bengal Atlas* with maps on a scale of 8 miles to 2.5 cm was published in London in 1779, two years after Rennell's retirement from the Survey of India. Gradually London became the leading centre of cartographic production. Perhaps the most important among the British cartographers of this period was Arrowsmith who prepared a chart of the Pacific Ocean in nine sheets in 1778. When mounted, the dimensions of this chart exceed 1.8 × 2.16 m.

FIRST HALF OF TWENTIETH CENTURY

The chief characteristics of this period are:

1. Organization of national topographic surveys,
2. Mass production and use of maps,
3. Increase in the influence of science and technology; and
4. International cooperation mapping and related issues.

During this period a number of countries established national survey institutions, which carried out the great national surveys based on the triangulation method developed by Giovanni Domenico Cassini (1625-1712). Gradually, the surveying instruments were improved and the errors resulting from such causes as refraction, curvature of the earth's surface and climate changes were removed. Topographical surveys were organized in seven stages:

1. Determination of mean sea level to which all altitudes are Referred,
2. A preliminary plane table reconnaissance to determine suitable points for triangulation,
3. Determination of initial latitude, longitude and azimuth;
4. Measurement of the baseline,
5. Triangulation with the help of theodolite,
6. Calculation of the triangulation and heights, and transfer of Trigonometrical points on to the sheets used by plane tabler, and
7. Filling in of the details by the plane tabler.

The accuracy and efficiency in surveying considerably improved with the use of theodolite, wireless and other mechanical aids. With the introduction of cameras in 1920 it was possible to use aircraft for surveying. The aerial camera advanced the mapping of the visible face of the earth at a speed and with a fullness of detail never obtained before. It became possible to photograph even those areas which could not be seen by the earth-bound observers. Development of several electronic devices further revolutionized the cartographic processes. Almost all the countries of the world have establishments charged with surveying of land. Those which do not have, rely on other countries for this purpose. In United Kingdom, the Ordnance Survey of Great Britain was established in 1791. In France, the Institute Geographique National is responsible for National Surveys. The U.S. Geological Survey is now the chief agency for topographical mapping in the United States. And in India it is the Survey of India which is charged with the responsibility of carrying out topographic surveys. All these organizations produce 1:1 million maps in addition to several other series of large scale maps.

The use of maps as tools for research was demonstrated by such early geographers as Alexander Von Humboldt and Karl Ritter. Humboldt, in particular, demonstrated that a great variety of facts could be represented in a readily intelligible manner. Later *Physikalischer Atlas* was issued by Heinrich Barghaus in 1838. Its next edition (1852) contained several thematic maps. At the same time John Bartholomew (father and son duo) in England published several atlases. The Soviet Union produced a series of atlases of the world as well as the U.S.S.R.

Mass production and use of maps of various types is another distinguishing feature of this period. Type and quality of maps produced, methods and techniques to represent quantitative as well as the quantitative data to suit the varying needs of map users, and the construction of thematic maps to meet the needs of specialized groups are some of the special features of this period.

POST-WORLD WAR TWO DEVELOPMENTS

Computer technology has brought revolutionary changes in cartographic processes during the last four decades. It represents a completely new and different technology for map making. The process of adoption of this technology can broadly be outlined in three stages:

1. In the first stage (until 1960) computer manufacturers and Governmental research and development units used cartographic algorithms.
2. In the second stage (1970's) remote sensing and satellite imageries entered the cartographic arena; despite initial reluctance on the part of established cartographic units, it spread fast its product was comparatively better.
3. The third stage (post-1980) which is still on the horizon for computer cartography further expanded and is on the verge of replacing all the old and traditional techniques. This phenomenal change is, perhaps, due to the speed, accuracy, automation, low cost, data processing, preservation of data and map design in memory, possibility of increasing output and freedom from tedious production tasks that characterize computer cartography.

The cameras have revolutionized the cartographic processes. They not only help surveyors get quick and better photographs but also assist cartographers design and develop better maps. Special cameras help them compile, draft and reproduce maps.

Publication of *National Atlases* arising from general advance in geography has been a special contribution of this period. The Royal Scottish Geographical Society's Atlas of Scotland published by J.G. Bartholomew in 1895 was perhaps the first such atlas. The *Atlas of Finland* was published in 1899, and *Atlas of Canada* in 1906. Czechoslovakia, Germany and France also published National Atlases. The French Atlas had maps on 1:1,000,000 scales. *Atlas of American Agriculture* prepared by A.E. Baker was published in 1936. The *Soviet Atlas of the World* was published in two volumes in 1937 and 1939. India published its first *National Atlas* in 1957.

An outstanding feature of this period is the ease with which many of the mapping processes could be accomplished. Today we have satellite imageries for the whole world at different scales which can be used as base maps to build new maps both general and thematic with the help of computers. The days of negatives and positives are on the way out. It is now possible to produce and print maps in multi-colours much easily and inexpensively than ever before.

Cartography as a science got further fillip by its international recognition. The United Nations established an International Cartographic Organization in 1961 that helps national governments in improving their spatial data collection and mapping activities. The cartographic department of the UNO produces 1:1 m series of International Maps of the World (IMW). In addition to this, the International Council of Aeronautical Organization (ICAO) produces World Aeronautical Charts on 1:2 million scales.

As mentioned earlier, there is an International Cartographic Association (ICA) similar to International Geographical Union (IGU). The first General Assembly of the Organization was held in 1960. The Association has been holding conferences once in four years in different parts of the world. The ICA holds Regional meetings too almost annually. The UN Regional Conference for Americas is held at the interval of four years; for Asia and Pacific Region at the interval of three years, and for Africa every fourth year. These Conferences offer opportunities for the geographers and cartographers to share their knowledge and experiences periodically. In addition to this, several sister organizations like British Cartographic Society, American Congress on Surveying and Mapping and Australian Cartographic Society, Indian National Cartographic Association, etc. have come up in various countries to discuss problems cartography encounters in serving the people. The Indian National Cartographic Association (INCA) was established in March 1979 with headquarters at Hyderabad. Geographers, professional cartographers and officials of the Survey of India were responsible for the formation of this association.

HISTORY OF CARTOGRAPHY PROJECT

The History of Cartography Project is a research, editorial, and publishing venture drawing international attention to the history of maps and mapping. The Project's major work is the multi-volume *History of Cartography* series. Its inter-disciplinary approach brings together scholars in the arts, sciences, and humanities. By considering previously ignored aspects of cartographic history, the Project encourages a broader view of maps that has significantly influenced other fields of study.

Organized by region and time period, *The History of Cartography* looks at maps in the context of the societies that made and used them. The volumes integrate existing scholarship with new research, examining an unprecedented range of artifacts from local maps to those of the cosmos. The books are extensively illustrated and contain detailed footnotes, appendixes, and reference maps. The award winning series, founded by J.B. Harley and David Woodward, is now edited by a team of scholars and published by the University of Chicago Press.

Volume 1 : *Cartography in Prehistoric, Ancient, and Medieval Europe and the Mediterranean (1987)*

Volume 2.1 : *Cartography in the Traditional Islamic and South Asian Societies* (1992)
Volume 2.2 : *Cartography in the Traditional East and Southeast Asian Societies* (1995)
Volume 2 : Book 3: *Cartography in the Traditional African, American, Arctic, Australian and Pacific Societies (1998)*
Volume 3 : *Cartography in the European Renaissance* (2007)
Volume 4 : *Cartography in the European Enlightenment*
Volume 5 : *Cartography in the Nineteenth Century*
Volume 6 : *Cartography in the Twentieth Century*

REFERENCES

Bagrow, Leo and R.A. Skelton (1985), 'Select Bibliography of the History of Cartography' in *History of Cartography*, Chicago: Precedent Publishing, pp. 201-300

Bagrow, Leo and R.A. Skelton (1985), *History of Cartography*, 2nd ed. Chicago, Ill, USA; Oxford, UK: p. 312.

Berthon, Simon, and Andrew Robinson (1991), *The Shape of the World: The Mapping and Discovery of the Earth*, Chicago: Rand McNally.

Blakemore, M.J. and J.B. Harley (1980), *Concepts in the History of Cartography: A Review and Perspective*, Cartographical Monograph., 26. Canada: University of Toronto Press, p. 120 .

Bricker, Charles and R.V. Tooley (1969), *A History of Cartography: 2500 Years of Maps and Mapmakers*, London: Thames and Hudson, p. 276.

Brown, Lloyd Arnold (1949), *The Story of Maps*, Boston: Little, Brown, p. 393.

Campbell, Tony (1981), *Early Maps*, New York: Abbeville Press, p. 148.

Cortesao, Armando. *History of Portuguese Cartography*, Agrupamento de Estudos de Cartografia Antiga, Seccao anexa a Universidade de Coimbra, (Pubs.) 6 and 8. Lisbon: Junta de Investigaçoes do.

Crone, G.R. (1968), *Maps and Their Makers: An Introduction to the History of Cartography*, 4th (revised) ed. London: Hutchinson, p. 184.

Dilke, O.A.W. (1958), *Greek and Roman Maps*, Ithaca, NY: Cornell University Press, p. 224.

Fordham, Herbert George (1927), *Maps, Their History, Characteristics and Uses: A Hand-Book for Teachers*, 2nd ed. Cambridge England: The CUP, p. 83.

Goss, John (1993), *The Mapmaker's Art: A History of Cartography*, London: Studio Edition, p. 376.

Harley, J.B. and David Woodward (1987), *The History of Cartography*, Chicago: University of Chicago Press.

Harvey, P.D.A. (1980), *The History of Topographical Maps: Symbols, Pictures and Surveys*, London: Thames & Hudson, 199p.

Harvey, P.D.A. (1991), *Medieval Maps*. Toronto: University of Toronto Press, p. 96.

Hodgkiss, Alan G. and A.F. Tatham.(1986), 'Annotated Bibliography of Reference Sources: History of Cartography,' in *Keyguide to Information Sources in Cartography*, part 2, pp. 79-127. New York: Facts on File Publications.

Hodgkiss, A.G. (1981), *Understanding Maps: A Systematic History of Their Use and Development*, Folkestone, Kent: Dawson, p. 209.

Jäger, Eckhard (1978), *Bibliographie Zur Kartengeschichte Von Deutschland Und Osteuropa: Eine Auswahl Des Kartographischen Schrifttums Mit Einem Exkurs Uber Landkartenpreise Im 18. Jahrhundert Im Vergleich Zu Anderen Kosten*. Lüneburg: Nordostdeutsches Kulturwerk, p. 92.

Library of Congress, Geography and Map Division, *The Bibliography of Cartography*, Boston: G.K. Hall, 1973-1980.

Mollat, Michel and Monique de La Ronciere (1984), *Sea Charts of the Early Explorers: 13th to 17th Century*, New York: Thames and Hudson, p. 298.

Monmonier, Mark S. (1985), *Technological Transition in Cartography*. Madison: University of Wisconsin Press, p. 282.

Raisz, Erwin. (1948), 'The History of Maps', In *General Cartography*, 2nd ed., p. 7-70. New York: McGraw-Hill Book Co.

Ristow, Walter William (1985), *American Maps and Mapmakers: Commercial Cartography in the Nineteenth Century*, Detroit: Wayne State University Press, p. 488.

Ristow, Walter William (1973), *Guide to the History of Cartography: An Annotated List of References on the History of Maps and Mapmaking*, 3d ed. Washington: Library of Congress, p. 96.

Sobol, Dava (1995), *Longitude: The True Story of a Lone Genius Who Solved the Greatest Scientific Problem of His Time*, New York: Walker & Co.

Thrower, Norman J.W. (1972), *Maps and Man: An Examination of Cartography in Relation to Culture and Civilization*, Englewood Cliffs, N.J.: Prentice-Hall, p. 184.

Thrower, Norman J.W. (1978), *The Compleat Plattmaker: Essays on Chart, Map, and Globe Making in England in the Seventeenth and Eighteenth Centuries*, Berkeley: University of California Press, p. 241.

Wilford, John Noble. (1981), *The Mapmakers*, New York: Knopf, p. 414.

Winchester, Simon (2001), *The Map that Changed the World: William Smith and the Birth of Modern Geology*, New York: HarperCollins.

Woodward, David, (1974), 'The Study of the History of Cartography: A Suggested Framework,' in *The American Cartographer* 1, pp. 101-115.

4
Earth as a Cartographic Problem

Had the earth been flat, Cartography would have developed into an elaborate scientific discipline. Much of the subject matter of cartography centres on earth being spherical. The earth derives its energy and light from sun as it revolves around the sun once in 365 days 6 hours, 9 minutes and 9.54 seconds. Like other planets of the solar system, it has its own magnetic power that far exceeds that of any other planet. It is this magnetic power that keeps it and other planets in position and at a certain distance while revolving around the sun. Earth is also the reference point for time. The earth rotates on its own axis once in the mean solar day which consists of 86,400 mean solar seconds. There are some periodic variations but for our purpose we can proceed further without entering into those finer details.

The moon, the only satellite of the earth also has magnetic power. Being closest to the earth, its magnetic power produces tides in oceans and has influence on all that exists on the earth including humans. Many other planets and the stars act as reference points to determine locations on the earth surface.

The solar system is helio-centric i.e. all planets, including the earth revolve around the sun (Fig. 4.1). This model was first mentioned in India's Vedas more than 5000 years ago. "The Sun never sets or rises and it is the earth, which rotates around it (*Sama-Ved*, 121). The gravitational effect of solar system makes the earth stable (*Rig.Ved.*, 1-103-2, 1-115-4 and 5-81-2). The axle of the earth does not get rusted and the earth continues to revolve on its axis (*Rig.Ved.*, 1-164-29)." In the west, it was first mentioned in *De Revolutionikus Orbium Coelestium* written by Niklas Koppernigk (Copernicus) in 1543. Prior to Copernicus, people in the west believed in a geo-centric model of the planetary system wherein earth was the centre around which other planets and stars revolved (Fig. 4.2). The heliocentric theory was further refined by Newton and Kepler who said that the orbits of the planets were slightly elliptical and that the stars do not revolve round the sun.

The period required by a planet to revolve round the sun is called planetary year. During the course of its revolutionary journey, the distance of the earth from the sun varies from 147 m km in January to 152 m km in July. In July it is said to be at perihelion (nearest to the sun) while in July it is at aphelion (*ap* means away and *helios* means sun). The average distance of the earth from the sun is 150 m km.

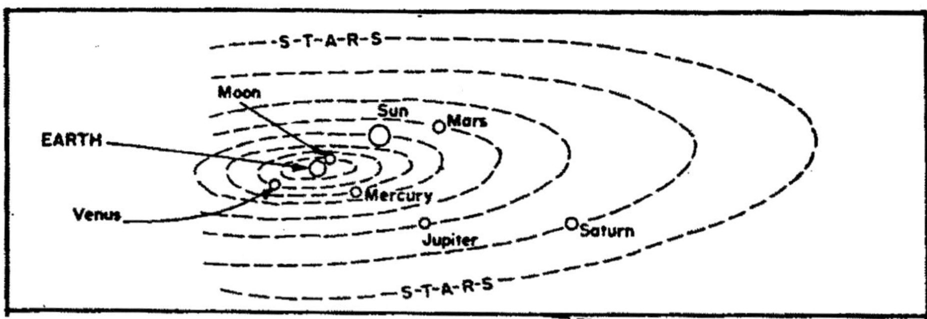

Fig. 4.1: Geocentric (top) model of the Solar System

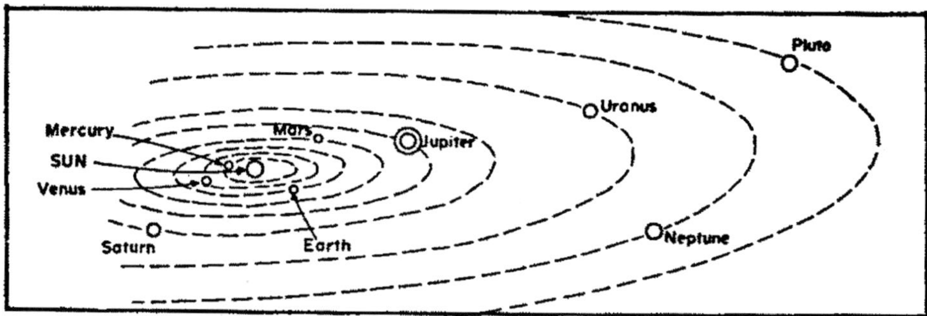

Fig. 4.2: Heleocentric (bottom) Model of the Solar System

The earth revolves round the sun counter-clockwise completing one revolution in a sidereal year which is a few minutes longer than the calendar year. Similarly, the moon completes one revolution around the earth in one sidereal month which is equal to 27.3 days. The earth itself rotates on its own axis in one sidereal day which is equal to 23 hours, 56 minutes, and 4.09 seconds. As a complete rotation of the earth on its axis is equivalent to 360 degrees, the earth traverses 15 degrees every hour. Thus the local apparent solar time, changes one hour for every 15 degrees of longitude. It decreases as one move to the East. By an international agreement, the day begins at the International Date Line, which roughly follows the 180° longitude. One gains a day while crossing this line from west to east, but loses a day when crossing it from east to west. The zero degree meridians passing through Greenwich in UK is called Greenwich Meridian. Civil Time on two sides of this meridian changes by an hour for every 15 degrees of longitude. So, if it is mid-night at Greenwich meridian, it would be 2 AM at 30° E meridian, and 10 PM at 30° W meridian.

SHAPE OF THE EARTH

Six thousand and four hundred years before Isaac Newton, the Rig Veda asserted that the shape of Earth is like an Oblate Spheroid. (*Rig Veda*, XXX. IV. V). It also said that gravitation held the universe together. Greeks believed in a flat earth. The Indians of the fifth century A.D. calculated the age of the earth as 4.3 billion years; scientists in 19th century England were convinced it was 100 million years old. In this regard, Pythagoras (570-495 BC) was an exception in the west. He was a mathematician and to him the most perfect figure was a sphere. He reasoned that God will create only a perfect figure and, therefore, the earth must be spherical in shape. The Pythagorean idea was apparently based on the phases of the moon. About two hundred years later, Aristotle was led to the same conclusion. In the third century B.C. Eratosthenes demonstrated that the earth was spherical by measuring the angle at which the rays of the sun fell at two points located on the same longitude but on different latitudes.

Eratosthenes had observed that on the day of the summer solstice, the mid-day sun rays were directed to the bottom of a well in the town of Syene (now Aswan). At the same time, he observed that the sun was not directly overhead at Alexandria. It rather cast a shadow with a vertical equal to 1/50th of a circle i.e., 7° 12′ (Fig. 4.3). Knowing that:

1. Syene was situated on the tropic of cancer because the mid-day Sun was overhead on the day of summer solstice;
2. Linear distance between Alexandria and Syene was 805 km; and
3. Alexandria and Syene lay on a direct north-south line;

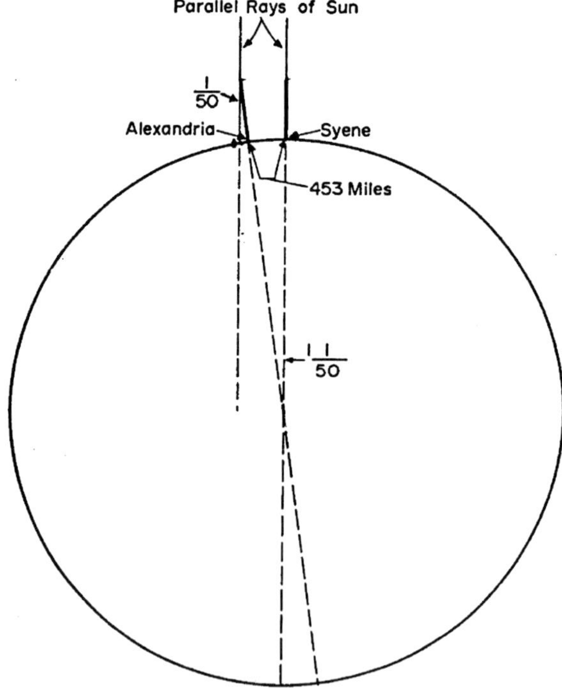

Fig. 4.3: Eratosthenes's Method of Determining Earth's Circumference

He concluded that the circumference of the earth at the equator must be 40,070 km. It is even more remarkable that such accuracy was obtained even though the facts which were made the basis for the calculations were not correct. As we know Syene is not at the tropic of cancer, it is about 59.5 km north of it; the distance between Alexandria and Syene is not 805 km, it is only 729 km, and Syene and Alexandria are not on the same longitude. The latitudinal difference between the two places is not 7°12′. It is only 7°5′.

The next significant advance in geodesy came in 1617 when Snellius (Willebrord *Snell*, 1591-1626) calculated the circumference of the earth from astronomical measurements. Later in 1671 Richer in French Guiana discovered that gravity was weaker near the equator than on the higher latitudes. After fifteen years, in 1786, Isaac Newton gave the correct description of the shape of the earth as being an oblate spheroid with its polar radius shorter than the equatorial radius. The theoretical proof of the spherical shape was given by Clairaut in 1743. Since then only minor changes have been made in Clairaut's model. The recent satellite measurements have resulted in two small modifications in the shape of the earth. One is the discovery that the southern hemisphere has a slightly larger circumference than the corresponding latitudes in the northern hemisphere, making the earth some-what pear-shaped. The other is that the equatorial cross-section of the earth is not an exact circle; it is slightly elliptical and tri-axial.

Since the earth is flattened slightly at the poles and some-what bulged at the equator, the geometrical shape used in geodetic computations is an ellipsoid of revolution (Fig. 4.4). This is a figure which one would obtain by spinning an oval disc on its flat side or by rotating an oblate ellipse on its north-south axis.

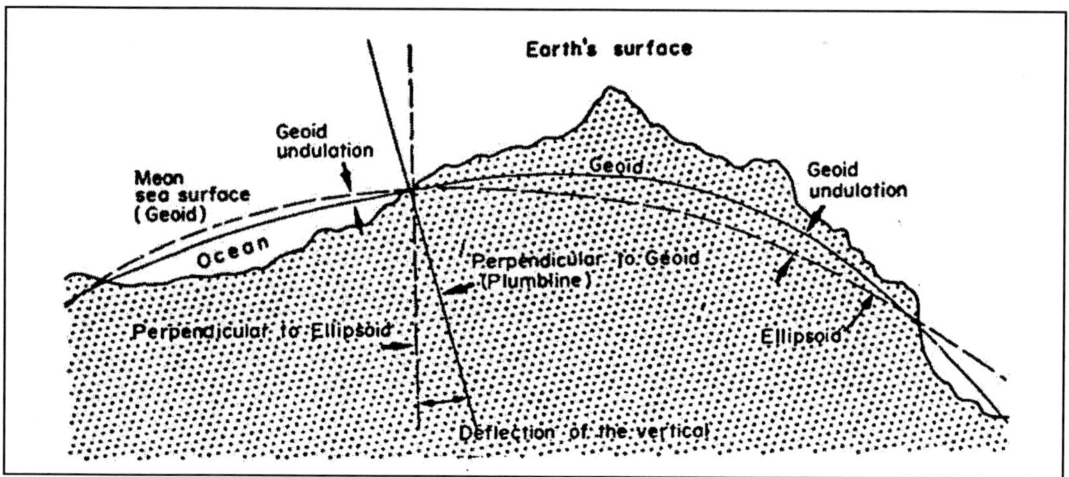

Fig. 4.4: Difference between an Ellipsoid and a Geoid

At present 11 ellipsoids are being used in different parts of the world. Seven of these are given in Table 4.1.

The ellipsoid of revolution differs from reference surface called geoids (Fig. 4.4).

Table 4.1: Ellipsoid of Revolution Used by Different Countries/Regions

Country	*Name of the Ellipsoid*	*Year*	*Equatorial Flattening*	*Radius*
India	Everest	1830	6,377,276	1/300
UK	Airy	1830	6,376,542	1/299
Japan	Bessel	1841	6,337,397	1/299
N. America	Clarke	1866	6,378,206	1/295
France	Clarke	1880	6,378,249	1/293
Europe	International	1924	6,378,388	1/297
Russia	Krassowsky	1940	6,378,249	1/298

Geoids represent a level surface, i.e., one that is perpendicular to the direction of the force of gravity. To visualize geoids, imagine the surface of the earth to be completely covered with water. The surface of this endless sea will correspond to geoids. Because of the irregularities in the distribution of the earth's mass, this surface slightly departs from the reference spheroid or the ellipsoid of the revolution.

GRAVITY OF THE EARTH

The concept of gravity is one of the fundamental cornerstones of geo-physics. In cartography, gravity is of special significance mainly because it helps us in determining the shape of the earth. The idea that the earth attracts everything towards its centre had been prevalent even in ancient times. In India the concept of '*Gurutwakarshan*' was developed during the Vedic and Puranic times. In Greece, Aristotle believed that the speed of a falling body depended on its mass. Early in the 17th century Galileo refuted this theory and proved that all bodies, irrespective of their masses, fall with the same constant acceleration. The acceleration of a body is the rate at which its velocity changes with time. Half a century later, Newton formulated the law of gravitation and thus explained the rotation of the planets and their satellites. He calculated the magnitude of the gravitational pull of the earth being proportional to the mass of the body falling on the earth multiplied by the mass of the earth (a constant) and inversely proportional to the square of the distance between the centre of the mass of the body and the centre of the mass of the earth.

It is the force of gravity which is largely responsible for the present shape of the earth. To understand this let us imagine a soft plastic sphere rotating about its axis. Because of the centrifugal force (inertia) the mass is pushed outward from the axis of rotation. As this force increases with distance from the axis of rotation, it is more effective at and near the equator. As a result, the sphere bulges outward around the equator and an oblate spheroid comes into being. This is the present shape of the earth (Fig. 4.5).

An object on the rotating earth is subject to outward-directed centrifugal force, which increases from zero at the poles to a maximum at the equator. The gravitational force due to the earth's mass is directed toward the centre of the earth almost everywhere but the centrifugal force detracts it from the centre. The measured gravity of the earth is the result of the combination of mass gravitation and centrifugal force. Gravity is, therefore, weaker at the equator than at the poles. It is about 0.5 per cent greater at the poles than at the equator.

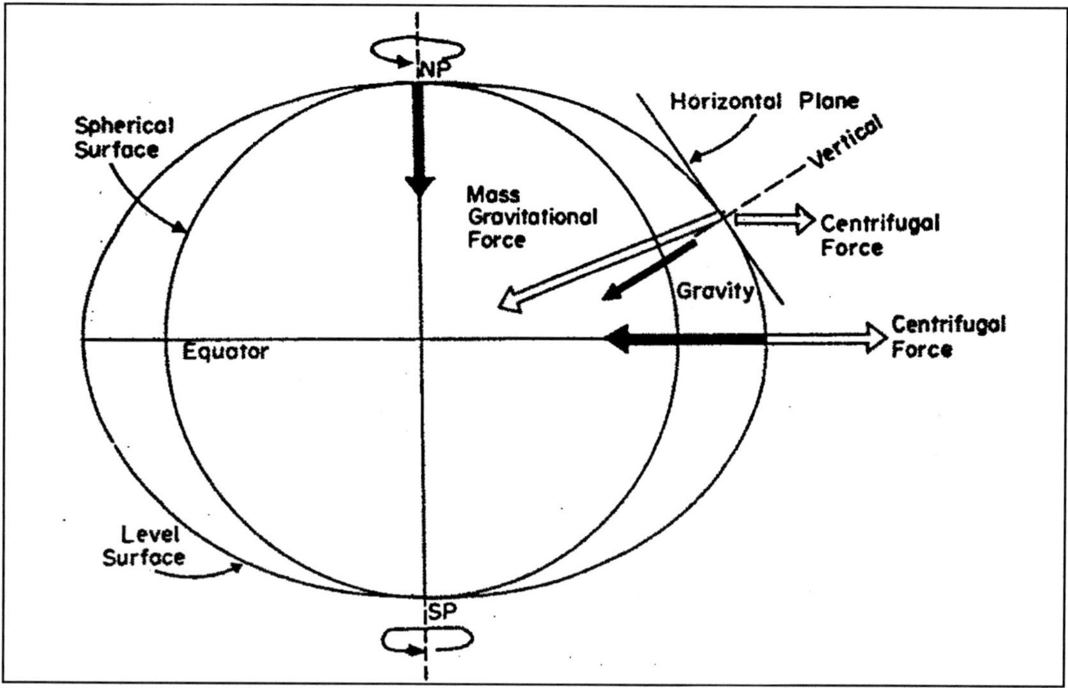

Fig. 4.5: Effects of the Rotation of the Earth and its Gravity on its Shape

DIMENTIONS OF THE EARTH

The International Union of Geodesy and Geophysics (IUGG) have accepted the length of the radii of the earth at different latitudes. These are:

Polar radius	6357 km
Equatorial radius	6378 km
Mean radius	6371 km

The surface area of the earth can be determined easily if we know the radius of the earth. Area of a sphere is 4 π times its radius ($4\pi r^2$). The area of the earth is about 510,900,000 sq. km. The total mass of the earth is determined by the gravitational force. It is almost 5.98×10^{27} grams. When we divide the mass (5.98×10^{27} grams) by the volume (1.08×10^{27} cu cm) we get the average density of the earth (5.52 grams per cu cm). This is 5.52 times greater than the density of pure water.

One of the basic problems is cartography is to reduce the earth to a convenient size and to convert its spherical surface into a plane surface.

REFERENCES

Grotzinger, John, *et al.* (2007), *Understanding the Earth,* New York: W H Freeman & Co.
Kump, Lee R. *et al.* (2009), *The Earth System*, 3rd ed., Englewood Cliffs, N.J.: Prentice Hall.
Mackenzie, Fred T. (2010), *Our Changing Planet: An Introduction to Earth System Science and Global Environment Change* (4th ed.), Englewood Cliffs, N.J.: Prentice Hall.
Skinner, Brian J. *et al.* (1999), 2nd ed., New York: John Wiley.
Stanley, Steven M. *Earth System History*, New York: W H Freeman & Co.
http://ebooks.bfwpub.com
www.whfreeman.com/understandingearth5e
www.whfreeman.com/earthinquiry

5
Scales and their Functions

Maps can be considered as a scale model of the world. The world is too big and complex; it has to be greatly reduced to fit on a piece of paper. Map scale is the means to do it. Map scale is a statement setting the relationship between map distance and distance in the real world. It tells how much the world has been reduced in size to fit on a sheet of paper. Scales can be expressed in many ways. More often than not, it is given as a ratio between one unit of distance on a map and the number of the same units of distance in the real word. Thus, on a map with the scale 1:633,600, one cm on the map would be equal to 633,600 cm in the real world. It can also be written in the form of statement like one inch = 10 km. One can represent scales graphically too. In this case, a calibrated bar is drawn to show map distances. Each of these representations of scales has its own advantages. The graphic scale is a must if maps have to enlarged or reduced in size photographically or through a computer. A map with ratio and written scales is easy to understand but if it gets reduced or enlarged by photocopying, its scale will become incorrect, since the relationship between map distance and real world distance has been altered.

For small areas like villages and towns, the scale may be taken as a constant ratio over the whole map. But for larger areas, one has to take into account the curvature of the earth too. And this is done with help of map projections discussed in chapter 8, which convert the sphere (or ellipsoid) into a plane. Projections, however, distort the scale and hence the concept of variable point scale or place/area specific scale has been developed. It can be defined as the ratio of the length of a small line on the map to the length of the corresponding length of a line on the surface of the Earth. The point scale varies with the position of the point and also the direction of the line.

While studying the point scale, it is better to so define the projection formulae that the scale is unity on some lines (or points) of the resulting map projection. For small areas it can be scaled down by a constant ratio called Representative Fraction (RF) or Principal Scale. Thus we have two uses of the word scale: the fixed ratio or constant scale is usually used to reduce a large scale map; and the variable point scale or particular scale related to map projections and referred to as small scale maps. In the second case, the curvature of the earth is taken into account, while in the first case, the curvature is ignored.

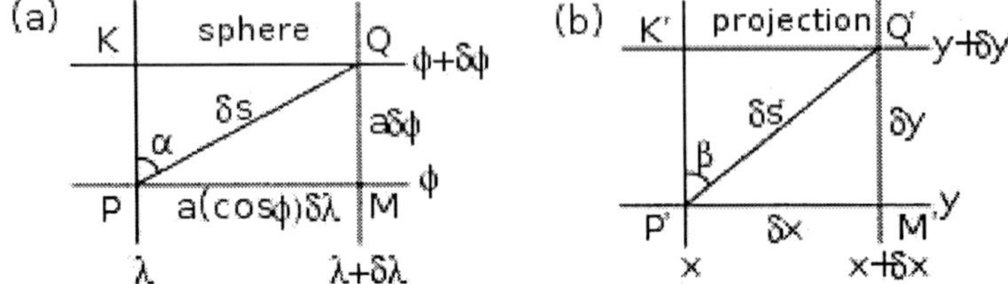

In the above figure, P is a point at latitude φ and longitude λ on the sphere (or ellipsoid) and Q is another point on latitude $\phi + \delta\phi$ and α, the angle between the element PQ and the meridian at P is the azimuth angle of the element PQ on the earth surface. When this space is put on a projection, the corresponding points P and Q become P′ and Q′. The angle between the direction P′Q′ and the meridian P′ is the bearing β. Generally, $\alpha \neq \beta$ because the azimuth on the Earth's surface and bearing on the map are not the same. The point scale at P is the ratio of distances P′Q′ and PQ and this ratio approaches zero as Q approaches P. That is:

$$\mu(\lambda, \Phi, \alpha) = \lim_{Q \to P} \frac{P^1 Q^1}{PQ}$$

Thus the point scale is a function of the position of P and also the direction of the element PQ. If the point scale is tied only to position, not to direction, it is isotropic i.e. *small* elements are stretched equally in all directions, the shape of those element is preserved. This is the property of orthomorphism (from Greek 'right shape'). The Mercator projection is conformal because it preserves angles and its scale factor is a function of latitude only.

CALCULATING THE MAP SCALE

To find out the scale of a map, we proceed as follows:

$$\frac{\text{Distance on Map}}{\text{Distance on Ground}} = \frac{1.27 \text{ inches}}{5 \text{ miles}} = \frac{1.27 \text{ inches}}{316800 \text{ inches}} \frac{1}{250{,}000} = 1{:}250{,}000$$

On the basis of details shown, maps are often referred to as large scale, intermediate scale and small scale. A large scale map shows a limited amount of space with considerable amount of details about that area. City and village maps showing the location and dimension of all the buildings, trees, streets or the location of temples, parks, and agricultural fields are examples of large scale maps. As against this, the small scale maps represent extensive areas, but few details. Maps with small scales are appropriate for showing large countries, continents, and the world. In between stand the medium scale maps. Usually, large scale maps vary in RF from 1:50,000 to

1:100. The small scale maps are those which are made on scales smaller than 1:250,000. Maps ranging in scales from 1:50,000 to 1:250,000 are categorize as intermediate scale maps.

The potential for accuracy of details drops as the area mapped grows larger and the scale becomes smaller. Small scale maps are thus likely to distort details. This is less likely to be so with large scale maps. Hence, distances on world and continental maps need to be viewed with great caution, while those having limited spatial extents like villages, urban neighbourhoods, towns, cities, and states and drawn on the large scale in general have higher levels of accuracy.

A mentioned in the very beginning of the chapter, it must be repeated here that the scale of a map does not express the ratio of actual distances between the corresponding points on the map and the ground as most books on Cartography and Practical Geography say. It is only the horizontal straight line distance that is taken into account. So if the airline distance between two points is 100 km and the actual distance is 120 km (due to the presence of mountains, for example) we will take only 100 km into account for the purpose of the scale (Fig. 5.1).

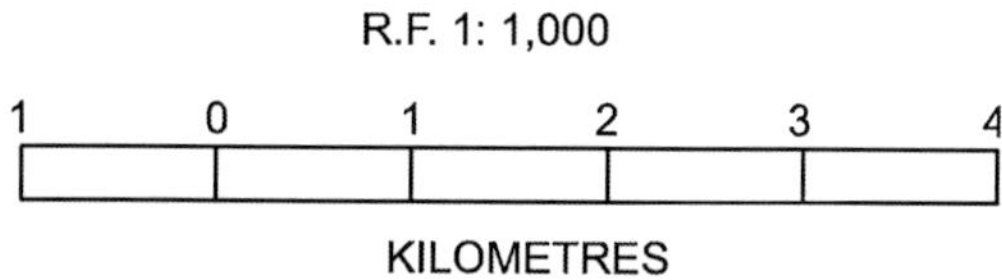

Fig. 5.1: Relation between Ground Distance and Map Distance

Objects can be reduced only so far and then they must be eliminated because it becomes impossible for the human eye to see them. Important details to be retained on a map have, therefore, to be exaggerated. These are no longer drawn to scale. So we should not expect a map to give every detail to scale. What we should be sure of is that all maps give areas covered by spatially distributed details to the scale shown in the map.

NOTATION OF SCALES

The Representative Fraction (RF) or Principal Scale

Most of the printed maps are based on projection map by a *constant* scaling denoted by a ratio. So if one cm on the map represents 1000 cm on the ground, the scale of the map is: 1: 1000. It also means that 1 inch on the map is equal to 1000 inches on the ground. If 1 cm on a map represents 1 km on the ground, the scale of the map will be 1: 100,000. These examples are enough to explain how scales are the ratios between the distances on a map and on corresponding crow flight distances on the earth surface. On a scale of 1:20,000, twenty thousand ground units are be compressed into one map unit.

To avoid confusion in the use of the word 'scale' this constant scale fraction is called the Representative Fraction (RF) of the printed map and it is to be identified with the ratio printed on the map.

We have already noted the ways to show scales on maps. One of the commonly used scales on the map is Representative Fraction (RF). The numerator of the RF represents the map distance while the denominator represents the actual crow flight distance on earth surface. The distances may be measured in any unit—mm, cm, or inches. One must keep in mind that a map showing RF scale will not be true to scale if it is reduced or enlarged in size photographically or by printer.

Verbal Notation

Another way of expressing scale is to make a simple statement giving the number of map units as a fraction of the corresponding earth units. So instead of saying 1:100,000 in the form of RF, we can say that one cm on the map represents one km. on the ground. To further shorten the expression we can say,
One cm to one km, or
1 cm = 1 km

Plane or Bar Scale

By far the most commonly used scale is bar graph. The bar scales are calibrated to express visual equivalents of the representative fraction or verbal scale. They can be used directly without the preliminary conversion needed in representative fraction. Further, they do not get affected if the map is reduced or enlarged because the graphic scale notation automatically conforms to the change in scale. Beginners often commit the mistake of using representative fraction or verbal scale on original drawings for being reduced in size for printing. If such scales do not adhere to the scale of the printed map, they will be incorrect.

To convert the RF 1:100,000 into a plane scale, all we have to do is to read it as 1 cm or inch on the map = 100,000 cm or inch on the ground. Taking cm as the unit of measurement we can write it as 1 cm = 100,000 cm or 1 cm = 1000 m or 1 cm = 1 km.

For graphical scales, a horizontal line or bar of appropriate length is drawn and divided into desired primary parts, let us say five parts. The first part should be earmarked for secondary divisions. Our starting point is zero (see Fig. 5.2) to the left are the secondary divisions and to the right the primary divisions. The portion left of the zero is divided into fractional parts such as a km being divided into ten parts each representing 100 meters. The secondary divisions enable the map user to read the distance in fractions, if need be. Conventionally, a bar scale should not be longer than 20 cm and shorter than 7.5 cm. If we have to prepare a bar scale of 1 cm = 25 km, a base line of 20 cm is drawn and then divided into five equal parts, each representing 2 km. We leave the first division on the left hand side for secondary divisions and start with zero at the end of the first primary division from the left. In this case the line of 10 cm is divided into five equal parts, so each division represents 1 km (Fig. 5.2).

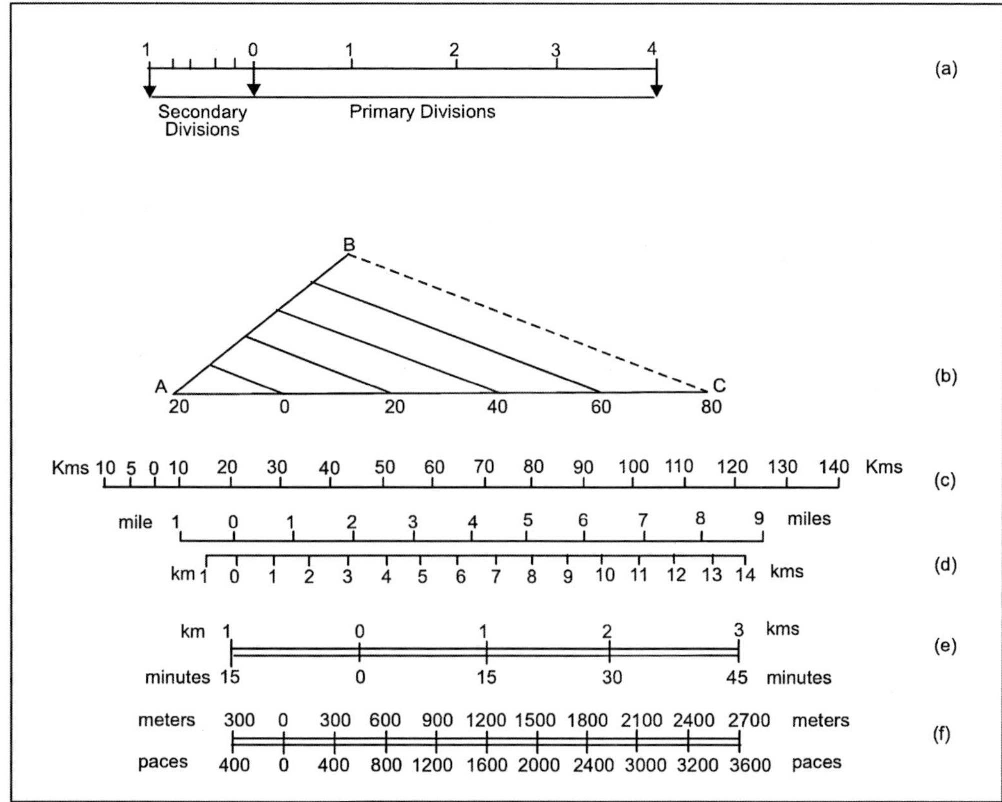

Fig. 5.2: A Graphical Scale

The graphical scale can also be drawn to represent not only the absolute but also the comparative values. Even the scales of cube roots and square roots can be presented. Rudiments of the methods of constructing these scales are given in Fig. 5.3. The first of these comparative scales shows distances in km and miles with the help of two bars one for miles and the other for km. Both the bars should be approximately equal.

The second example illustrates the rates of movement. Suppose a contingent of police trainees is moving at the rate of 10 km an hour. If we have to draw a time scale for a map of inch to a mile, we will proceed as follows: if 1 mile on the ground is represented by 1 inch on the map, 4 miles on the ground will be represented by 4 inches on the map. Given the speed of 4 miles an hour, we draw a bar of 4 inches; divide the upper line of the bar into 4 parts each part representing a mile, and the lower line into 4 parts representing 15 minutes.

At times we may have to measure distances by paces or revolutions of wheels with comparable distances in meters or yards. If we know the circumference of the wheel and the number of revolutions made by it, we can be calculated the distances covered by it. The distances covered and the number of revolutions made can then be shown on the same bar scale (Fig. 5.3).

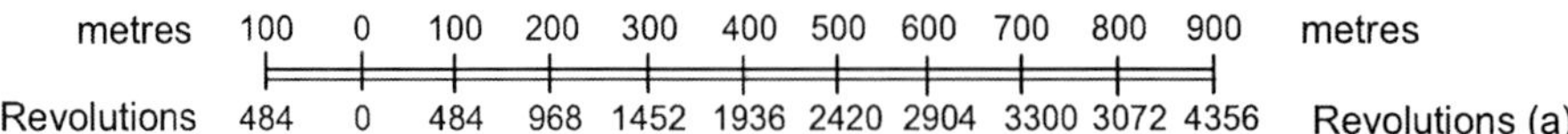

Fig. 5.3: Comparative (Miles and Km), Movement (Miles and Minutes), Yards and Paces), and Furlongs and Revolutions Scales

While preparing a scale, we have to keep in mind that the scale of a given map stands for the whole map area. Maps drawn on projections (see chapter 8) that are not equal area cannot have a single scale no matter how it is presented. For example, a single bar scale for Mercator's projection is applicable only for a given parallel of latitude. As we move towards the poles, distortions increase. In such cases multiple scales have to be adopted to get correct distances along succeeding parallels.

Diagonal Scale

A diagonal scale gives greater precision to measurements. Suppose we have to draw a scale of 4 inches to measure 2.56 inches. We draw a line AB of 4 inches and divide it into 4 equal parts (Fig. 5.4). Draw perpendiculars AC and BD from A and B each measuring 1 inch. Divide AC and BD into ten equal parts and draw lines parallel to AB. To get the secondary divisions, we divide the top and bottom lines AA' and CC' into ten equal parts. Join zero of the bottom lines with one of the top lines, one of the bottom with the two of the top lines, two of the bottom with the three of the top line, and so on. To measure 2.56 inches, place one leg of the divider at X and the other at Y. To get 1.33 inches measure the distance MN.

The XY distance consists of Xa+ab+bY = 5/10 + 6/100 + 2 inches = 0.5+0.06 + 2 inches = 2.56 inches.

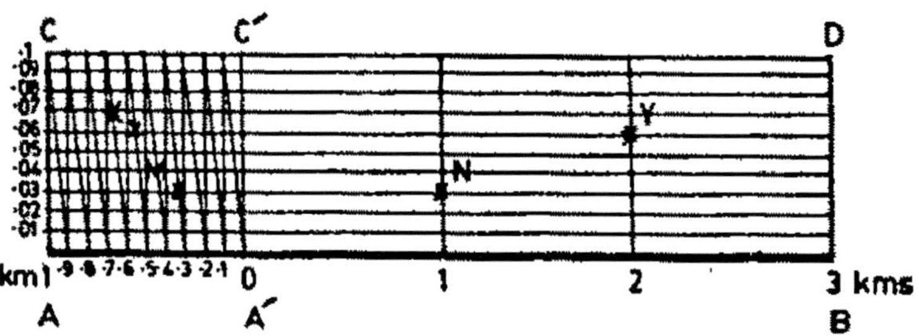

Fig. 5.4: Example of a Diagonal Scale

Vernier Scale

Vernier Scale (devised by B.P. Vernier) helps estimate with greater accuracy the fraction of a division of line or angle. It consists of a moving scale which slides along a primary scale. On the basis of the length and numbering system it is divided into three types:

Direct or Positive Vernier, Retrograde or Negative Vernier, and Double Vernier. Here are the details of the Direct Vernier Scale which is the more common than the other two. In this scale, the graduation on both the scales (primary and Vernier) run in the same direction and the zero of the Vernier is marked by an arrow.

Suppose we have to construct a Vernier with a least count of one hundredth of an inch, we follow the following procedure. Draw a line AB and divide it into inches and tenths of inches as shown in Fig. 5.5. Now draw a Vernier CC on AB equal to 9 small divisions of AB. Divide CC′ into 10 equal parts so that a small division of the Vernier scale is 1/100 of an inch shorter than a small division of the primary scale. Here, the zero of the Vernier scale coincides with the zero of the primary scale and hence, the first Vernier division lags behind by 1/100 inch, the second by 2/100 inch and the third by 3/100 inch and the tenth by 10/100 inch or 1/10 inch or 1 small division of a primary scale. To read 2.56 inches slide the Vernier so that the zero of the Vernier coincides with the 2.5 inch mark of the primary scale. Now shift the Vernier to the right, so that the 6th division of the Vernier coincides with the primary scale graduation. The length between the zero of the Vernier scale and the zero of the primary scale is 2.56 inches. Similarly we can have Vernier scale showing degrees and minutes.

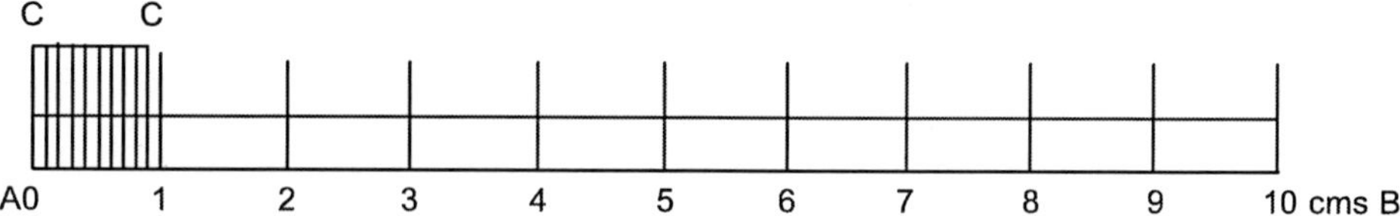

Fig. 5.5: Example of a Vernier Scale

Determining the Scale of a Map

To determine the scale of a map (if it is not given), first find out the ground distance between any two points shown on the map and then measure the map distance between the same points and compare the two. If the ground distance is two km and the map distance is two cm, then 2 cm on the map represent 200,000 cm on the ground. Since, the scale ratios are always expressed with one as the base, we divide both, 2 and 200,000 by 2 to obtain the RF 1:100,000. RF can then be converted into statement or bar scale.

Reduction and Enlargement of Maps

In the process of compiling maps cartographers are often required to reduce or enlarge maps. Reduction or enlargement involves change in the size. One simple way to illustrate what happens to the size of a map when it is reduced or enlarged is to fold a sheet of paper. Take a sheet of ordinary notebook and assume it to be a map of a given scale. To show the same area reduced to 1/2 of the original scale, fold the paper in half each way since the both length and breadth of the map are being reduced to half. Now we have one-fourth of the paper area of the original, while the scale is 1/2 of the original. Fold the paper once again in each direction to illustrate four times reduction which gives a paper having 1/16 the size of the original.

This paper folding can also be depicted mathematically. The ratio between the area of a map on one scale and its area on another scale is equal to the square of the ratios between the scales of the original and enlarged maps.

Example 1:

Reduce 1:10,000 to 1:50,000 scale.

$$10{,}000 : 50{,}000 = \frac{10{,}000}{50{,}000} \text{ Or } 1/5 \text{ of the original line or scale.}$$

$$\text{And } 1/5)^2 = \frac{1}{25} \text{ of the area of the original}$$

Example 2:

Enlarge 1: 100,000 to 1: 20,000

$$100{,}000 : 20{,}000 = \frac{100{,}000}{20{,}000} = 5 \text{ times of the original line or scale and}$$

$(5)^2 = 25$ times the area of the original.

If we know the original scale of a map and want to find the new scale of its reduced or enlarged version, we should use the principle of ratios.

Example 1:

Given scale 1: 100,000 is to be enlarged 10 times

$$\frac{1}{x} : \frac{1}{100{,}000} \text{ or } \frac{100{,}000}{x} = 10 \text{ or } 100{,}000 = 10 \times \text{ or } \times = 10{,}000$$

The enlarged scale is 1: 10,000

Example 2:

Given scale 1: 10,000 is to be reduced 5 times.

$$\frac{1}{100{,}000} : \frac{1}{x} = 5 \text{ Or } \frac{x}{10{,}000} = 5 \text{ Or } \times = 5 \times 10{,}000 = 50{,}000$$

The reduced scale is 1: 50,000

G-Scale

Haggett, Chorley and Stoddart have proposed a standard of geographical measurement which indicates the relative size of the area studied with the total area of the earth as the base. They call it G-scale. G-scale of an area can be determined by the following formula:

$$G = \log Ga/Ra$$

Where

G = G-scale of the area studied
Ga = Earth's surface area
Ra = Area under investigation

The area can be measured in any of the several known units such as acres, hectares, square km, etc. Formulae for computing G values from different conventional area standards are given below:

Conventional Standard	Formula
Square miles	8.2941-log Ra ($mile^2$)
Square kilometres	8.7074-log Ra (km^2)
Acres	11.1003-log Ra (ac)
Hectares	10.7074-log Ra (ha)

On the basis of this scale, the following will be the G-scale for a few selected areas.

Earth — G	
Asia (Excluding USSR)	G = 1.1
India	G = 2.2
Karnataka state	G = 3.4
Mysore district	G = 4.2
Mysore taluk	G = 5.8
Mysore city	G = 6.9
University of Mysore campus	G = 8.30
Geography Department and its Surroundings (10 acres)	G = 10.10

CONCLUDING REMARKS

The beginners often interpreted map scales incorrectly, largely because the smaller the map scale, the larger the reference number and vice versa. For example, a 1:1,000,000 sheet produced by the Survey of India is considered a large scale map and a toposheet at 1:250,000 scale is considered a small scale map. It should be kept in view that small scale map is one which shows large areas or distances on the ground by small areas or distances on the map.

Most geographical areas of the world have been mapped but at different scales. While preparing a map, its intended use should be considered before deciding on which scale to use. To get a close view that gives the most detailed information of a small geographical area, one should choose a map with a small second number, i.e. 1:100. For a broad general view of a large geographical area, a map with a larger second number in its RF, i.e. 1:100.000 should be selected.

Following is a short list of map scales.

RF Scale	*1 Inch on Map Represents*	*1 Centimeter on Map Represents*
1:24,000	2,000 feet on ground	240 meters on ground
1:50,000	4,166 feet	500 metres
1:63,360	1 mile	633.6 metres
1:100,000	1.6 miles	1 kilometre
1:250,000	4 miles	2.5 kilometre
1:500,000	8 miles	5 kilometres
1:1,000,000	16 miles	10 kilometres
1:2,000,000	32 miles	20 kilometres

REFERENCES

Gonzales, Doren (2008), *Are We There Yet? Using Map Scales*, Capstone.
Monmonier, Mark (2005), 'Lying with Maps', *Statistical Science*, vol. 28, no. 3, pp. 215-22.

6
Directions and Their Functions

Direction has been defined as an imaginary line on the map or ground showing a common direction by an arrow with respect to which the position of all other points are determined. The line pointing to the north is regarded as the zero direction or base direction line.

The two ends of the axis of the earth's rotation are known as north and south poles. These are also called geographical poles. The line that joins these two poles is the zero direction line. Any line cutting this line at right angles is the east-west line. North, South, East and West are the four main directions. These are also called cardinal points. In between the cardinal points, one can have several intermediate directions as given in Fig. 6.1.

A map must have the base direction represented on it to enable the user to locate different features with respect to each other. Such a direction is often shown by an arrow pointing to the north. At times the base direction is not indicated on a map. This generally means that the top represents the true north or that the longitudes are indicative of the north-south direction.

As we can see in Fig. 6.1, we can represent only a few intermediate directions by means of arrows. If we have to show a direction in between N E and E N E, we will be hard put to express it. It is for this reason that directions are often given in terms of angles. North is regarded as the angle of zero direction. All directions are expressed as angles, measured clockwise from north throughout the full range of the directional circle. These angles are called azimuths. In this system we have no problem with regard to expressing the angles between N E and E N E. All we have to do is to measure the angle between the zero direction line and the line pointing to the object whose direction is needed.

Directions are generally measured in terms of degrees but they are also sometimes measured in terms of 'Mil' and 'grad'. The sub-divisions of a degree are called minutes; and minutes are divided into seconds. When a circle is equally divided into 360 angles each of the angles represents a degree. In the same way each degree is equally divided into 60 minutes, and each minute into 60 seconds. This plan establishes over one million seconds in a circle. With the help of these sub-divisions, it is possible to measure

Fig. 6.1: Cardinal and Intermediate Directions

directions with utmost accuracy. For the sake of convenience, degree, minutes and seconds are represented by conventional signs so that if we want to write 168 degrees, 56 minutes and 43 seconds, we will write 168° 56′ 43″. (see Fig. 6.2)

A mil is the angle subtended by 1 meter across distance of 1000 meter from the centre of a circle. It can be expressed as an angle of tangent of which is approximately .001. It is 1/6, 400th part of the circumference of a circle, so that 90° = 1600 mils. Fig. 6.2 shows the relationship between a mil and a degree. Mil is used more for military purposes because change of one mil in the direction of the tube of a weapon changes the centre of the impact of a round 1 yard at a range of 1000 yards, 2 yards at a range of 2000 yards and so on.

A third system of measuring directions is called 'grad'. 400 grads form a circle. Being divisible by 10, it can be more conveniently expressed in metric system. Its use is, however, not very common.

TRUE, MAGNETIC AND GRID NORTH

True north is the direction pointing to the North Pole. On a map, true north is represented by meridians or lines of longitudes. Magnetic north is the direction which points to the magnetic pole. Magnetic attractions in earth's core cause the needle of a compass to be pulled away from the north. Magnetic attraction differs from locality

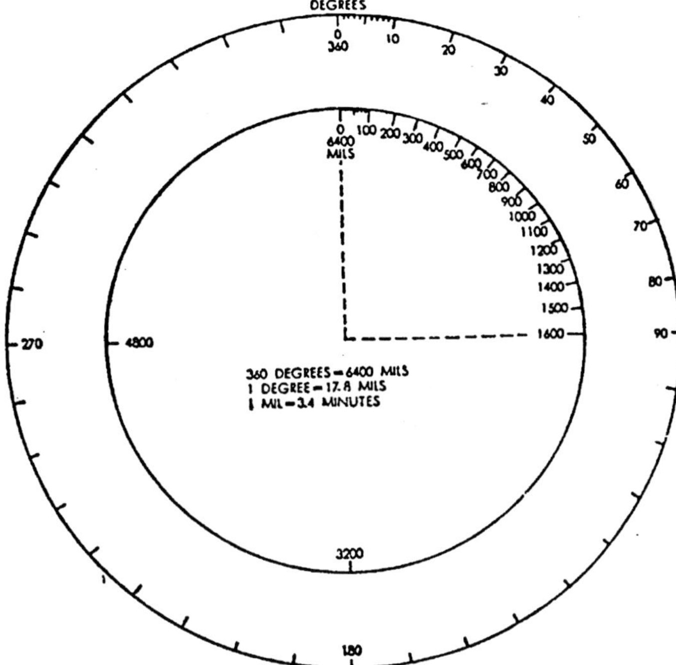

Fig. 6.2: Relationship between a degrees and a mail

to locality, depending upon the position of a point with respect to the magnetic pole. Fig. 6.3 shows the difference between the magnetic and the true north.

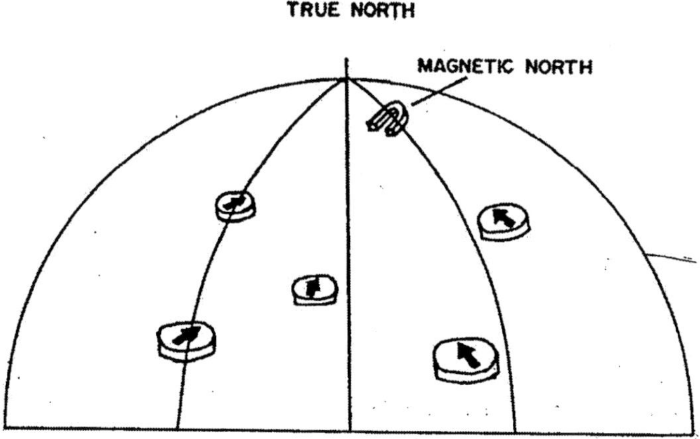

Fig. 6.3: Magnetic North and True North

Grid directions also vary from place to place. These variations are illustrated in Fig. 6.4 which shows a grid super-imposed on a projection with meridians pointing to true north. The angular difference in magnetic or grid north and true north are respectively known as magnetic and grid declinations. These values are often given on the margins of the topographical sheets. At times they are also expressed diagrammatically (Fig. 6.4). As the magnetic declination also give the amount of annual change and the year to which the declination pertains.

There are certain points at which the compass needle pointing to magnetic north also points to true north. At these points the true and magnetic north coincides. The line which joins these points is called 'agonic line' (Fig. 6.5). On two sides of an agonic line, are the lines connecting the points of same magnetic declination. Such lines are called isogonic lines. Maps covering the entire world with isogonic lines are used in determining the magnetic declinations. It is from these maps that we obtain information to construct declination diagrams.

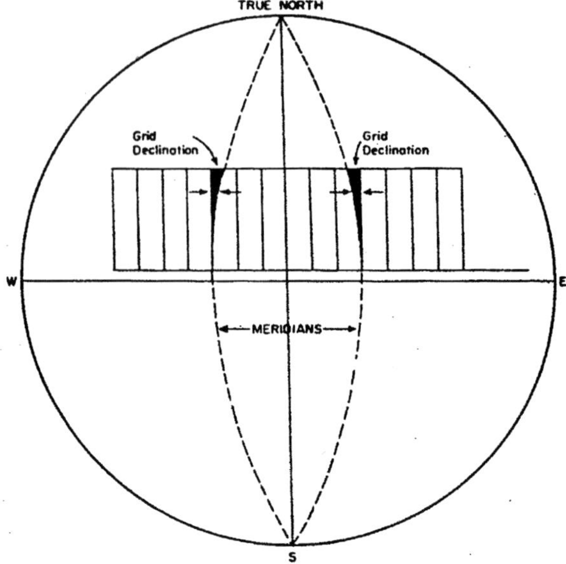

Fig. 6.4: Grid North and True North

DETERMINATION OF TRUE AND MAGNETIC NORTH

Using the Azimuth

As stated earlier, a direction angle is always measured clockwise with north as the zero direction. For the sake of convenience, the zero angle is called 'azimuth'. There are three types of azimuths. These are:

1. True azimuths,
2. Magnetic azimuth, and
3. Grid azimuth.

Fig. 6.5 gives the three types of azimuths for the same geographic position. To determine the magnetic azimuth we use a magnetic compass. The details of this instrument are given in Chapter 9. This instrument gives angular values in degree with reference to the magnetic north. One end of its needle always points to the magnetic north. These angles are known as bearings.

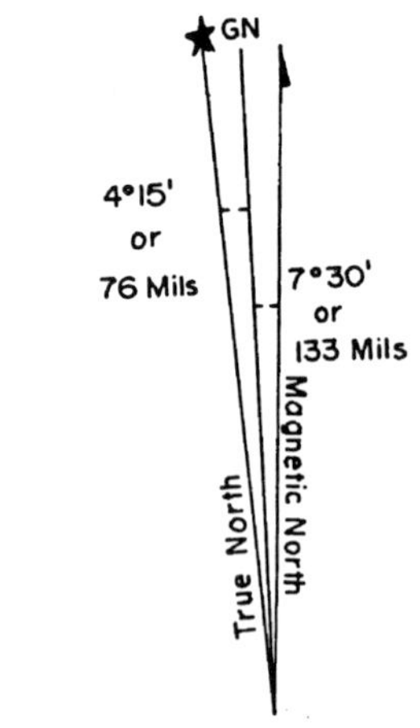

Fig. 6.5: Representation of True, Magnetic and Grid North on a Map

Using the Stars

To determine the true north all that we have to do is to look at the North Star. This star always points to the North Pole. A similar method is used to determine the true south in southern hemisphere. These techniques are illustrated in Fig. 6.6.

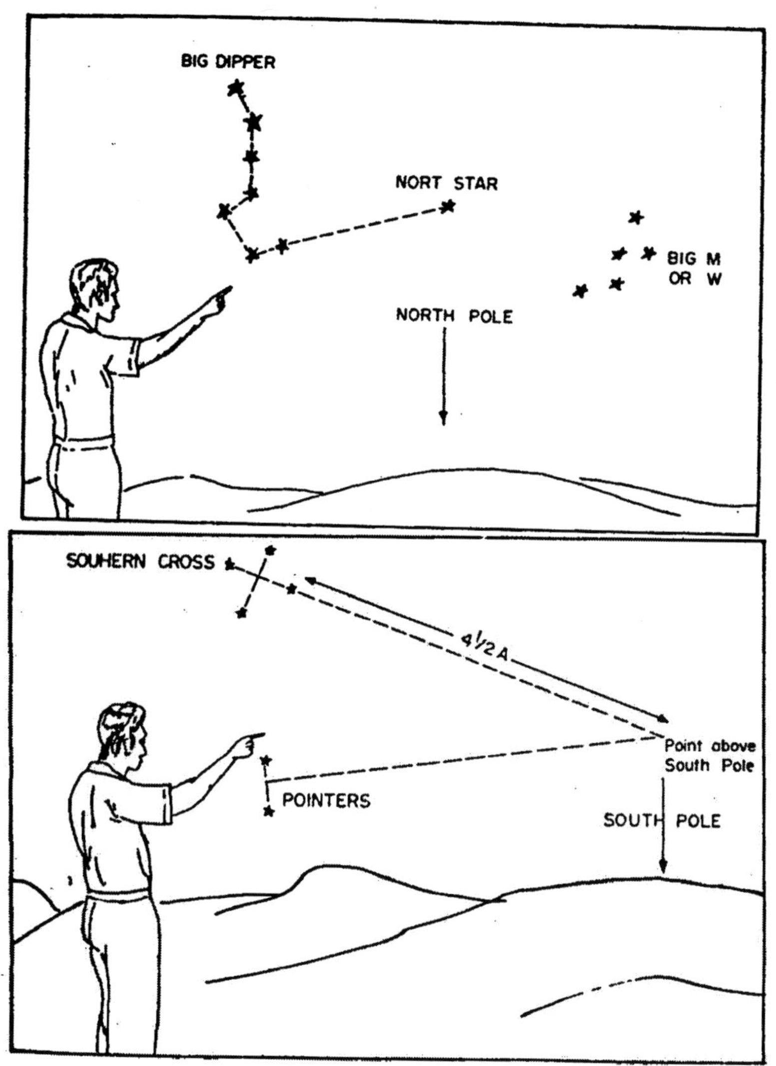

Fig. 6.6: Determination of the True North and South with the Help of Stars

7
Coordinates and Their Functions

We have already noted that a map is a working model of the earth or its parts. To know whether this model represents the earth truly or not, we need a frame of reference to compare the details of the real earth with that of its model. Geographic coordinates are the most commonly used frame of reference. Another frame of reference is known as 'grid system'. The latest method of locating one's position while moving of the land or navigating in the ocean or air is the GPS. It is discussed in chapter 25.

Geographic coordinates provide the convenient reference points for determining location, distance and direction relationships on the ground as well as on the map. Had the earth been flat, the development of the networks of geographical coordinates would have been a simple matter. Because of its spherical shape the principle of spherical geometry involving complicated trigonometric calculations have to be used for this purpose.

The principle underlying the division of the earth by a network of geographic coordinates is the same as the preparation of a line graph with Y and X axis. In all graphs we must have a point of origin and two reference lines: a horizontal line or X axis or abscissa, and a vertical line or Y axis or ordinate. As shown in Fig. 7.1, we can draw innumerable lines parallel to OX and OY. The intersection of a pair of these lines will give the location of a point. If we want the location of point m, we can say it is at 2X and 2Y. When a similar set of lines are drawn on the spherical earth, the horizontal lines are called parallels or latitudes and the vertical lines are called meridians or longitudes. The network of these parallels and meridians is called geographic coordinates.

PARALLELS OR LATITUDES

There are three types of parallels of latitudes. They are:

1. Astronomical,
2. Geocentric, and
3. Geographical.

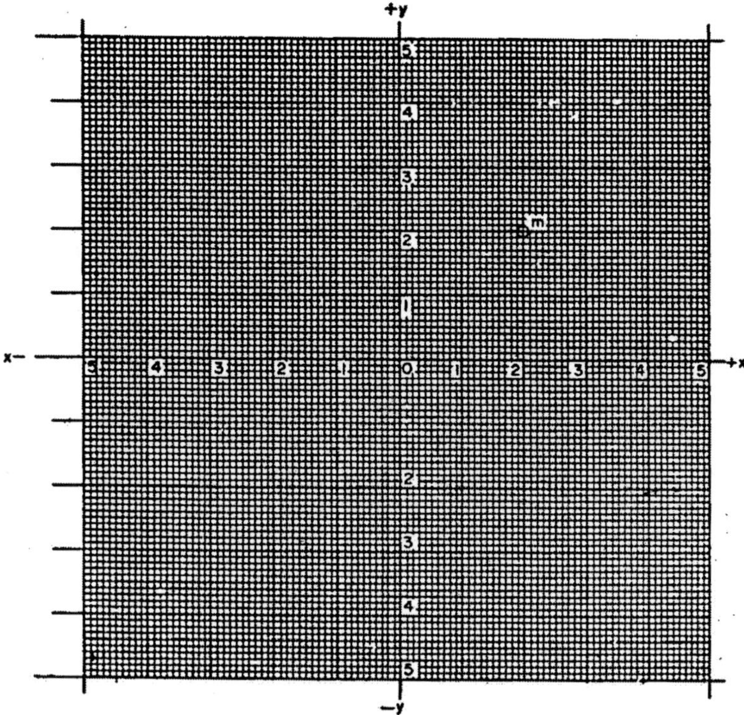

Fig. 7.1: Locating a Point on an Arithmetic Graph Paper

It is the geographical latitudes that we use in our maps, and geographical studies. We should, however, know that there are other types of latitudes which form the bases for the geographical latitudes.

Astronomical and Geocentric Latitudes

Before we move on to longitudes, we should know that the latitudes that we discussed above are derived from a compromise between the Astronomical and Geocentric Latitudes. As illustrated in Fig. 7.2a the Astronomical latitudes constitute the angles between the direction of a plumb line and the plane of the equator. A plumb line indicates the direction of the gravity of the earth. The Geocentric latitude is the angle between a radius drawn from the centre of the earth to a point on the earth's circumference and the equatorial plane, as shown in Fig. 7.2b. If the earth were a true sphere the astronomical and geocentric latitudes would have been the same.

The geographical latitudes discussed hereinafter constitute a compromise between the astronomical and geocentric latitudes. They are the astronomical latitude corrected for the so-called 'station error' i.e., local gravity anomalies. The lines of astronomical latitudes are wiggly whereas those of the geographic latitudes are perfect circles. The difference between the geocentric and geographic latitude is less than 12 minutes of an arc.

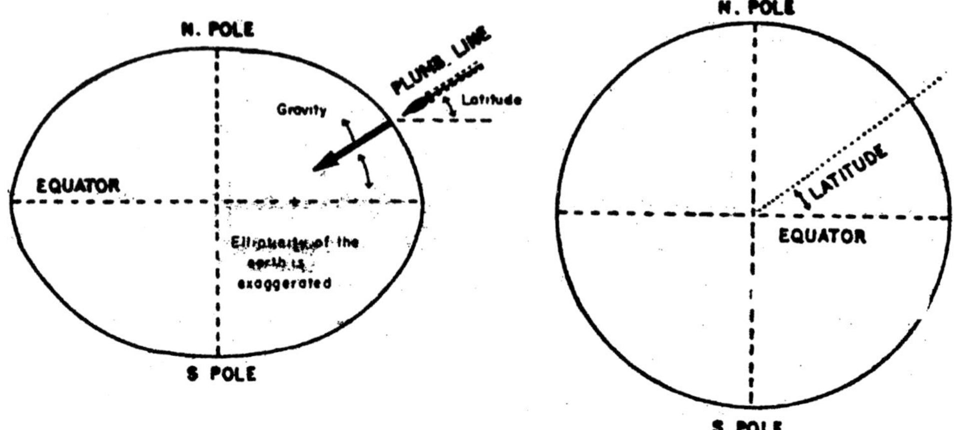

Fig. 7.2 (a): Astronomical and Latitudes

Fig. 7.2 (b): Geocentric Latitudes

Geographical Latitudes

In a spherical earth, a line drawn mid-way between the northern and southern ends of the axis, will divide it into two halves (Fig. 7.3). This mid-line is called equator. The angular distance from the centre of the earth north or south of the equator is called latitude. As the equator is the line of origin from which latitudes are measured, it is called 0° latitude. The angular distance between the equator and the poles is 90° i.e. one-fourth of a circle (360°). As such the latitudes can never be numbered beyond 90°. Latitudes, being the lines parallel to the equator, are also called 'parallels'.

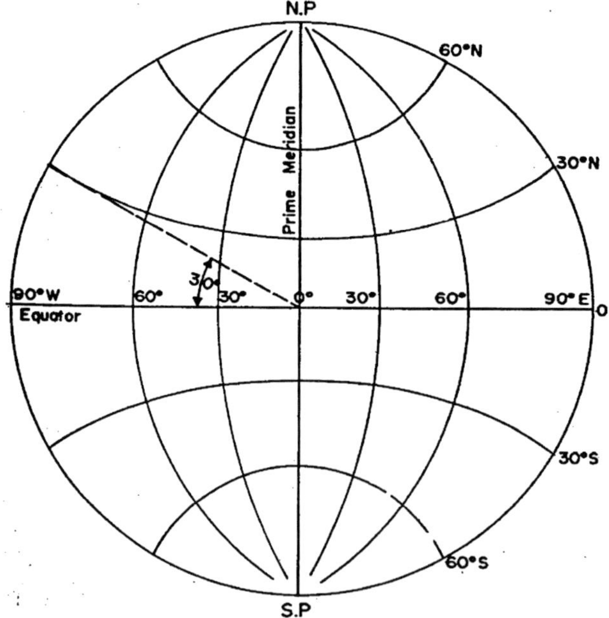

Fig. 7.3: System of Latitudes and Longitudes

The earth being spherical, determination of angle of latitudes is a bit cumbersome. To know the latitude of a place, we have to take the help of the Pole star or sun. In the former case, let us cut the earth north of equator into two halves along the polar axis (Fig. 7.3). The polar circumference of the earth is drawn as a circle on OE equatorial axis. OE and the polar axis OP intersect at O. Another line MN is drawn parallel to the equator (OE) to represent the plane of latitude. If we draw OM from the axial intersection O to a point M and extend it outwards to L, line OL will become a diagonal. When two parallel lines are cut by a diagonal, the opposite angles are equal. In this case the equator OE and the plane of the latitude MN are parallel lines and the extended radius OL is the diagonal line cutting the two parallel lines. Hence, angle MOE = angle LMN. Angle LMN is 30° and hence the latitude is 30°. And since this latitude is in the northern hemisphere, it is called 30° N.

In this way any number of parallels or latitudes can be drawn on both side of the equator. And any line drawn parallel to the equator on a globe is a latitude or parallel. To account for fractions of degrees, a degree is divided into 60 minutes and minute into 60 seconds. For an approximate determination of latitude of a place we can take the help of the polar star. We can measure the elevation of the North Star (Polaris) above the horizon. The line of sight to Polaris is nearly parallel to the axis of the earth and the elevation of this line above the horizon is equal to the latitude as shown in Fig. 7.4.

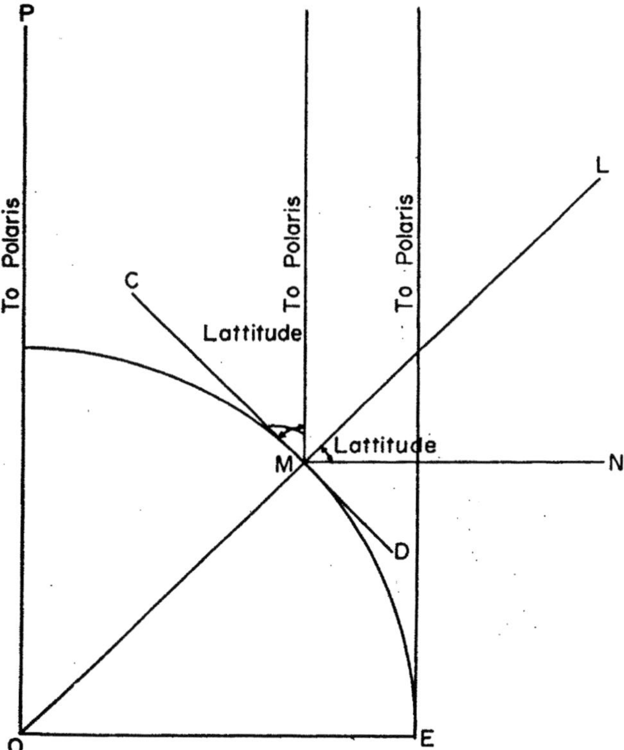

Fig. 7.4: Determination of Latitude with the Help of the North Star

We can also get the latitude of a place by measuring the declination of the sun at the local apparent noon. This technique is illustrated in Fig. 7.5. The elevation of the sun above the equatorial plane is called the sun's declination. This varies with season from + 23.5 degrees on the summer solstice to zero at the equinoxes (March 21, and September 22), and –23.5 degrees at the winter solstice. The angle between the vertical and the sun is called the zenith-angle of the sun. It is equal to 90° minus the angle of the elevation of the sun above the horizon. The sum total of the zenith-angle and the declination is the latitude. On the equinoxes, the declination angle is 0 so that the latitude is equal to the zenith-angle.

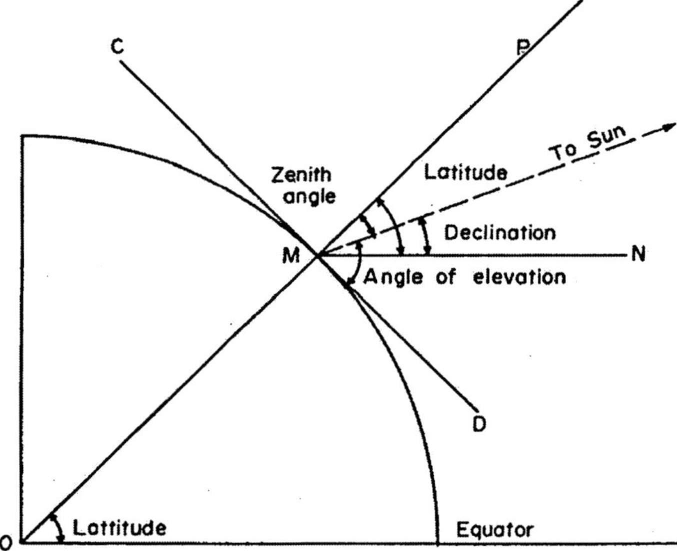

Fig. 7.5: Determination of Latitudes with the Help of Declination of the Sun

All observers standing at a given distance from the equator but located at different points will record the same angle of declination. If we join these points by a straight line, we will get the plane of that particular latitude or parallel. As the earth is spherical we can get a complete circle for each of the innumerable latitudes.

MERIDIANS OR LONGITUDES

Latitude tells us how far from the equator an object is, but to pinpoint the exact location of the object on the earth surface, we have to state where exactly the object lies on the given latitude? This can be done with the help of the longitudes. If we know the latitude as well as the longitude along which the object lies then we can pinpoint its exact location, which would be none other than the point of intersection of the two.

Let us now see how longitudes are determined? We take the help of the sun. The earth revolves around the sun once a year (about 365 ¼ days), and rotates on its axis once a day (nearly 24 hours). If we line up observers along a straight line running from North to South Pole, all of them will see the sun reach its highest point in the sky at the same time (12 noon) on any day. If we had another set of observers along another straight line east of the first line, they would have already seen the sun overhead. If there are some other observers on a third line west of the first line, they will have to wait before they watch the sun overhead. All such lines drawn from pole to pole are called meridians or longitudes.

A meridian is about half of the equator for the later circles the earth and the former makes only half of a circle. Latitudes are parallel to each other that is why they are also known as parallels but the meridians are, not parallel to each other; the distance between the meridians along succeeding parallels decreases pole ward from the equator. The meridians are, however, equally spaced along any one parallel. In all there are 360° of longitudes. Each degree is further sub-divided into minutes and each minutes into seconds.

POINTS OF ORIGIN

To draw the network of latitudes and longitudes we have to have common point of origin. For latitudes, equator is the line of origin. Equator is, therefore, called 0° latitude. For longitudes, the Greenwich meridian is considered to the prime meridian (0 meridian).

There has been no dispute over treating equator as a line of origin for latitudes but the point of 0 degree longitude has been long contested. Earlier, almost every country wanted to have its own 0° meridian passing through a prominent place in it. This naturally caused lot of cartographic confusion. The earliest attempt to settle this issue was made by an international group of people meeting at Paris in 1634. They agreed to treat 19° 55′ 03″ of Paris as the prime (zero degree) meridian. It gradually came to be called Paris meridian, which the other countries did not like. An international conference held in Washington D.C. in 1884 decided to use Greenwich meridian as the prime meridian. Many of the older maps are based on the prime meridians passing through Ferro (western most of Canary Islands), the Pantheon in Paris (2° 20′ 13.95″ east of Greenwich) Madrid, Oslo, Monte Mario (in Rome), etc.

NOTATION OF DEGREES

There are two systems of representing geographical coordinates: hexadecimal and centesimal. The former is based on the divisions of sixty whereas the latter of hundred. The hexadecimal system is the most commonly used system now. It evolved from the experience of our ancients who watched the moon wax and wane every 30 days. One moon round was called a month. It took a cycle of 12 month from one spring to another. They, there-fore, conceived of a cycle or circle having 30 × 12 = 360 days or degrees. As 60 was an even multiple of 12 and at the same time even divider of 360, it was considered appropriate sub-division of degree. That is why we have 60 minutes in a degree and 60 seconds in a minute.

According to the centesimal system, a circle is divided into 400 grads. Each grad in turn is divided into 100 minutes and each minute into 100 seconds. In hexadecimal system, 4 degrees 10 minutes and 15 seconds will be written as 4° 10′ 15″ but in centesimal system they would be expressed as 4° 10′ 15″.

Numbering of Meridians

Numbering the latitudes is easy far they start with 0° and end with 90° on both sides of the equator. In case of the meridians, it is different. The Greenwich is accepted as the 0° meridian a full circle around the globe equals 360 degrees. If the numbering is continuous, 0° and 360° meridians will overlap. In order to get over this problem, we have to stop mid-way and reverse the order of numbering. That is why we have the meridians starting from 0 to 180 degrees only. While moving eastward the numbering increases to 180° and then decreases back to 0°. Latitudinal values from 0 to 180 are labeled E (east of Greenwich) and those from 180 to 0 are labeled W (west of Greenwich).

We can locate any point on the surface of the earth in terms of its latitude labeled N (North) in the Northern Hemisphere, and as S (South) in the Southern Hemisphere. The meridians or longitudes are labeled E (East) to indicate east of the Greenwich and W (West) to indicate west of the Greenwich.

The East or West labeling of the meridians should not be confused with the direction of movement. One can follow latitude and go due East or West and reach the same spot without changing the direction. But this will not be the case if one has to go in North-South direction. Poles are the points from which directions change. One may go along a meridian to north or south to come back to the same spot but the direction of the journey would change on the way.

ASTRONOMICAL AND GEOCENTRIC LATITUDES

Astronomical latitudes (Fig. 7.6) are angles between the direction of the plumb line and the plane of the equator. A plumb line points to the direction of the gravity of the earth. Geocentric latitude is the angle between the radius drawn from the centre of the earth to a point on the circumference of the earth and the equatorial plane (Fig. 7.7). Had the earth been a true sphere, there would have been no difference between the astronomical and geocentric latitudes.

The latitudes that we use in our maps constitute a compromise between the astronomical and geocentric latitudes and are called geographical latitudes. When the astronomical latitudes are corrected for the so-called 'station error' i.e. local gravity anomalies they become geographical latitudes. The lines of astronomical latitudes are wiggly whereas those of the geographic latitudes are perfect circles. The difference between geocentric and geographic latitude is less than 12 minutes of an arc.

INTERNATIONAL DATE LINE

The earth rotates on its axis from west to east causing the apparent movement of the sun from east to west. The full circle (360°) rotation of the earth is completed in

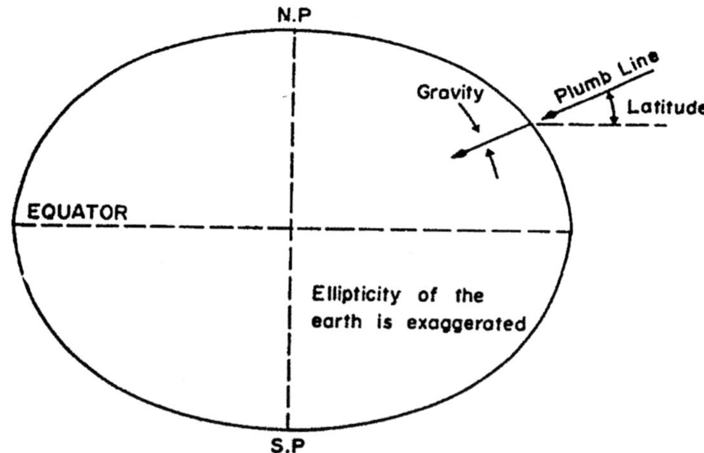

Fig. 7.6: Astronomical Latitude

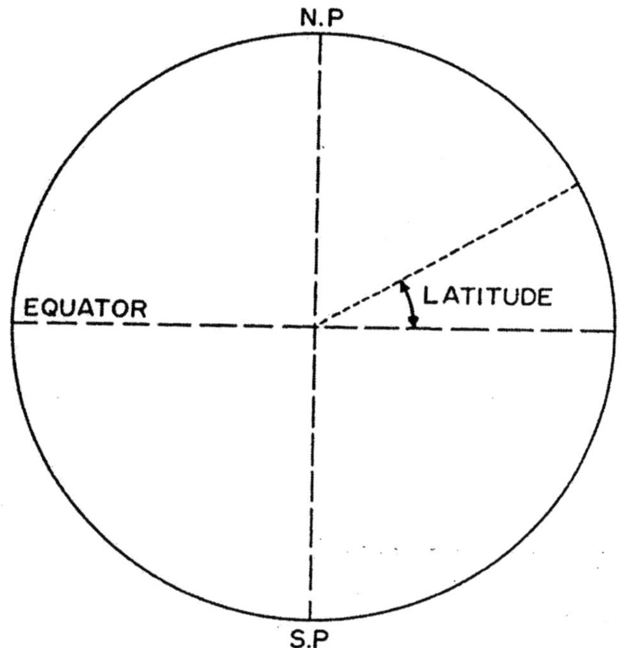

Fig. 7.7: Geocentric Latitude

approximately 24 hours or 15 degree per hour. In other words the sun traverses 15 degree of longitudes from east to west every hour or 1 minute of longitude every 4 seconds. Thus the apparent noon time changes by one hour for ever 15 degrees of longitude. The time decreases as one goes to the west, as the sun also moves with the traveler and increases as one moves eastward. By international agreement the day begins at about longitude 180°, which passes through the Pacific Ocean. This is called the International Date Line. The International Date Line does not follow 180° longitude all the way in order to avoid passing though land. This avoids a country or island having two different dates on any particular day. When crossing the International Date Line from East to West one gains a day. Thus if it is 6 a.m. on January 10 on the eastern side of the line, it will be 6 a.m. on January 9 on the other side. Similarly while crossing the line from West to East one loses a day (Fig. 7.8).

For the purpose of common time keeping, it is necessary to define time zones within which all clocks read the same time. If it is not done, people living at different longitudes within the same country will have different times. The time zones are approximately 15 degrees of longitude wide. The central meridian of the zone defines the standard time within the zone. India has only one time zone whereas the USA has four.

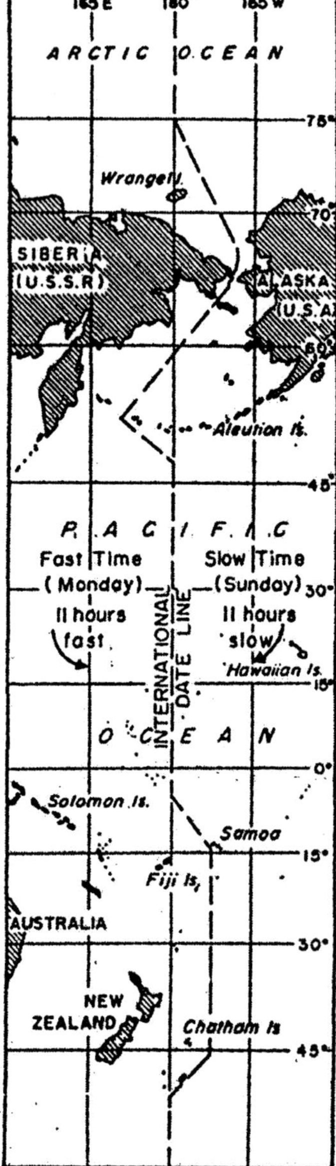

Fig. 7.8: International Date Line

THE GRID SYSTEM

A third reference system in common use is called Grid System. It has been developed as a common-language means of locating points on a map.

A grid is a network of a series of vertical and horizontal lines constructed perpendicular to each other. One series of lines runs from east to west and the other from north to south. Together they form square of same dimensions within a given map. Each of the lines forming the square is given a value so that the position of an object on the map can be easily identified.

Grids are generally used on maps with 1:250,000 or larger scale. On small scale maps, point locations can be better done with reference to longitudes and latitudes. Maps on 1:250,000 or smaller scales have grid markings at an interval of 10,000 units, but those having larger scales have 1000 unit intervals. The units may be yards, feet or meters.

Every grid line on a map is identified by a number printed in the margin of the map, directly opposite the relevant grid line. The numbers for the vertical lines are given on the bottom margin and for the horizontal lines on the right hand margin. In some cases the numbers are given on all the four sides. The point of origin of the grid line is always the South-West corner of the sheet. The numbers, therefore, increases in northerly and easterly directions.

Grid numbers have several digits. To accommodate these digits and at the same

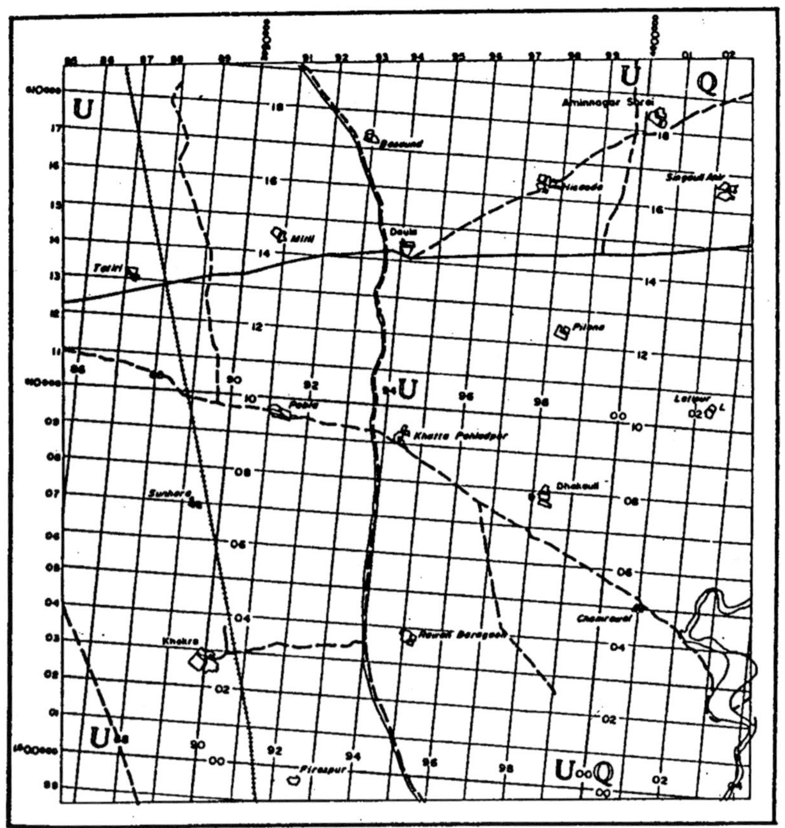

Fig. 7.9: Reading the Grids (From Map Reading)

time to keep them legible, some digits are skipped. On maps of larger than 1:250,000 scale, the last three digits are omitted. On maps with 1:250,000 scale, the last four digits are omitted.

The rule for reading the grid is read 'right up'. That is to say that first read eastward to the right across the bottom of the grid and then northwards up the side. For example let us make the grid reference for Dhakauli from an Indian topo-sheet (Fig. 7.8). It is U 983078. To read it, identify the letter i.e., 'U'.98 first; it represents the grid number towards east. The particular point is located exactly at the third division (3) from the south-west corner of the grid square. Similarly, from the south the point is located after '07' grid line. Further, the exact location in this case is found on 8th division from the gridline '07'. Therefore, the reference for 'Dhakauli' is U 983078.

REFERENCES

Defense Mapping Agency (1983), *Geodesy for the Layman*, Washington DC. This book is a must for all those who want to know about cartography and who want to make it a career.

Read, HH; Watson Janet (1975). *Introduction to Geology*. New York: Halsted. pp. 13-15.

8
Map Projections

The work of cartographers would have been simple had the earth been flat. Because of the spherical shape of the earth and the plane surface on which this shape has to be represented, cartographers have to devise complex graphical, geometrical and mathematical methods to accomplish this task. The devises used to transform to accomplish this are collectively known as map projections.

Map projection means reflection of an idea or image on something. Map projection means reflecting the network of latitudes and longitudes of a globe on a plane surface. It is, however, only in few cases that we get map projections by this method. In most other cases we use geometrical and mathematical methods to draw the networks of longitudes and latitudes. Nevertheless, the graticules so derived are called projections. In view of this, we will define map projection as the transformation of the spherical network of latitudes and longitudes on a plane surface, irrespective of the method of transformation.

IS THERE AN IDEAL MAP PROJECTION?

An ideal map projection is one which represents the globe truly. The network of meridians and parallels of the globe has the following characteristics:

1. The equator divides the globe into two halves: northern hemisphere, and southern hemisphere;
2. The equator is perpendicular to the polar axis;
3. All the latitudes are parallel to the equator;
4. The spacing between any two parallels is the same. Equator is the only latitude that makes a Great Circle. All other latitudes are shorter than the equator; and none of them makes a great circle;
5. Each meridian is one half of a great circle in length. It is the shortest line between the two poles;

6. All meridians coverage at the north and south polar points;
7. All spacing between meridians is equal along a given parallel, but the space between meridians decreases polewards;
8. The parallels and meridians intersect at right angles; and
10. All areas are correct to the scale in relation to earth.

Because of the above characteristics of meridians and parallels, many features of the earth surface can be measured independently of its location but this cannot be done on maps. Some of the properties of the earth surface which cartographers try to preserve are:

1. Area (equal area, equilateral or equivalence);
2. Shape locally (conformal or orthomorphic);
3. Direction (azimuthal), a trait possible only from one or two points to every other point;
4. Distance (equidistant), a trait possible between one or two points and every other point;
5. Shortest route, a trait found only in gnomonic projection; and
6. True Bearings.

EQUAL AREA OR EQUIVALENCE

A projection that is equal area, cannot be conformal i.e. it can not maintain true shape. And *vice versa* is also true. An equal area map preserves the ratio of earth area to corresponding mapped area but it cannot preserve shapes of corresponding features. An equal area projection has two properties:

1. The parallels and meridians are drawn to scale, and
2. Spacing between the parallels and meridians is true. Measurements of most equal area projections are mathematically derived.

Conformality

If the shape of the map surface at any given 'spot' is identical to the shape of the corresponding spot on earth, the projection is called conformal. Here, the use of the word spot should be noted. Implications of conformality are sometimes wrongly enlarged to cover only the shapes of large areas like continents. A correct interpretation of the term implies that the shape of small areas and features would also be correctly maintained.

The angle at which the parallels intersect the meridians governs the shape of areas. On a globe, the meridians cross the parallels at right angles. Preservation of the right angle and the maintenance of the same scale both along the parallels and meridians make a projection conformal.

True Direction or Azimuth

The *directions are truly represented* on a globe i.e. the directions of points on the globe from a given point are the same as the directions on the surface of the earth. True direction is the horizontal angle of the observer's bearing (in surveying), measured clockwise from a referent direction, as from the north or from a referent celestial body, usually Polaris.

The conformal projections show directions correctly. The Mercator's projection was developed for the purpose of retaining true directions, not only along projection line but also diagonally between them. This could be achieved by a uniform distortion of the network intervals. Another way of maintaining correct direction is to evolve a projection, which is constructed on a plane from one central point. In such projections, directions of all point on a map, as taken from the central point, will be the same as between corresponding points on the ground.

True Distance (Equidistance)

The geodetic lines on the earth, which give the shortest distance between any two points appears as shortest lines on the globe. Distances are true along the parallels drawn to scale and spaced correctly. Similarly, the distances along the meridians drawn and spaced correctly are also true.

True Bearings

In marine navigation, a bearing is the direction of an object from one's own vessel. In aircraft navigation, a *bearing* is the actual (corrected) compass direction of the forward course of the aircraft. In land navigation, a *bearing* is the angle between a line connecting two points and a north-south line, or *meridian*. Bearings can be measured in Mils and Degrees.

True Scales

Similarly, the scales are true on globe in all directions. A map drawn on a plane surface can never achieve that property for any area no matter how small it may be. No map projection can ever maintain all the above characteristics of the earth surface or globe. And all projections are compromises with respect to the characteristics of the globe discussed above.

It has been the endeavour of cartographers since early times to develop a projection which has the same properties as a globe. This could be achieved only if a projection possesses all the ten properties of a globe. Since, it is impossible to truly represent a globe on a flat surface without losing one or more of the above noted characteristics, there is no projection which satisfies all the properties of a globe. It is, however, possible to develop projections which have some of the five characteristics but not all the characteristics.

TYPES OF MAP PROJECTIONS

It has been the endeavour of cartographers since early times to develop a projection, which adheres to the properties of the globe. This could be achieved, only if a projection conforms to a globe that possesses ten above mentioned characteristics. In view of the fact that it is not possible to truly represent the globe on a flat surface without losing one or several of the ten characteristics, it has not been possible to develop a projection which conforms to the globe in area, distant, direction, and simplicity at the same time. All projections are, therefore, short of one or more of the characteristics of the globe.

There can be as many types of map projections as the criteria used for classification. Table 8.1 gives the most frequently used types of Projections:

Table 8.1: Types of Projections and the Criteria Used to Derive Them

S. No.	*Criteria Used*		*Types*
1.	Method of Derivation (Source of Light)	I.	Perspective projection
		II.	Non-perspective projection
		III.	Mathematical or Conventional
2.	Developable Surface	I.	Conical
		II.	Cylindrical
		III.	Azimuthal or Zenithal
		IV.	Mathematical or Conventional
3.	Global Properties	I.	Homolographic or Equivalent or Equal Area
		II.	Orthomorphic or Conformal
		III.	Azimuthal or Correct Bearing

Perspective projections are those which can be derived by projecting the image of the network of meridians and parallels of the globe on any developable surface. Many projections have been evolved by slight modifications in developable surfaces. The lines forming the network are straightened or curved and the spacing between the parallels and the meridians are reduced or enlarged to make a perspective projection equivalent, orthomorphic or azimuthal. Such projections are called non-perspective projections. All non-perspective projections are also non-developable projections because they can never be transferred by means of a source of light casting shadows on surfaces capable of being flattened. They are, nevertheless, considered to be developable because they can be derived by simple modifications in perspective projections. Mathematical or conventional projections are those which are derived by mathematical computations and formula and have little, if any, relation to a projected image.

The developable surface projections are classified as:

1. Conical,
2. Cylindrical, and
3. Zenithal.

A developable surface is one which can be flattened and which can receive lines projected or drawn directly from an assumed globe. To derive a projection on a come, we will have to wrap a cone around a globe. If we place a light in the centre of the globe, it will cast shadows of the geographic network on the inner surface of the cone. When the shadow net is drawn and the cone is cut open and laid flat, we get a projection to work on. Cylindrical projections can be derived the same way. When a plane touches the globe at a point, we get a zenithal projection. The method of deriving projections from developable surfaces is illustrated in Fig. 8.1.

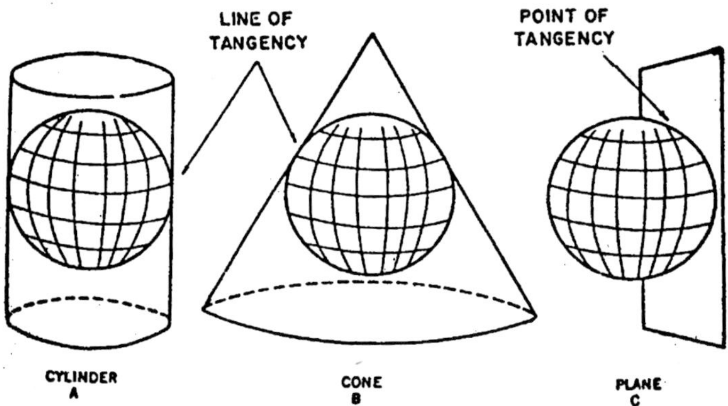

Fig. 8.1: Methods of Deriving Map Projections from Developable Surfaces

Projections can also be classified on the basis of the global criteria they satisfy. These criteria are:

1. Equivalence,
2. Orthomorphism, and
3. Correct bearings.

Such projections can be developable as well as non-developable. They can be conical, cylindrical, zenithal or mathematical.

The above typology of map projections reveals that there is no way of classifying projections into mutually exclusive classes. Most projections fall into two or more of these classes.

CONSTRUCTION OF MAP PROJECTIONS

Construction of a map projection involves two steps:

1. Selecting a model for the shape of the Earth or any planetary body for which a map is to be prepared knowing that the Earth's actual shape is irregular, and hence information is lost in this step. Usually, the model selected falls between a sphere or ellipsoid.

2. Transforming the geographic coordinates (longitude and latitude) to Cartesian (x y) or polar plane coordinates. Cartesian coordinates normally have a simple east-west and north-south grids superimposed on the projection.

The first preliminary step in the construction of map projections is to select the shape of the earth one envisages. It is usually assumed to be spherical but it is only partly true. The shape of the earth is closer to ellipsoid, bulging around the equator. Spherical models can be used for small-scale maps made for world atlases because the error at that scale does not distort the reality. The ellipsoidal model is commonly used to construct large and medium scale maps that aim to accurately depict the land surface. Topographical maps use ellipsoidal model.

Geoid is the third model of the shape of the Earth. It represents the global mean sea level surface of the earth obtained through a combination of terrestrial and satellite gravity measurements. It is, however, not used for mapping due to its complexity. It is mainly used for control purposes in the construction of geographic datum. It takes into account the large scale features like mantle convection patterns in the Earth's gravity field, and the gravity impacts of very large geomorphic features like mountain ranges, plateaus and plains while constructing a datum. Thus the geoids help in improving Earth's actual shape. Historically, datum has been based on ellipsoids that best represent the geoids within the region the datum is intended to map. Each geographic region of the world has it own datum. A few modern datum, such as WGS84 (the one used in the Global Positioning System or GPS), are optimized to represent the entire earth. Some inaccuracies do enter in smaller regions if used for a single ellipsoid.

The simplest map projections are those which are obtained by keeping the source of light at a particular point relative to the globe and projecting its reticules onto to a developable surface placed on the globe. A developable surface is one that can be unfolded or unrolled into a sheet or plane such as cylinder, cone, and flat plane. The sphere and ellipsoids are not developable surfaces. A sphere or ellipsoid projected onto a developable surface will necessarily distort the image, in the same way as football cannot be made into a plane without tearing it apart and distorting its shape, area, directions, etc. Most projections are, however, defined only by mathematical formulae and hence they are not the reflected images of the globe. The distortions, however, remain in one form or the other.

Naturally then, no projection is perfect i.e. true representation of the globe. Some maintain correct area, while others maintain correct shape, direction, bearings, etc.

Before we embark on constructing a map projection, we have to reduce the true earth to a scale we want to have the projection. We know that the equatorial circumference of the earth is about 40,075.4 km (24,901.7 miles) and the mean radius is 6371 km (3981.9 miles). The radius of the earth when converted into inches comes to about 250,000,000. Thus, if we plan to make a globe of 1″ radius it will be 1/250 millionth the size of earth. The length of the equator on the globe will be equal to the circumference $2\pi R = 2 \times 22/7 \times 1'' = 6.3$ inches $\left(2\times\frac{22}{7}\times1''6.3\right)$ or $2\times\frac{22}{7}\times22.5$ cm = 15.5 cm.

The network of latitudes and longitudes is then drawn from the equator. One degree of latitude is equal to about 69.6 miles (111,895 m) at the equator; it decreases slightly to 100,607 m near the poles. A degree of longitudes, however, has far more variations (Table 8.2).

Table 8.2: Length of a Degree of Longitude along Various Latitudes

Latitude	*Length of a Degree of Longitude*	
	Miles	*Kilometers*
0	69.60	111,321
10	68.30	109,641
20	64.80	104,649
30	59.90	96,448
40	52.90	83,063
50	44.60	71,698
60	34.80	55,802
70	23.70	38,183
80	11.80	19,349
90	0	0

We can find out the length of each parallel using graphical or mathematical method. The choice of the method would depend upon the degree of accuracy desired. For approximate measurements we can use graphical method as illustrated in Fig. 8.2. To find the length of 30 degree parallel, let us draw a circle with the earth radius of our choice. Let EQ be its diameter. Draw OB making QOB equal to 30 degrees. Draw line AB parallel to line EQ. With AB as diameter, draw another circle A′B′D. From its centre C draw angle DCB′ using the desired interval. The arc distance DB′ is the required longitudinal interval along 30 degree parallel. To get the length of 30° parallel, multiply DB′ by $\frac{360}{\text{interval}}$.

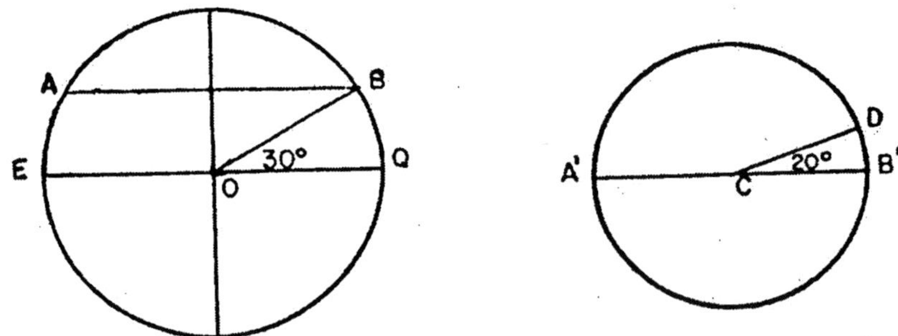

Fig. 8.2: Graphical Method of Determining the Length of a Parallel.

If greater accuracy is desired, a mathematical method as illustrated in Fig. 8.3 should be used. In this figure OE is the radius (R) of the circle, and AB is a parallel drawn at 45 degree latitude.

Angle OAC = angle EOA = latitude (being alternate opposite angles).
Angle AOC = 90° – Lat. = co-lat.

In triangle OAC, $\frac{AC}{AO}$ = cos. lat. or AC = AO cos. lat. = R cos. lat.

Because AO = R

Since $\frac{AC}{AO}$ = sin.cos-lat.; AC = AO sin. co-lat.

= R sin. co-lat.

$$AC = \frac{AB}{2}$$

Hence, the length of the parallel AB = 2πR sin. cos-lat. = 2πR cos. lat.

The length of any other parallel can be determined the same way. Thus when R = 1″ (2.5 cm) the length of the arc of the parallel AB for an interval of 10° will be

$$\frac{2\pi\,R\cos.\,lat.\times d}{360} = \frac{2\pi\,R\cos.\,45^\circ\times 10}{360}$$

$$= \frac{2\times 22\times 1\times 0.71\times 10}{7\times 360} = .12''.$$

$$\frac{2\pi\,R\cos.\,lat.\times d}{360} = \frac{2\pi\,R\cos.\,45^\circ\times 10}{360}$$

$$= \frac{2\times 22\times 2.5\times 0.71\times 10}{7\times 360} = 0.3\text{ cm.}$$

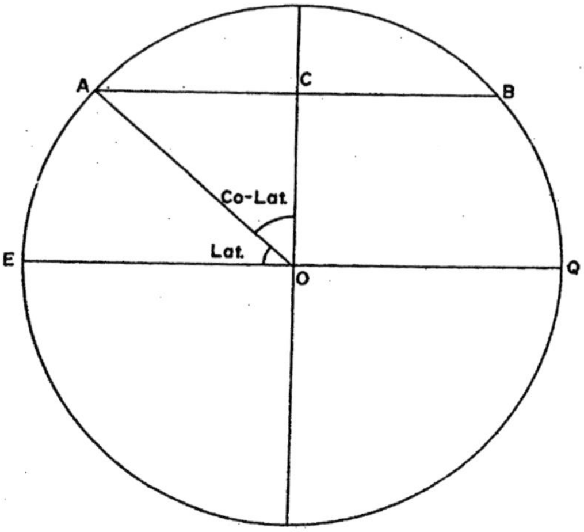

Fig. 8.3: Mathematical Method of Determining the Length of Parallels

CONICAL PROJECTIONS

Perspective Conical Projection

When the image of the network of parallels and meridians is projected on a developable cone, we obtain a conical projection. In Fig. 8.4, PQ is the circle or contact of tangency. Such a circle (parallel) along which the cone is tangent to the globe is called the standard parallel.

A perspective conical projection is one which is derived by projecting the image of the network of parallels and meridians of a globe on a developable cone which touches the globe along the latitude called standard parallel (PQ in Fig. 8.4). The light is located at the centre of the globe. The cone touches the globe along PQ, so that the position of this parallel on the globe coincides with that on the cone. But the position and length of other parallels on either side of the standard parallel are distorted. For example, parallel RS on the globe is pushed up as R′S′ on the cone. Its length is also considerably increased. The North Pole is shifted from N to V. The distortion of shape and area away from the standard parallel is so progressive and so much that this projection cannot be used for anything except representing areas along the standard parallel.

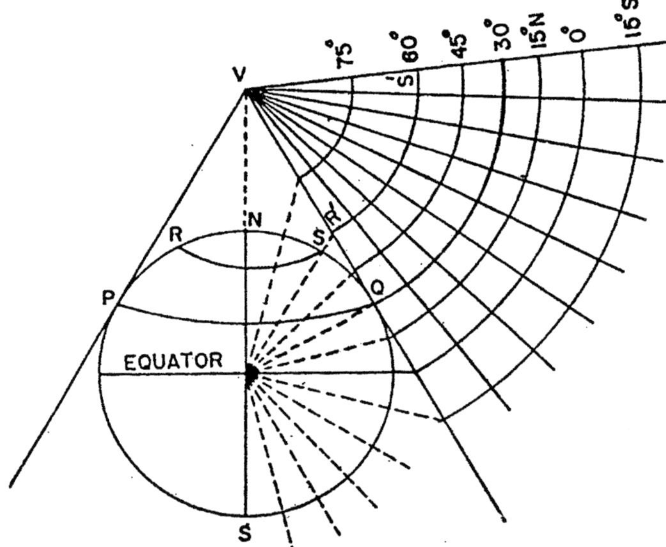

Fig. 8.4: Derivation of a Perspective Conical Projection

To minimize the distortion in shape and area a number of non-perspective conical projections have been developed. A few of the important ones are discussed below.

Simple Conical Projection with one Standard Parallel

A simple conical projection is an improved version of the perspective conical projection.

This projection also has only one standard parallel which is true to scale. All the parallels are drawn as concentric circles. It differs from the perspective conical projection in the spacing of the parallels. In the perspective conical projection the distance between the parallels increase progressively as one move to north or south of the standard parallel; in simple conical projection the parallels are drawn at equal intervals. As a result the distortion in shape and area from the standard parallel are considerable minimized.

For the graphical construction of this projection the following procedure may be adopted. Determine the radius of the earth from the given scale. If the scale is 1=250,000,000, the radius of the reduced earth will be 250,000,000/250,000,000 inches = 1 inch. Draw a circle with a radius of 1 inch to represent the earth (Fig. 8.5). Now with the help of given latitudinal extension of the projection, select the Standard Parallel which should be, as far as possible, the central latitude. Draw the angle for the standard parallel (in this case AOT = 45°) and of the given interval (in this case AOB = 15°). Tangent TP is drawn at T to meet the polar axis of the reduced earth at P. with AB as radius draw an arc of a circle with O as the centre. Draw XY parallel to EA. PT is the radius of the standard parallel; AB is the interval between the parallels along the central meridian and XY is the interval between the meridians along the standard parallel.

Now draw a straight line VV′ to represent the central meridian. With O′ as the centre and PT as the radius, draw an arc of a circle to represent the standard parallel. From the intersection of the standard parallel and central meridian, mark intervals on both lines (central meridian and the standard parallel). With O′ as the common centre, draw arcs to circles through the points marked on the central meridian to represent other parallels. Join O′ with the points marked on the standard parallel. This gives the required graticules for the Simple Conical Projection.

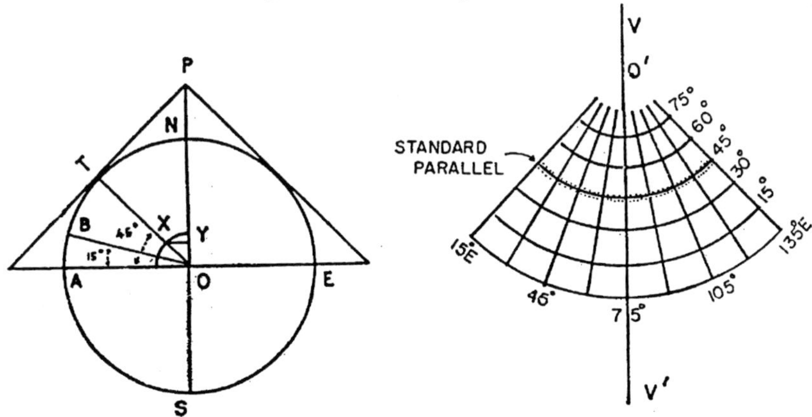

Fig. 8.5: Simple Conical Projection with One Standard Parallel

The Projection can be drawn mathematically adopting the following procedures. Determine the length of the radius of the standard parallel = R cot. latitude where R = radius of the reduced earth. The length of the standard parallel equals $2\pi r$ cos. lat.

and the intervals between the meridians along the standard parallel $= \frac{2\pi \text{ R cos. Lat.} \times \text{Interval}}{360}$.

The intervals between the parallels along the central meridian $= \frac{2\pi \text{ R Interval}}{360}$.

This projection is a modified version of the perspective conical projection. Its parallels form arcs of concentric circles and are equally spaced. The standard parallel is the only parallel which is truly divided and hence is the only parallel along which the scale and the distances are correct. The meridians converge toward the pole and are equally spaced along the parallels. The vertex of the cone does not represent the pole. Distance between the parallels is the same along all the meridians. Distance along the meridians is fairly accurate. Shape and area represented are generally well maintained near the standard parallel. Great circle can be represented by near straight lines anywhere on the projection and are straight lines along the meridians. It is not used for world maps because there are extreme distortions in the hemisphere opposite to the one in which the standard parallel is selected. Even within the same hemisphere, distortions both toward the pole and the equator are large enough to discourage its use for representing a hemisphere. It is most suited for mapping meddle latitudes areas of limited latitudinal and relatively larger longitudinal extent.

CONICAL PROJECTION WITH TWO STANDARD PARALLELS

The simple conical projection with one standard parallel can be further improved by selecting two standard parallels. The standard parallels should be so selected that they enclose about two-third of the longitudinal extent of the area to be represented. For example, if the area to be represented extends from 20°N to 60°N, the selected standard parallels would be 30°N and 50°N. This should not, however, be treated as a rule. The selection will partly depend upon the purpose of the map and the area to be emphasized. By making a judicious choice of standard parallels, one can represent much larger area fairly accurately than one can do on the simple conical projection with none standard parallel (Fig. 8.6).

The following procedure can be used to draw this projection. Draw a circle or a quadrant of a circle with a given earth radius. And then, draw the angles of latitudes for the two standard parallels and the selected interval along the parallels and the meridians. As a cone resting on a sphere cannot touch it along two parallels tangentially, it is made to cut across the sphere keeping the strips of the sphere between the two standard parallels outside the cone.

The procedure for getting the measures for graphical construction is the same as in simple conical projection. To get the vertex of the cone, draw a straight line CD as given in Fig. 8.7. Take the distance EF and mark the required number of divisions on CD. EF can be calculated by formula 2 R × Interval/360. The intervals in Fig. 8.7 are 10 degrees apart. From H and L which represent the intersections of 20° and 50° degree parallels and the central meridian, draw perpendiculars HG and LK equal to XY and MN respectively. Join GK and extend it to meet CD at C. Point C will form the vertex of the cone.

To construct the projection, draw a vertical line C′D′ representing the central meridian. Divide the line in the same way as CD. Among the divisions, locate points H and L to represent the intersections of 20° and 50° parallels and the central meridian. Take CH and CL as radii and draw arcs of concentric circles with C′ as the centre passing through H′ and L′. These will constitute the two standard parallels. Draw the other parallels also as concentric circles with centre C′. Take distances GH and KL and divide the respective standard parallels on both sides of the central meridian as illustrated in Fig. 8.7. Join the corresponding points on two standard parallels by straight lines and extend them pole-wards as well as equator-wards.

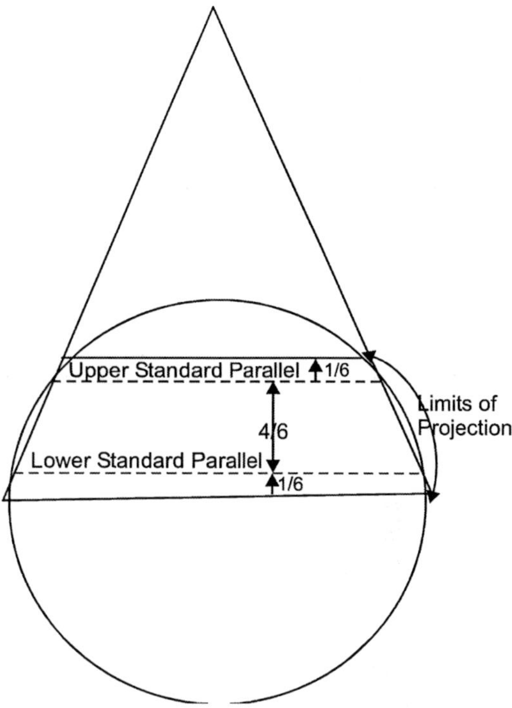

Fig. 8.6: Selection of Standard Parallels for Conical Projection with Two-standard Parallels

The mathematical method of construction of the projection is easier than the graphical method. The following formulae can be used to get various measurements:

$$\text{Intervals along the standard parallels} = 2\pi\,\text{R}\cos.\text{Lat.}\times\frac{\text{Interval}}{360}$$

$$\text{Intervals along the central meridian} = 2\pi\,\text{R}\times\frac{\text{Interval}}{360}$$

Radii of the standard parallels

$$\text{First parallel (y)} = 2\pi\,\text{R}\frac{\text{y}-\text{x}}{360}\cdot\frac{\cos.\text{x}}{\cos.\text{x}-\cos.\text{y}}$$

$$\text{Second parallel (x)} = \text{Radius of y} + 2\pi\,\text{R}\frac{\text{y}-\text{x}}{360}$$

As in the case of the simple conical projection, in this projection too the parallels are drawn as arcs of concentric circles and are equi-spaced. The meridians appear as straight lines and coverage at the vertex. The scale is true along the standard parallels.

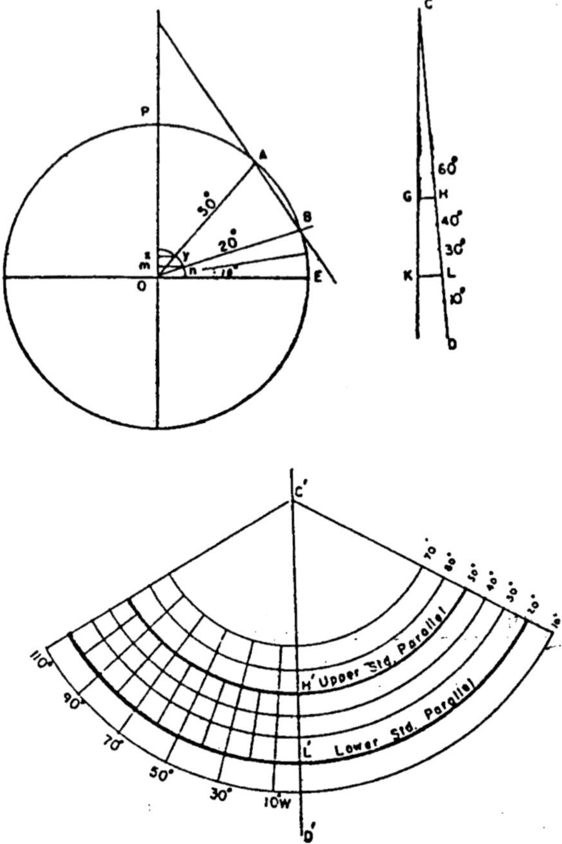

Fig. 8.7: Conical Projection with Two Standard Parallels

Distances are reduced between the standard parallels and are increased outside them. The pole is represented by an arc of a circle. It does not coincide with the centre of the circles representing the parallels. The intervals between meridians on any given parallel are equal, but they are progressively reduced as one goes pole-wards.

Being an improved version of simple conical projection, this projection is suited to areas extending over 50° latitudes such as Canada, Russia, and to some extent China.

BONNE'S PROJECTION

This project is also a modified version of the simple conical projection. It has only one standard parallel but each parallel is truly divided. It is an equal area projection. The method of finding out the radius of the standard projection is almost the same as in the case of the simple conic projection. All the parallels are equi-spaced and drawn as arcs of concentric circles from a common centre. The central meridian cuts all the parallels at right angles and the standard parallel cuts all the meridians at tight angles. The length of the parallel is true with reference to the globe. The distance between any two given parallels along the central meridian is true and constant. It must, however, be noted that no other parallel or meridian except the standard parallel and central meridian intersect each other at right angles. As a result the shape is distorted as the distance increases from the standard parallel and the central meridian.

The graphical construction is similar to that of the simple conic with one standard parallel except that in this case each parallel is truly divided. The intervals along the parallels can be found out from Fig. 8.8.

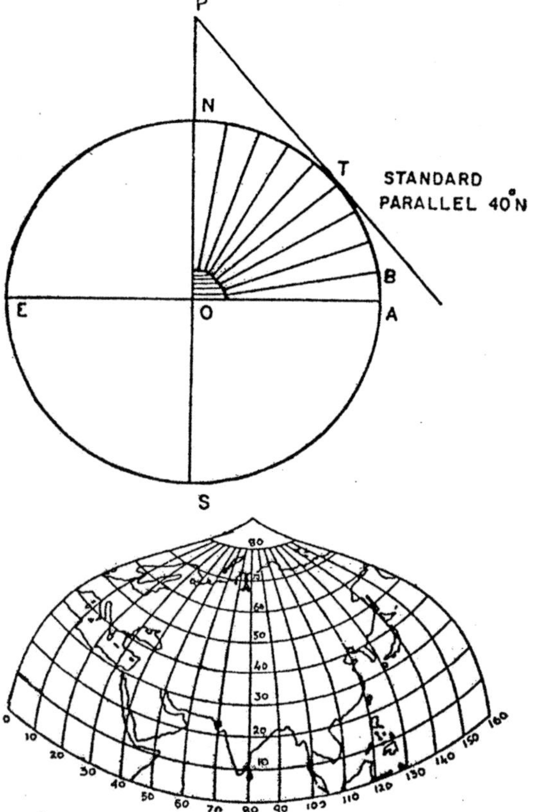

Fig. 8.8: Bonne's Projection

The mathematical formulae for construction are as follows:

Radius of the standard parallel = R cot. Lat

$$\text{Interval along the central meridian} = 2\pi R \frac{\text{interval}}{360}$$

$$\text{Interval along the parallels} = 2\pi R \text{ cos. lat. } \frac{\text{interval}}{360}$$

$$= \text{interval along the central meridian} \times \text{cos. lat.}$$

This projection has the following properties: its parallels are arcs of concentric circle; they are equi-spaced and truly divided; distances along them are true, but they are not true between them except along the central meridian; except for the central meridian, all meridians are curved lines; the central meridian intersects the parallels at right angles; away from the centre of the projection intersections become progressively oblique, and, therefore, the shapes of area represented get distorted. It is not an orthomorphic projection. Each quadrangle of the projection is equal in area to its corresponding quadrangle on the globe made on the same scale.

It is often used to represent continents like Europe, North America and Australia. It is less suited for representing Asia because large chunks of areas falling on the margins away from the central meridian get distorted in shape. It has been used in topographic mapping by several European countries like Netherlands, Belgium and Switzerland.

POLYCONIC PROJECTION

This projection is also a modified version of the simple conic projection. The very name of projection implies the concept of using several cones at the same time. It was developed by Prof. Ferdinand Hassler, the first Superintendent of the Coast and Geodetic Survey of the U.S.A.

In the simple conic projection, the radius of the standard parallel is equal to the radius of the corresponding parallel on the globe of the same scale. Meridians intersect the parallels at right angles. It is, therefore, possible to fit maps of adjacent areas lying east and west of each other. But in the absence of bounding parallels of constant length (the length of parallel will depend upon the selected standard parallel) maps of areas north or south of one another cannot be fitted together. The polyconic projection is a device to minimize this drawback.

In the construction of this projection it is presumed that there are as many cones covering a globe as parallels to be drawn. Each of these cones is tangent to its corresponding latitude, thereby making each parallel a standard parallel. The representation along all the parallels is, therefore, correct. This projection divides the map surface into a number of east-west running belts. These belts fir each other only along the central meridian and hence an element of error enters as one moves away from the central meridian. Only along the central meridian the areas as well as the shapes are correct.

The construction of this projection is similar to that of the simple conical projection. Draw tangents at the latitudes in the circle (or its quadrant) representing the earth as shown in Fig. 8.9. Then draw a straight line to represent the central meridian. With respective tangents as radii, the parallels are drawn through the equally spaced points marked on the central meridian. The arcs of circle so drawn are, however, not concentric. It this case, instead of starting from the apex, we start from the points marked on the central meridian. Fix the pencil-bearing hand on the compass on a point marked on the central meridian and let the needle of the compass fall anywhere on the central meridian in the direction of the pole. Non-concentric arcs are thus drawn to represent the parallels. The intervals to be marked on the parallels can be determined in the same way as in the case of Bonne's Projection.

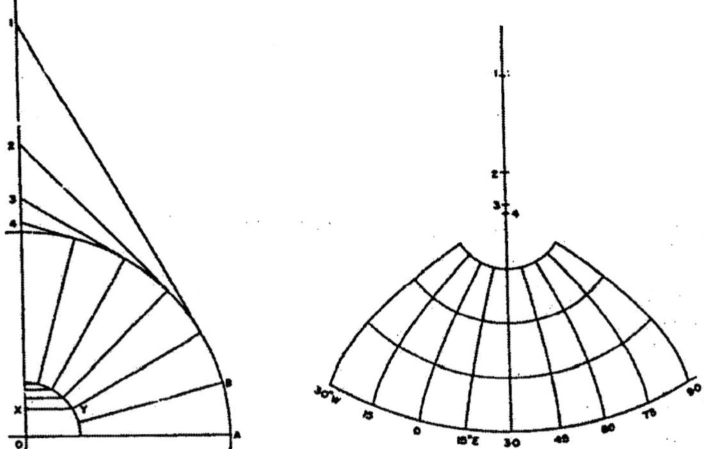

Fig. 8.9: Polyconic Projection

In Polyconic Projection all the parallels are treated as standard parallel and they are not arcs of concentric circles. The equator is shown as a straight line. The parallels are equi-spaced along the central meridian but they diverge away from it. The scale is true along every parallel. The central meridian is also truly divided and hence the scale is true along it. The other meridians are curves diverging toward the equator. The meridians, except the central meridian, do not intersect the parallels at right angles but the obliqueness is less than that in the Bonne's Projection. The scale is not true along the meridians except the central meridian. The projection is neither equal area not orthomorphic. Both these properties are, however, found near the central meridian.

The projection is not suited for representing large areas. It is useful in representing areas on small sheets designed to fit together on their northern and southern ends. Areas with large latitudinal but limited longitudinal extent in the middle latitudes can be shown satisfactorily on this projection.

INTERNATIONAL MAP PROJECTION

This projection is a modified form of Polyconic Projection and is suited for world coverage on 1:1m scale. The idea of preparing 1:1m sheets to cover the whole world on a unified network was first advanced by Prof. Penk at the International Geographical Congress held at Berne in 1891. After a long controversy the proposal was ultimately accepted by the International Map Committee of the International Geographical Union in 1909. The committee recommended that modified Polyconic Projection should be adopted for the purpose.

On this Projection, one can prepare a map of the world in sheets which can fit together on all the sides. This is done by making certain modifications in the Polyconic

Projection. To begin with, the curved meridians of the Polyconic Projection are drawn straight. This enables the sheets to fit together along the meridians if drawn on the same scale. Parallels are drawn true to scale, as in the Polyconic Projection. This enables the sheets to fit along the parallels too (Fig. 8.10).

The Polyconic Projection has only one central meridian; in International Map Projection, there are two other meridians, one on either side of the central meridian. The central meridian is shorter than its true length. The other two meridians are true to scale. These meridians are located at a distance of 2° to the east and west of the central meridian between 60° N and S, and 4° east and west of the central meridian between 60° and 88° N and S. This arrangement reduces and better distributes the exaggeration than is the case in Polyconic Projection.

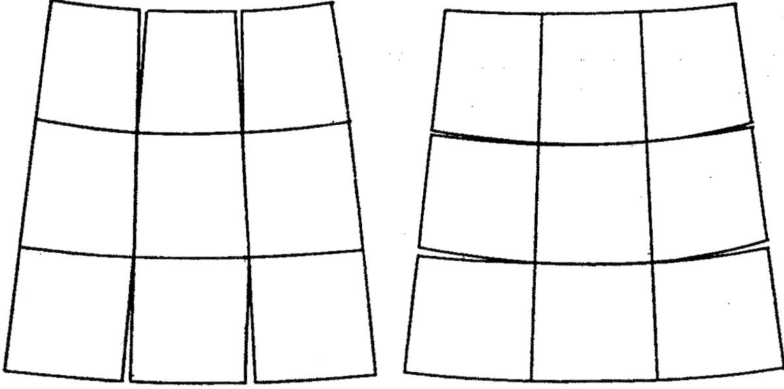

Fig. 8.10: Fit of Modified Polycone Sheets of International Maps of the World

This projection is drawn on 1:1 m scale. Each sheet covers 4° latitude and 6° longitude between 60° N and S and then 4° latitudes and 12° longitudes between 60° and 88° N and S. The two polar areas are represented by two circular sheets each having a diameter of 4 degree of latitude. The sheets falling within 60° and 88°, number 420. Including two polar sheets, the total number of sheets is 2222.

On this projection errors do not exceed one hundredth of an inch at any place on the map. This is because of the better distribution of the error as well as the accuracy.

The construction of this projection involves the use of Tables 8.3 and 8.4. Table 8.3 gives the corrected lengths between intervals of 4° latitudes, while Table 8.4 gives the X and Y coordinates of the points of intersection of parallels and meridians east and west of the central meridian.

Take the corrected length of the relevant meridian from Table 8.2 and draw a vertical line. Take the coordinates of the top and bottom parallels east and west of the central meridian from Table 8.4 and join them to get the two parallels. Join the coordinate on two parallels to get the six straight line meridians. Divide all the meridians into four equal parts to get the remaining parallels.

Table 8.3: Length of the Central Meridian in International Projection

Latitude	*Natural Length (mm)*	*Corrected Length (mm)*
0-4	442.27	442.00
4-8	442.31	442.04
8-12	442.40	442.14
12-16	442.53	442.28
16-20	442.69	442.45
20-24	442.90	442.67
24-28	443.13	442.91
28-32	443.39	443.19
32-36	443.68	443.50
36-40	443.98	443.81
40-44	444.29	444.14
44-48	444.60	444.47
48-52	444.92	444.81
52-56	444.22	445.13
56-60	445.52	445.44

Table 8.4: Co-ordinates of Intersection of Parallels and Meridians in International Projection

Latitudes	*Coordinates*	*Longitude from Central Meridian*		
		1°	*2°*	*3°*
0°	x	111.32	222.64	333.96
	y	0.00	0.00	0.00
4°	x	111.05	222.10	333.16
	y	0.07	0.27	0.61
8°	x	110.25	220.49	330.74
	y	0.13	0.54	1.21
12°	x	108.91	217.81	326.73
	y	0.20	0.79	1.78
16°	x	107.74	214.08	321.13
	y	0.26	1.03	2.32
20°	x	104.65	209.31	321.13
	y	0.31	1.25	2.81
24°	x	101.76	203.52	305.31
	y	0.36	1.45	3.25
28°	x	98.37	196.75	333.96
	y	0.40	1.61	3.63
32°	x	94.50	189.01	283.56
	y	0.44	1.75	3.93
36°	x	90.17	180.36	270.50
	y	0.46	1.85	4.15
40°	x	85.40	170.82	256.29
	y	0.48	1.92	4.31
44°	x	80.21	160.45	240.73
	y	0.49	1.95	4.38
48°	x	74.63	149.29	224.00
	y	0.48	1.94	4.36
52°	x	68.69	137.40	206.16
	y	0.47	1.89	4.25
56°	x	62.40	124.83	187.31
	y	0.45	1.81	4.06
60°	x	55.81	111.64	167.52
	y	0.42	1.69	3.80

It is neither an equal area nor an orthomorphic projection. Both these properties have been sacrificed to make the adjoining sheets fit together. Each parallel is treated as a standard parallel; it is curved and is plotted according to its corrected length. Meridians are straight lines. The parallels intersect the central meridian at right angles; they intersect the other meridians at slightly oblique angles. The two meridians on either side of the central meridian are drawn correct to scale. Distance along the bounding meridians are slightly longer, whereas along the other meridians slightly shorter.

CYLINDRICAL PROJECTION

When a globe is enclosed in a developable cylinder and light placed at the centre of the globe, casts the image of the graticules of the transparent globe on the cylinder, we get cylindrical projections.

Perspective Cylindrical Projection

This projection is also known as Natural or True Cylindrical Projection. Draw a circle as per scale to represent a globe. Produce the equatorial diameter to the extent of $2\pi R$. This produced line will be equator of the projection. Draw two perpendiculars with half the length of the equator on either side of equator. One of these perpendiculars is tangent to the circle. Radii representing the angles of various latitudes are then drawn from the centre of the circle and produced to meet the tangent line as shown in Fig. 8.11. From the points of intersection draw lines parallel to the equator. Divide the equator into desired number of divisions to get the meridians.

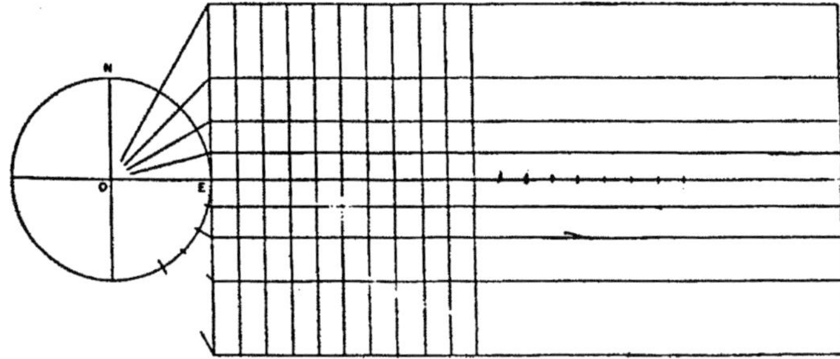

Fig. 8.11: Perspective Cylindrical Projection

In this projection, the parallels and the meridians are drawn as straight lines cutting each other at right angles. The meridians are equi-spaced but the intervals between parallels increase toward the poles. All the parallels are equal in length. Both the east-west and north-south scales are increasingly exaggerated but the exaggeration is not the same in both the directions. Distances are correctly represented along the equator. This projection is rarely used as the exaggeration of area and shape is too much and only a narrow strip along the equator gives tolerable results.

Simple Cylindrical Projection

To avoid the shortcomings of the perspective cylindrical projection, a number of non-perspective cylindrical projections have been developed. One of these is the simple cylindrical projection. On this projection, the meridians are drawn true to scale and they together with the parallels are equally spaced. They cut each other at right angles to form a network of squares.

For graphical construction of the Projection, draw a circle with the given earth radius. The equatorial diameter is extended out equal to the required length of the equator. If a network for world map is desired, the length of the equator will be $=2\pi R$. Draw an angle showing the latitude of the given interval and making an arc at the circumference of the circle. This arc distance is the interval along the parallels as well as the meridians. Mark these intervals along the equator and the two perpendiculars erected at the two ends of the equator. Join the corresponding points marked on the perpendiculars to get the parallels. Draw meridians through the points marked on the equator (Fig. 8.12).

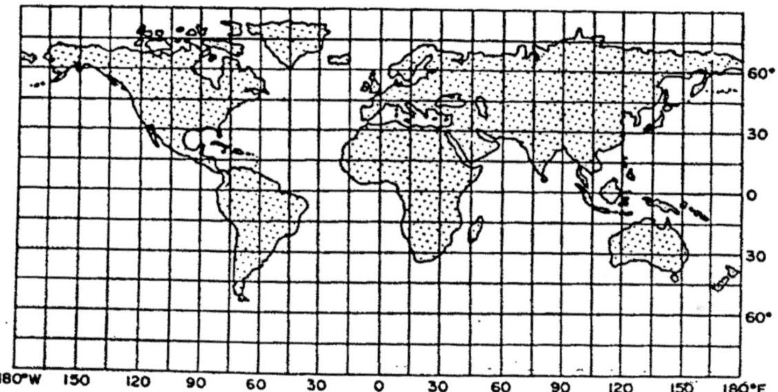

Fig. 8.12: Simple Cylindrical Projection

For mathematical construction, the following formula can be adopted:

1. The length of the equator $= 2\pi R$
2. The length of the interval along the equator and the meridian $=2\pi R \dfrac{\text{Interval}}{360}$
3. The length of the meridian $= \pi R$

Having worked out the length of the equator, meridians and the interval along equator and meridian, construct the graticules as it is given in Fig. 8.12.

Since the latitudes are drawn parallel to the equator, the scale is exaggerated at increasing rate away from the equator, so much so that the pole which is a point is represented by a line equal to the equator and parallel to it. The parallels and meridians intersect each other at right angles, and form a series of squares. This projection is, therefore, not good for representing areas in higher latitudes. It is neither an equal area nor an orthomorphic projection. It can be used for areas which extend along or near the equator as Africa and South-America.

EQUAL AREA OR LAMBERT'S CYLINDRICAL PROJECTION

To represent areas correctly, the simple cylindrical projection has been modified to get Equal Area or Lambert's Cylindrical Projection. It is so designed that the areas of the respective zones are equal to their counterparts on the globe. The shape of the area represented gets exaggerated.

The procedure for constructing this projection is illustrated in Fig. 8.13. All that is needed is to draw a circle with a given radius and the equator equal to $2\pi R$. The equator is divided into required number of parts = $2\pi R \frac{\text{Interval}}{360}$ meridians. Lines parallel to the equator are drawn from the points where the radii of latitudes touch the circumference of the circle.

For mathematical construction, the following formulate may be used:

Length of the equator = $2\pi R$

Intervals along the equator = $2\pi R \times \frac{\text{Interval}}{360}$

Distance of parallels from the equator = R sin lat.

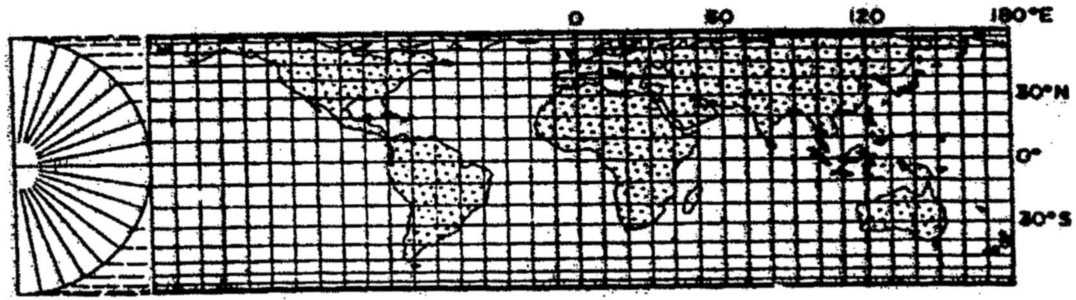

Fig. 8.13: Equal-area or Lambert's Cylindrical Projection

This is an equal area projection because the exaggeration in areas caused by the increases in the lengths of the parallels towards the pole is equalized by the decreasing distances between the parallels. The scale is, however, correct only along the equator. The parallels and the meridians intersect each other at right angles. The projection is not orthomorphic. The above deficiencies render it unsuitable for areas which extend to higher latitudes. Africa can be shown on this projection better as the equator almost cuts it into two halves.

GALL'S PROJECTION

Gall's projection is an improved version of cylindrical projections discussed earlier. In this projection the cylinder is assumed to cut across the globe at 45° N and S, i.e., halfway between the equator and the poles (Fig. 8.14). The length of the equator is equal to the length of the latitude along which the cylinder cuts the globe. Thereupon, it is given a perspective effect by supposing the light being placed stereographically (at the opposite end of the equator).

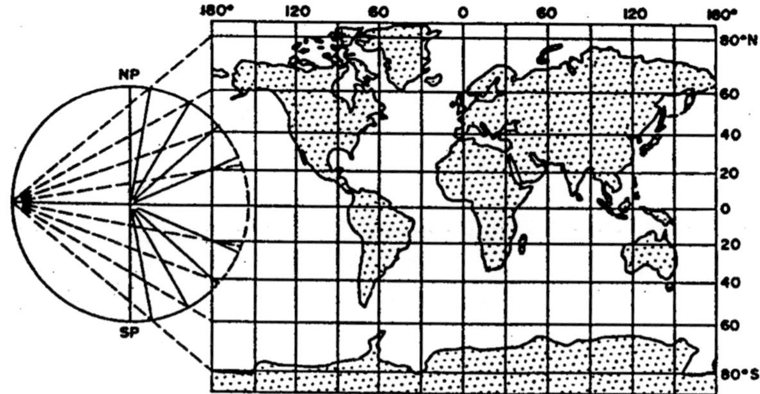

Fig. 8.14: Gall's Projection

To construct this projection, draw a circle with the given radius of the reduced earth. Draw radii for latitudes from the centre of the circle to meet its circumference. Through the points marked by radii of 45°N and S, draw a vertical line to represent one of the sides of the cylinder. Now if the light is placed at the opposite end of the circle the points on the circumference of the circle will be reflected on the side of the cylinder. These reflected points serve as the starting points for the parallels which are drawn parallel to the equator. The length of the equator, and hence of all the parallels is equal to $2\pi R \cos. 45°$. Divide the equator into required parts and draw perpendiculars to get the projection as illustrated in Fig. 8.14.

For mathematical construction, the following calculations are required:

Length of the equator $= 2\pi R \cos 45°$

Intervals along the equator $= 2\pi R \cos 45° \dfrac{\text{Interval}}{360}$

Distance between the equator and the parallels $= 1.7071\, R \tan \text{lat}/2$.

This projection differs from other cylindrical projections in the sense that only the parallels of 45° and 90° N and S are drawn true to scale. As a result of this east-west scale is shortened between these two parallels and exaggerated between 45° and 90° N and S. The meridians are equi-spaced but not correctly divided. Parallels are not equi-spaced; they get closer towards the equator between 45° N and S and farther apart toward the poles between 45° and 90° N and S. It is neither an equivalent nor an orthomorphic projection. Because of the lessened distortion in the higher latitudes it is often preferred for world maps.

MERCATOR'S PROJECTION

The *Mercator Projection* was invented by Gerardus Mercator (1512-1594), a Flemish mapmaker in 1569. His name is a latinized version of Gerhard Kramer. It was modified and improved by Edward Wright—a British cartographer in 1599 to give it its present form. Projection is very useful for seafarers as it shows correct directions by straight lines. There are four main reasons for its early popularity:

1. Britain being the greatest maritime power and shipping being the chief mode of international transportation during the 19th century, its wide use was inevitable.
2 On this projection the major portions of the then British Empire was Highly exaggerated in area giving the impression to the average map reader that the British islands, and its middle latitude colonies were large. It is interesting to note in this regard that on this projection India does not appear to be more than 5 times the size of the British Isles and Greenland appears to be larger than the world of South America.
3. The countries of the world under the political control of the Europeans appeared smaller than what they were.
4. The shape of the countries are well maintained on this projection.

This projection has two outstanding features. These are:

1. Orthomorphism, and
2. Constant Bearing.

Being an orthomorphic projection small areas represented on this projection maintain their shape accurately. This does not mean that it does not suffer from distortions. The distortions in length and breadth are not at uneven rate; they are at the same rate.

As in most other cylindrical projections, in this projection too, all the parallels are equal to the equator. This projection differs from other cylindrical projections in that the distance between the successive parallels increase at the same rate at which the distance between meridians increase towards the poles. As a result the amount of exaggeration of scales both north-south and east-west is the same at any given point (Fig. 8.15).

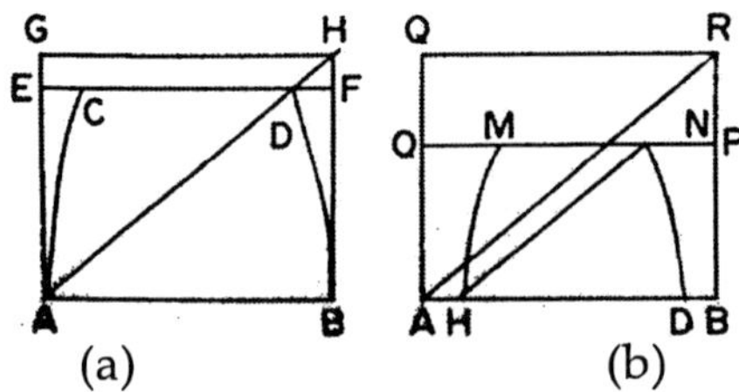

Fig. 8.15: How are Directions Maintained on Mercator's Projection?

In Fig. 8.15 (a) AB is a part of the equator and CD a part of a corresponding parallel. CD has been increased in length east-west to get EF = AB. But EF has also been shifted simultaneously to the position of GH. Here the amount of north-south expansion is of the same order as the east-west expansion. As a result the directions AD and AH remain the same. In Fig. 8.15b HD and MN are any two parallels. Both the lines have been increased in length to equal the equator, because in all cylindrical projections the parallels are equal. To keep up with the increase in the length of the two parallels the interval between the two has been increased, so that MN is represented by QR and not OP. Now the figure MHDN has assumed the shape of figure QABR wherein the direction of R from A is the same as that of N from H.

To construct this projection, first determine the length of the equator with the help of the usual formula of equator being = $2\pi R$. The distances of other parallels from the equator can be found out as = 2.3 log tan (45+lat/2). So that the interval between the equator and 20° N or S (with R=1") = 2.3×1×1548 =356". This can also be found out with the help of Table 8.5.

Table 8.5: Distances from Equator in Terms of R in Mercator's Projection

Degree of lat.	*Distance from equator*
5	0.087 R
10	0.175 R
15	0.265 R
20	0.356 R
25	0.450 R
30	0.549 R
35	0.652 R
40	0.763 R
45	0.880 R
50	1.011 R
55	1.153 R
60	1.317 R
65	1.505 R
70	1.736 R
75	2.025 R
80	2.437 R
85	3.132 R

The meridians are obtained by drawing perpendiculars from equally spaced points marked on the equator, as in the case with other cylindrical projections (Fig. 8.16).

Mercator's projection is perhaps the best known projections the world; it is most commonly used projection for the portrayal of the world as whole. As already mentioned, it has certain merits, but it has also been responsible for giving a distorted picture of the world to the average map user. The areas of the higher latitudes shown on this map are highly distorted. It, nevertheless, continues to be popular.

Owing to the convergence of the meridians of the globe at the poles, the shortest distance between any two points on a parallel except the equator, is not along that parallel of latitude on which the points are located, but to the north of that hemisphere in the northern hemisphere and to the south in the southern hemisphere. This

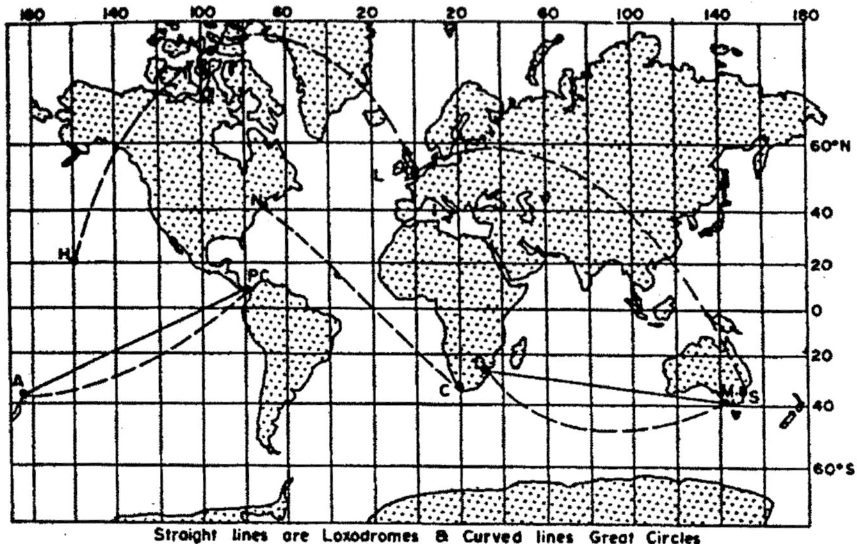

Fig. 8.16: Mercator's Projection

applies even to the points which do not lie on the same parallel and are located east and west of each other. This shortest distance is an arc of a circle and the completion of the circle gives us a great circle. On this projection the shortest distance between any two points on the earth's surface is along the arc of the great circle passing through them. The navigators' therefore, follow the great circle routes.

If a sailor has to follow a great circle, other than the equator or one of the meridians, he will have to change directions at every bit of the distance he sails. This happens because great circles are not shown by a straight line on this projection. As against this on the Gnomonic chart, which we propose to discuss in a subsequent section of this chapter, the shortest distance between any two points is given by a straight line joining the two points. This drawback in the Mercator's projection is amply compensated by the fact that a straight line joining any two points on this projection gives a constant bearing, i.e., it cuts all the meridians and the parallels through which it passes at the same angle. This line is called *Loxodrome* or *Rhumb* line. This amounts to saying that all that a navigator has to do to reach a given point, is to follow a single bearing all through his course. This naturally gives considerable advantage to the navigator.

We have seen that there are two factors which make the job of a navigator easier. The first is the great circle route and the second is the *Rhumb line or Loxodrome.* Unfortunately, the arcs of great circles and the *Loxodromes* joining any two points do not coincide with each other on the Mercator's projection. Ways and means have, therefore, been found to combine the two properties and get the maximum advantage. This has been done by dividing the arc of the great circle into convenient number of sectors as shown in Fig. 8.16. These sectors are connected by straight lines. And since any straight line joining any two points on this projection is a Loxodrome, these sectoral lines become a set of Loxodromes the navigator has to change his directions at only a few places to pass through the shortest route. This is the great merit of Mercator's projection and the reason why it is used so widely by the navigators.

The Mercator's chart is superior to the Gnomonic chart in several ways. In the former straight lines are lines of constant bearing, whereas in the latter they are lines giving the shortest distances. To many the latter may seem to be a more useful a property, but it is not so because a Gnomonic projection has excessive exaggeration as one goes farther from the centre of the projection. Although poles cannot be shown on the Mercator's projection, it does not matter much because polar areas are not so important for human activities. Moreover, it is not possible to map the whole world on a Gnomonic chart.

ZENITHAL PROJECTIONS

To construct a Zenithal Projection light is thrown on a plane is placed tangently on a transparent globe. Each point within and out side the globe from which light is focused on the plane produces different images of the graticules of parallels and meridians. The projections thus formed have their directions from the centre of the map projection to different points are correct. The projections so formed are also called Azimuthal.

If the light is placed at the centre of the globe, we get a zenithal projection which is called gnomonic. If it is placed at a point diametrically opposite to the point where the plane touches the globe, we then get a stereographic projection. And, if the light is placed at the infinity to cast parallel rays, we get an orthographic projection. Thus we have three types of perspective zenithal projections. These are:

1. Gnomonic,
2. Stereographic, and
3. Orthographic.

Each of these three types of Zenithal projections can be further sub-divided on the basis of the position of the plane surface on the globe. The plane can be placed at three different positions. These are:

1. North or South Pole,
2. Equator, and
3. Anywhere between the equator and the poles.

Thus each of the three types of Zenithal projections can have three sub-types:

1. Polar,
2. Equatorial, and
3. Oblique.

Thus we get nine types of perspective Zenithal projections.

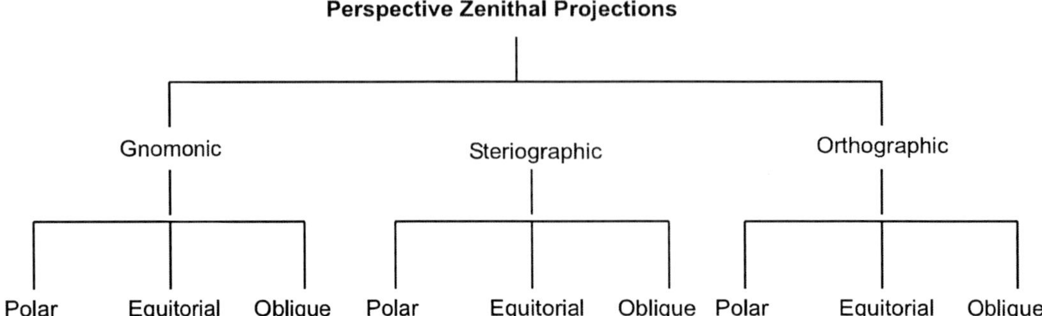

In addition to these nine types of perspective Zenithal projections, one can create a number of non-perspective Zenithal projections. Two of these are discussed in this chapter. These are:

1. Zenithal equi-distance projection;
2. Zenithal equal-area projection.

GNOMONIC PROJECTION: (POLAR CASE)

This projection is obtained by casting the shadow of the network of the parallels and meridians on a tangent plane touching the globe at one of the poles. On this projection a straight line joining any two points gives the shortest distance between those points. In other words the great circles are represented by straight lines.

To construct this projection, draw the angular lines for the latitudes to be represented from central point C in a circle as shown in Fig. 8.17. A plane surface LP is made tangent to this globe at the pole. To determine the radii of the parallels of the projection produce the angular lines drawn within the circle to meet the plane surface at M, N, O, P, etc. The required radii are LM, LN, LO and LP, etc. Now with L as the centre, draw concentric circles with these radii. Divide the circles into required parts to show the meridians.

We can draw this projection by calculating the radii of each parallel mathematically using the fomula:

Radius of the parallel = R cot lat.

In this projection the parallels are concentric circles; they are not equi-spaced; the distance between them increases progressively towards the equator. The meridians are straight lines radiating from the centre. All great circles are shown as straight lines, hence the shortest distance between any two points on the earth can be easily found out on this projection. This distance along the parallels and the meridians increase away from the centre and hence the areas are exaggerated. The shapes of the areas represented are also distorted as on goes away from the centre.

This projection is suited small areas in Polar Regions. The equator can not be shown on this projection.

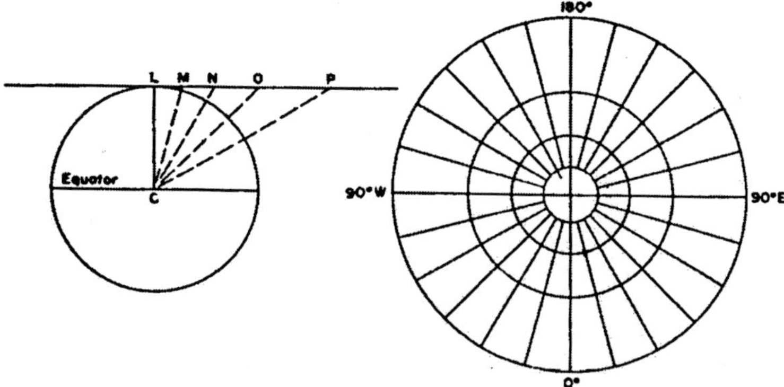

Fig. 8.17: Gnomonic Projection (Polar Case)

GNOMONIC PROJECTION (EQUATORIAL CASE)

In this case the plane is assumed to be tangent at the equator. Draw a circle with the given radius. Produce the angular lines to the plane surface EQ to get the intervals along the equator as well as the central meridian. To get the intersections of parallels and meridians, draw perpendiculars on angular lines showing the degrees of latitudes at the points these lines EQ. Thus s perpendicular Aa is drawn on OA at point A. this perpendicular is intersected by other angular lines like OB, OC, etc. at points a_1, a_2, a_3, etc. Aa_1 is the first interval along its meridian, a_1, a_2 is the second one and so on. After we sub-divide all the meridians this way, we get the network of the Equatorial Gnomonic Projection (Fig. 8.18).

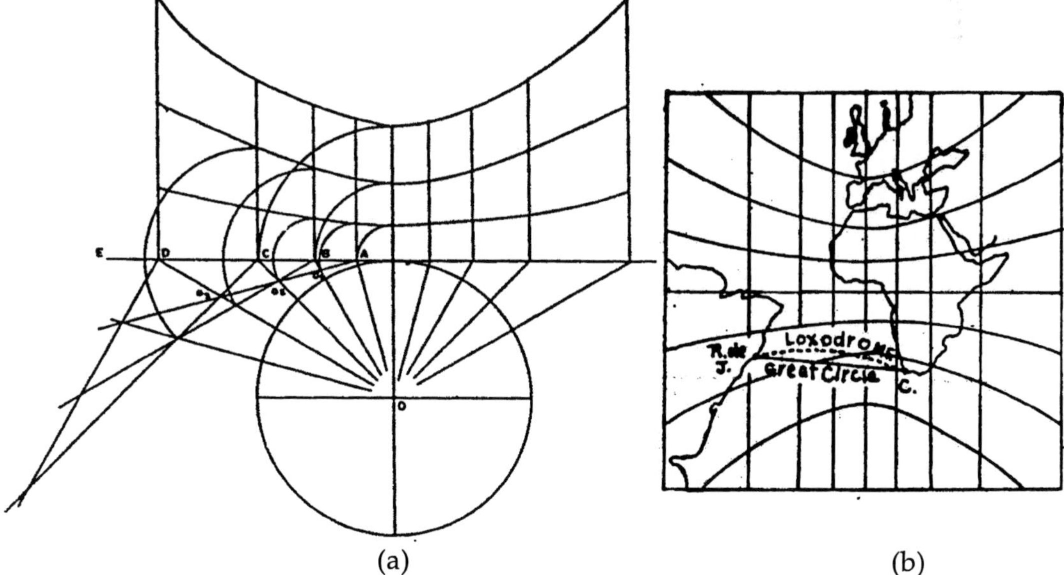

(a) (b)

Fig. 8.18: Gnomonic Projection (Equatorial Case)

In this projection the equator is represented by a straight line. All other parallels are curves. Distances between the parallels increase away from the equator on either side. Meridians are great circles and are projected as straight lines. Distances between the meridians also increase away from the central meridian. Both the area and the shape are distorted away from the centre of the map. It can be used for showing small areas in the equatorial region.

STEREOGRAPHIC PROJECTION (POLAR CASE)

In the construction of this projection the light is assumed to be placed at a point diametrically opposite the point where the plane touches the globe tangentially (Fig. 8.19). To begin with a circle is drawn to scale to represent the earth and then the required angles to show the latitudes are drawn. Let the rays emanating from O project the latitudes on the plane surface NF. The required radii for the parallels are: NA, AB, BC, CD, DE and EF. Using these radii draw concentric circles from a common centre N. The meridians can be drawn with the help of a protractor.

The formula for calculating the radii of the parallels mathematically is:

$$\text{Radius} = 2\text{ R tan } \frac{90 - \text{lat}}{2}$$

As apparent from Fig. 8.19, the parallels in this projection are concentric circles but they are not equidistant. The distance between parallels increases as one moves poleward from the centre, but the rate of increase is slower than what it is in the case of Gnomonic Polar Projection. As the equator can also be represented on this projection, it can be used for hemispheric maps too. The meridians are straight lines radiating from a common centre. The direction from the centre to any point on the projection is correct. As the scale along the meridians increases away from the centre at the same rate at which it increases along the parallels, the shape of areas shown on this projection is not distorted. In other words, it is orthomorphic projection. Since, a circle on the sphere remains a circle on this projection, it is very useful for astronomical purposes.

Like its polar version, the equatorial case of this projection is used to represent a hemisphere. The polar case is used to show the northern and southern hemispheres whereas the equatorial case is used to show the eastern and western hemisphere. The same procedure is followed in getting the intervals along the equator and the central meridian as in the polar case. Intervals along the two are the same.

Draw two straight lines perpendicular to and intersecting each other to represent the equator and the central meridian. Mark the intervals along them (Fig. 8.20). Then draw the bounding meridian. With the help of a protractor mark angles of the given interval along the circumference of the circle which is also the bounding meridian. join the corresponding points (two on the bounding meridian and one on the central meridian) by arcs of circles. To draw an arc of a circle passing through any three points, the following procedure should be adopted. Suppose A, B and C is the points through which the arc should pass. Join AB and BC and locate their mid-points. From

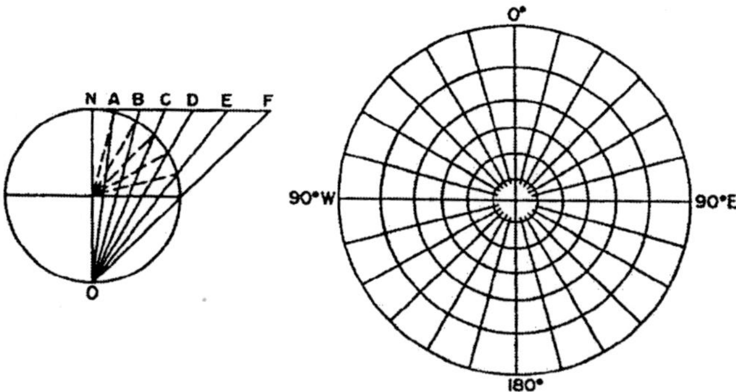

Fig. 8.19: Gnomonic Polar Projection

the mid-points, draw perpendiculars which meet at a common point O. OA = OB = OC will be radius of the circle.

For mathematical construction, the following procedure is suggested. Draw the bounding meridian (circle) with the given radius. Draw the equatorial and polar axes and produce them outside the circle (Fig. 8.20). As the meridians and parallels are arcs of non-concentric circles, we have to determine the radii and the centres of these arcs. The formulae of these are:

With the point of intersection of the polar and the equatorial axes:

Centre of a parallel = 2 R cosec. lat.
Radius of a parallel = 2 R cot. lat.
Centre of a meridian = 2 R cot. long.
Radius of a meridian = 2 R cosec. long.

All the parallels and the meridians excepting the equator and the central meridian are arcs of non-concentric circles. Neither the parallels nor the meridians are equi-spaced. Distances between them increase away from the centre. The projection is commonly used for maps of hemispheres.

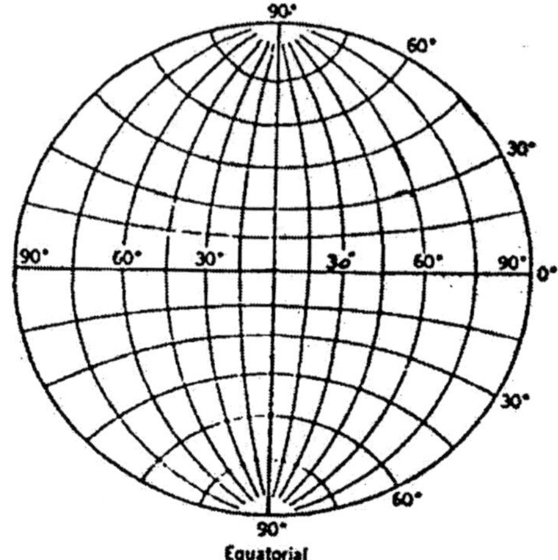

Fig. 8.20: Stereographic Projection (Equatorial Case)

ORTHOGRAPHIC PROJECTION (POLAR CASE)

In this projection the light is presumed to be focused from infinity, so that the rays passing through the globe come parallel to each other. The plane surface touches the globe at the pole. To construct this projection, first, draw a circle with the given radius. Draw angles at given intervals. Draw perpendiculars on the plane surface NF passing through the points where the angular lines intersect the circumference of the circle. NA, AB, BC, CD, DE, and EF are the radii of the parallels. With the radii draw the projection as shown in Fig. 8.21.

For mathematical construction the radii of various parallels can be found out by using the formula: Radius of a parallel = R cos. lat.

On this projection the parallels are concentric circles. They are not equi-spaced. The distance between them decreases away from the pole. Meridians are straight line radiating from the pole. As is the case with all Zenithal projections, the direction from the centre to any other point on this projection also is correct. Scales along the parallels are correct and hence distances along them are correctly represented. This is, however, not the case with respect to the meridians. Because of the excessive shortcomings of scales along the meridians, both the area and shape are excessively distorted. It is particularly so at places away from the centre of the projection.

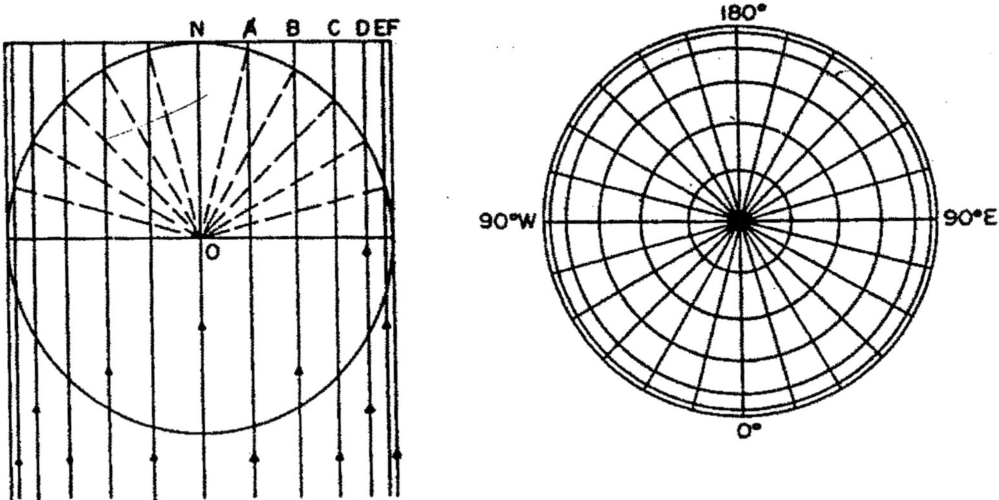

Fig. 8.21: Orthographic Projection (Polar case)

ORTHOGRAPHIC PROJECTION (EQUATORIAL CASE)

The construction of the equatorial case is either to that of the polar case. A circle of a given radius is constructed and intervals along them marked as in the polar case. The meridian is then drawn as ellipses joining the poles and the relevant interval along the equator. The parallels are drawn as straight lines (Fig. 8.22).

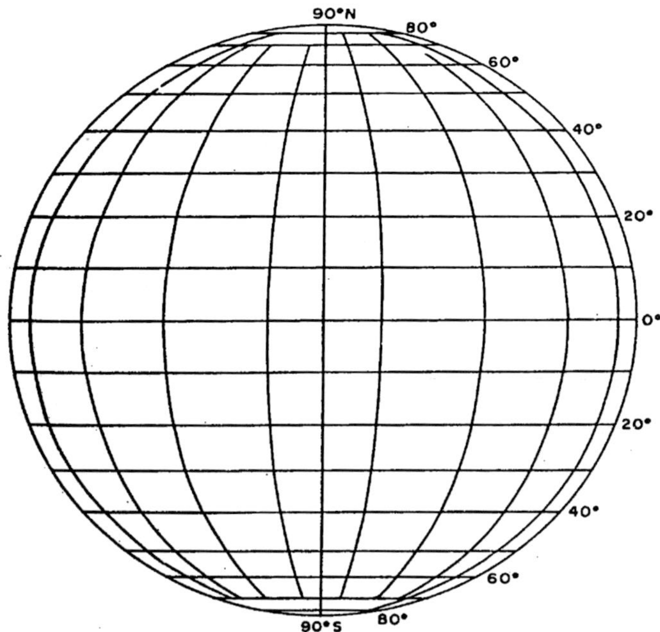

Fig. 8.22: Orthographic Projection (Equatorial Case)

ZENITHAL EQUIDISTANT PROJECTION (POLAR CASE)

This is non-perspective projection. The parallels are drawn at their true distances from one another. Because of this characteristic, it is called equi-distant projection.

For constructing this projection, all that is needed is the true distance between parallels along a meridian. This distance is = $2\pi R \times$ interval/360. Having gotten the interval, take a meridian and mark the intervals from the centre. These markings give the radii of various parallels (Fig. 8.23). Draw the meridians as straight lines radiating from the centre.

On this projection one of the pole forms the centre of the projection. The parallels are concentric circles equally spaced from one another. Meridians are straight lines. Scales along the meridians are correct. Hence, directions as well as distances from the centre to any point along the meridians are correct. The scales along the parallels are not correct and hence the distances along them are exaggerated. The excessive divergence of the meridians outwards distorts the area. It is neither an orthomorphic nor an equal area projection.

This projection can be faithfully used for a hemisphere if the objective is to show correct distances and directions from the centre of the projections. Large areas cannot be shown satisfactorily. It is generally used for polar areas extending from 90° to 50°, although it can be extended to cover the entire globe.

The equatorial case of the projection can also be drawn very easily. The radius of the circle drawn as the bounding meridian of the projection will be 1.57 times that of the reduced earth radius.

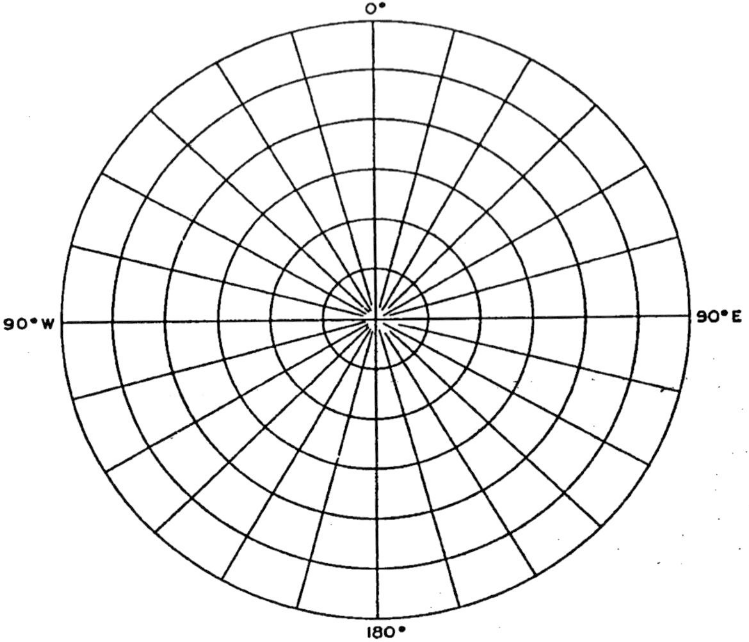

Fig. 8.23: Zenithal Equidistant Projection (Polar case)

ZENITHAL EQUAL AREA PROJECTION (POLAR CASE)

It is also called Lambert's Azimuthal Equivalent Projection and Lambert's Equal Area Projection. It is also a non-perspective projection. This projection is made equivalent by compensating the increase in scale along parallels with a proportionate decrease in scale along meridians. The area of a zone of the projection is equal to the area of the corresponding zone on the globe.

The radii of various parallels of the projection are found out by joining the pole with the points where lines showing the angles of latitudes cut the circumference of the circle. Meridians are drawn plotted as straight lines (Fig. 8.24).

The radii of the parallels can also be found out mathematically. The formula for this is—

$$\text{r (radius)} = 2\text{R} \sin \frac{\text{co} - \text{lat.}}{2} = 2\ \text{R} \sin \frac{90 - \text{lat.}}{2}$$

On this projection the parallels are drawn as concentric circles, but they are not equi-spaced. The spacing between them decreases outwards from the centre. The scale along the parallels is increasingly exaggerated away from the centre. The radial compression of the meridians is proportionate to the enlargement of scale along the parallels. This makes it an equal area projection. The directions from the centre are also correct.

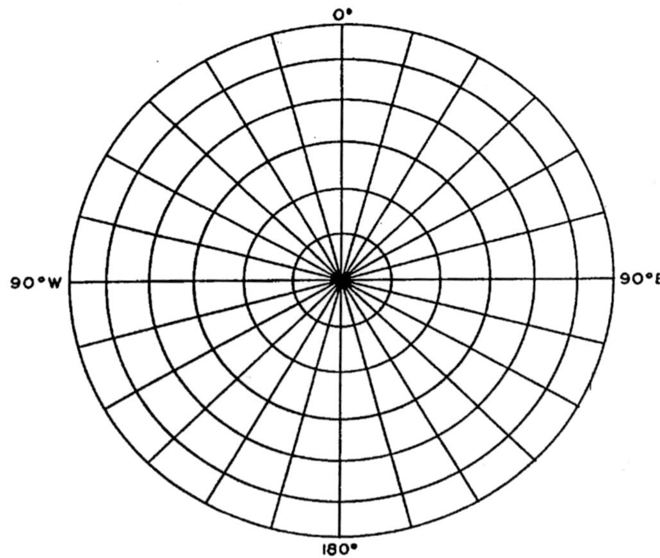

Fig. 8.24: Zenithal Equal-area Projection (Polar Case)

ZENITHAL EQUAL AREA PROJECTION (EQUATORIAL CASE)

For constructing the equatorial case of the projection draw a circle with a radius = 2 R. The distances of various latitudes from the intersection of the polar and equatorial axes can be had from the Table 8.6 given for Mollweide's projection. The intervals between the intersection of the two axes and various meridians and parallels are the same.

CONVENTIONAL PROJECTIONS

The following conventional projections are discussed here:

1. Globular
2. Mollweide
3. Interrupted Mollweide
4. Sinusoidal
5. Interrupted Sinusoidal
6. Hammer

Globular Projection

This projection is one of the easiest and most popular ones. On this projection a hemisphere is represented by a circle. The radius of this circle is equal to 2 R. To show the whole world we need a pair of circles (Fig. 8.25).

Draw a circle with the radius = 2R. Also draw the equator and the central meridian

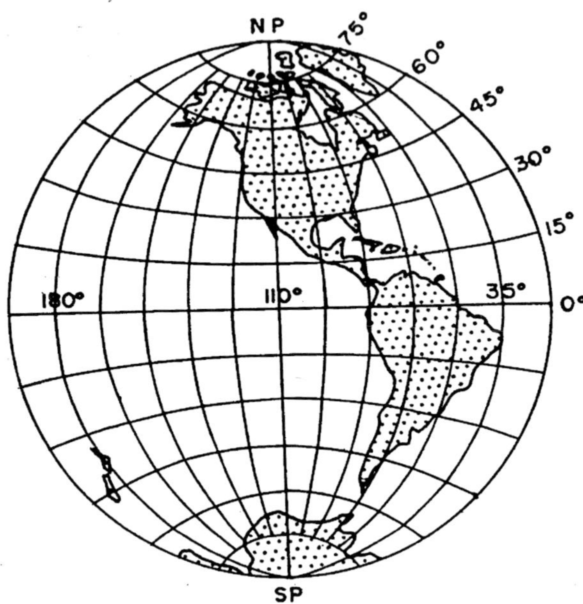

Fig. 8.25: Globular Projection

as straight lines intersecting each other at the centre of the circle. Divide the central meridian and the equator into equal parts. The number of parts will be determined by the interval selected for the projection. Divide the bounding meridians (the circles) also into some number of parts. Join the corresponding points on the central meridian and the bounding meridian on both sides of the equator by arcs of circles. These arcs will form parallels. Similarly join the north and south poles through the various divisions on the equator to get the meridians.

The projection is highly arbitrary in matters of cartographic details. The parallels and the meridians (except for the equator and the central meridian) do not intersect each other at right angles. Because of the inequality in scales both along the parallels and meridians, shapes get distorted away from the centre of the projection. As distances between the parallels tend to increase away from the central meridian, areas also get distorted toward the margins. It is neither an equal area nor an orthomorphic projection. Nevertheless, it is often used for showing hemisphere in atlases for the simple reason of its simplicity.

Mollweide's Projection

Unlike the globular projection, the Mollweide's Projection can represent the whole world within one network. This is an equal area projection and maintains the shapes fairly well in the equatorial, tropical and mid-latitude areas. It is quite a popular projection these days.

The network of the Mollweide's projection is drawn in an ellipse the area of which equals the area of the globe. The ellipse of the projection is so drawn that it is double the area of the circle encompassing 90°E and W of the central meridian. First of all draw a circle with the radius = 2 R. Now produce the equatorial axis both sides equally so that its total length is four times the radius of the circle. This horizontal line measuring twice the diameter of the circle will represent the equator. The central meridian forms the polar axis of the sphere. Tables 8.6 and 8.7 give the intervals along the parallels and the meridians.

Table 8.6: Mollweide's Projection: Distance of the Parallels from the Equator

Latitude	*Distance*	*Latitude*	*Distance*
5°	0.0970 R	50°	0.9209 R
10°	0.1936 R	55°	1.0014 R
15°	0.2896 R	60°	1.0783 R
20°	0.3846 R	65°	1.1508 R
25°	0.4787 R	70°	1.2189 R
30°	0.5714 R	75°	1.2814 R
35°	0.6621 R	80°	1.3370 R
40°	0.7508 R	85°	1.3835 R
45°	0.8372 R	90°	1.4142 R

Table 8.7: Mollweide's Projection: Intervals along the Parallels

Latitude	*Interval*	*Latitude*	*Interval*
5°	0.9977×I.E.	50°	0.7589×I.E.
10°	0.9906×I.E.	55°	0.7062×I.E.
15°	0.9788×I.E.	60°	0.6411×I.E.
20°	0.9623×I.E.	65°	0.5811×I.E.
25°	0.9410×I.E.	70°	0.5071×I.E.
30°	0.9148×I.E	75°	0.4232×I.E.
35°	0.8836×I.E.	80°	0.3259×I.E.
40°	0.8474×I.E.	85°	0.2680×I.E.
45°	0.8059×I.E.	90°	0.0000×I.E.

Note: IE = Interval along the Equator

Divide the equator and the parallels into equal parts. Mark the intervals along the central meridian and draw parallel to the equator through the points marked on the central meridian. Join the corresponding points marked on various parallels, including the equator, to get the meridians (Fig. 8.26).

To determine the length of intervals along the various parallels we can use the following formula.

$$\text{Length of one interval along the equator} = 4\sqrt{2}\,R\frac{\text{Interval}}{360}$$

$$\text{Length of one vertical along other parallels} = 4\sqrt{2}R\cos.y\frac{\text{Interval}}{360}$$

where y = elliptical equivalent of the latitude.

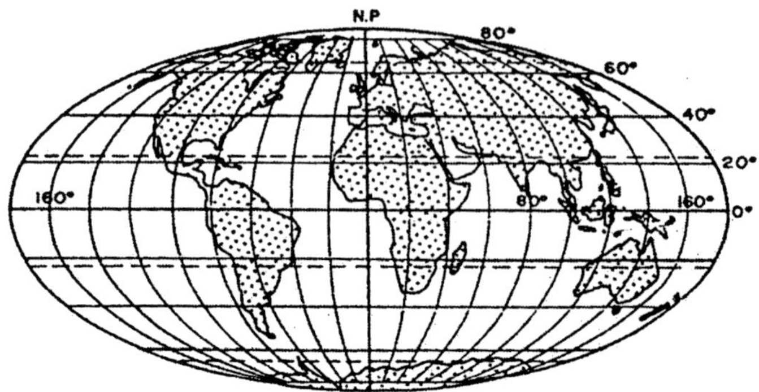

Fig. 8.26: Mollweide's Projection

The parallel are straight lines parallel to the equator. They are not equi-spaced. Except for the central meridian, all meridians are ellipse. The 90° E and W meridians form a circle. This projection is an equal area projection. This scale along the parallels differs from parallel to parallel. The shape of the area shown gets too exaggerated near the margins. It is, therefore, not a conformal or orthomorphic projection.

Being an equal area projection, it is used for showing distribution of various things, especially those which are found in the equatorial, tropical things, especially those which are found in the equatorial, tropical or mid-latitude areas. It represents the area better than the Sanson-Flamsteed's projection does. It is, therefore, preferable.

Interrupted Mollweide's Projection

This projection is a modified form a Mollweide's projection. It was developed by Prof. J.P. Goode in 1916. He called it as homolographic projection. The chief aim of this projection is to minimize the distortion in the shape of marginal areas by selecting a number of central meridians instead of only one central meridian (Fig. 8.27).

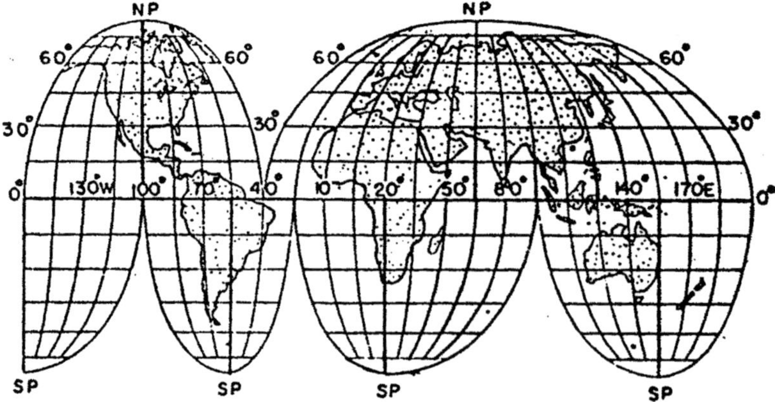

Fig. 8.27: Interrupted Mollweide's Projection

To construct this projection select the meridians to be treated as central meridians and then mark the meridians where the projection has to be interrupted. The selection should be so made that no meridian on either side of the equator, central meridians, and meridians of interruptions alternate. Divide the equator into the required number of divisions. The procedure for division is the same as in the case of the Mollweide's projection.

At selected meridians, erect perpendiculars to get the central meridians. From the Table 8.7 given for Mollweide's projection, mark the intervals between the parallels along all the central meridians. Draw lines passing through these points and parallel to the equator. Now mark intervals along all the parallels with the help of the table given for Mollweide's projection. These intervals need to be marked on both sides of each of the central meridians upon the meridian of interruption. Join the relevant points marked on various parallels to get the other meridians.

While selecting the central meridian and the meridians of interruption, care should be taken to see that the interruption do not occur along areas which are sought to be given importance in the map. For example, if the objective is to show land areas of the world, the meridians of interruption should occur in the water areas.

Sinusoidal Projection

This projection is also known as Mercator-Sanson-Flamsteed Projection because it is supposed to have been used by Mercator in 1605; by Sanson (a French cartographer) in 1650; and by Flamsteed (a British astronomer) in 1729. Name Sinusodial has been derived from the fact that sine curves are used in its construction.

This projection is a special case of Bonne's projection with equator as the standard parallel. The equator is represented as a straight line. Since all other parallels are parallel to the standard parallel, they are also straight lines. The central meridian and parallels are correctly divided as in the case of the Bonne's projection. It is an equal area projection.

For the construction of this projection one has to go through the same process as used in the construction of the Bonne's projection. The length of the equator can be found out by the formula 2πR. The central meridian is half the equator and is drawn perpendicular to it and its mid-point. The divisions along the equator or the central meridian can be found out the same way as in the case of Bonnes projection. Through the points dividing the central meridian, draw parallels to the equator. Join the corresponding points on the parallels by smooth curves to get the meridians (Fig. 8.28).

One can derive its measurements mathematically too by using the following formula.

$$\text{Equator} = 2\pi R$$

$$\text{Central Meridian} = \frac{2\pi R}{2} \text{ or } \pi R$$

Interval along the equator and the central meridian = $2\pi R \frac{\text{interval}}{360}$

Intervals along other parallels = Intervals along Equator x cos lat.

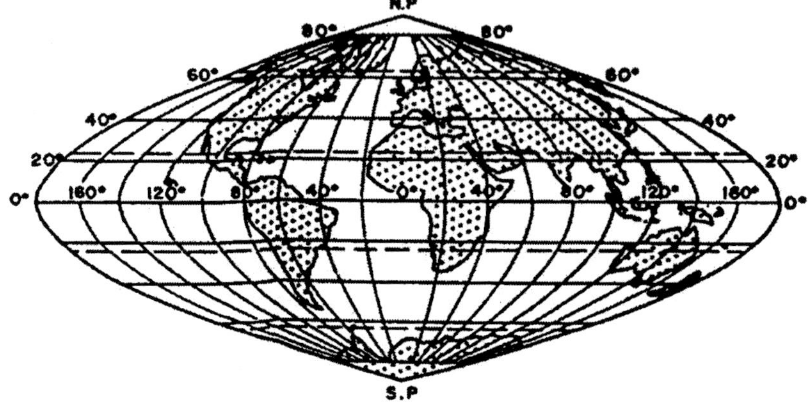

Fig. 8.28: Sinusoidal Projection

Equator is the standard parallel and is drawn as a straight line. Other parallels are also drawn as straight lines. Parallels are equi-spaced and correctly divided. Except the central meridian all meridians are sine curves. The scale is true along all parallels and the central meridian. But the scale along other meridians is not correct; it gets increasingly exaggerated away from the central meridian because of the varying obliquity of the intersections of meridians with parallels.

The projection, being an equal area one, is useful in representing distributional data. It is not suited for continents extending to equatorial regions and middle latitudes such as Australia, Africa, and South America. A World Map is not well represented on this projection because the increasing distortion of shape on the margins reduces the legibility of the map.

Interrupted Sinusoidal Projection

This projection too was developed by Goode. He had named it as Homolosine projection. His aim was to improve the shape of marginal areas, which get too distorted in a Sinusoidal projection.

The construction of this projection is easier than that of the interrupted Mollweide's projection. First select the meridians which are to be treated as central meridians and the meridians of interruption. Draw the equator = $2\pi R$. Divide the equator into required divisions. Draw each central meridian equal to one-fourth of the equator. Mark intervals along the central meridians. Draw lines parallel to the equator along the points marked on the central meridians to get the parallels. Mark intervals along each of the parallels on both sides of the central meridians. The intervals along the parallels are equal to 2π R cos lat. interval/360 while making these intervals on the parallels, the interrupted parts should be left unmarked. Draw smooth curves through the relevant points on the equator and the parallels, to get the meridians (Fig. 8.29).

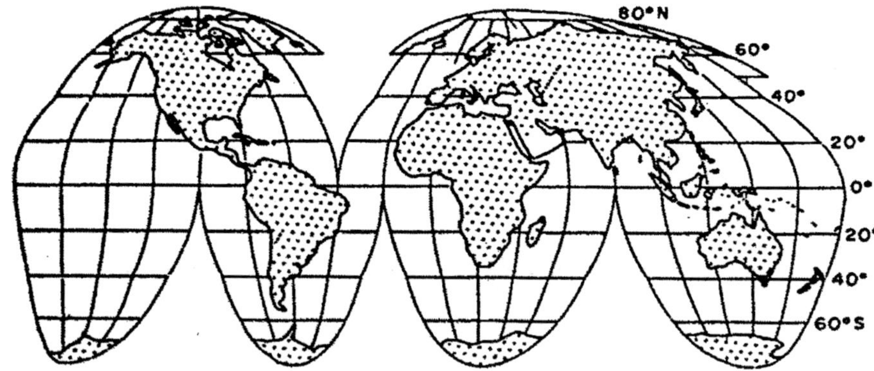

Fig. 8.29: Interrupted Sinusoidal Projection

Hammer's Projection

This projection is often referred to as Aitoff's projection. It is, however, different from Aitoff's projection. Hammer's projection is an equal area projection but Aitoff's projection is not so. The graticules of Aitoff's projection are derived from Zenithal equidistant projection while that of Hammer's projection are derived from Zenithal equal area projection. Hence, to equate Hammer's projection with Aitoff's projection is not correct.

To construct this projection, draw the equator = $2\pi R$. Draw a perpendicular to divide the equator as well as itself into two equal parts. The full length of this line should be half that of the equator and will mark the North and South Poles at its two ends. This line will function as the central meridian. To mark the intervals along the equator and the central meridian, follow the procedure adopted in the construction of the equatorial case of the Zenithal equal area projection. As there is no change in the dimensions of the central meridian, the table used in the Zenithal equal area projection (equatorial case) to determine the intervals along the central meridian, can be used in this projection also.

In this connection it may be recalled that this table is the same as used for the Mollweide's projection. The equator can also be divided the same way keeping in view that it is double the length of its counterpart in the Zenithal equal area projection. The intersections of all other meridians and parallels in the Zenithal equal area projection can be determined by treating the equator as x axis and the central meridian as the y axis, and the intersection of the two as the point of origin. Same measurements can be used to locate the corresponding intersections on the Hammer's projection, if the x axis is doubled. The completed projection is illustrated in Fig. 8.30.

It is an equal area projection and is superior to the Mollweide's projection because the distortions of shapes in the marginal areas are less than in Mollweide projection.

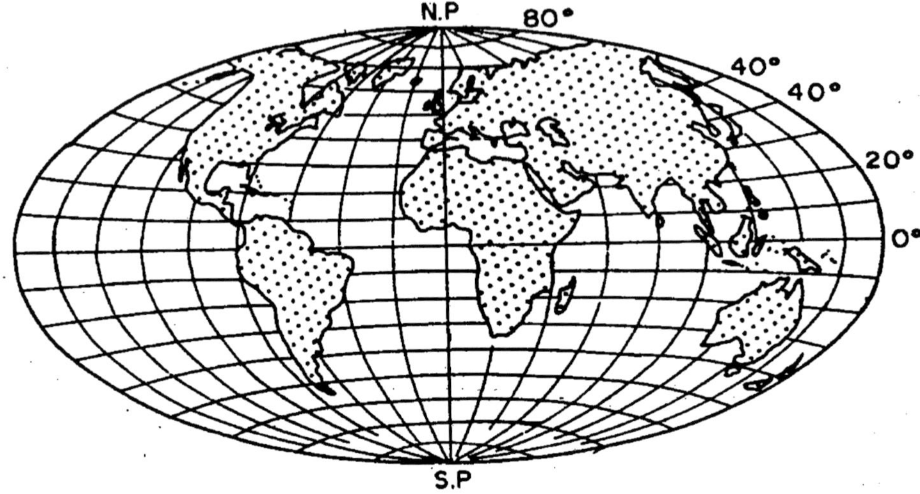

Fig. 8.30: Hammer's Projection

Choice of Projections

In this chapter we have so far discussed over 25 projections. These, however, do not exhaust all available and conceivable projections. There can be as many projections as our cartographic needs. With the availability of computer services, one can develop hundreds and thousands of projections. But many of these projections will be so complicated and will give such a confused and meaningless outline of the areas represented, that they will not serve any purpose. It is for this reason that we usually do not make use of more than 50 projections. Some of the most relevant of these projections have been discussed here.

Projections like any other cartographic device are only tools to achieve certain ends. They can help fulfil our objectives only if we use them judiciously. The correct use of projections involves their right selection for the purpose in view. We know that there is no projection which can portray all the characteristics of a globe. Some of them show areas correctly, some others give correct shapes, while still others give correct bearings (directions). Some projections are meant to show only small areas, other are meant to represent larger areas like continents while still others are good for representing the whole world. Some are good for equatorial areas, others for polar areas. There are few other characteristics which individual projections possess. But there is no projection, which possesses all the characteristics simultaneously. A judicious choice of projections, therefore, forms an important step in map making.

Compatibility of geographic information (data sets) is one of the major concerns that determine the choice of a projection. As each projection assigns slightly different coordinates to the same location, these data have to be related to conform to the selected projection. In small areas (large scale) metric distortions are minimal and hence data compatibility is more important but in very large areas (small scale), the critical factor is distortion of shape.

Look at Fig. 8.31. It shows Alber's Projection with two standard parallels. On this projection areas are shown accurately but the shapes are deformed as distance from the parallels increases.

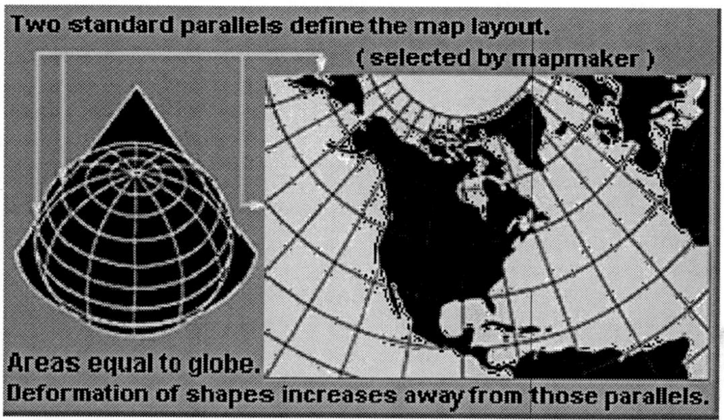

Fig. 8.31: Alber's Projection

Cartographers design maps and non-cartographers use them. Here lies the problem. All those who use maps are not conversant with the fact that the spherical earth cannot be truly represented on a flat surface. All maps, therefore, give a distorted picture of the earth. This distortion may be in scale, shape, or direction. But most of the map users do not know this. They build the image of the earth as it appears on the map they see. If they see a rectangular world map, they think the world to be rectangular. Even educated people commit this mistake. Cartographers have, therefore, tried to dissuade publishers from the inept use of map projections and data representation.

According to a *New York Times* editorial published in 1943 "*...The time has come to discard (the Mercator) for something that represents the continents and directions less deceptively... Although its usage... has diminished... it is still highly popular as a wall map apparently in part because, as a rectangular map, it fills a rectangular wall space with more map, and clearly because its familiarity breeds more popularity.*" The American Cartographic Association (now Cartography and Geographic Information Society) Committee on Map Projections, 1986 published a series of booklets to educate public about map projections and distorted images of the earth they conveyed. One of these booklets was titled: *Which Map Is Best?* In 1989 and 1990, after some internal debate, seven North American geographic organizations adopted the following resolution, which rejected all rectangular world maps, including the Mercator and the Gall projections:

"1. *Whereas*, the earth is round with a coordinate system composed entirely of circles,
2. *Whereas*, flat world maps are more useful than globe maps, but flattening the

globe surface necessarily greatly changes the appearance of Earth's features and coordinate systems,

3. *Whereas,* world maps have a powerful and lasting effect on people's impressions of the shapes and sizes of lands and seas, their arrangement, and the nature of the coordinate system,
4. *Whereas,* frequently seeing a greatly distorted map tends to Make it 'look right', and
5. *Therefore,* we strongly urge book and map publishers, the media, and government agencies to cease using rectangular world maps for general purposes or artistic displays. Such maps promote serious erroneous conceptions by severely distorting large sections of the world, by showing the round Earth as having straight edges and sharp corners, by representing most distances and direct routes incorrectly, and by portraying the circular coordinate system as a squared grid. The most widely displayed rectangular world map is the Mercator (in fact a navigational diagram devised for nautical charts), but other rectangular world maps proposed as replacements for the Mercator also display a greatly distorted image of the spherical Earth."

The following are a few suggestions which may be kept in view while selecting map projections.

For A World Maps

General Purpose: For general purposes we can use the equatorial cases of the zenithal equidistant and equal area projections. Other projections which can be used are: Mollweide's, Hammer's, Simple Cylindrical, Cylindrical Equal area, Mercator's, Gall's and Sinusoidal.

Distributional Purposes: Equal area projections are ideal for showing the distribution of various geographic phenomena. Among the equal area projections are cylindrical equal area, sinusoidal, Mollweide and Hammer. For showing the distribution of those things which are found in the equatorial or sub-tropical areas, cylindrical equal area projection is most suited. On this projection the shape is not too much distorted unless we go beyond 50° N and S of the equator. Equally good, for such purposes, is the Sinusodial projection. The Hammer and Mollweide also can be used. In case the objective is to show the distribution of something that is found in the mid-latitudes, the cylindrical equal area projection should not be selected. The Hammer projection will be the best choice. But the selection of Mollweide will not be bad. If the distribution has to be shown both in the mid-and high-latitudes, Gall's projection would be ideal.

Navigational Purposes: Mercator's is the best projection for navigational purposes.

For A Continental Maps

General Purposes: The conical projection with two standard parallels is very useful for such purposes. Asia, Australia and North America can be shown well on this projection. For South America and Africa, we may use Sinusoidal projection.

Distributional Purposes: To show the distribution of something in a continent the best projection is Bonne's. It is, however, less suited for Asia. The zenithal equal area projection should be the next choice. For Africa Sinusoidal projection can also be fruitfully used. At times Mollweide's and Cylindrical equal area projections are also used.

For Polar and Sub-Polar Regions

The best projection to show these areas is the polar stereographic projection. For distributional purposes Zenithal equal area projection should be selected. If correct distances are to be maintained, Zenithal equidistant projection should be selected.

For Countries and Regions

A number of projections are suitable for general purpose maps. For polar and sub-polar areas, the choice should go to one of the Zenithal projections. For the mid-latitudes, selection would normally be made from among one of the conical projections. And for tropical and equatorial areas, a cylindrical projection may be the best choice.

To show linear phenomena such as boundaries, railways or roads which pass through a limited latitudinal extent, the selection of the conical projection with one or two standard parallels is ideal. For example, the US-Canadian boundary can be better shown on a simple conic with one standard parallel. Similarly the Trans-Siberian railway will be better represented on conical projection with two standard parallels. If a strip of land with narrow longitudinal extent but wide latitudinal extent has to be represented on a map, the projection to be selected should be zenithal equidistant or sinusoidal. For example, to show the Cape to Cairo railway route, the zenithal equidistant will be an ideal choice.

Recent Developments in Map Projections

A wide range of projections are readily available for mapping parts of a country, continent or a portion of the Earth. Yet Geographers often need suitable projections to show certain distributions throughout the globe. A large number of projections of the world are in use today. Yet there is scope to develop new projections, which can help map the globe with minimum distortion in area, shape, and direction. A few such projections developed recently are discussed below.

The Sinusoidal and the Bonne's projections have been combined and transformed into a set of new projections called Improved Re-centreed Sinusoidal projections. These projections retain the advantages of the Sinusoidal projection while at the same time greatly reduce the angular distortions. Likewise a set of modified Pseudo-Cylindrical projection have been developed. These projections generally show less angular distortions than the Sinusoidal projection and have symmetrical wings.

The launching of space-crafts has added a new dimension to the problems of projections. Almost daily we get new satellite imageries. How to convert them into accurate maps is a baffling problem. In 1976 John P. Snyder, Alden Patridge Colvocoresses and John L Junkins developed a new projection called Space Oblique Mercator Projection (SOM). This is a modified version of the oblique Meracator projection, and can be applied with equal accuracy to mapping landsat imageries taken anywhere on the globe, with minimum distortions. It overcomes the problem of curvature of the spacecraft ground track which varies from latitude to latitude.

The story begins when the US Geological Survey (USGS) needed to develop a system for reducing the amount of distortion caused when satellite pictures of the ellipsoidal Earth were printed on a flat surface. They appealed for help at one of their annual conferences. Snyder worked on the problem and devised the mathematical formulas needed to solve the problem. The USGS has since been using his formulas to produce maps from Landsat 4 images launched in 1978.

The SOM projection offers continual conformal mapping of the swath sensed by a satellite. The scale along the ground track is true in the sense that varies just 0.01 per cent within the normal sensing range of the satellite. Conformality is also correct within a few parts per million for the sensing range. Distortion is essentially constant along lines of constant distance parallel to the ground track. SOM also takes into account the rotation of Earth.

REFERENCES

American Cartographic Association's Committee on Map Projections (1986), *Which Map Is Best,* p. 12, Falls Church: American Congress of Surveying and Mapping.

American Cartographer, 1989, 16(3), pp. 222-223.

Bauer, H.A. (1942), 'Globes, Maps, and Skyways, *Air Education Series'*, New York. p. 28

Bugayevskiy, Lev. M and John Parr Snyder (1995), *Map Projections: A Reference Manual*, Taylor & Francis.

Canters, Frank Canters and Hugo Decleir (1989), *The World in Perspective: A Directory of World Map Projections,* New York: John Wiley & Sons.

Fenna, Donald (2006), *Cartographic Science: A Compendium of Map Projections, with Derivations,* Canmore, Canada: University of Alberta.

"http://www.gis.psu.edu/projection/chap6figs.html"Snyder, J.P. (1989).

Album of Map Projections, United States Geological Survey Professional Paper, United States Government Printing Office.

http://pubs.er.usgs.gov/usgspubs/pp/pp1453.

Miller, Osborn Maitland (1942), "Notes on Cylindrical World Map Projections". *Geographical Review* 43 (3): 405-409.

Raisz, Erwin Josephus (1938). *General Cartography*. New York: McGraw-Hill, 2d ed., 1948. p. 87.

Robinson, Arthur Howard (1960), *Elements of Cartography*,.second edition. New York: John Wiley and Sons. p. 82.

Snyder, John Parr (1987), *Map Projections: A Working Manual*, US Geological Survey Professional Paper, United States Government Printing Office, Washington, D.C. 1395.

http://pubs.er.usgs.gov/pubs/pp/pp1395.

Snyder, John Parr (1982), "Map Projections Used by the U.S. Geological Survey", 2nd edition, *Geol. Survey Bulletin* 1532, 313 p., U.S. Government Printing Office, Washington, D.C., 1982.

Snyder, John Parr (1987), *Map Projections: A Working Manual*, USGS Professional Paper, 1995.

Snyder, John Parr (1997), *Flattening the Earth: Two Thousand Years of MapProjections*, Paperback, 366 pages, Published by University of Chicago Press.

Part II

Cartographic Information: Collection, Processing and Interpretation

9
Geodetic and Plane Surveying

Surveying is commonly considered to be the science of making measurements needed to locate accurately the details of the earth's surface. Map making in its complete sense involves surveying, for surveying provides the accurate measurements that go into map making. The earth surface undergoes constant changes because of the natural as well anthropogenic activities. Surveying helps record the changes. Cartographers use these records to update the maps.

Surveying involves extensive field work and elaborate laboratory work. In addition, it involves the knowledge of the equipments used. Field work includes: the measurement of distances and angles, recording of the numerical values, preparation of sketches of the areas and features surveyed, and the explanation of the recorded data.

Surveying can be classified into several types. If we take the purpose of the survey as the criterion, we can have cadastral surveys, topographical surveys, aerial surveys, geological surveys, military surveys, and so on. But if we consider the accuracy desired and the technique used, we can divide land surveying into two: geodetic and plane.

In geodetic surveying, the curved surface of the earth is taken into consideration and, therefore, the measurements are done in three dimensions. In plane surveying, the land is considered to be flat and the measurements are done in two dimensions. Thus, plane surveying is less accurate but when surveying small and nearly flat area s, it is ideal.

GEODETIC SURVEYING

The main purpose of geodetic surveying is to establish a set of control points to which the surveys of lower orders can be tied up subsequently. A control point is merely a spot on the earth surface which has been precisely located. It is a point of reference from which measurements are made to locate other points. Greater, the number of control points in a given area, better is the accuracy with which earth surface can be mapped.

Before any definitive geodetic surveying covering an extensive area can be done, two critical factors have to be decided. The first factor is the *Datum Point* which is the origin for further comprehensive linear measurements and the other is the *Datum Plane* which establishes the base level for determining elevation. *Datum Point* gives mean earth gravity and the *Datum Plane* the zero elevation.

To understand the importance of these points let us imagine ourselves having been sent to a newly found island to map it and also to establish its position on a world map. Before carrying out our assigned job we must know the longitude and latitude of at least one point. For this we will have to look to the sky and make use of one of the celestial bodies as explained in the Chapter seven. To determine the elevation of the selected spot we will have to assume a certain point to be at the zero elevation and then measure the elevation of the proposed control point with respect to the zero level point. We will also determine the position of our point with respect to true or magnetic north.

Once the latitude, longitude, elevation and true north of a *Datum Point* are established, it becomes a prime control point of reference from which other control points can be determined without astronomical observations. Finding a true *Datum Point* is difficult, because levelling of instruments used in stellar observations is a difficult process. Had the earth been spherical and made of same types of rocks, a plumb bob would have pointed directly towards the centre of the earth. In most places, however, it does not do so because it is deflected by earth masses like mountains; it actually leans towards the pulling force. Although this pull is known to exist, there is no known way to determine its force and thus to correct the error. We, therefore, select a station (point) that seems to be least affected by these disturbing elements. It is with reference to this point that the latitudes and the longitudes of other points are determined.

The *Datum Plane* is equally difficult to determine. It is supposed to be the sea level. But we know that the sea level is not the same always and everywhere; it fluctuates. As a convention, it is considered to be an average of the hourly water-height seen for a period of a year or more at a tidal gauge. This has given rise to controversies over the height of such peaks as Mount Everest. The height depends on where the measurer thinks sea level would be under Mount Everest is a sea where there.

Horizontal and Vertical Controls

Horizontal Controls

Two types of controls are established from the *Datum Point* and *Datum Plane*. The first is the horizontal control. It consists of points from which horizontal distances and azimuths can be measured in linear units discounting elevation on the earth's surface. Such points are derived by Triangulation which is a method of establishing control stations by means of a set of mutually connected triangles.

The process of triangulation begins with a reconnaissance survey to select a number of prominent points which constitute the vertices of a set of triangles. These stations are known as trigonometric stations. In between these points, the elevation of several other points is also determined by primary horizontal control levelling. These are called Bench Marks. The trigonometric stations are the primary horizontal control stations, and the Bench Marks the primary vertical control stations.

The next step commences with the determination and measurement of the base line (Fig. 9.1). Base line is selected on a ground which is free from obstructions to accurate measurements. It is, usually, one of the sides of the first triangle in the triangulation. An object is sighted from both ends of this line. When rays are drawn from each end of the line to the object, we get a triangle. Each angle or Azimuth is carefully measured several times and checked by adding the three angles of the triangle to see if they total 180°. Another object is sighted and angular measurements are repeated for the next triangle whose base is formed by a side of the previous triangle. This process is repeated until all triangles of the system are complete. It must be noted that in a triangulation the base line is the only side of a triangle which is measured. All the other sides of the triangles are calculated trigonometrically. Given a side and two angles a triangle, the remaining sides and angles can be easily determined. These triangulation stations and Bench Marks are then plotted on a projection selected for the purpose. The topographical and other details are incorporated after the plane and socio-geographic surveys are also completed.

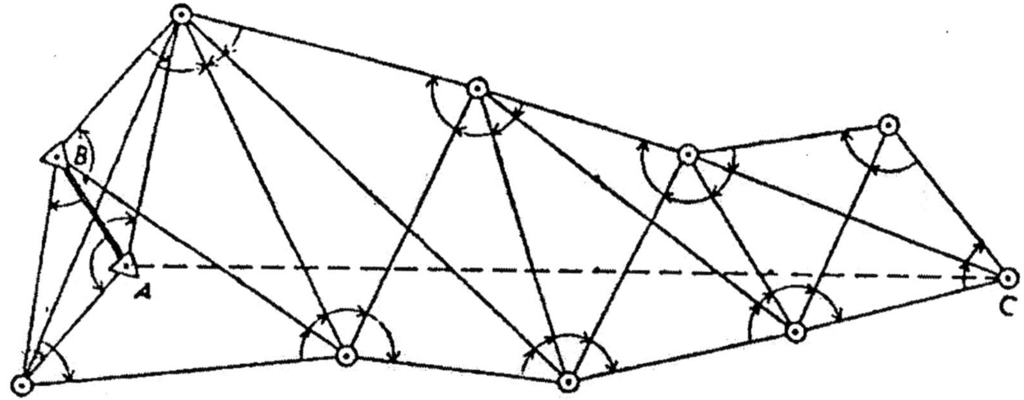

Fig. 9.1: A Simple Triangulation Net (Reproduced from Geodesy for the Layman)

One can get three types of data from Fig. 9.1: These are:

Known Data

1. Length of the base line AB.
2. Latitude and Longitude of points A and B.
3. Azimuth of line AB.

Measured Data

Angles to new control points

Computed Data

1. Latitude and longitude of point C, and other new points.
2. Length and azimuth of line AC.
3. Length and azimuth of all other lines.

There are four orders of triangulation. The basis for this classification is the degree of accuracy desired. The first order triangulation requires that all triangles must close within 1 second of error and the side lengths must agree to earth measurements within 30 cm in 7620 meters or 1 foot in 25,000 ft. The second order triangulation permits discrepancy of 3 seconds and the horizontal discrepancy of 30 cm in 3048 meters or 1 foot in 10,000 feet. The respective figures for third and fourth order triangulations are 5 seconds and more than five seconds and 30 cm in 1524 meters or 1 foot in 5,000 feet and 30 cm in 152.4 meters or 1 foot in 500 feet respectively. The accuracy required in the first and second order triangulations can be had only if geodetic adjustments are made to conform to the earth's ellipsoid and geoid. The triangulations of these two orders are, therefore, the results of geodetic surveys. The third and the fourth order triangulations are used in Plane Surveying.

Normally, triangulation is carried out by parties of surveyors from preplanned stations along the arcs of the spherical triangles. If there are obstacles in movement from one station to another, a method known as flare triangulation is employed. Parties locate themselves at all the stations from which measurements have to be taken and magnesium flares are parachuted from aircraft or 'shot' into the air from ships at suitable points between them. Intersections of lines are marked simultaneously at all the stations and thus reasonably accurate 'bridges' are established.

In cases where the distances between the points are too long to be inter-visible, the above noted methods of triangulation are not helpful. In such situations Trilateration (Fig. 9.2) is used. Trilateration involves the use of radar and other electronic devices. Instruments used in trilateration are geodimeter and tellurometer. In the forer, light is focused from one point to another. The focused light is received and a mirror installed in the receiving station. The mirror reflects the light back. The time taken by the light to travel back gives the horizontal distance between the points. Distance between stations lying 60 km apart can be measured this way. Tellurometer involves the use of radar and other electronic devices. Instruments used in trilateration are geodimeter and tellurometer. In case of the former, light is focused from one point to the other. The receiving station has a mirror which reflects the light back. The time taken by the light to travel back gives the horizontal distance between the points. Distance between station lying up to 60 km apart can be measured this way.

Tellurometer is less accurate and hence it is used for shorter distances only. This instrument uses radio waves for measuring distances.

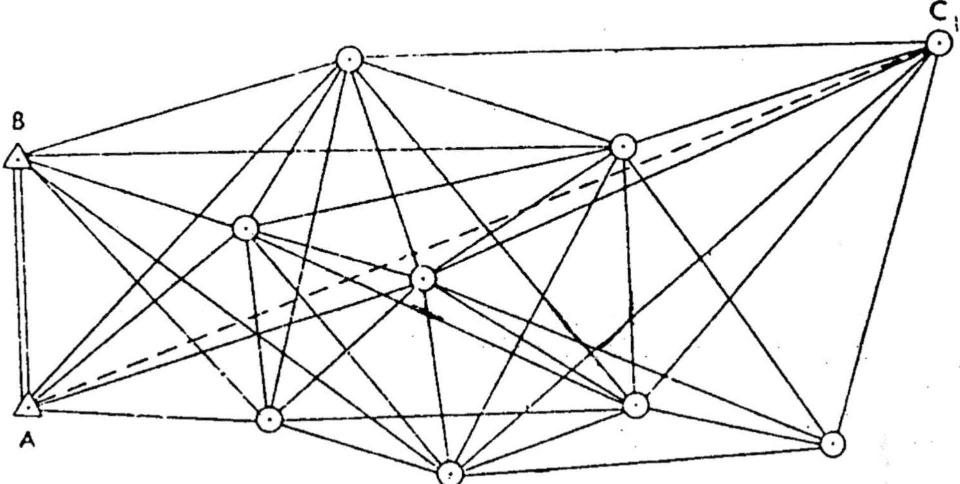

Fig. 9.2: A Simple Trilateration Net (Reproduced from Geodsey for the Layman)

In Fig. 9.2, the available data are of three types:

Known Data

1. Length of base line AB.
2. Latitude and longitude A and B.
3. Azimuth of line AB.

Measured Data

Length of all triangles sides

Computed Data:

1. Latitude and longitude of point C, and other new points.
2. Length and azimuth of line AC.
3. Length and azimuth between any two points.

To get over the distance constraints, Short Range Navigation (Sharon) and High Precision Range Navigation (Hiran) electronic distance measuring systems are also used in geodetic surveys. Since, lines of about 800 km (500 miles) can be measured by this system, geodetic triangulation network could be extended over vast areas in comparatively shorter time. Surveys of islands and isolated areas can be carried out by these techniques. In these systems, radio-waves are sent and received by a plane flying between two points. The time taken by the radio waves determines the distance of two points from the plane. When the height of the plane is known the horizontal distance between those two points can also be known.

The celestial techniques of geodetic triangulation involve the use of solar eclipse, star occultation, moon position, camera method, and observation and tracking of artificial earth satellites. Among these the last one has proved most promising. A satellite is photographed from several observation points. The small angular distance between the images of the satellite and the star background are measured and the direction of the satellite from the observation point is computed. High altitudes of the satellites enable us to triangulate across the oceans and determine relative positions between continents.

Vertical Control

Triangulation nets provide the control for locating geographic co-ordinates and for measuring plane distance on the earth or map surface. For the interpretation of terrain, we, however, need vertical control. This deals with the third dimension. Horizontal control supplies the 'where' 'how big' and the vertical control 'how high'. Obviously, the question 'where' must be answered first so that the location of an earth feature can be pin-pointed. Once, the spot is located, its physical form, which depends partly on its height is determined. This is done by establishing vertical control.

Any measurement length, breadth, height or depth has to begin from somewhere. Some measures are relative. For example, height is relative to some accepted 'o' level. In our case it is the sea level. Abstract elevations in themselves are difficult to visualize because human mind comprehends things only by comparison. The measurement of vertical control, therefore, should logically be related to a common base, Datum Plane, which is the Mean Sea Level at a selected point. Elevations of selected points are measured progressively inland from the Datum Plane to complete a network of vertical control that coincides with points of the horizontal triangulation net. The technique of determining the height is called levelling.

Precise geodetic levelling is used to establish a base network of vertical control points. With reference to these, the elevation of other points in the areas to be surveyed can be determined. There are three levelling techniques in common use (Fig. 9.3). These are:

1. Differential levelling
2. Trigonometric levelling, and
3. Barometric levelling.

The differential levelling is the most accurate of the three techniques. The instrument used in this type of levelling is called Level. It includes a bubble tube which helps in adjusting the level parallel to the geoid. When set properly, the telescope of the level is in horizontal position. Readings are done on two calibrated staffs held in an upright position ahead of and behind the level. The difference between the readings is the difference in elevation between any two points.

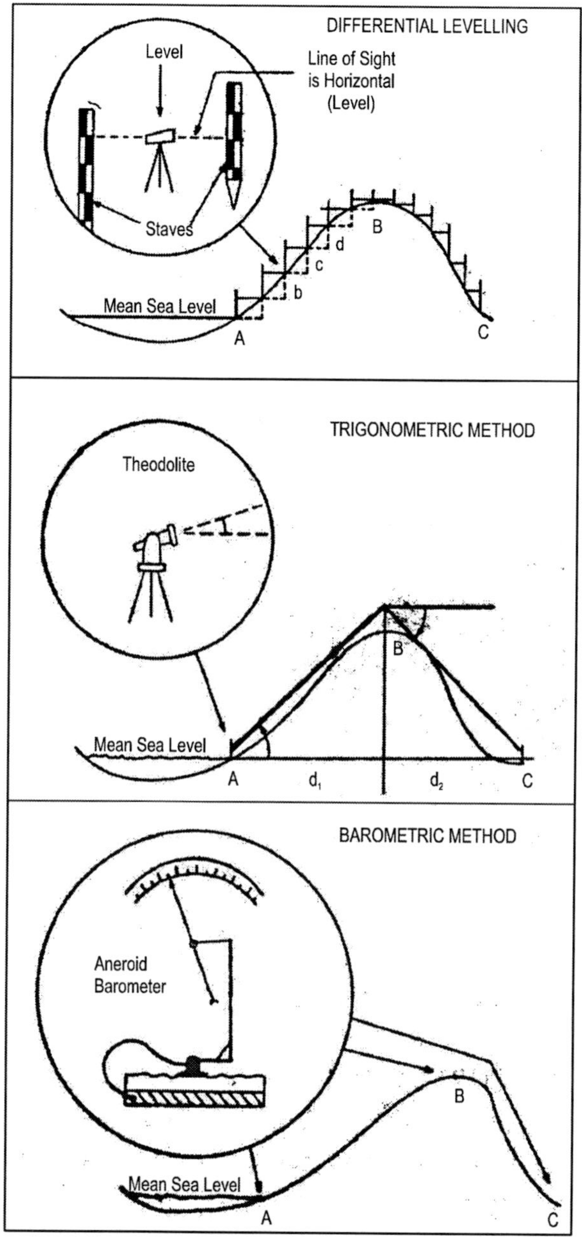

Known Data: Elevation starting point, A.

Measured Data: Elevation differences, a, b, c, d, etc.

Compared Data: Elevation of B, C, and all other points.

Known Data: Elevation starting point, A. Horizontal distances, d_1, d_2, between points.

Measured Data: All vertical angles.

Compared Data: Elevation of B, C, and all other points.

Known Data: Elevation starting point, A. Gravity and density data.

Measured Data: Air pressure at A, B, C, and all other points.

Compared Data: Elevation of B, C, and all other points.

Fig. 9.3: Levelling Techniques (From Geodesy of the Layman by Burkard, Richard, K, 4th revision, 1984)

Trigonometric levelling involves measuring a vertical angle from a known point with the help of a theodolite and then computing the elevation of the selected point. It is a relatively less accurate method although somewhat more economical. It is most suited to mountainous terrain.

Barometric levelling is based on calculations of height from the pressure data.

Barometer is an instrument which registers atmospheric pressure. The rule is: the higher a point, the lesser the atmospheric pressure. The pressure at the sea level is about 1013.25 millibars or about 14.5lbs per sq. inch. At an altitude of 18000 ft 1013.25 (5500 meters) above sea level the pressure is about 500 millibars. Table 9.1 gives the relationship between height and pressure.

Table 9.1: Relationship between Elevation and Pressure

Barometric Readings		*Boiling Point of Water (100° C/212° F) at Different Heights*	
Millibars	*Inches Hg*	*Meters*	*Feet*
473	13.970	6063	19,887
541	15.964	5004	16,412
616	18.199	3960	12,988
701	20.690	2938	9,638
794	23.461	1937	6,354
898	26.521	961	3,151
1013	29.921	0	0

So if the barometer registers pressure of 20.69 inches, we can assume it to be at a height of 2938 meters or 9638 feet.

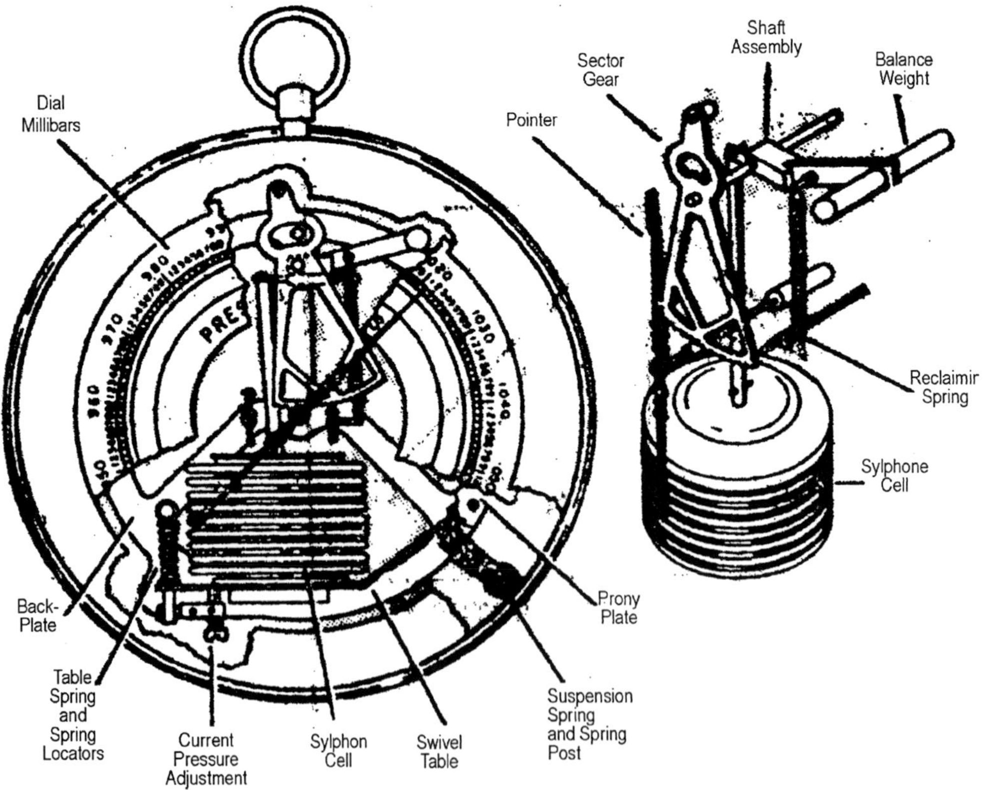

Fig. 9.4: Barometer and Its Parts

PLANE SURVEYING

Plane Surveying, unlike geodetic surveying assumes the earth to be flat. The reason behind this assumption is that plane surveying is used only for relatively small areas. Usually it is used to fill up the details within the framework of the control points established by geodetic surveying. Plane surveying is also used for such purposes as demarcation of various types of boundaries, property lines, and alignment of transport lines.

All horizontal and vertical measurements in plane surveying are tied to the measurements given by the trigonometrical stations and bench marks.

Plane surveying is of five types all related to the instruments used. These are:

1. Chain surveying
2. Plane-table surveying
3. Prismatic compass surveying
4. Theodolite surveying
5. Levelling

Chain Surveying

This method is employed for the purpose of small surveys. At times it is also referred to as chain and tape surveying. The name 'Chain and Tape' has been derived from the two main measuring instruments used in this type of survey. It is especially suited to areas which are relatively level, and whose horizontal distances can be measured manually. It is still being used for all types of land based surveys especially cadastral surveys.

The instruments used in Chain Surveys (Fig. 9.5) are discussed below.

Chain: It is made of steel wire. It consists of 100 links. An engineer's chain is 100 ft. long and a Günter's chain is 66 ft. long. Gunter's chain used to be common in India because it is convenient in measuring acreage and mile-distances. 80 chains equals 1 mile and 10 sq. chains make an acres. As against, the engineer's chain has the advantage of easy counting in decimals. In India where metric system of measurement is now in vogue, chains of 10, 20 or 25 meters are used. These chains are also divided into 100 links.

A chain has two brass handles at its two ends. To facilitate the counting of links, a brass tag is attached at every tenth link. The tags have 1, 2, 3 or 4 teeth to indicate the number of links.

One tooth	10 or 90 links
Two teeth	20 or 80 links
Three teeth	30 or 70 links
Four teeth	40 or 60 links
Zero teeth	50 links

Tapes: There are many varieties of tapes in length and make. The length of tape varies form 1 metre to 300 metres. Tapes are made of cotton, nylon, metallic, or steel. The first two tend to shrink with changes in weather. These are not suitable for precision jobs. Metallic tapes are cloth tapes with metallic threads twisted with tinsel. These are better than cotton or nylon tales. Steel tapes are made of special steel which does not expand or contract easily. These are sturdy and suitable for rough use.

Off-set Staff: It is a round or square wooden staff of 10 feet (3 metres) length with a pointed iron shoe on one end. It has 10 divisions, each division being marked by alternate colours (red, black or white).

Ranging Rods: These are used to indicate the position of an object or point in the field. These are made of bamboo or wood and coloured to aid visibility.

Arrows or Pins: These are iron nails 15″ (35 cm) to 18″ (45 cm) long with a loop at the top. Arrows are used to mark the chain length on the ground. In the process of measuring horizontal distances, the surveyor fixes an arrow each time he measures one chain.

Wooden Pegs: Arrows are removed after the full length of a line has been measured. But in certain cases the surveyor may like to refer back to some of the points on the line already surveyed. To keep such points distinct and visible wooden pegs are fixed.

Magnetic Compass: It is used to determine the magnetic north.

Optical Square: While measuring the distances of objects or points close to the chain line an optical square helps in setting out right angles. Two mirrors fitted vertically on a brass plate and inclined to each other by 45° from the basic parts of the square (Fig. 9.5).

When and How to Do Chain Surveying?

Chain Surveying has the advantage of being relatively simple, inexpensive and moderately accurate. This can be used for:

1. Showing property lines,
2. Surveying a small area,
3. Surveying a level area.

Chain Survey is a slow process; it is, therefore, not suited for quick surveys, unless the area to be surveyed is very small or when time factor in surveying is not crucial.

Chain surveying is of two types:

1. Open traverse, and
2. Closed traverse.

In open traverse, the surveyor does not come back to his starting point. This often happens in the case of route surveys. In the case of the closed traverse, the surveyor comes back to the starting point.

The following steps should be taken before starting the survey:

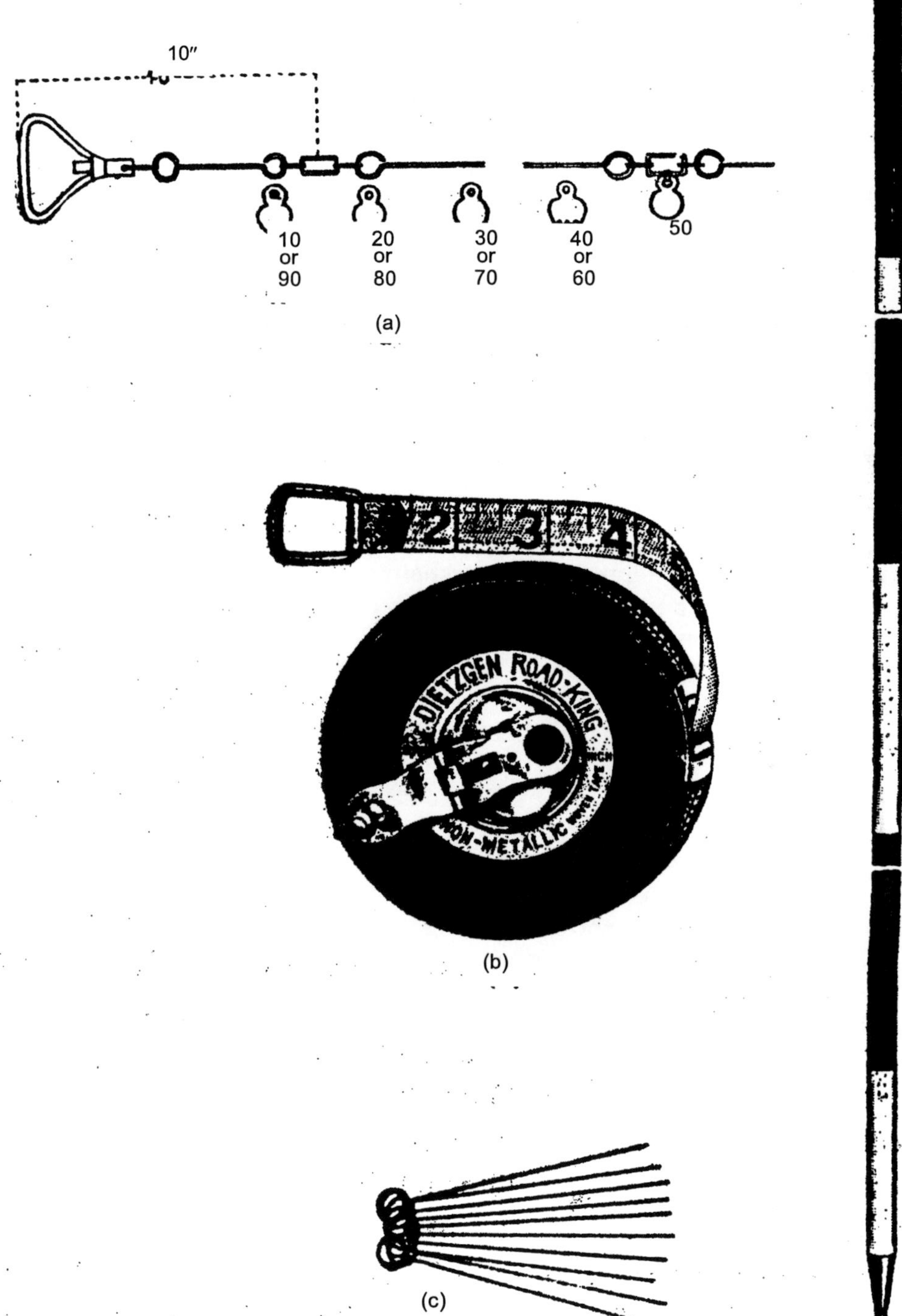
10″
10
or
90
20
or
80
30
or
70
40
or
60
50
2
3
4
DIETZGEN ROAD-KING
NON-METALLIC
(a)
(b)
(c)
(d)

Fig. 9.5: Some Instruments Used in Chain Surveying: (a) Chain, (b) Tape, (c) Arrows, and (d) Ranging Rod

1. Make a sketch of the area to be surveyed and locate all the prominent objects in it.
2. Divide the area included in the sketch into convenient number of triangles and mark their vertices on the ground by wooden pegs.
3. Determine the magnetic north with the help of a magnetic compass.

While dividing the area into triangles, make sure that the vertices of the triangles or station points are free from obstacles and are easily accessible.

And that:

1. The number of stations are as small as possible,
2. The triangles are equilateral or near equilateral,
3. The sizes of triangles are as large as possible, and
4. The sides of triangles, as far as possible, run along the bounding of the area and close to the objects so that the offsets are not too far away.

After these preliminary steps, the chaining can be started from any of the several points selected as stations. A ranging rod is fixed at the opposite end of the line. The chain is stretched along the line: Before the chain is pulled forward, off-sets to all the objects falling within the range of the first chain length are taken with the help of a tape and an optical square. All off-sets must be vertical to the chain line. The distances of the offsets from the chain line and from the starting point where the off-set line meets the chain line are measured and recorded in a field book, an example of which is given later in this section. The data so recorded are then taken on a map of suitable scale. The preparation of the map completes the chain survey.

In a Chain Survey at least five persons are needed. Two persons handle the chain, two other measure the off-sets and one person prepares the field book. The front man holds one of the handles of the chain and walks from the starting point to the end point. He carries 10 arrows with him. When he walks approximately a chain length, the rear man instructs him to stop by shouting 'chain'. The rear man then instructs the front man to be on the straight line by aligning him with the object or the ranging rod fixed at the end of the line. Kinks, if any, in the chain are removed and the chain is straightened. The front man then fixes an arrow at the end of the chain. The men deputed to take off-sets finish their work and the field man records the data in the field book. After all this is done the front man pulls the chain forward. The whole process is repeated until the front man reaches the end of the line. The rear man goes on collecting the arrows fixed at the end of the chain length. The number of arrows in the hands of the rear man gives the length of the line measured. The party then moves to the next point and repeats the process.

Let us take an example of an area to be surveyed as given in Fig. 9.6. We start from point A and move towards point B. the chain is stretched on line AB with one end of the chain being at A. The chain is straightened and aligned. An off-set is measured. The distance between Aa and aa are recorded in the field book as shown below.

Similarly distance Ab, Ac, bb and dd are also recorded. It may be restated here that all off-set distances are measured along straight lines drawn as perpendiculars from the objects to the chain line.

For Fig. 9.6, There will be five field book records, one for each of the lines AB, BC, CD, DA and EA in the same way as given for AB. The data given in the field book are reduced to scale and the map of the area is drawn.

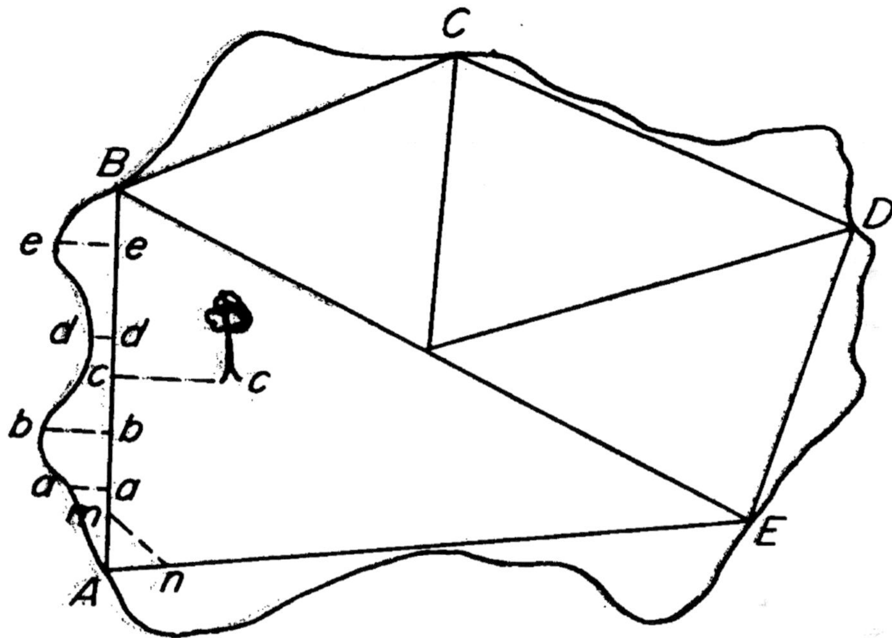

Fig. 9.6: Procedure for Chain Survey

The field book in which all the measurements are recorded has three columns. The central column is enclosed and is intended to record the measurements along the chain line. Starting with the bottom of the column, the starting point is noted and then the distances of all points at which off-sets are recorded upwards. The side columns are utilized to record the length of off-sets. Off-sets on the right side of the chain line are recorded in the right hand column and those on the left side in the left hand column.

	B	
	180′	
Boundary 30′	156′	
Boundary 10′	114′	
	96′	55′ tree
Boundary 36′	68′	
Boundary 18′	40′	
	A	
	From A to B	

Problem in Chain Surveying

In chain surveying we measure only horizontal distances. This is easy to do if the ground is not sloppy or if there are no obstacles in the way of the surveyor. If the ground is sloppy we can use one of the following formulas to convert the actual distance into horizontal distance.

1. $X = L \cos A$,
   where
   X = Horizontal distance
   L = actual length along the slope
   A = slope in degrees
2. $X = L - \frac{h^2}{2L}$ $X = L - \frac{h^2}{2L}$

   where
   h = the height
   L = actual length along the slope.

Those who cannot work with the above formula can use what is commonly called measurement-in-steps. The method is illustrated in Fig. 9.7. A ranging rod is fixed at B. With the help of tape, the horizontal distance Ab is measured and recorded. Same is done between B and C, and C and D.

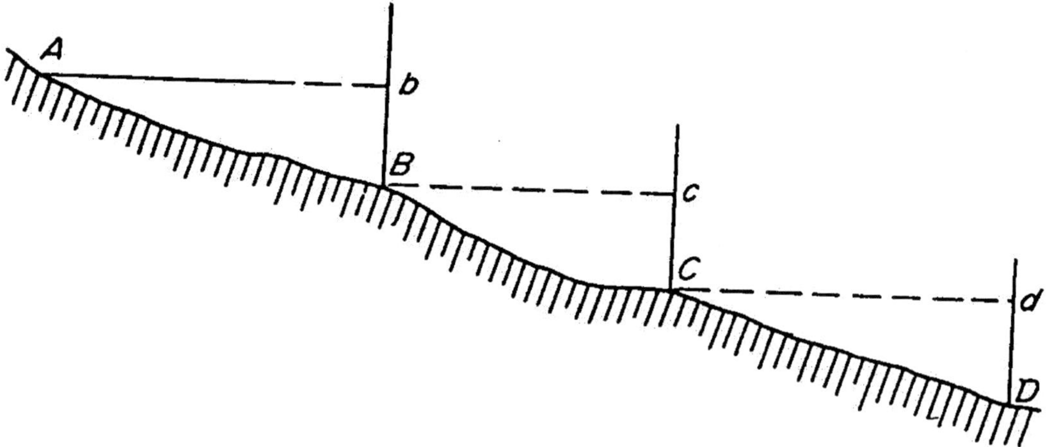

Fig. 9.7: Measurements in Steps along a Sloppy Ground

If there are obstacles in the way of measurement the following procedure is suggested. Let us suppose that we have to measure the distance between two objects located on the two opposite sides of a river, or on two sides of a pond. If latter is the case, we can measure three perpendicular distances as shown in Fig. 9.8a. In this figure PQ = XY. In case the object lie on two sides of a river, the method illustrated in Fig. 9.8b can be used. In this case point B is fixed on the other side of the river. Both A and B fall on the same chain line, AO is drawn perpendicular to AB from A and produced to Q, making AO = OQ. OP is drawn parallel to AB. It is also perpendicular to AQ at Q. Point P lies in a straight line POB. <AOB and QOP being opposite, are equal and hence AB = PQ.

Another problem in a chain surveying is the exact measurement of the angles that different sides of a triangle make. This may become a real problem if one has to make an open traverse survey. In this case there are no triangles. Even in the case of closed traverse one may come across some unavoidable obstacles in measuring one of the lines of a triangle. In chain survey we must know the length of all the sides of the triangles to prepare a map, for we do not measure the angles.

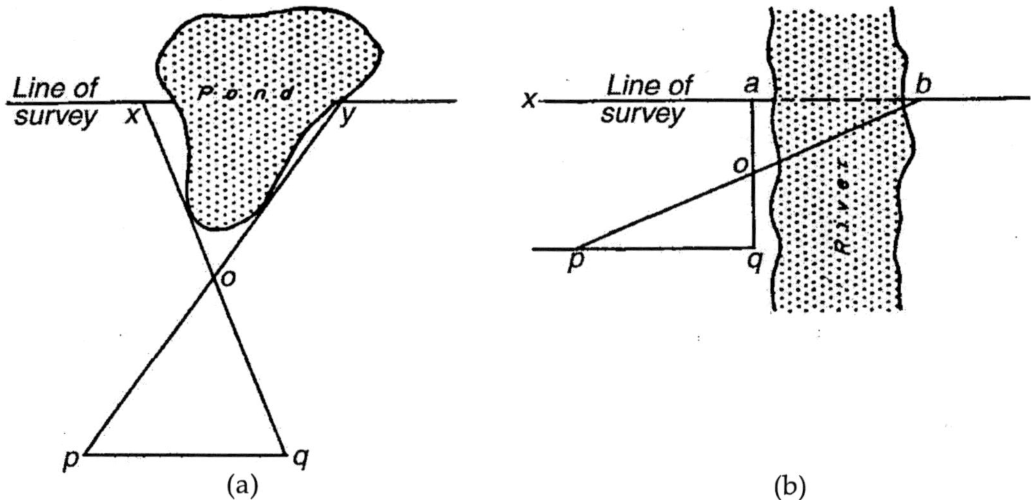

Fig. 9.8 (a): Horizontal Measurements across a Pond

Fig. 9.8 (b): Horizontal Measurements across a River

To solve this problem we can measure what we call tie lines. A tie line is one that connects two sides of a triangle to make a small triangle in one of the corners. This is shown in Fig. 9.6. In this figure there is an obstacle between points A and E which forbids the measurement of the line. We, therefore, establish points 1 and 2 on AB and AE respectively at convenient distances and measure the tie line mn. With the help of this tie we can draw lines AB and AE without knowing either angle BAE or line AE.

Errors in Chain Surveying

Most of the errors in chain surveying result from the carelessness of the surveyor in handling the equipment or from the defects in instruments. If the measurement is defective, errors will be repetitive and will grow cumulatively as work proceeds. Some of the common causes of errors are mentioned below:

1. Length of the chain is not correct,
2. Chain is not aligned and stretched properly,
3. Chain or the tape is not horizontal,
4. Ranging rods or arrows are misplaced,
5. Fieldname did not record the data correctly.

Prismatic Compass Surveying

We have seen how chain surveying is based on the measurement of horizontal distances. Even where measurement of angles becomes necessary we resort to the measurement of tie lines. Suppose we have to construct a quadrilateral ABCD. If we know the angles, A, B, C, D and any two of the sides, we can make this figure. Another way we can draw this figure is by knowing the angles each of the lines AB, BC, CD, DA, DB and AC make with a north-south line. In the latter-case, we will need to know the length of only one line. The prismatic compass surveying is based on the principle of measuring the angles that different lines make with respect to an N-S line. Only one line measured to maintain the scale relationships between the ground and map details.

Prismatic Compass

To determine the angles each line makes with a north-south line, we make use of a prismatic compass. Figures 9.9, 9.10, and 9.11 show a prismatic compass and the bearing marked in it. It is like any ordinary magnetic compass except that it has a prism, an eye vane at one end, and an object vane with a hair-line at the other.

Following are the main parts of the compass.

1. Needle,
2. Ring,
3. Pivot,
4. Glass cover,
5. Object vane,
6. Adjustable mirror,
7. Spring brake,
8. Brake pin,
9. Lifting pin,
10. Lifting lever,
11. Eye vane,
12. Prism,
13. Prism Dust Cap,
14. Focusing stud for prism,
15. Hinged sunglasses,
16. Hinged strap, and
17. Screw for Jacob staff.

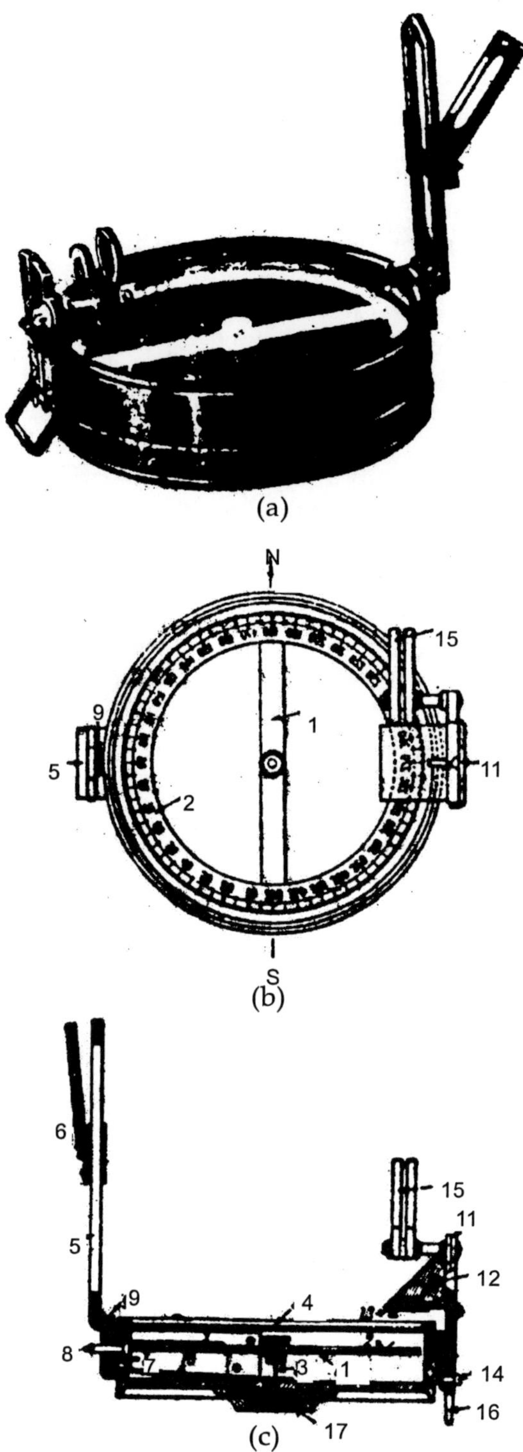

Fig. 9.9: A Prismatic Compass and its Parts

While using the compass the eye vane, the centre of the compass, and thin wire fixed vertically in the object vane make a straight line. Both the prism and the straight line are hinged so that they can be folded horizontally. A magnetized needle is pivoted at the centre. One end of this magnetized needle is coloured or made distinct. The coloured end points to the magnetic north. Usually, there are two rows of regulations on the disc. The inner row has its north point marked 0° or 360° and the south point 180°. In the outer row the 0° or 360° are marked at the south point and 180° at the north-point and the figures are given in reversed image. In many compasses the graduations run only from 0° to 90°. The reversed figures how up through the prism in a magnified and up-side down form (correct form) while sighting an object. The magnetized needle is fixed to the disc on the 0° to 180° line. As the needle moves to the point the magnetic north, the disc also moves along with it. So that when we look through the prism, we read the angles that the object sighted makes with the magnetic north.

To facilitate the sighting of luminous objects, the prism is fitted with a pair of coloured glasses which can be fixed on the line of sight. In addition to these a sliding plane mirror is also fitted to the frame of the sight vane. This mirror can be adjusted to any angle and thus the bearing of the objects which lie too far below or high up the plane of the compass can be conveniently read.

The angles read with the help of a prismatic compass are called bearings. A bearing is an angular measurement with respect to a reference direction. If the reference direction is magnetic north, the bearings are called magnetic bearings; if it is true north, it would be called true bearings. In the case of the prismatic compass survey, we deal with magnetic bearings only.

There are two systems of expressing bearings:

1. Whole Circle System (WCS), and
2. Quadrantal System.

The whole circle system runs from 0° to 360°. This system is the one which is used in most of the prismatic compasses and also one which has been explained earlier. In the quadrant system the bearings runs from 0° to 90°; they are preceded by N or S and followed by E or W to indicate the quadrant to which they belong. The first quarter of the circle is NE. If 50° bearing is to be shown, it will be shown as N 50 E. The second is SE, the third SW and the last NW.

Surveying Procedure

The first step in prismatic compass survey, as in all other surveys of lower order, is to prepare a sketch showing the objects for which bearings have to be recorded. Then a base line is selected. The base line should be such that its two ends are inter-visible and easily accessible. The object to be located should also be clearly visible from both ends. In Fig. 9.10, AB is the base line and C, D, E, F and G are the points for which bearings have to be noted.

To start with, the compass is set at A. The compass must be in the horizontal position to enable the magnetic needle to swing freely. Lift the prism and allow the needle to come to rest. Fix a ranging rod at B. Look through the prism and make the hairline of the object vane intersect the ranging rod. Read the reading of line AB and record it on the field book as shown below. Similarly, read the bearings of all other objects and record them in the field book.

The next step is to measure the length of AB. It should also be recorded in the field book. Now place the compass at B. After levelling, read the bearing of A and B. This bearing is called back bearing and that from A to B (90°) is called forward bearing. The difference between the two readings should always be 180°. Read the bearings of other objects also and note them in the field book.

The next step is to measure the length of AB. It should also be recorded in the field book. Now place the compass at B. After levelling, read the bearing of A from B. This bearing is called back bearing. Whereas the bearing A to B (90°) is called forward bearing. The difference between two bearings should always be 180°. Read the bearing of other objects also and note them in the field book.

In many cases the objects selected for representation on map are so close to the traverse lines that the recording of their bearings is not necessary. We can use such objects as off-sets and note down the perpendicular distances from the line of sight to the object as in chain surveying. The distances at which the off-set lines meet the line of sight can be recorded in the same way as in the case of chain surveying.

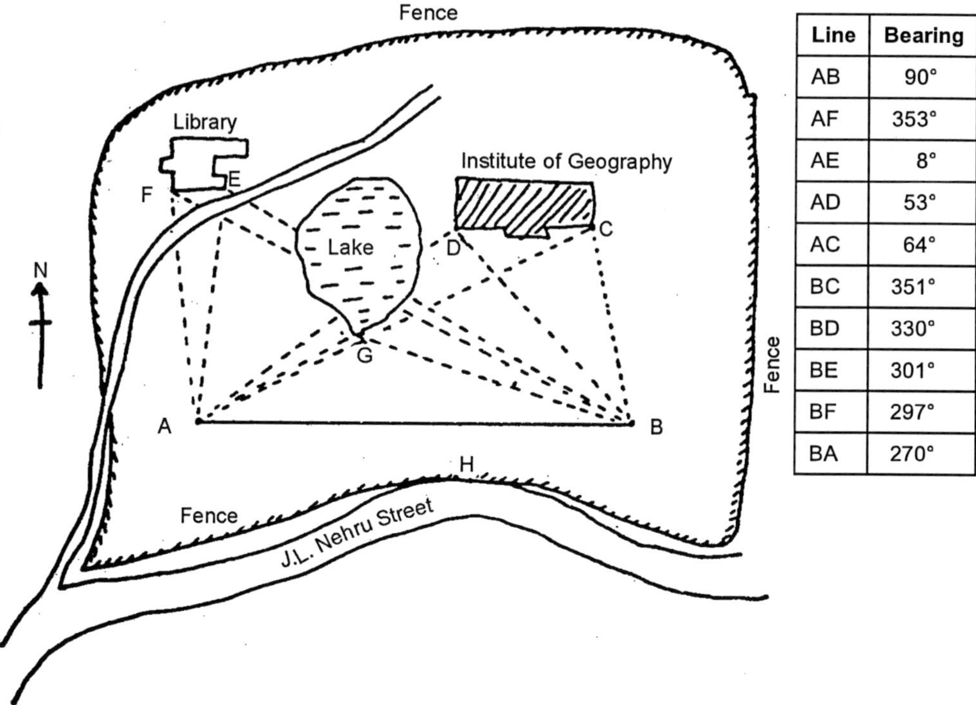

Line	Bearing
AB	90°
AF	353°
AE	8°
AD	53°
AC	64°
BC	351°
BD	330°
BE	301°
BF	297°
BA	270°

Fig. 9.10: Procedure for Compass Surveying

Like chain surveying, compass surveying is also of two types:

1. Open traverse; and
2. Closed traverse

Open traverse is one in which the surveyor does not come back to the starting point as in the case of route survey (Fig. 9.11). The surveyor starts from A, reads the forward bearing of B and moves on to B after measuring line AB. He then takes the back bearing of A and the forward bearing of C and repeats the process of measuring the line and moving to C, D, etc. While measuring the lines, he also takes the off-sets whenever necessary. In a closed traverse the surveyor comes back to the starting point. If he starts at A he comes back to A after passing through other points as shown in Fig. 9.11.

After the field work is over, the surveyor comes to the laboratory to plot the data recorded in the field book. The first thing that he does is to mark a magnetic north-south line at the starting point. From that point and with reference to N-S line he marks the respective bearings recorded in the field book. For example, in Fig. 9.12 our starting point is A. If the forward bearing of AB is 10°, then we will mark this and draw line AB to its proper length. We then move to point B, and repeat the processes for other lines until we come back to A.

If the bearings are taken and recorded correctly and there are no mistakes in plotting, BA and EA would meet at a common point A. But, more often than not, they would not meet and may deviate as shown in Figs. 9.11 and 9.12.

Sources of Error and Methods of Correction

The errors in prismatic compass surveying may be due to one of several of the following reasons:

1. Presence of iron in the area surveyed
2. Defects in the compass
3. Errors in observation
4. Wrong recordings in the field book
5. Wrong plotting

Magnet attracts iron, so if iron pipes or rods are present in the area surveyed, the bearings will not be correct. Errors may also result from the use of defective instruments. Very many errors result from wrong observations. At times the bearings are wrongly recorded in the field book, at others the plotting is not done correctly.

There are three methods of correcting the errors noted above. They are:

1. Proportional distribution
2. Re-checking
3. Plotting by included angles.

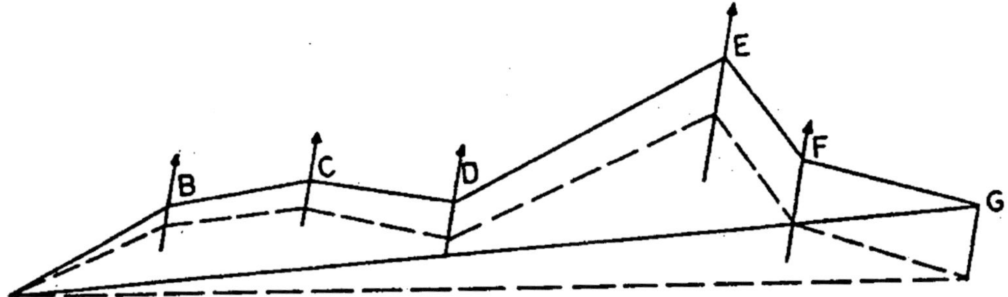

Fig. 9.11: Open Traverse in Compass Surveying (Dotted Lines Show the Corrected Traverse)

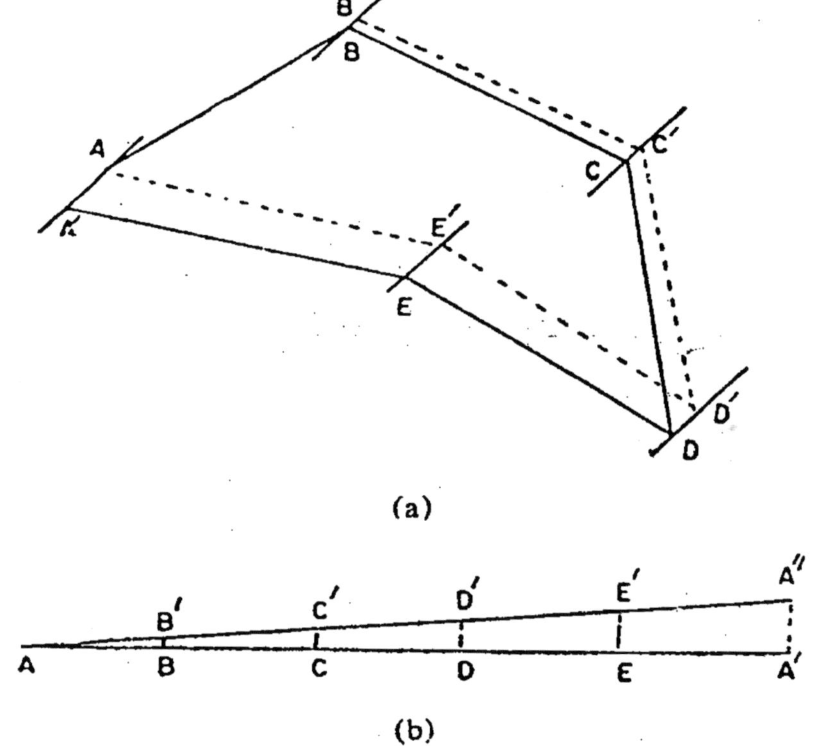

Fig. 9.12: Close Traverse in Compass Surveying and the Method of Correcting the Error (Dotted Lines Show the Corrected Traverse)

Proportional Distribution

If we happen to observe a closing error as shown in Fig. 9.12a, we can distribute the total error in such a way that points A and A' will meet at a common point. The principle of proportional distribution implies that the total error AA' should be distribution at all the points, B, C, D, E and A, to make the whole plan graphically balanced. The method of distribution is as follows. Draw a straight line AA′ = the parameter (total length of all lines). At A′ draw perpendicular A′ A″ equal to A A′ of the traverse. Join A and A″. Mark distances equal to AB, BC etc. on AA′ and erect perpendiculars to meet AA″ at B′C′D′ and E′.

Now in the traverse draw line at all points parallel to AA′ and form these mark distance BB′, CC′, etc. at their respective places. Connect these points. This new figure AB′ C′ D′ E will be the corrected figure (Fig. 9.12a).

Re-checking

In those cases where the existence of iron in the area to be surveyed distorts some of the bearings, we can correct them without resorting to the graphical method given above.

Suppose that following bearings were recorded in a survey:

Line	*Forward Bearing*	*Back Bearing*
AB	14°	192°
BC	58°	238°
CD	137°	313°
DE	218°	40°
EA	292°	116°

We know that the difference between the forward and backward of a given line should always be 180°. From the data given above, we find that this difference is found only in case of BC where 238°-58°=180°. This means that the bearings taken at points B and C were not affected by the presence of iron. This being the case, the back bearing taken by B to A (192°) is also correct. The mistake lies in the forward bearing of A (14°), which apparently should be 12°. Similarly, the forward bearing at (137°) is also correct and the discrepancy lies in the back baring D (313°) which should be 317°. If the discrepancy at D is equivalent to +4°, then forward bearing of D should be 222°, and not 218°, and hence the back bearing should be 42°. The corrected bearings are given below:

Line	*Forward Bearing*	*Back Bearing*
AB	12°	192°
BC	58°	238°
CD	137°	317°
DE	222°	42°
EA	294°	114°

Plotting by Included Angles

If the compass is defective, the bearings it recorded on it would also be defective and the difference between the forward and back bearings will not amount to 180°. Since a defective compass registers the same amount of discrepancy at all points, it is easy to adjust or correct this type of error. If we subtract the back bearing of a point from its forward bearing, we will get the included angle. In case the forward bearing is less than 180°, add 360° before deducting the back bearing. The included angles so obtained run clockwise and are measured with reference to the line joining the previous station to the station at which the bearings are observed.

Figure 9.13 explains the principle underlying this method. Let us suppose that X is the amount of discrepancy caused by the malfunctioning of the compass. The back bearing at point B is <a, BN (lower) is the magnetic north line shown by the compass and BN (upper) is the correct magnetic north line. Forward bearing of B is b. If we deduct < a from < b, we get <ABC which is an included angle. Hence, < a is the back bearing and < b the forward bearing. If we add < X to both the bearings and deduct < a from < b, the value of the <ABC will remain unchanged. Knowing <ABC and all other included angles of the figure we will have no difficulty in drawing the figure.

A few examples of the method of calculating the included angles are given below.

Included angle

$$
\begin{aligned}
A &= (14° + 360°) - 116° = 258° \\
B &= (58° + 360°) - 192° = 226° \\
C &= (137° + 360°) - 238° = 259° \\
D &= (218° + 360°) - 313° = 265° \\
E &= (292° - 40°) - 313° = 252° \\
& \qquad\qquad\qquad\qquad\quad \overline{\;1260°\;}
\end{aligned}
$$

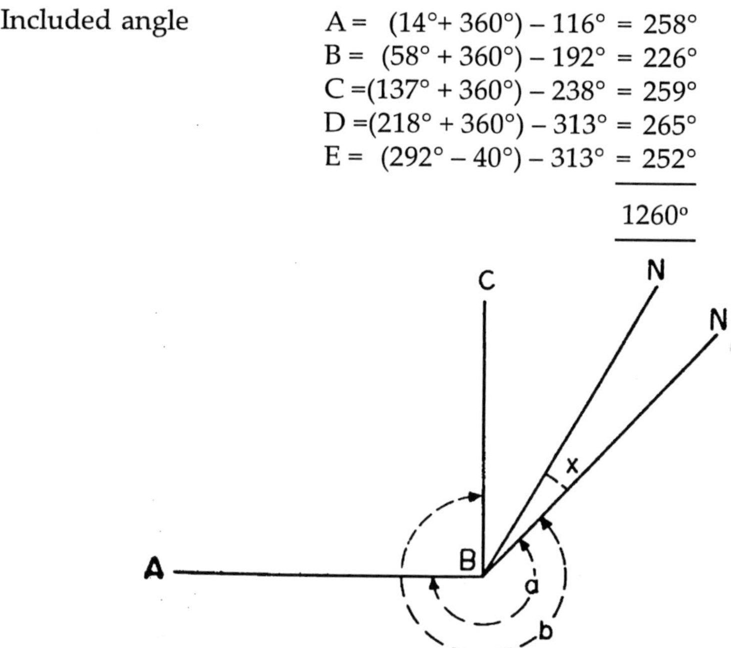

Fig. 9.13: How Included Angles are not Affected by Wrong Bearings Resulting from Magnetic Influences

To check the correctness of the included angles one of the following two methods can be used:

1. The total of the included angles should be 2n + 4 right angles; where n is the number of angles. 4 will be added if the angles are exterior, and subtracted if they are interior. In our example, all the angles are exterior ones. Hence, the total of the included angles should be 2 × 5 + 4 right angles = 14 right angles = 1260°

2. The other formula is a variant of the first. According to this, the sum of the included angles should be 2n × 90 + 360, where n stands for number of angles. According to this formula the total in our example should be 2 × 5 × 90 + 360 = 1260°

Advantages

1. Instruments used in the prismatic compass surveys are light and portable;
2. Wrong bearings at one point do not affect bearings at other points. As such the error is not cumulative;
3. It is most suitable to areas which are very congested; and
4. It is quicker than chain surveying.

Disadvantages

1. Its results are vitiated by the presence of metals like iron near the surveying place; and
2. A magnetic compass is not a precision instrument. Like ordinary watches, difference compasses may record different bearings for the same object.

PLANE TABLE SURVEYING

In the chain and prismatic compass surveying, the sides and bearings are measured on the ground, noted in a field book and finally plotted on a sheet of paper. Now we come to another type of plane surveying in which surveying and plotting go together in the field itself. The laboratory work does not involve plotting of the data, it involves only fair drawing.

Plane table surveying is a graphical method of surveying. It is based on triangulation done geographically. It is one of the simpler methods of surveying. It is speedy also. The errors are easily detected in the field itself. The horizontal distance of only a base line is measured. The plotting is so done that the measurements of angles are not necessary. Plane table survey, unlike other types of surveys, can be completed single handed. There are no changes for discrepancies resulting from the presence of magnetic materials. Further, plane tabling can be stopped temporarily and taken up subsequently without having any adverse effects on survey results.

With the advantages listed above, plane table survey has certain disadvantages also. First, even small obstruction makes plane tabling difficult. It cannot be used in wooded areas, for example. Secondly, it is not suited to areas having frequent rains. The sheet of paper on which the results are plotted has to be used in the open. With sudden changes in weather the paper contracts or expands and the accuracy of the work is marred. It is not suitable for large scale surveys covering extensive areas.

Equipments

Plane Table: It is a drawing board fitted to a tripod. It can be rotated freely and fixed in any desired horizontal position. The table has a brass plate with a bossed head which fits into the hole of a clamp to keep the table in a fixed position (Fig. 9.14).

Altitude: It is a ruler with parallel edges. There are two flaps at the two ends of the alidade. One flap has a slit and the other has a vertically fixed wire. The method of observing the objects is the same as in the prismatic compass surveying. A telescopic alidade (Fig. 9.15) is used to observe distant objects.

Chain and Tape: A chain or a tape is used to measure the base line.

Trough Compass: A trough compass is a rectangular compass used to determine the magnetic north.

Spirit Level: It is used to set the table in a horizontal position. There is an air bubble in the level. If the table is fixed horizontally, the bubble rests in the centre. Some of the alidades are fitted with a spirit level.

Ranging Rods and Wooden Pegs: These are needed for making the stations or objects to be included in the survey.

Plumb Bob: A fork or U-shaped frame with a plumb bob is used to centre the table exactly at the station.

Field Glasses: Those are used to get a clearer view of distant objects.

Thumb Pins: Thumb pins are used for fixing the paper to the table.

Fig. 9.14: Equipment of Plane Table Surveying

Fig. 9.15: Telescopic Alidade

Procedure

Preliminaries: Before starting the actual survey, the surveyor should see that the equipments he is going to use are in perfect working order. The sheet of paper to be mounted on the table should be larger than the table so that it can be thumb pinned underneath the board of the table. It should be pinned on the board because the pins will hinder the free movement of the alidade. They will also make the surface of the table rough.

Measuring the Base Line: The first step constitutes in plane table surveying constitutes the selection of a base line in the area to be surveyed. The area should be level so that some prominent objects at its other end can be easily seen. The base line should be measured and then reduced to scale and drawn on a drawing paper mounted on a board. The position of the base line on the paper should approximately correspond to the position of the base line on the ground (with respect to the features to be plotted on the map).

Centreing and Levelling: After the base line is drawn, the table should be placed at one of its two ends on the ground, in such a way that the starting point on the paper is exactly over its corresponding point on the ground. This can be done with the help of the plumb bob. To use the plumb bob the fork should be placed across the table in such a way that the upper blade of the fork is fixed at the selected point on the paper and the thumb bob attached to the lower fork points directly to the corresponding point on the ground. This is called centreing the table.

While centreing, the levelling of the table should also done with the help of a spirit level. Place the spirit level on the board between any two legs of the tripod. When the bubble comes to stay in the centre, remove and the level and place it in between the third and any first two legs. The table should now be adjusted by moving the third leg only in order to bring the air bubble of the spirit level in the centre taking care to see that the centreing is not disturbed. A more simple alternative method is place the spirit level among any one edge of the table and brings its bubble in the centre with the help of the two legs of the tripod. Now remove the spirit level and put it at right angles to the earlier position or along the opposite edge of the table and bring the bubble of the spirit in the centre by moving up and down the third leg only. Now the table is levelled in both the directions and it should be leveled everywhere. While doing so, one should keep in mind that the bubble will move in a direction opposite to the one in which the legs of the tripod are moved. A final checking can be done by placing the spirit level at each corner of the table.

Orienting the Table: Once the table is levelled, the fiducial edge of the alidade is aligned to the baseline drawn on the paper. The board should be unclamped and turned to allow the surveyor to see the ranging rod fixed at the other end of the base line on the ground through the alidade placed along the base line drawn on paper. The hairline of the alidade and the ranging rod should be in the same line when seen through the eye vane. When this is done, the table is oriented. Orientation of the table means the alignment of the base line drawn on the paper with the one marked on the ground. It is done to determine the map direction.

Laying the Table in Azimuth: Once the table is oriented, a trough compass is placed at one of the corners of it. A line showing the magnetic north is drawn with the help of this compass. With this marking, the table is said to be lying in azimuth.

Locating Points by Intersection

With the marking of the direction, the surveyor is ready to draw rays to the objects he wants to locate on the map. Before starting the work, he should re-check the levelling of the table. Without moving the table, the alidade is so placed on the table that the selected point and the object on the ground are on the same line when seen through the eye vane. A line should be drawn in the direction of the object. This process should be repeated with respect to each object. And thus will emerge an array of rays radiating from one end of the base line as shown in Fig. 9.16. As the rays to different objects are drawn, the names of the objects are jotted down along them to distinguish the objects from each other.

After completing the work at one end, the table is placed at the other end of the base line. In Fig. 9.16 the other end is B. Now orient the table which includes the whole process of levelling, centreing and orienting towards north and along base line A B. After this put a middle at point B and sight all the objects sighted from A. The location of the object sighted is at the intersection of the rays drawn from the two points. After the locations are marked, the rays can be rubbed off.

The process of plane table surveying as discussed above is a simple one. Here rays are drawn only from point A and B; but there are reasons why an object is visible from point A, but not from point B. If such objects are also to be located, the surveyor will have to go to a third or even a fourth point to repeat the process of drawing the rays. The process of locating the objects by means of drawing rays is called intersection.

Locating Points by Resection

Resection is a method to locate and plot the position of the plane table during surveying. In the process, the plane table also gets orientated. This method is based on principles

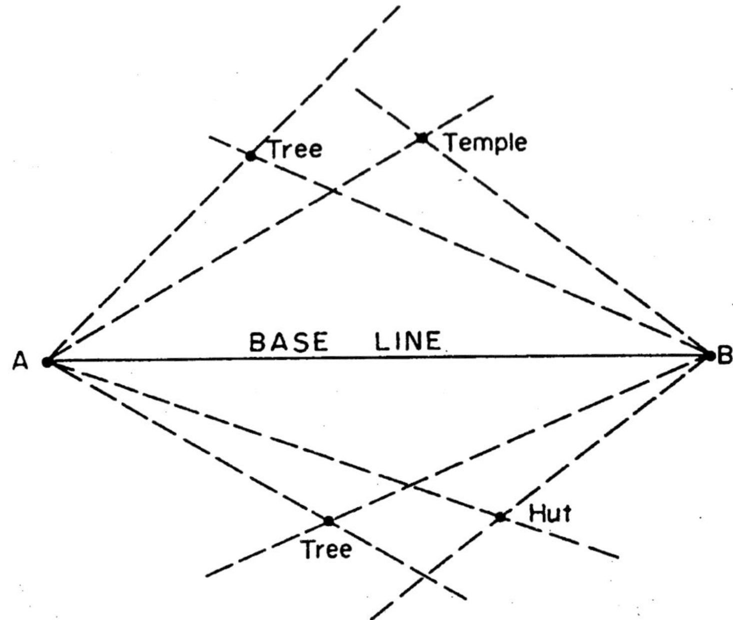

Fig. 9.16: Locating Points by Intersection: The Two-Point Problem

quite opposed to those underlying the method of intersection. Resectors are then drawn from already plotted points to determine the exact position of the plane table. There are different methods for locating plane table by method of resection and are primarily based on the type of orientation which precedes resection. Of these, resection after orientation by three points is widely prevalent.

The surveyor locates his own position with respect to the objects already shown on the map. This he does by drawing rays to the known points from his unknown position. This amounts to following procedures just opposite to the one followed in intersection. That is why the process is known as resection (Fig. 9.17)

Resection problems are of two types:

1. Two-point problems, and
2. Three point problems.

Two-point problem: Let us suppose that there are two well defined points A and B on the ground, which are visible from station P over which the plane table is set. The position of these two points has already been plotted on the plan or map as a, and b. We are required to locate other points on the plan. To do this, we have to determine the ground position P of the surveyor on the plan.

This can be done by adopting the following procedure:

1. Choose an auxiliary point C on the ground in such a way that <PAC and <PBC

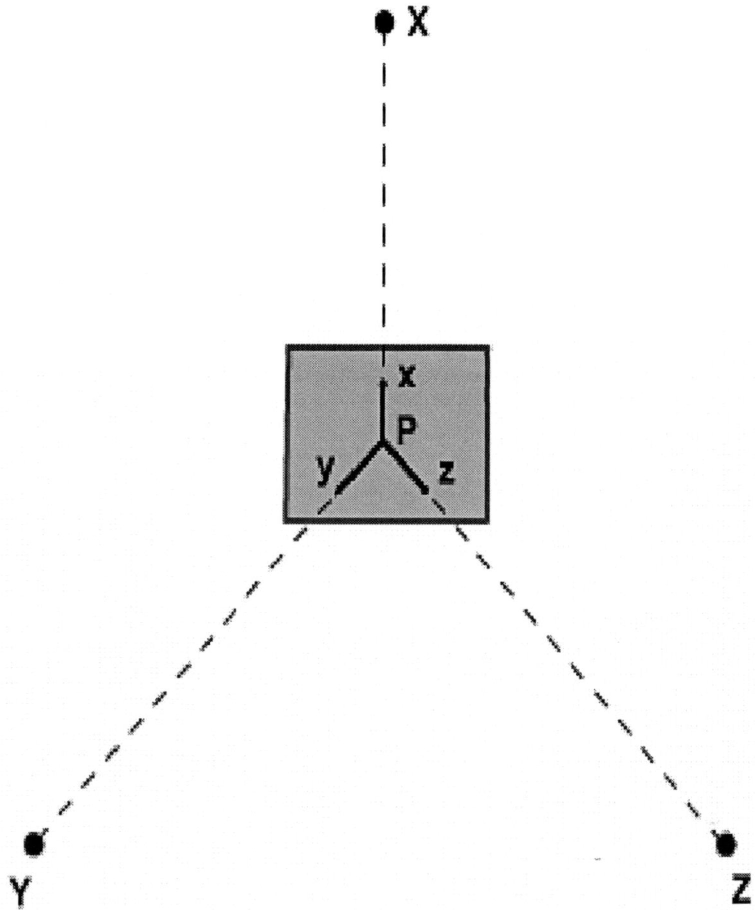

Fig. 9.17: Principle of Resection

are not too acute (Fig. 9.18). In other words these angles should not be less than 30°

2. Set the plane table at C; level the orient it with the help of a spirit level and magnetic compass. The orientation can also be done by keeping line a b parallel to AB.
3. Placing the alidade at a, sight A and draw a ray. Similarly sight B from b and draw a ray. These two rays intersect at c′. c′ is the approximate location of C.
4. Sight P from c′ and draw a ray c′ p′.
5. Now shift the table to P and level it. Orient the table by back sighting c′. Clamp the table. Assume a point p′ on p′ c′ as the position of P on the plane.
6. Sight A from p′ and draw a ray p′ A. Sight B from p′ and draw a ray p′ B. The rays will pass through a and b, provided that the initial orientation of the table at C and P was correct. If it was correct, p′ A may intersect c′ a at a′ and p′ B may intersect c′ b at b′ as illustrated in Fig. 9.18.

To eliminate the error, the following procedure is suggested:

1. Fix a ranging rod R so that a′ b′ and R are in the same line.
2. Place the alidade along a′ b′ and bisect R and extend the ray up to r. Place the alidade along ab and turn the table until R is bisected. Now the orientation is maintained. Clamp the table. Now a b is parallel to AB.
3. To determine the correct position of C, sight A through a′, and draw a ray.

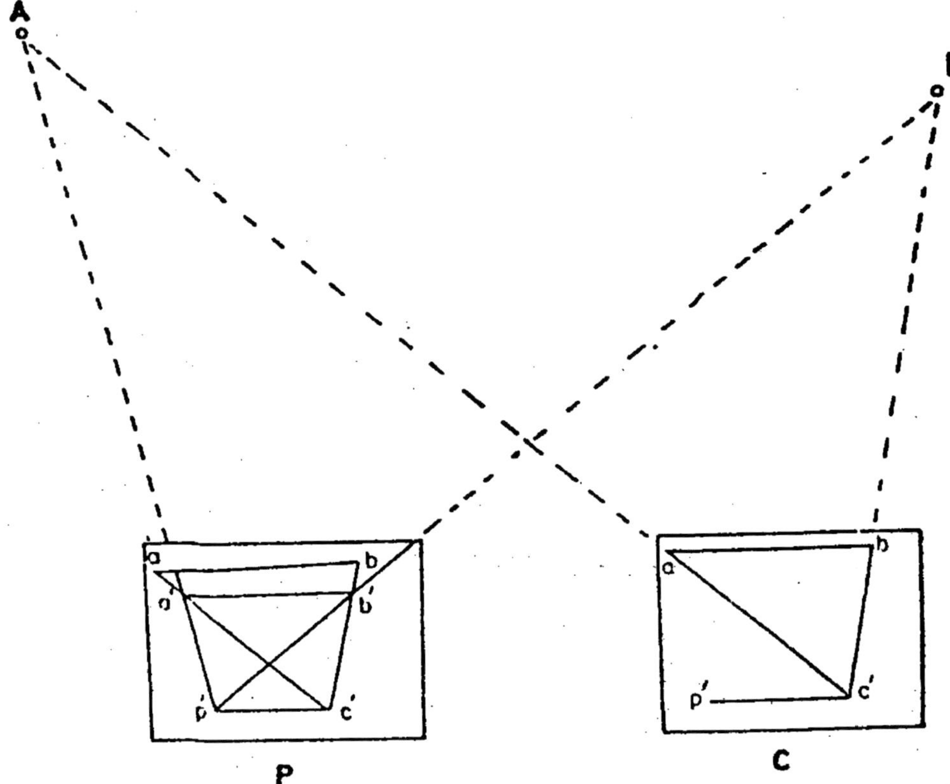

Fig. 9.18: Locating Points by Resection: A Two-Point Problem

Sight B through b′ and draw a ray as given in Fig. 9.17. The intersection of A c′ and B c′ gives the true position of ground station C (as c′ on the plan). Similarly to determine the position of P, sight A through a′, and B through b′. The intersection of the two (p′) gives the correct position of P.

Three point problem: There are three methods of solving a three point problem. These are:

1. Graphical method
2. Mechanical method, and
3. Lehmann's method.

Graphical Method: If three ground points A, B, C are represented on a plane as a, b, c and the location of a few more ground objects is to be determined on the map, the surveyor would have to first determine his own location on the map. Once he has done it, he can draw rays to the required points by intersection method. There are

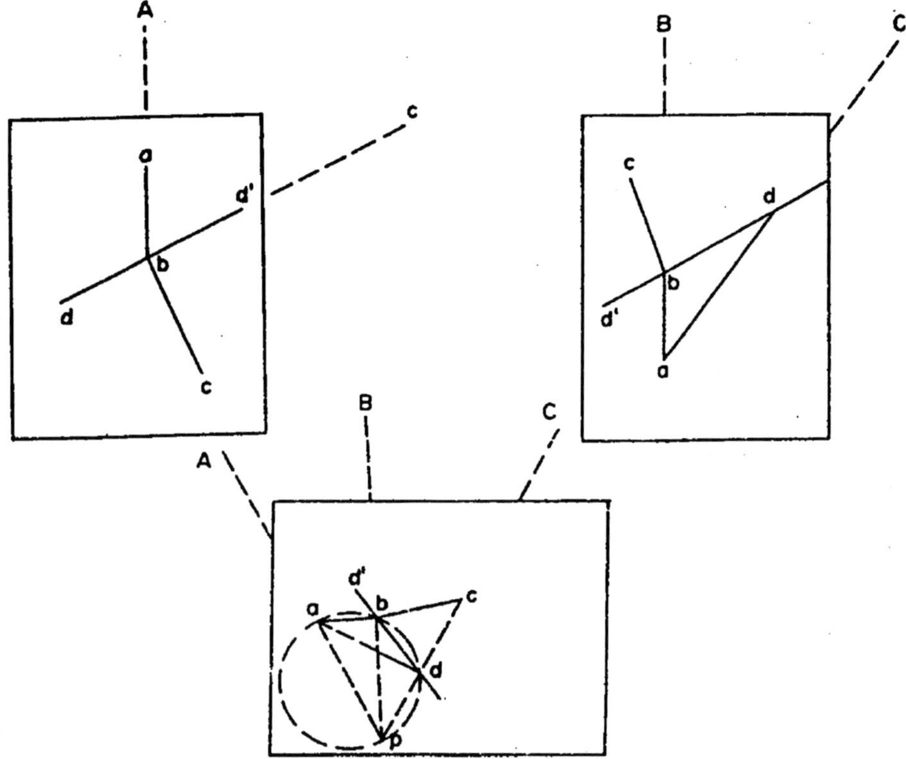

Fig. 9.18(a): Graphical Method

several graphical methods to locate the position of the surveyor on the plane but here only the most commonly used Bessel's Method is discussed.

In this method, the procedure used to locate the surveyors position is the following:

1. Set the plane table at a ground station T. Level the table and turn the table so that a is toward A. Place the alidade along a b. Sight A. Clamp the table. Sight C through b and draw d b d (Fig. 9.18a).
2. Unclamp the table and place the alidade along a b. Turn the table until the ground point B is bisected, a being towards B. Clamp the table. Draw a d towards C through a meeting d b d′ at d (Fig. 9.18b)
3. Unclamp the table and place alidade along ac. Turn the Table, sight C and clamp it. The table is now oriented. It must be on D c, or dc produced (9-19c). It must also be on A a and B b. To get the point T, bisect A through a. The meeting point p between A a produced and d c produced is the location of point T.

4. To be doubly sure, check the orientation of the table and bisect B through b. If the location of T is correctly determined, this ray will pass through p. Otherwise it will form a small triangle which can be eliminated by trial and error method.

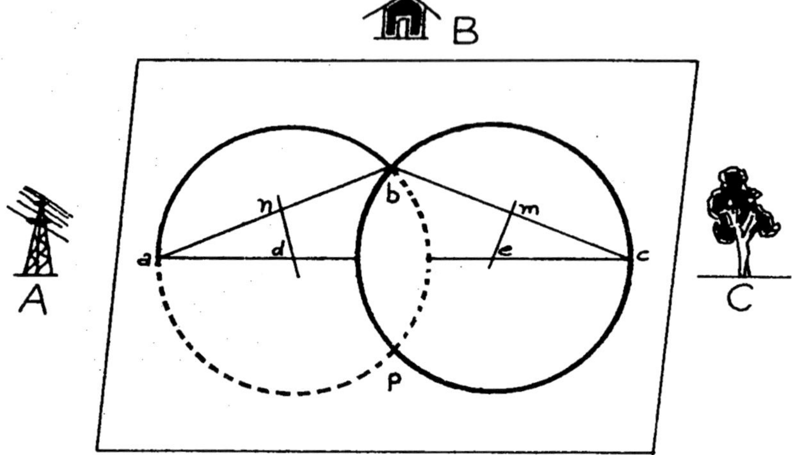

Fig. 9.18(b): Locating Points by Resection: A Three-Point Problem

Mechanical Method: To illustrate this method we assume to have a map which represents three points, A, B and C on the ground, as a, b and c respectively on the plan. Fasten the map on the plane table so that it is not allowed to change its position. Pin a tracing paper on the plan. Now level the table and select a point p on the tracing paper to represent your ground position. Draw rays from p to A, B and C. Release the tracing paper and move it on the plan without altering the position of the board to make the rays drawn on it pass through a, b and c on the plan. The point just below point p on the tracing paper will be ground position P on the plan. If owing to careless manipulation or unequal stretching of the tracing paper, a small triangle of error is formed, it can be eliminated by trial and error method.

Lehmann's Method: This method is also known as trial and error method. To use the previous example, we assume to have three known point A, B and C on the ground. These points are represented by a, b and c on a plan. After levelling the table draw rays from A, B and C, passing through their respective positions a, b and c on the plan. If every thing is all right, the three rays will meet at a single point to give the location of the surveyor. In case the table was not property oriented or leveled, a triangle as shown in Fig. 9.19 will be formed. This triangle is called the triangle of error.

The position P of the surveyor will be either within the triangle of error, or on the right or left of the triangle of error.

The size of the triangle of error will depend upon the angular error in the orientation of the table. From the triangle the true position of p may be estimated and by second trial a new triangle of error may be obtained which will be smaller than the former. By successive trials this triangle may be made so small that it becomes almost a point.

In estimating the correct position of point P on the map, after the first triangle of error has been drawn the following geometric relations will be found useful:

1. If the table is inside the triangle ABC, the point p is inside the triangle of error.
2. If a circle is drawn through a, b and the intersection of the corresponding resection lines it will pass through the true position of p, for the angle made by any two resection lines from a and b is the angle APB. Similarly, a circle drawn through a, c and the intersection of the corresponding resection lines will pass through p. A corresponding circle through b and c would also pass through the intersection of these two circles which is the true position of p. Hence, the intersection of these circles will give the position of point p [see Fig. 9.19(a) and 9.19(b)]. This is called Position circles method;
3. The distance of p from any resection line is proportional to the distance of the table from the point from which that line was drawn i.e., if that table is nearer B than the other points, the p will be nearer the resection line drawn through b than to others.

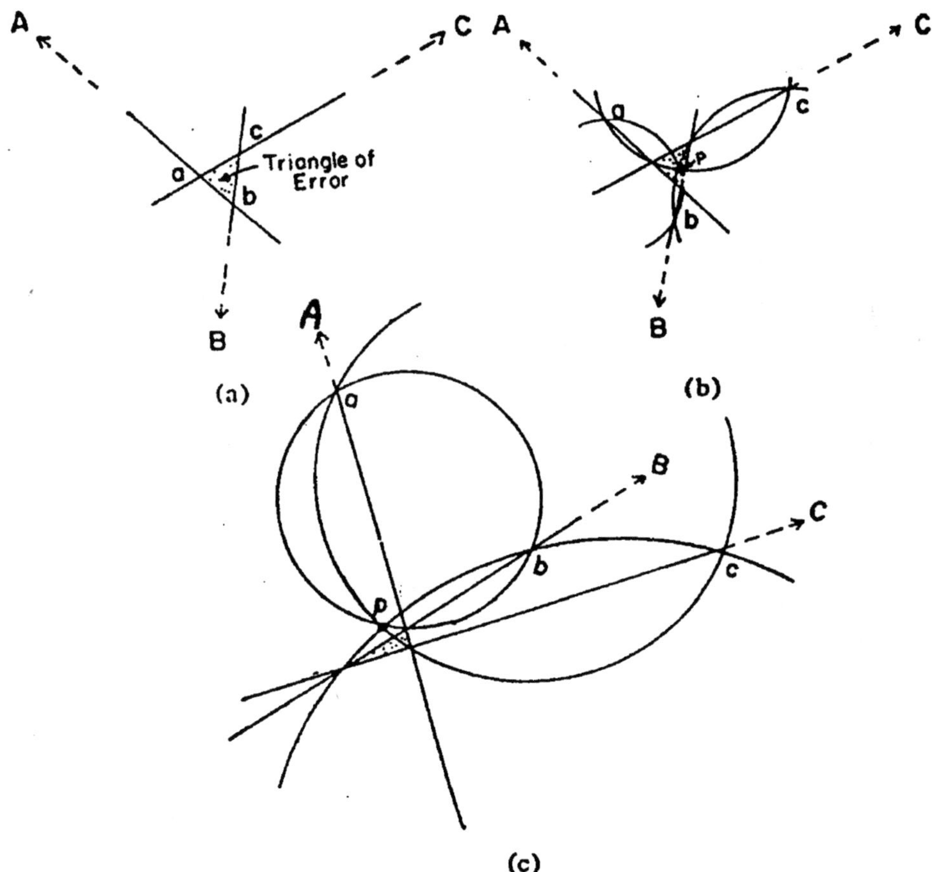

Fig. 9.19: Lehmann's Method of Solving : A Three Point Problem

To check the position of point p, correctness of draw rays from A, B and C passing through a, b and c. If they all intersect at point p the position is correctly determined. If a triangle of error is formed, use the method of position circles explained earlier to get the point.

Use of Telescopic Alidade

Telescopic Alidade (Fig. 9.15) is used in plane table surveying when the work has to be completed in a short time. Only one ray is needed to locate an object if a telescope alidade is used because we can read the horizontal as well as the vertical distances with the help of this instrument. We can see objects located at greater distances also with the help of instrument.

Both the vertical and the horizontal diameters of the diaphragm of the telescope are ruled over it. On the vertical diameter, there are two or four lines called stadia lines. An assistant stands at the object with a vertically held staff having graduation of feet, meter and hundredths of a foot, meter (Fig. 9.14). The observer looks at the staff through the telescope whose stadia lines intercept a distance marked on the staff. The intercepted distance bears a fixed ratio to the horizontal distance between the telescope and the staff. This ratio is usually 1:100. So if the intercepted if 1 foot, the staff lies at a distance of 100 ft. from the telescope. As the additive constant of the telescope is about 1 foot, this 1 foot is also added to the horizontal distance making it 101 ft. Some of the telescope has internal focusing system and hence the additive constant is less than 1 foot. It may vary from 4 to 8 inches.

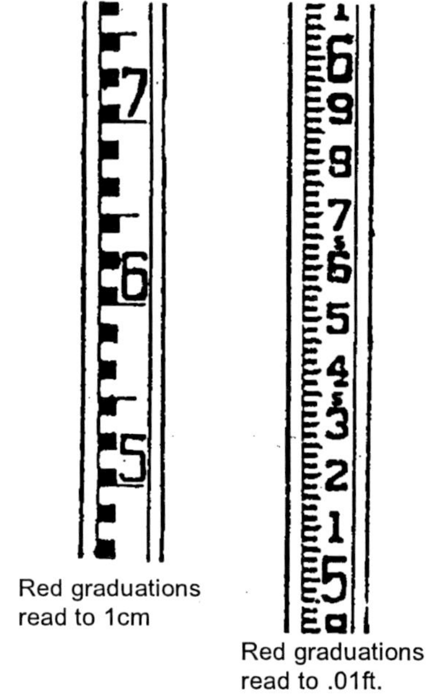

Fig. 9.20: Graduations of a Staff

THEODOLITE AND TOTAL STATION SURVEYING

The methods of plane surveying discussed so far give the technique of measuring only horizontal distances. All of them are relatively simple. Now we come to the method which can be used for both horizontal and vertical measurements and which is the most complex and accurate among the methods discussed so far. It is called Theodolite and Total Station Surveying.

Theodolite and its Parts

Theodolites are electronic instruments widely used in surveying to measure vertical and horizontal angles. Their use in topographic surveys heralded a great advancement in surveying in mid-20th century. Apart from measuring both horizontal and vertical angles, they can sight distant objects with the help of telescopes attached to them.

Theodolites have become indispensable instruments for topographic mapping. No wonder then, that they are widely used in the engineering and surveying disciplines, especially in remote locations. Accurate measurement of angles is essential for huge construction projects like dams, railways, highways, etc. Theodolites have been modified to numerous other functions such as rocketry and meteorology. The leveling systems have become more accurate and reliable, and the latest models employ infrared based technology. Despite, the fact that theodolites have been used for a long period, their basic function has not changed much.

Types of Theodolites

There are various kinds of theodolites. A few of them are discussed below:

Transit Theodolite

It is the older version of theodolite still in use for teaching and small scale construction projects (Fig. 9.21). It's telescope can make a complete revolution about its horizontal axis in a vertical plane.

Trivate Stage: It is the lower most triangular plate hinged to tripod. It consists of the following parts:

1. *Levelling Screws*: There are 3 or 4 screws fixed between trivate stages and the tribrach plate. In the case of a four screw instrument, there is uneven distribution of pressure on the screws. For stability, three points of support are sufficient. Besides, the three screw instruments can be more quickly leveled.
2. *Tribrach or upper parallel plate*: it is a triangular plate fixed above the leveling screws. It is also known as upper parallel plate.
3. *Spindles*: There are two conical spindles fixed one inside the other. They form the vertical axis of the transit. The inner spindle is attached to a vernier (upper horizontal) plate and the outer one to the lower horizontal plate.
4. *The lower horizontal plate*: The outer axis is attached to the lower plate also called the scale plate. Its edge or limb is silver plated and graduated from 0° to 360° in a clockwise direction. The horizontal circle may be graduated in (1) degree and half of a degree, (2) degree and thirds of a degree, or (3) degrees and sixths of a degree, depending upon the size of the instrument. The lower horizontal plate is provided with (a) A clamp, and (b) tangent or slow motion screw, by means of which it can be fixed accurately at a desired position. When the clamp screw is tightened, the lower plate is fixed to the upper tribrach or parallel plate and on turning the tangent screw, the lower plate and the upper part of the instrument can be rotated slightly.
5. *The upper horizontal plate*: This is also called the vernier plate: and is attached to the inner axis. A clamp and a tangent or slow motion screw are provided for the purpose of accurately fixing the vernier plate to the scale plate. This

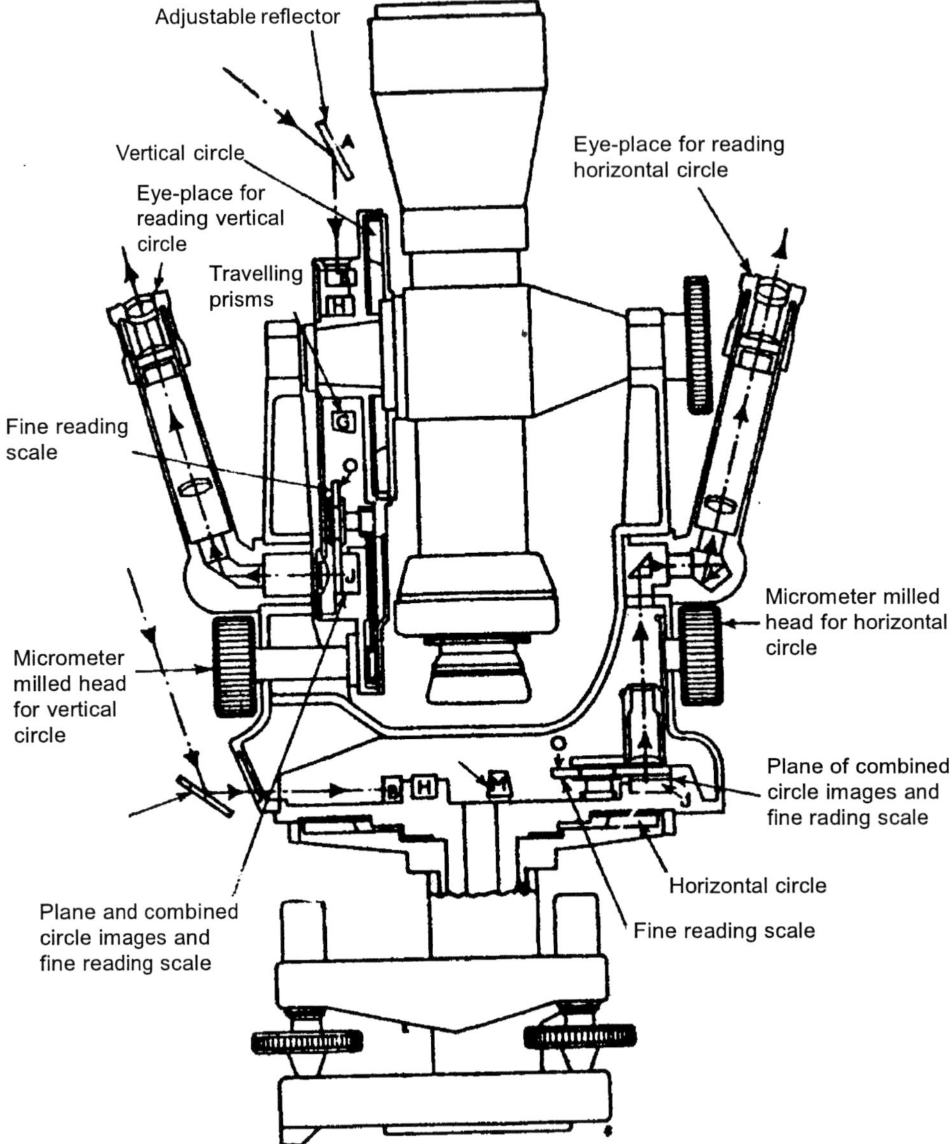

Fig. 9.21: Transit Theodolite and Its Parts

upper horizontal plate carries two verniers A and B with magnifiers placed 180° apart for reading horizontal angles to a minute or 20 seconds.

6. *Spirit levels*: Two spirit levels called plate levels placed at right angles to each other are fixed on the upper surface of the vernier plate for levelling the instrument.
7. *Standards*: There are two frames called standards to support the horizontal axis.

8. *Tube compass*: The compass box may be either of tubular type or of a trough type. The former is mounted on the vernier plate between the standards, while the latter is either fixed below the lower plate, or screwed to one of the standards. This compass is used in directing the telescope towards the magnetic north.
9. *The telescope*: The telescope is fixed at right angles to the horizontal axis. The telescope consists of (i) body, (ii) object glass, (iii) eye piece, (iv) diaphragm, and (v) focusing screw.
10. *The Vertical circle*: It is attached to the inner axis and is placed vertically. It has two verniers called C and D. It is graduated in four quadrants of 0° to 90° each. There are magnifiers to facilitate the reading of the scale. It is also provided with a clamp screw and a slow motion tangent screw. This plate, combined with the telescope of the theodolite, is called the alidade of the theodolite. It is used in measuring the vertical angles.
11. *The Index Bar (or T. frame)*: The index bar is T shaped and centreed on the horizontal axis of the telescope in front of the vertical circle. It carries two verniers at the extremities of its horizontal arms or limbs, called the index arm. The vertical leg called the clipping arm is provided with a fork and two screws called clips or clipping screws at its lower extremity. The index arm and the clipping arm are together known as the T-frame. A long sensitive bubble tube is attached to the top of this frame. It can be centreed by means of the clip-screws.

Repeating Theodolite

When readings of angles are recorded on a graduated scale, the average angle measurement is obtained by dividing the sum of the readings by the number of readings. They are most suited for locations where the support base is unstable, or space for using other such instruments is limited.

Direction Theodolites

In these theodolites, the circle is kept fixed, while the telescope aims on a number of signals. The circle gives the readings for every direction. Direction theodolites are ideal instruments for triangulation.

Total Station

A total station is an electronic theodolite of transit variety. It is widely used in surveying. It is an Electronic Distance Meter (EDM) to read slope distances from the instrument to a given point. The robotic total stations allow the operator to control the instrument from a distance via remote control. Thus the operator can hold the reflector and control the total station from the observed point without the aid of an assistant.

If the direct line of sight between any two points is established, coordinates of an unknown point relative to a known coordinate can be determined. All angles and distances to points under survey are measured from the total station, and the coordinates and elevation of surveyed points relative to the total station position are calculated using trigonometry and triangulation.

To determine an absolute location a Total Station requires line of sight observations and must be set up over a known point or with line of sight to 2 or more points with known location. For this reason, some total stations also have a Global Navigation Satellite System (GNSS) interface which help in determining coordinates without any direct line of sight. The GNSS measurements, however, may at times give relatively poor accuracy in the vertical axis. Since most modern total station instruments use electro-optical scanning of extremely precise digital bar-codes etched on rotating glass cylinders or discs within the instrument to measure angles, they can measure angles ranging from 5 or 10 arc-seconds to 0.5 arc-second depending on the quality of the instruments.

The distances are measured with the help of a modulated microwave or infrared carrier signal, generated by a small solid-state emitter within the instrument's optical path, and reflected by a prism reflector or the object under survey. The computer in the total station reads and interprets the modulation pattern of the returning signal. The emitting and receiving multiple frequencies determine the distances. Most total stations use purpose-specific glass Porro prism reflectors for the EDM signal. Reflectorless total stations can measure distances to any object that is reasonably light in colour, to a few hundred meters. A typical total station can measure distances with an accuracy of about 1.5 millimetres (0.0049 ft) + 2 parts per million over a distance of up to 1,500 metres (4,900 ft).

Total stations are mainly used by land surveyors, mining surveying, archaeologists to record excavations, by the police, investigators scenes of crimes, private accident reconstructionists, and insurance companies to take measurements of scenes. As the drifts of an underground mine are driven, the total station records the location of the tunnel walls (stopes), ceilings (backs), and floors. The data so recorded is then downloaded into a CAD programme, and used to maintain the layout of the tunnel. Small steel plugs fixed in pairs in holes drilled into walls or the back, function as control points. Same procedure is adopted for wall stations, two plugs are installed in opposite walls, forming a line perpendicular to the drift. For back stations too, two plugs are installed in the back, forming a line parallel to the drift.

Comparison of Theodolite and Total Station

Both the Total Station and Theodolite are devices for the measurement of vertical and horizontal angles during engineering projects and surveying. Each has specific characteristics due to which they are used. Normally, the factors of time, expertise available and cost will determine the use of either of these instruments. However, the major differences in these survey instruments are explained below.

Due to its digital integration and all-inclusive features, a total station is considered to be a superior surveying tool compared to the theodolite. A total station incorporates the functions of theodolite to determine angles and distances by an electronic distance meter. Total stations employ a combination of lasers and prisms to record digital readings of the measurements in a computer. This data can be used for further analysis. Robotic total stations have been developed that can be operated remotely. However, the total stations are expensive and need not only expertise in surveying, but also detailed training in software. Total stations are suitable for survey requirements over large distances, particularly over difficult terrain. The results of a total station on such environments are more accurate.

Procedure for Using Theodolites and Total Stations

Centreing

It means fixing of the theodolite exactly over a selected point on the ground. This is done by hanging the plumb-bob by the hook attached to the inner spindle. The centreing is done by slightly moving any two of the legs of the tripod, so that the plumb-bob points directly to the ground station.

Levelling

Level is used is used to keep the surveying equipment horizontal. Procedure is as follows. Bring the spirit level of the horizontal plate parallel to any two foot screws. Turn the foot screws until the bubble settles down in the centre. Turn the screw in such a way that the thumbs of both the hands move either outwards or inwards simultaneously. Then turn the instrument to let the level make an angle of 90° with its former position. Bring the bubble again in the centre by turning the third foot screw. Repeat the procedure twice or thrice to get an accurate levelling. Thereafter, clamp the lower plate and unclamp the upper plate. Turn the horizontal upper plate until 0° to 360° coincides with the vernier V. Clamp the upper plate also. Finer adjustments can be made with the help of the upper tangent screw. It may be pointed out, however, that such adjustments have to be made with reference to the vernier A. Unclamp the lower plate.

Reading Horizontal Angles

The horizontal angle is read with reference to magnetic north, which can be found with the help of a tubular compass attached to one of the standards. Turn the horizontal plate until the needle in the compass comes in between the lines of collimation. Clamp the lower plate. Release the clamp of the upper plate. Turn the telescope in a clockwise direction towards the object. Sight the object and align it with the gun-sight. Clamp the telescope. Focus the object by gently moving the eye piece, so that the cross wire coincides with the object. Now read the verniers A and B.

GRADUATED 30 MINUTES READING TO ONE MINUTE
DOUBLE DIRECT VERNIER

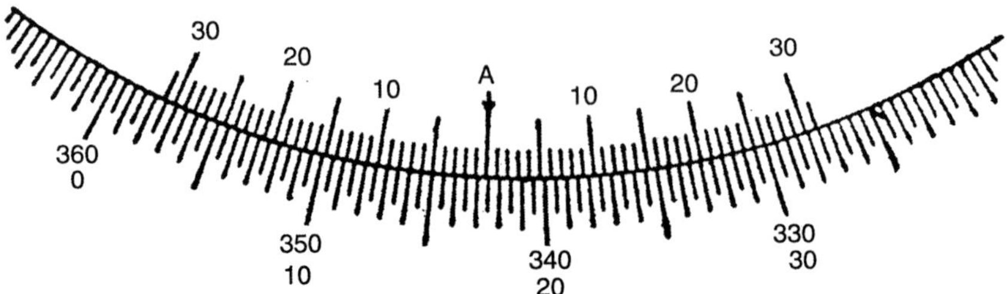

Fig. 9.22: Reading Angles with the Help of Vernier Scale

Reading the vernier scale is a bit problematic. In Fig. 9.22 is given a Vernier scale in which zero of the Vernier cuts the main scale at a point where the angle is more than 342°30′ but less than 343°. To know the exact measurement of the angle, one should move along the Vernier scale to the left until one finds a Vernier division coinciding with the main scale division. The number of divisions one has to move from 0 gives that many parts of 1 degree which should be added to 342°36′. In this case it is 5′, and thus the required angle is 342°35′. In some of the transits, angles can be read even up to 10 seconds.

Reading Vertical Angles

Unclamp the lower plate; bring the vertical spirit level parallel to any two foot screws, and then turn the screws outward to inward and inward to outward simultaneously in order to bring the bubble of the spirit level to its centre. Now turn the spirit level 90° from the previous position. With the aid of the third foot screw, adjust the bubble of the spirit level. Repeat the procedure twice or thrice for accurate levelling. Now, let the 0 of the main scale coincide with 0 of the Vernier scale in the vertical plate. Clamp the telescope and move it horizontally towards the object and unclamp it. Focus it at the foot of the object with the help of the eye piece. Read the angle in both C and D verniers. Now point the telescope to the top of the object and read the angle in both C and D verniers. If one of the readings is below 0 and the other above 0, add both the readings to get the vertical angle. If the readings are above or below 0, the first reading is to be subtracted from the second one in order to get the vertical angle.

Example:

I. Reading 3° 10′ 20″ (foot of the object: below zero)
II. Reading 9° 10′20″ (top of the object: above zero)
∴ *Vertical angle*: 12°20′40″
I. Reading 3° 10′20″ (above zero)
II. Reading 40° 10′20″ (above zero)
∴ *Vertical angle:* 37°0′0″

Plotting the Results

If the theodolite survey is based on triangulation method, the triangles can be plotted either by computing the angles or the sides. The angular measurements made by the theodolite are very precise. If one wants to find out the length of the sides of a triangle, one has to measure the base line and then get the measurements of other lines with the help of the following formula.

$$\frac{\text{Sin A}}{a} = \frac{\text{Sin B}}{b} = \frac{\text{Sin C}}{c}$$

Where A, B and C are the three angles of the triangle with three sides a, b and c facing respective angles. Verbally stated, the sides of a triangle are proportional to the sin of the opposite angles. If we know one of the sides, we can know the other two sides too.

LEVELLING

Levelling is a process of determining the relative heights of various features with the help of an instrument called a 'level'. Earlier in this chapter some of the levelling techniques have already been explained. Further, details of some of them are given here.

There are several makes of Levels; the ones in common use in India are:

1. Wye-Level,
2. Dumpy Level,
3. Watt's Self-Adjusting Level, and
4. Watt's High Way Level.

The telescope of the Wye-Level (Fig. 9.23) stands on two standards whereas that of the Dumpy Level (Fig. 9.24) stands on one pivot. Both are easy to operate. The Watt's Self-Adjusting Level is a simple modern reversible level. It has reflecting prisms, which show the bubble and an index line. There are two mirrors to reflect light through the level tube. In this instrument, the telescope can be reversed through 180° about its longitudinal axis so as to bring the bubble on the right of the instrument.

The Watt's Highway Level (Fig. 9.25) is a very delicate and precise instrument. Its operational procedure is given below. After screwing the level to the tripod, centre it at a convenient place somewhere between the turning points (first two stations). With the help of the foot screws bring the bubble of the circular spirit level in the centre. Thereafter, bring the main bubble in the mid-run. Focus the eye piece as well as the object glass. Direct the telescope toward the staff and clamp it. With the help of slow motion tangent screw bring the image of the staff on the hair of the diaphragm. While reading the graduations on the staff it should be kept in that readings at the intersection of the cross-wire indicate backsight (or + s) of the starting point. For horizontal distances

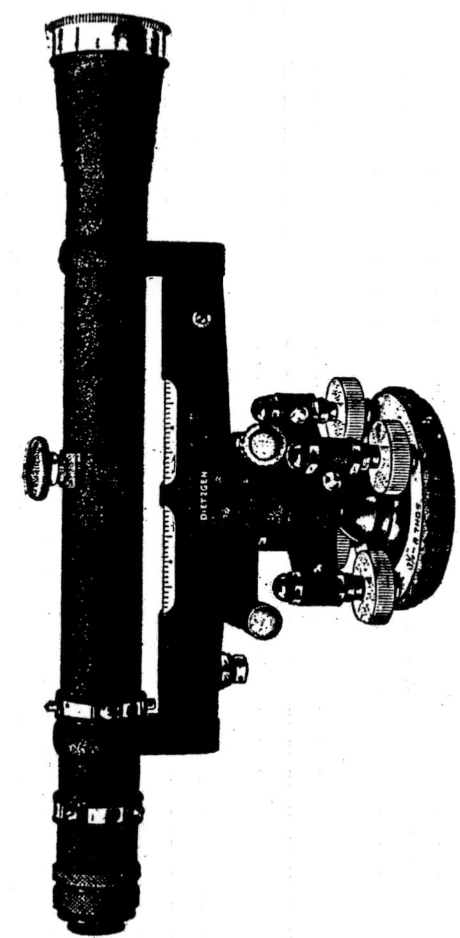

Fig. 9.23: Wye-Level

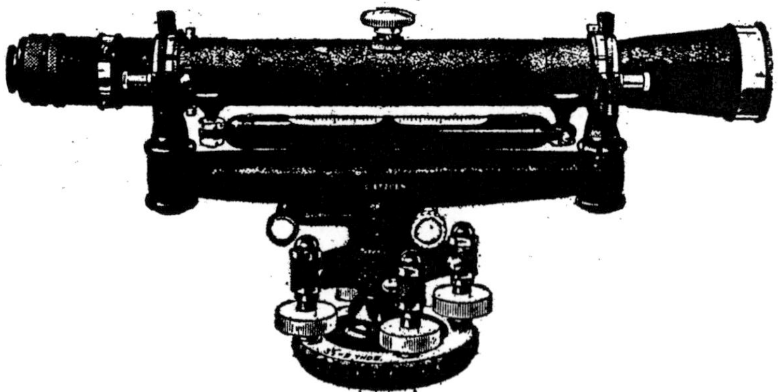

Fig. 9.24: Dumpy Level

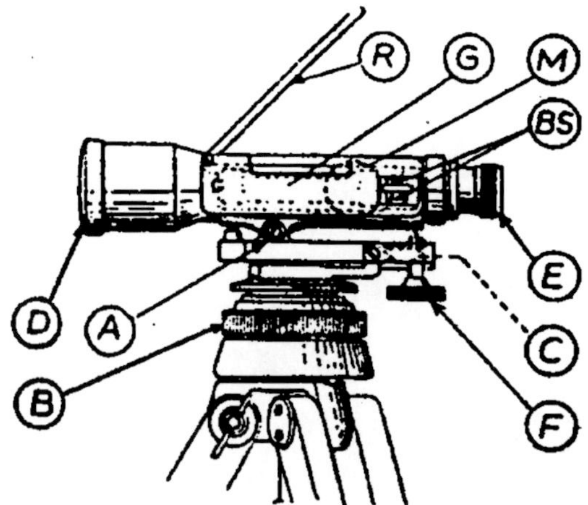

A—Horizontal fuleram about which the telescope and its attachments rotate.
B—A bolt-and-socket jointed head used for primary levelling.
C—Air bubble
D—Ray shade
E—Eyepiece
F—Levelling screw
G—Air bubble attached to telescope
B—Copstan screws used for adjusting the air bubble
R—Mirror reflecting the bubble.

Fig. 9.25: Watt's Highway Level

read the stadia. Then focus the telescope on the staff fixed at the second point. This time the reading will give the fore-sight (or-s). Nothing these reading in the field book, we get the results as given below:

Point	Backsight +S	Foresight -S	Reduced Level (R. L).	Remarks
A	3 ft.		500 ft.	Point a has bench mark giving height of 400 ft
T.P. (B)	2 ft.	6 ft.	497 ft.	
T.P. (B)	4 ft.	5 ft.	494 ft.	
T.P. (C)	3 ft.	6 ft.	492 ft.	
E		5	490	
Arithmetic check	12 ft	22ft	– 10ft	

Calculation of R.L. : 500+3 – 6= 497
497 + 2 – 4 = 494
494 + 4 – 6 – 492 +3 – 5 = 490

While finding out the slope of a ground, the levelling instrument is placed in between the points for which elevation has to be determined. It is not placed on the stations for which readings have to be taken. That is why seeing the first point is called the back sight reading and the next point is called the foresight reading.

Abney Level and Indian Clinometer are also used in measuring elevations. Abney level consists of a telescope attached to a protractor (Fig. 9.26). The middle point of the protractor is marked by a zero. Both sides of the zero, graduations increase to

60° or 90°. Left side shows rise and right sides shows fall. The slopes are marked on the inner margin of the protractor. They are read by the vernier. A spirit bubble at the top keeps a check on the horizontal positioning of the level.

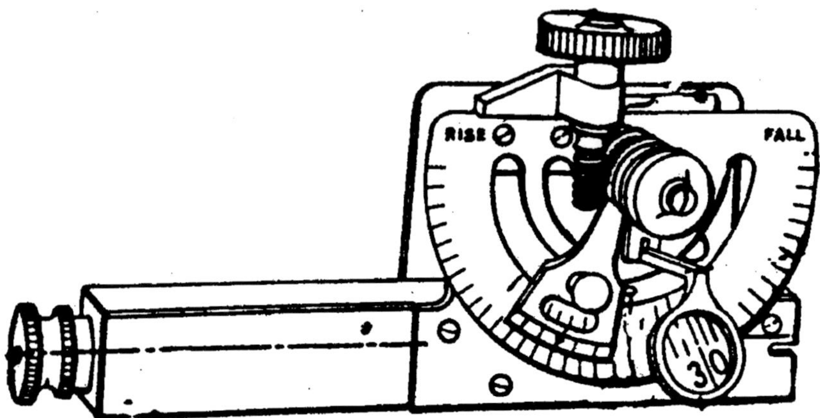

Fig. 9.26: Abney Level

The Indian Clinometer (Fig. 9.27) was developed by the Survey of India. It is used along with the plane table. It consists of a brass plate with an air bubble and two flaps at its ends. One of the flaps has a for-sighting hole and the other has a long slit and is graduate along both of its edges giving trigonometric tangents and degrees of slope. In the middle is a common zero which is at the same horizontal level as the Eye-Hole.

Determination of height with the help of an Indian Clinometer involves some arithmetical calculation. To determine the height of an object C from a point A in Fig. 9.28, the surveyor will go to the field with point B already marked. He will determine his position A by resection and then find out the ratio between b c and A b by measuring the height of c. Suppose, the ratio comes to bc/Ab = .5. As the slope of A c is the same as that of AC, the ratio between BC and AB will also be the same. Since, the map gives the horizontal distance AB and we know that BC/AB = .5.

We can find out the length of B c.

HYDROGRAPHIC SURVEYING

Hydrographic surveying is the term applied to the processes used in surveying of water bodies. In the case of oceans or lakes this may include determination of shore lines, soundings, characteristics of the bottom, location of buoys, etc. The survey of a river may also include the determination of the velocity and characteristics of the flow. In its broad sense the term may be applied to the survey of drainage areas and proposed reservoirs for the storage of water.

The surveyor of shore lines of oceans, lakes and rivers are done in almost the same way as the land survey. In addition to the instruments employed in land surveying, sextant is frequently used in hydrographical surveying. It is an instrument

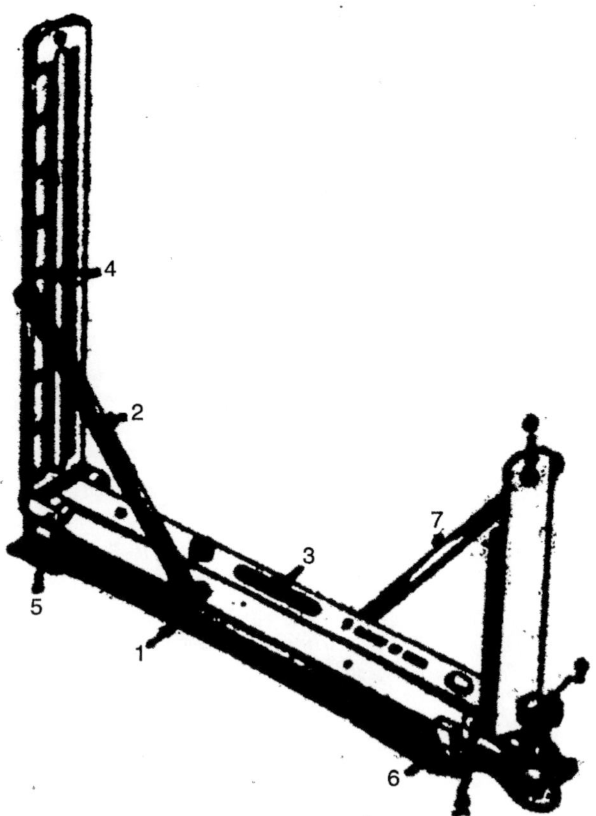

Fig. 9.27: Indian Clinometers

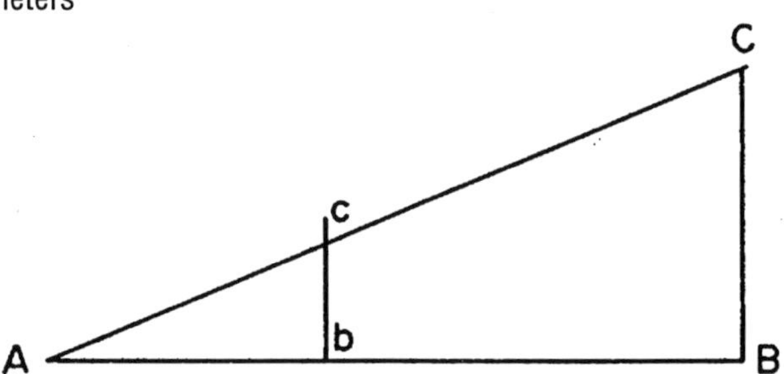

Fig. 9.28: Measuring the Height with the Help of Indian Clinometers

which, unlike the transit, is adapted to measuring angles in any plane. It is constructed in such a way that angles as large as 120 degrees can be measured. Owing to the fact that it can be used by an observer who is on a moving object, such as boat, it is especially valuable for hydrographic work. Hydrographic surveys are also done with reference to the control points fixed by geodetic surveys.

The determination of the topography of the bottom of the sea, lake, or other bodies of water is one of the common problems in hydrographic surveying. Surveys of this kind are made for a variety of purpose, such as to prepare charts for navigations, to determine where dredged material may be dumped, or to measure the quantity removed. They are also done for the purposes of discovering what changes are taking place in the bed of water body for scientific or constructional purposes.

This work is usually done by first establishing certain points on the shore (by triangulation or traverse) to which the hydrographic survey may be referred to, and then measuring, usually from a boat or hydrographic survey ship, the depth of the water at various points and determining the positions of these points. The measurements of depths are called soundings. Since, the sub-aqueous surface is not visible, it is evident that for given degree of accuracy a great many more points must be located to obtain the shape of the surface that would be necessary in an ordinary topographic survey of equal area.

REFERENCES

Andrews, John Harwood (1985), *Plantation Acres: an Historical Study of the Irish Land Surveyor and His Maps*, The U.H.F. historical series 8, Belfast: Ulster Historical Foundation, p. 462.

Arora, K.R. (1998), *Surveying*, Vol. I and II, Delhi: Standard Book House.

Bedini, Silvio A. (1964), *Early American Scientific Instruments and Their Makers*, U.S. National Museum, Bulletin 231, Washington DC: Museum of History and Technology, p. 184.

Cary, Edward Richard (1916), *Geodetic Surveying*, New York: John Wiley & Sons.

Clark, David and James Clendinning (1958), *Plane and Geodetic Surveying for Engineers:* Vol. I. *Plane Surveying:* Vol. 2. *Higher Surveying*, London: Constable & Co.

Davis, R.E., F.S. Foote and J.W. Kelly (1966), *Surveying: Theory and Practice*, New York: McGraw Hill Book Company.

Davis, R.E. (1967) *Elementary Plane Surveying* (4th ed.), New York: McGraw Hill Book Company.

Higman, B.W. (1988), *Jamaica Surveyed: Plantation Maps and Plans of the Eighteenth and Nineteenth Centuries*. Kingston: Institute of Jamaica, p. 307.

Hughes, Sarah S. (1979), *Surveyors and Statesmen: Land Measuring in Colonial Virginia*. Richmond, Va.: Virginia Surveyors Foundation and Virginia Association of Surveyors, 196p.

Ingram, Edward L. (1911), *Geodetic Surveying and the Adjustment of Observations: Methods of Least Square*, New York: McGraw Hill book Co.

Kiely, Edmond Richard (1947), *Surveying Instruments: Their History and Classroom Use*, National Council of Teachers of Mathematics, 19th Yearbook. New York: Teachers College, Columbia University, p 411.

Mansfield, Merriman (1892), *Introduction to Geodetic Surveying, Part 1. The Figure of the Earth; Part 2. Principles of Least Square, Part 3. Fieldwork of Triangulation*, New York: John Wiley & Sons.

Nugent, Paul Cook (1902), *Surveying*, New York: John Wiley & Sons.

Punmia, B.C., A.K. Jain and A.K. Jain, (1996) *Surveying*, Vol. 1 and II, New Delhi: Laxmi Publications (P) Ltd.

Richeson, Allie Wilson (1966), *English Land Measuring to 1880: Instruments and Practices*, Cambridge, Mass.: Society for the History of Technology and the M.I.T. Press, p. 214.

Stewart, Lowell O. (1976), *Public Land Surveys: History, Instructions, Methods*, Minneapolis, MN.: Meyers Printing Co., p. 202.

Taylor, Eva Germaine Rimington (1956), *The Haven-Finding Art: A History of Navigation From Odysseus to Captain Cook*, London: Hollis & Carter, p. 295.

10
Interpretation of Topographical and Geological Maps

We can learn a lot about the nature of the area mapped from carefully prepared topographical maps. To make an effective and correct use of these maps we must, however, know the technique of map interpretation. Map is a model of the earth or its part. Real earth is not a jumble of bits of information which appear on the earth surface independent of each other. Nothing on the earth stands alone; everything forms part of a totality that is complex and interdependent. In fact, the earth forms a system with innumerable subsystems. We cannot understand the whole truth about any single subsystem unless it is seen as an integral part of the larger system. Map interpretation thus involves a synthesis of facts represented on a map and the interrelations among them to see from the map the real earth or the part there of it represents. This point is important to note because many a times map interpretation is considered to be an exercise in describing the individual components of the map.

By isolating a single component of a topographical map, say highways, we can know the direction and distribution of the roads. If we go a step further and take up two components, say relief and highways, we would not only know the details of these components but also the relationship between the two. But even this will not give the whole truth; it will give only a partial truth that the map in question illustrates. The study of individual components of a map separately should be done only as a means to understand the relationships underlying the various components.

For scientific interpretation of maps prior knowledge of the physical and cultural geography of the area represented on the map being interpreted helps.

INTERPRETATION OF TOPOGRAPHICAL SHEETS

The following steps are suggested in the interpretation of topographical maps:

1. Identifying the map

2. Interpreting the physical details
3. Interpreting the cultural details
4. Prepare a cartographic appreciation

Identifying the Map

The topographical sheet being interpreted should be identified by the details given in its margins. Among these details are the area to which the map belongs, time of the survey, magnetic north, scale, administrative index, contour interval and other details like the longitudes and the latitudes.

Interpreting the Physical Details

The following procedure is suggested for the interpretation of physical landscape:

1. Give the broad physical features first and then point out the general lay of mountains, plateaux and plains. Note the major rivers, if any and the general slope as indicated by the main contours.
2. After noting the broad features, look carefully the details of each of them. Mention the landforms like peaks, ridges, hills, spurs, escarpments, knolls, cirques, cols, flood plains, valleys, morainic deposits, etc. The number of sectional profiles should be drawn at this stage to indicate the nature of relief. For the method of drawing profiles see chapter 15. These profiles help in identifying the nature of the landforms mentioned above. For example, they will indicate whether a hill is residual, morainic or volcanic. Spacing of contours should get special attention. Several of the techniques discussed in Chapter 15 can be fruitfully used in identifying various landforms.
3. The next stage is to discuss the drainage pattern and the physical features associated with them. The catchment area of each river or stream should be marked. The glaciers and their morainic deposits should also be marked. A correlation between the contour pattern and the drainage pattern will be very revealing. Longitudinal profiles along a few rivers should be drawn to point out the stage at which different stretches of the streams are. The cross-sectional profiles will give data on the nature of the valleys such as gorge, V-shaped, U-shaped, broad, narrow, symmetrical, asymmetrical, hanging, transverse, longitudinal, etc. The nature of the banks of the rivers and the formation of flood plains, meanders, oxbow lakes, incised meanders, terraces, swamps, fans, deltaic flats and tidal deposits, if any, should be noted. The longitudinal profile will give information about the presence of water falls, cataracts, rapids, etc.
4. The type of drainage pattern visible in the map should be noted. It will be one or several of the following types (Fig. 10.1):

   (i) Dendritic pattern
   (ii) Trellis pattern
   (iii) Rectangular pattern

(iv) Radial pattern
(v) Deranged pattern
(vi) Barbed pattern
(vii) Karst pattern

5. Next step is to study the vegetation. The natural vegetation in Indian toposheets is shown by green colour. Certain types of trees are shown by symbols.

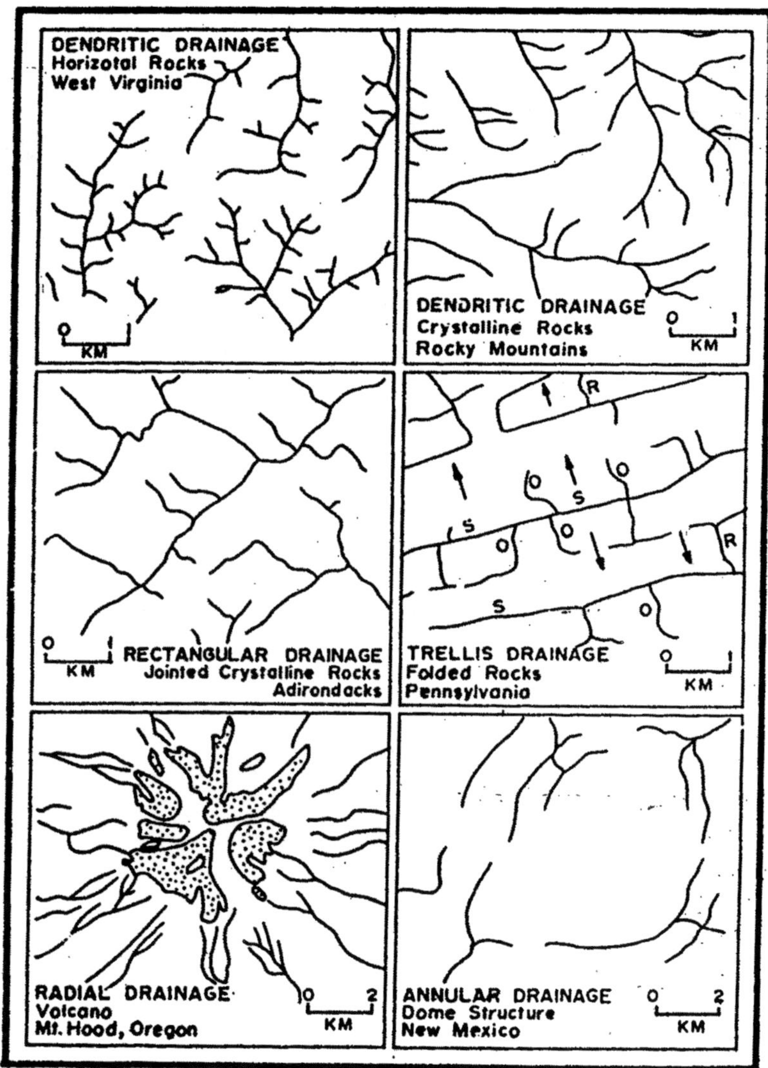

Fig. 10.1: Drainage Patterns

While interpreting the physical landscape, one has to keep in view W.M Davis' classification of landscape. According to him, physical landscape is a function of structure, process and stage. It is true that all the three elements cannot be fully interpreted from all the maps. As such no attempt is made here to suggest any stereotype procedure. The procedure will have to vary from area to area and map to map. In certain cases the main interest will have to be focused on the correspondence between structure and surface relief; in certain others, on the landforms developed in one of the special cycles; while in certain others on the erosional forms belonging to a certain stage of the cycle. While, structure, process and stage must form the basic framework for interpretation of physical landscape, they need not be discussed in that order or with same emphasis.

The problems to be discussed are:

- The Process
- The Structure

The Stage

To explain the *process*, discuss the erosion cycle currently in progress. It will be one of the following:

1. Normal Cycle
2. Glacial Cycle
3. Arid Cycle, and
4. Karst Cycle

In case of the normal cycle of erosion, physical evidences of former glacial erosion or deposition and former marine erosion should also be discussed. If the map has shorelines, the class, type and stage reached in the present shoreline cycle should also be taken note of and discussed.

Interpreting the Cultural Details

It is true that man's activities have not altered the major landforms, except locally, they have nevertheless altered certain surface details significantly. Look at how the earth looked like in its natural form or it would look like if untouched by man. There would be grasses, plants and trees all around. Wild animals would be roaming around everywhere providing fertilizers to the virgin land. Anthropogenic activities during the last ten thousand years have considerably altered the land. In place of grasses, there are domestic plants. We call them food, fiber and oil crops. The study of cultural features is, therefore, an essential part of topographic map interpretation. It involves the abstraction of a certain data such as roads, settlements, mines, etc., from a given composite map on the one hand and the perception of the underlying relationships among the various sets of cultural details as well between the cultural and physical features, on the other hand.

For the interpretation of cultural details the following outline can be used:

(a) Land Use

(i) Agricultural (with types of crops),
(ii) Forest,
(iii) Other uses,
(iv) Wasteland,
(v) Irrigation systems,
(vi) Land tenure, if given.

(b) Transportation and communication

(i) Railways
(ii) Roads
(iii) Cart tracks, Camel tracks, mule paths, etc.
(iv) Aerodromes
(v) Ports and harbours
(vi) Telegraph and telephone lines, post offices, etc.

(c) Settlements

(i) Rural settlements
(a) Units
(b) Forms
(c) Size
(d) Density
(e) Patterns

(ii) Urban settlements
(a) Site
(b) Situation
(c) Form
(d) Size

(d) Other details

(i) Political boundaries
(ii) Archaeological details
(iii) Cultural institutions
(iv) Others

(e) Synthesis

Cartographic Appreciation

Geographic interpretation of a topographical map is not complete unless the interpreter has also critical appreciation of the map. Cartographic appreciation involves the study of cartographic methods and techniques, and the estimation of their success or failure in the representing the map details. There are two aspects of cartographic representation which need special attention:

1. Suitability of symbols used, and
2. The design and aesthetic beauty of the map as a whole.

Both these aspects involve personal liking and disliking on the one hand and expertise and experience of the judges on the other. The taste, experience and aesthetic susceptibilities of the judges matter a lot. That is why cartographic appreciation has always been a controversial issue. Cartographic appreciation cannot be learnt by a rule of thumb. Long practice and considerable patience are needed. Comparisons should be made between maps on different scales, of single surveys, of different series of single surveys, and of different surveys on a single scale before passing a judgment.

A topographical map must omit a great deal of potential material suitable for mapping and should achieve some kind of balance in the material presentable and actually presented. Many map series achieve praiseworthy success with certain classes of information but fail miserably in dealing with others. At times an interpreter may also be biased. He may take his own national maps as standard ones and criticize foreign maps having unfamiliar symbols and representation techniques. To give an example, the 1/50,000 maps of Sweden use black hachure lines instead of contours for representing relief. One who is used to seeing relief through contours may not like it. But the technique is very suited to the terrain of Sweden. At times a few details are shown too prominently such as waterways in Netherlands. For an arid zone dweller it may appear to be unnatural. But it is quite inappropriate in a country where waterways are so important.

It must also be kept in view that the design of topographic maps differs from country to country. It is possible to make wholesale changes in series mapping. What is possible in a short run is the creation of a mixture of the old and the new. Although many modern surveys use common techniques, map styles do not become identical. So long as the perfect map remains unmade, so long as there is no agreement on cartographic perfection, and so long as improvements in surveying, data collection, and printing processes continue, varying styles in mapping will persist.

While attempting a cartographic appreciation, it must be kept in mind that the scale of mapping puts many limitations on the amount and type of information that can be given. The larger the scale, the greater the amount of detail that can be given, at the expense, however, of a larger coverage and also of wider and more synoptic view of the major features of the country side which can be gained from smaller-scale maps.

The balancing of the physical and cultural details marks success or failure of a map. The cultural patterns must be clear in themselves and clear in their relationship to the underlying terrain if correlations are to be understood and the basic purpose of map interpretation is to be achieved. The success achieved in clear representation of non-landscape, features like names, boundaries, etc., should also be kept in view. In general the relative importance of these features increases with the scale of the map, for on larger scales there is both more room and greater need for that class of information which is of use particularly to the indoor map reader. In the assessment of non-landscape features one must first consider how far the information provided suits the scale of the map and the type of terrain. Thus the representation of the least significant non-landscape feature would be justified in a map of a part of sparsely populated Rajasthan but the same details would be quite out of place on a map of densely populated East U.P., through drawn on the same scale.

INTERPRETATION OF GEOLOGICAL MAPS

A geological map suggests the nature of the underlying rocks and is extremely useful in determining whether landforms reflect control by lithology and structure or bear more the imprint of geomorphic processes and history.

A geological interpretation of a map should deal with following aspects:

1. Succession of beds,
2. Structure,
3. Relationship among rock groups,
4. Topography and its relation to structure, and
5. Geological history.

In this section certain methods are indicated which can be used in understanding these aspects. Further, some of the commonly used terms are defined below to facilitate the interpretation of geological maps.

Geological Beds

These beds occur in the earth's crust resting upon one another as successive layers. The surface of separation between two adjoining beds is known as a "bedding plane".

Geological Outcrop

These are the exposed portions of geological beds and formations on the earth surface.

Strike Line

It is a line joining two or more points which lie on the same surface at the same elevation from the mean sea-level. It is also called "structure contour" because it is analogous to the contour lines of the ground surface.

Dip

Dip of a surface is its inclination with horizontal plane. The direction of dip is the direction of inclination of a surface. The amount of dip is the angle between inclined surface and the horizontal plane. It is measured in the direction of the dip (Fig. 10.2).

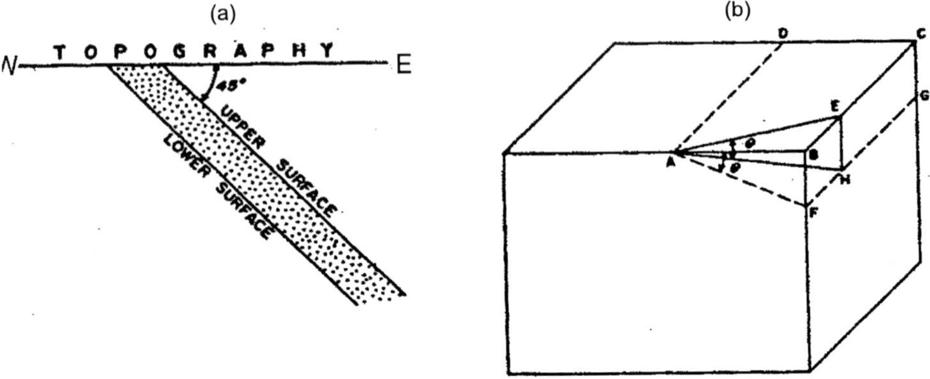

Fig. 10.2: Determination of True and Apparent Dips

True dip is the amount of dip measured along the direction of maximum slope of an inclined plane. It is clear from the Fig. 10.2b that FB is the direction in which the inclined plane has the minimum slope. < BFG is the true dip.

The apparent dip can be measured in any direction other than the direction of maximum slope. Moreover, it is always less than the true dip. < BPI is the apparent dip.

Method of Drawing Strike Lines and Determination of Dip

In Fig. 10.3 the outcrop cuts the same contour at two points. The line joining these two points is the strike line and the value of the strike line is that of the contour line (in this case, the 300′ ft contour line). If another contour is cut by the same outcrop, a strike line can be drawn through the point parallel to the first one and the value of the strike line will be the value of the second contour. All such strike lines will be parallel and equidistant. The direction of dip can be found out by drawing a line at right angles to any strike line towards the strike line of lower value: xy in Fig. 10.3.

The amount of dip can be calculated by using the following formula:

$$\text{Dip (in } 0^\circ) = \tan^{-1}\frac{\text{V.I.}}{\text{H.E.}}$$

where,

V.I. = vertical interval of the strike lines

H.E. = horizontal equivalent.

METHOD OF COMPLETING A GIVEN OUTCROP

Fig. 10.4 shows the contours and portions of beddings at certain places. First draw a strike line, when the same contour cuts the bed at two places. A second strike line can

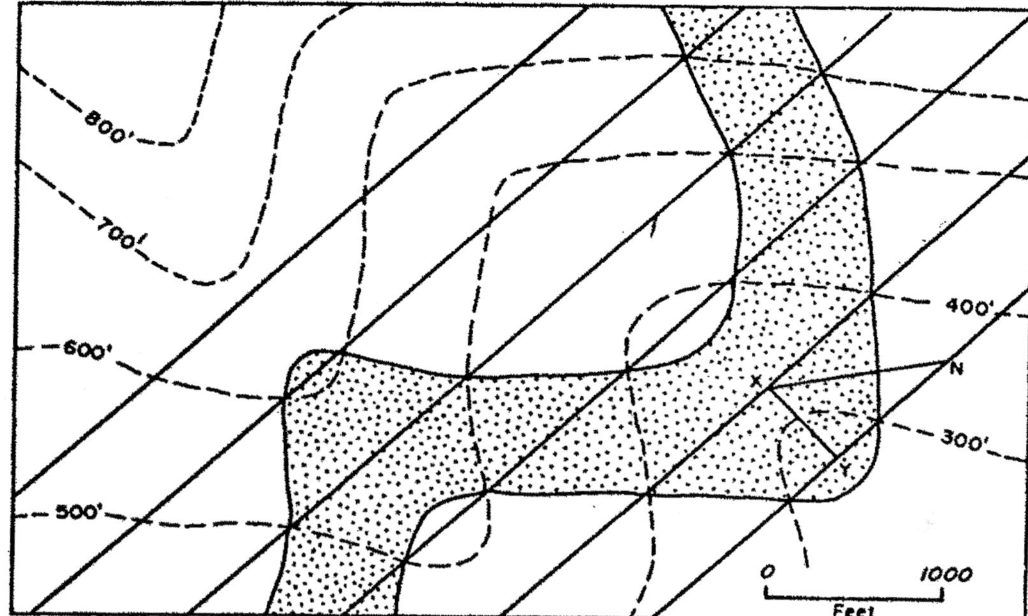

Fig. 10.3: Drawing the Strike Lines

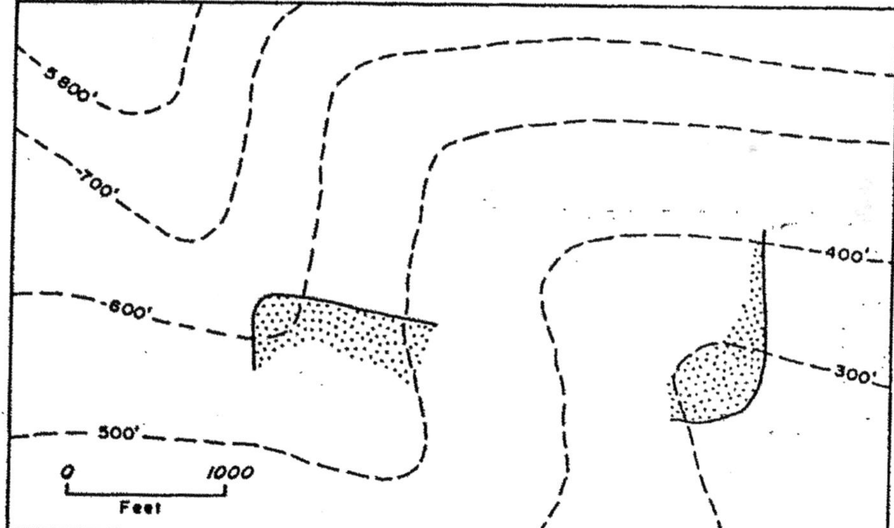

Fig. 10.4: Completing a Given Outcrop

also be drawn through a point where another contour of a higher or lower value cuts the same outcrop. This strike line should be drawn parallel to the previous one. One these two strike lines are drawn, the other strike lines can also be drawn following the same principle. Now, each point where a contour line is cut by a strike line of the same value can be marked. The outcrop can be completed by joining these points by a smooth line. Similarly, one can complete a series of outcrops in any map. Following rules may be followed while completing an outcrop:

1. The outcrop line can cut only a strike line or a contour where equal values intersect.
2. An outcrop line must not be taken across any strike lines or any other contour, except at the points of intersection proper.
3. Outcrop should go right through the point in opposite angle.

Calculation of Bed Thickness

The vertical distance between the upper and lower surface of the bed is called the "vertical thickness of the bed" (Fig. 10.5). The true thickness of the bed is the perpendicular distance between the upper and lower surfaces of that particular bed.

The vertical thickness of a bed can be determined by comparing the strike lines for its upper bedding plane with those of the lower bedding plane in a given geological map. Suppose the upper value is 600′ and the lower value 400\ the difference of 200′ is the vertical thickness of the bed.

The true thickness can be calculated from the following formula:

T = t x cos of the angle of dip

where t = vertical thickness.

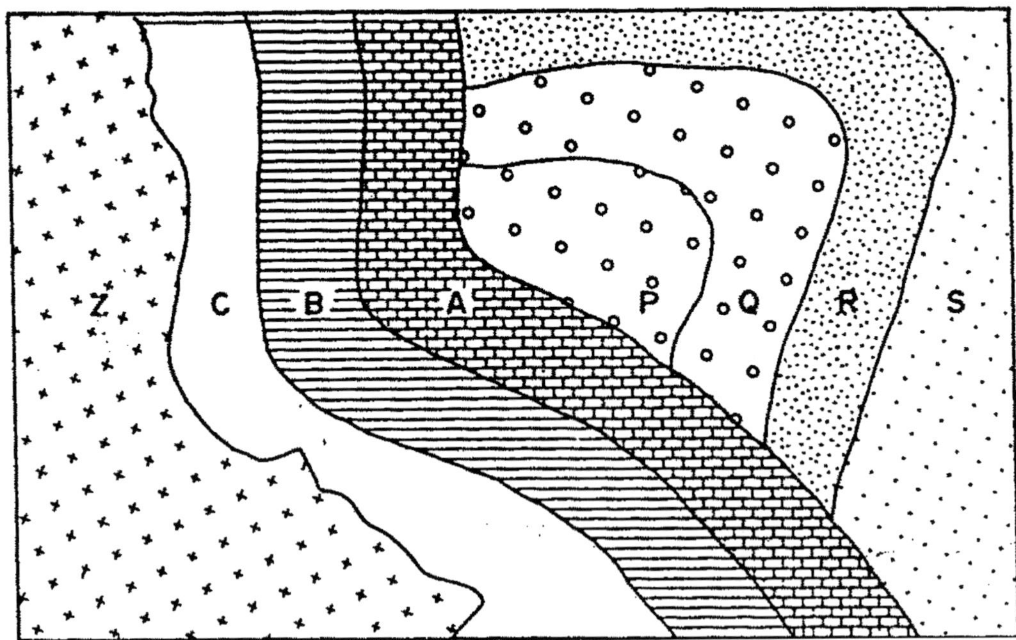

Fig. 10.5: Determining the Thickness of the Bed

Conformable Beds

A conformable series of rocks is one in which the beds are laid in a period of continuous sedimentation. In other words there is no break or interruption in the deposition of the different beds. Beds which are conformable to one another have the same dip and strike (Fig. 10.5). But the dip and strike of the beds of the first series are different from the dip and strike of the beds of the second series.

The relative ages of the beds can be determined from the direction of the dip i.e. in the direction of the dip younger bed comes after an older bed. Thus in the given map, under the first series, bed A is the oldest, followed by beds B and C. Similarly, in the second series, bed P is the oldest followed by beds Q, R, and S.

Unconformity on a Geological Map

Unconformity means a break in a stratigraphic sequence; it represents an interval in sedimentation and a period of erosion of the older rocks. It is usually indicated by a wavy line in geological sections. It can be identified by observing the following characteristics:

1. *Variations of structure:* Two series of beds will leave different altitudes and structures.
2. *Truncation of outcrop:* Against the line of unconformity the outcrops of the beds of the old series are abruptly ended or truncated.
3. *Trijunction points:* Three beds may meet at a point on a line of angular unconformity. By finding such points, the line of unconformity can be traced on the map.

METHOD OF STUDYING FOLDS AND FAULTS

Folds: The map in Fig. 10.6 shows series of beds viz. U, Q, S, T, R, W, P and a dyke denoted by V. It is obvious from the map, that there are three structural features:

1. Unconformity
2. Fold and conformity, and
3. Intrusive contact.

The fold can be identified on the map with the help of following characteristics:

(i) Repetition of Outcrops

This happens because of the erosion of the upper part of the fold. The beds of the series A are folded. One can easily observe the repetition of the outcrops of S, Q, U and T and the reversal of the dips of the beds. It is also clear from the map that in the central part of the area, the beds T, S and Q on both sides of the bed R are dipping

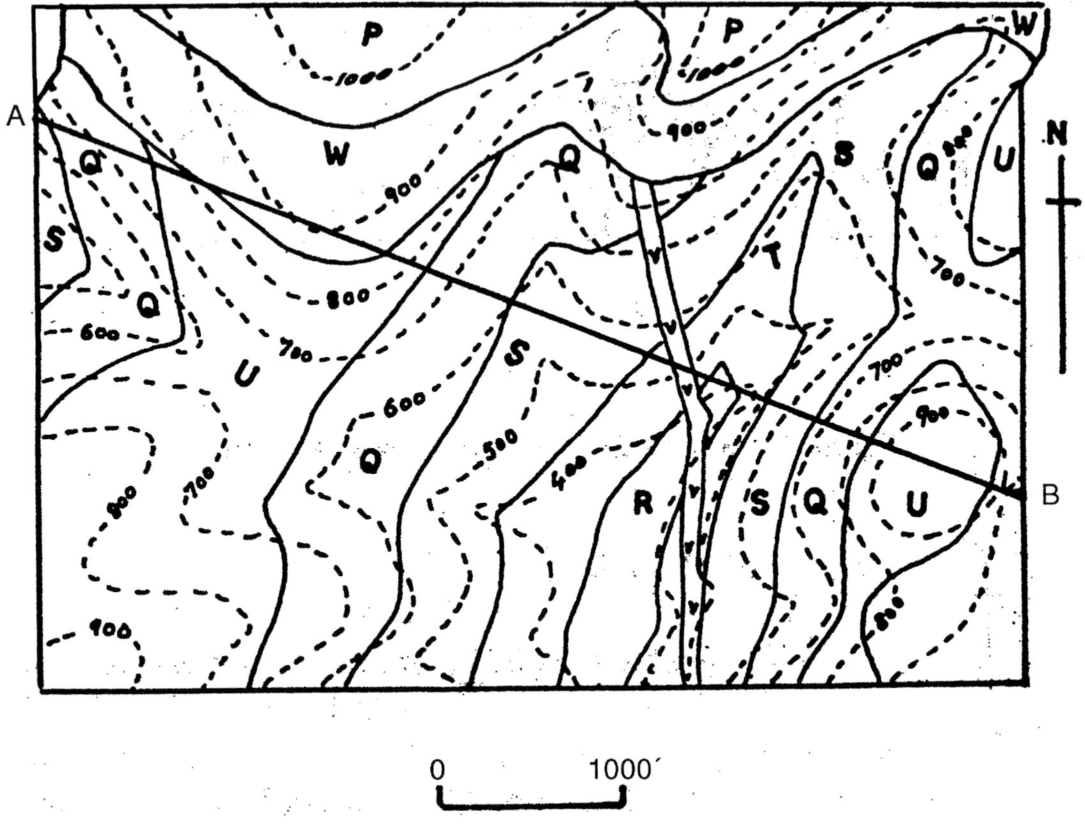

Fig. 10.6: Identification of Folds and Faults

away from bed Q. Here, the core of an antiform is R, and T, S and Q are the limbs. Similarly, in the western part of the area, U is the core of a synform. If the succession is normal, we can group beds of A, B, and C in the following order.

P Formation (B)
W unconformity
.... Unconformity
V Formation (C)
.... Intrusive Contact
U
Q
S Formation (A)
T
R
....

(ii) Reversal and Variation of Dip

The following criteria can be used to determine the characteristics of the fold:

(a) *Anticline*—the two limbs dip away from the axial plane.
(b) *Syncline*—the two limbs dip towards the axial plane.
(c) *Plunging fold*—the axis is not horizontal but dipping.

Given below are the definitions of other relevant terms:

Fault: It is a dislocation along a plane of fracture in rock strata. The plane of dislocation is called "fault plane". Before dealing with the geological maps, it is necessary to know the characteristics of a fault.

Slip: This is measured on the fault plane. Slip is the relative displacement of points on opposite sides of the fault which were formerly in contact with each other before faulting.

Pitch: It is the angle between a line on an inclined plane and the horizontal line (strike) on that plane.

Hade: It is the angle measured between the vertical planes that strike parallel to the fault plane.

Throw: Throw can be measured on a vertical section perpendicular to the strike of the fault plane.

Heave: The horizontal component of dip separation between the up throw and the down throw side is known as heave.

A fault can be recognized on a geological map with the help of the following criteria.

(a) Presence of a fault line
(b) Discontinuity of the outcrops
(c) Change in the strike on the opposite sides of the fault line, and
(d) Repetition of omission of strata.

METHOD OF DETERMINING A THROW

Fig. 10.7 illustrates a fault line F.F. Here, the conformable beds are affected by fault dip towards the south-east. Two strike lines a_1 b_1 and a_2 b_2 can be drawn for the upper surface of the bed P corresponding to 400 ft and 500 ft.

Then the distance xy viz., 400-500 ft. strike line should be measured. Now the 600 ft. strike line passes through a_3, the intersection point of the upper surface of the bed P and 300 ft. contour. This strike line is drawn in the direction opposite to the dip of 500 ft. strike line and at a perpendicular distance equal to *xy*. Now this strike line of the upper surface of the bed P actually becomes the 300 ft. strike line. Hence, the throw is 300 ft (600′–300′ = 300′) down in the north-western part of the fault. Thus it is obvious that the south-eastern part is the up-throw side of the series.

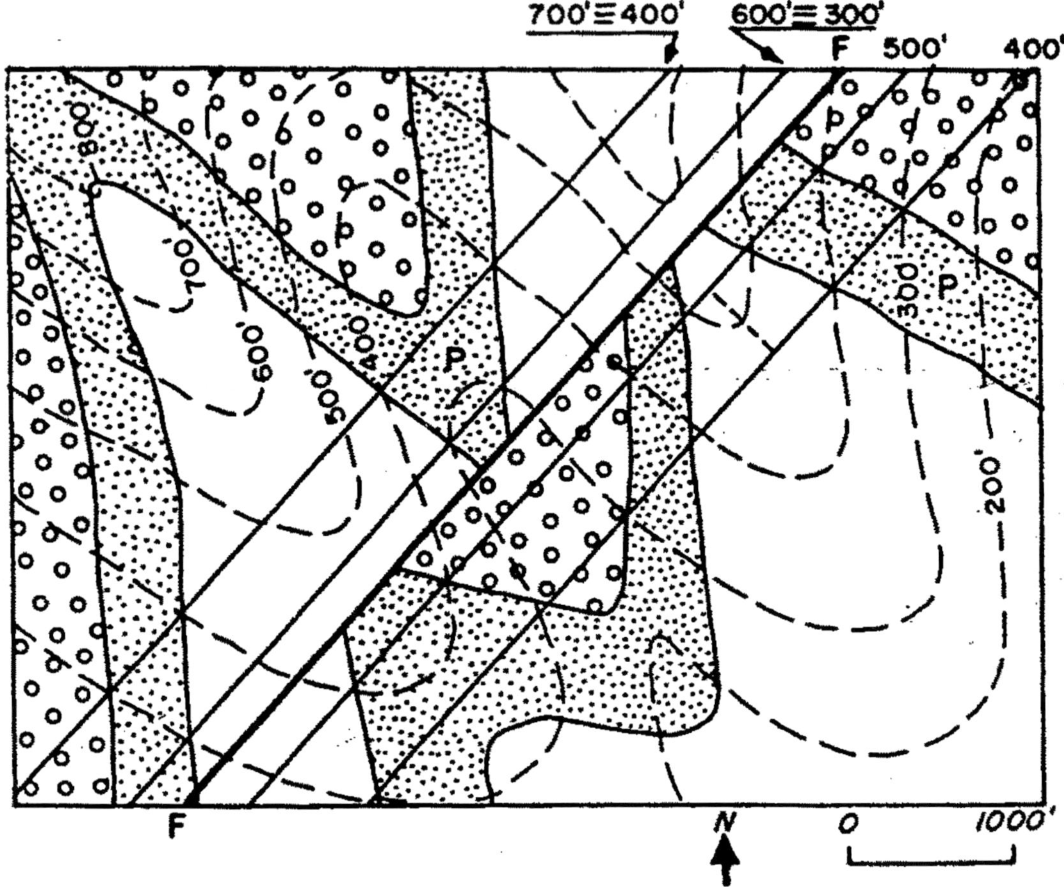

Fig. 10.7: Identification of Folds and Faults

METHOD OF DRAWING A GEOLOGICAL SECTION

To begin with, the profile should be drawn by projecting the points where the contours cut the lines of section of their corresponding altitudes. These points can be joined by a smooth line. The points of intersection of the different bedding planes with the line of section should be marked carefully on a piece of paper. Then these points are projected on the profile with the aid of the base of the section.

The apparent dip of the beds can be utilised to draw angles on the base in the direction to which they are dipping. Again, these points can be projected on the profile and the parallel lines drawn. One need not use the same scale for horizontal and vertical axes because the section should be a natural one.

Some of the symbols conventionally used to represent various types of racks are illustrated in Fig. 10.8.

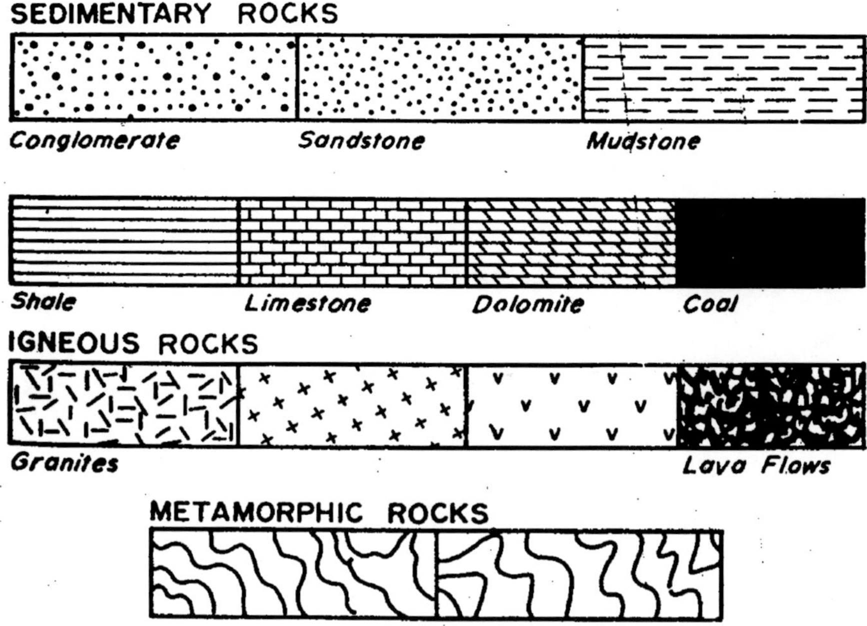

Fig. 10.8: Rock Symbols Used in Geological Mapping

REFERENCES

Dake, Charles Laurence, and J.S. Brown (1925), *Interpretation of Topographic and Geologic Maps with Spcial Reference to Determination of Structure*, New York: McGraw Hill.

Dwerryhouse, Arthur Richard (1911), *Geological and Topographical Maps, Their Interpretation and Use: A Handbook for Geologists and Civil Engineers*, London: Edward Arnold.

Harvey, P.D.A. (1980), *The History of Topographical Maps: Symbols, Pictures and Surveys*, London: Thames and Hudson.

Kraak, M.J. and F. Ormeling (1996), *Cartography: Visualization of Spatial Data*, London: Longman.

Salisbury, R.D. and W.W. Atwood (1908), *The Interpretation of Topographic Maps*, USGS Professonal Paper No. 60.

US Military Academy (1984), *Atlas of Land Forms*, New York: John Wiley & Sons.

11
Cadastral Maps and their Uses

What is a cadastre? According to the International Federation of Surveyors a cadastre is normally a land information system containing a record of interests (e.g. rights, restrictions and responsibilities) in each parcel of land. It includes length, breadth, area, ownership or control, value and the value and changes made therein. Each parcel is given a unique code or number and its exact location is recorded on a plan or map.

Cadastral maps are the graphic records of cadastral surveys; they show the relative location of all parcels of land in a given area or region on a plane paper or cloth. They are made on scales ranging from 1:500 to 1:10,000. Larger scale maps and diagrams showing more precise dimensions and features of the parcels are done by cadastral surveys for each parcel. These surveys may employ traditional surveying methods or depend on aerial surveying. With the help of the unique parcel codes or numbers, information on land value, ownership, or use, can be accessed.

Cadastral systems are not ends in themselves. They are created to support the land tenure systems which protect land rights, and support effective land use and management. They help in buying, selling, mortgaging, and leasing land rights. If these broad administrative, social and economic criteria are met, the system should be considered successful. It has to be kept in view that the system should not be too complicated for the common man who at times may not even be literate as is often the case in India.

Cadastral maps are used in all the countries of the world because they offer details of the land owned by different people and institutions and use to which it is put and can be put. Agricultural land in most countries cannot be put to non-agricultural uses without government clearance. Similarly, forestland cannot be put to any other use unless the change in the land use is permitted by relevant authorities. Cadastral maps are used to demarcate ownership boundaries and to settle other land ownership disputes.

In the context of India, a simple example of a cadastral map might be a map of a village, which shows the boundary of all the plots and lots within the village. Cadastral maps are also indispensable in city planning and land use planning. These maps are usually created and maintained by the government as a public record; anyone who wishes to go to the office which maintains the records can see or purchase a copy.

Apart from giving the number of each parcel of land and the name(s) of its owner(s), it also gives the kinds of use it is put to and the structures, if any, built on that. Detailed information about location of the parcels of land in the area is a key feature of a cadastral map. In many countries, cadastral maps use GPS locations as well, so that people clearly understand where everything on the map is in the real world. It also highlights the specific landmarks which people can use to orient themselves within the map, including buildings and natural features such as lakes and streams. Cadastral maps also show the details of the ownership of mines and forests. It gives the land use and history of the ownership.

In most countries, legal systems have developed around the original administrative systems and use the cadastre to define the dimensions and location of land parcels described in legal documentation. The cadastre is a fundamental source of data in disputes and lawsuits between landowners. In most countries of the world, an organization similar to the Cadastral Survey in the USA maintains records of all public lands. Such surveys often require detailed investigation of the history of land use, legal accounts, and other documents.

NATIONAL POLICY ON LAND RECORDS

Maintenance of proper land records has been emphasized in India ever since 1950 when the First Five-Year Plan was launched to develop the country. During the Seventh Five-Year Plan (1985-89), it was decided that 'the land records form the base for all land reform measures' and, therefore, regular periodical updating of the land records is essential in all States. This will necessarily have to include the scientific survey of unmeasured land and the recording rights of tenants and sharecroppers which have remained unrecorded until now.

Latest ICT for quicker storing, processing and retrieval of information database in land records have now been compulsory. During the last two decades the NIC has started computerization the of land records of the country by storing the records in simple file based database. Each State of the country has a computerized land records directorate with its office in each district. At the Tehsil/taluk levels too a Lekhpal/*Patwari* has been trained to maintain land records. The introduction of the computers has, however, enhanced the land records being tempered especially because the Lekhpals/Patwaris are known to be corrupt. There is a need to have well trained personnel in handling Relational Data Base Management System (RDBMS) and Geographic Informations System (GIS) packages at local levels. A number of organizations like DST, IITs, and NIC etc. are engaged in developing these packages.

The systematic land survey and land records are crucial for proper management of land and other natural resources. One of the major factors for delay in execution of land related projects is the lack of proper and correct information. There is a need to adopt modern methods of surveying and mapping to generate maps on a uniform scale and associated system of land records for the whole country.

The time has now come for the country to adopt latest technology for collecting cadastral data needs no emphasis. Computerization of land records and their updating in a consistent format is also needed for both macro and micro planning not only at the village level but also at Block and District levels. The government must open some training institutes to train local level officials in GIS, GPS, and LIS. Further, there is need to create national standards for cadastral surveys, methods, equipment, scale, data measurement, data structure, symbology, accuracy, and data exchange format. The cadastral map data base should integrated with national datum so that the individual land parcel and the rights of the land holders in the parcel get prime focus in all developmental activities launched by government or private agencies.

India is one of the largest and economically fastest developing countries of the world. Further, it is an agricultural country where two-thirds of its people eke their living from land. The cadastral system of the country requires that all land deeds be registered, the registration of land ownership is also compulsory. Let us have a look at the cadastral statistics of India. This would indicate how big is the task of maintaining records and training the professionals for making maps and keeping records.

Population of India
1020 million (projected for 2011)
Population Distribution
Rural 70 per cent; Urban 30 per cent
Total No of land parcels
210 million; 204500 land parcels for 1 million population

Total Number of Registered Strata Titles or Condominium Units

41 million
i.e. 39,900 strata titles per 1 million population

Land Parcels in Urban Areas

- Properly registered and surveyed: 99 per cent
- Legally occupied, but not registered or surveyed: 1 per cent
- Informally occupied without legal title: 0 per cent

Land Parcels in Rural Areas

- Properly registered and surveyed: 99 per cent.
- Legally occupied, but not registered or surveyed: 1 per cent
- Informaly occupied without legal title: 0 per cent

Number of Active Professional Land Surveyors

Total: 30,000
Proportion of time committed for cadastral matters: 30 per cent.
i.e. 9,000 full-time equivalents committed to cadastral matters
i.e. 8.8 full-time equivalents to per 1 million population

Number of Active Lawyers/Solicitors

Total: 100,000
Proportion of time committed for cadastral matters: 30 per cent
i.e. 30,000 full-time equivalents committed to cadastral matters
i.e. 29.2 full-time equivalents to per 1 million population

CADASTRAL SURVEYS

History of Cadastral Surveys

Cadastral surveys form the base element in Geographic (GIS) and Land Information Systems (LIS) used to assess and manage land and built infrastructures of various kinds. These systems are also needed to track the changes taking place in the land ownership and use. These surveys document the boundaries of land ownership through written documents, diagrams, sketches, plans, charts, and maps. They were originally planned to help the government levy taxes on the owners. In England, there was the Domesday Book; in France, Napoleon set up a detailed cadastral system now regarded as the forerunner of most modern versions. The Public Lands Survey System had developed in India even in ancient times but dependable records of it are there since Akbar's time (1556-1605). In the USA, it began in 1785 in the western parts of the country. In Canada, the Dominion Land survey was launched 1871.

History of land records is perhaps as old as the history of human kind, cadastral surveys and related maps are recent developments. Even in ancient civilizations like China, modern cadastre began only in 1930s. Australia was inhabited for over 40,000 years by the Aboriginal people their land rights did not gain legal recognition until 1993 when the Indigenous Native Titles Act came into force. Incidentally, the Australian cadastral system was not influenced by the land ownership systems or patterns of the indigenous peoples. Instead it was based on the English system of deeds registration for land transfer. As such the original inhabitants of Australia did not get the rights over their lands.

The King of Sweden established cadastral books for taxation purposes as early as 1553. These books list real properties of every village and give every unit a permanent number. This numbering system is still used for designation of real properties. During the 17th century village cadastral maps, showing all real parcels in the village were made part of cadastral books. These maps gave the area and value of each land parcel; they improved the taxation of land and make it more just and equal. National Land

Survey Organization (Lantmäteriet) was created in 1628 by the government to further improve this work.

Following the English and Western European systems, the British initiated revenue surveys in India nearly two century back. The Survey of India was established in 1767 as a Department of Government of India. It was the main agency for revenue surveying till 1904 when provinces (now states) were made responsible for cadastral surveys. Each state has evolved its own legal system of cadastral surveys for revenue collection. This separation of functions between the Centre and State governments has come in the way of modernization of cadastral surveys in India. Even today it has not taken full advantage of modern surveying, mapping and information technology.

We just do not know when was the first cadastral and revenue survey map prepared? The inscriptions and copper plates of donations made by Pallava and Chola Kings (9th to 13th century AD) show that their kingdoms had a widespread network of villages with carefully demarcations fields and land rights and their precise boundaries correct to virals. This could not have been done unless there were some sort of cadastral plans and maps of the fields in relation to visible features like rivers, wells, canals and tanks. May be, these plans and maps are now lost as they were made on palm leafs. Yet there are clear evidences of the existence of such maps. Indian geographers and cartographers should have a fresh look at the contribution of Indians to the field of cartography in ancient and medieval times and try to find out how people and governments decided on land rights.

The map given in Fig. 11.1 shows the gifted devadana lands to the Siva and Vishnu temples and the natural environment around on the southern bank of River Pennar close to Tirukoilur in Tamil Nadu. The map has been copied from an inscription. The plots of farmlands were parceled out during the Chola period but imprinted in the later day British cadastral plans. A number of Moghal period cadastral maps have been located by AMU historian Irfan Habib. Drawing his data from Abul Fazl's *Ain-I-Akbaree*, Habib mapped the revenue villages of Akbar's *subas* (provinces) as defined by Raja Todarmal on the present day map format. The Moghal Atlas is an authentic land record of Akbar's times.

Similarly, maps of the Peshwa period reflect the land reforms carried out by Chhatrapati Shivaji. It was difficult indeed to locate villages in the forested hills of the Sahyadri but the Maratha cartography was advanced enough to locate them in their graphic maps of valley heads. In Nepal, a cadastral map dated AD 1830 shows distinct boundaries of plots and their revenue estimates. Similarly, maps of the Peshwa period reflect the land reforms carried out by Chhatrapati Shivaji. It was difficult indeed to locate villages in the forested hills of the Sahyadri but the Maratha cartography was advanced enough to locate them in their graphic maps of valley heads. In Nepal, a cadastral map dated AD 1830 shows distinct boundaries of plots and their revenue estimates

Below, reduced lacsmile at a held map of a *** village (manua) in Saran D., Bihar-prepared by an indian measure (**) under the direction of a professional surveyor, c. 1845 Syntamatic revue surveys, beginnin in 1822, resulted in similar large scale cadastral maps for most of British hand. These maps, periodically revised, formed the basis for sand revenue assessments. From Phillimerem *op cit.* vol IV 1830 to 1842, map lacing p. 176

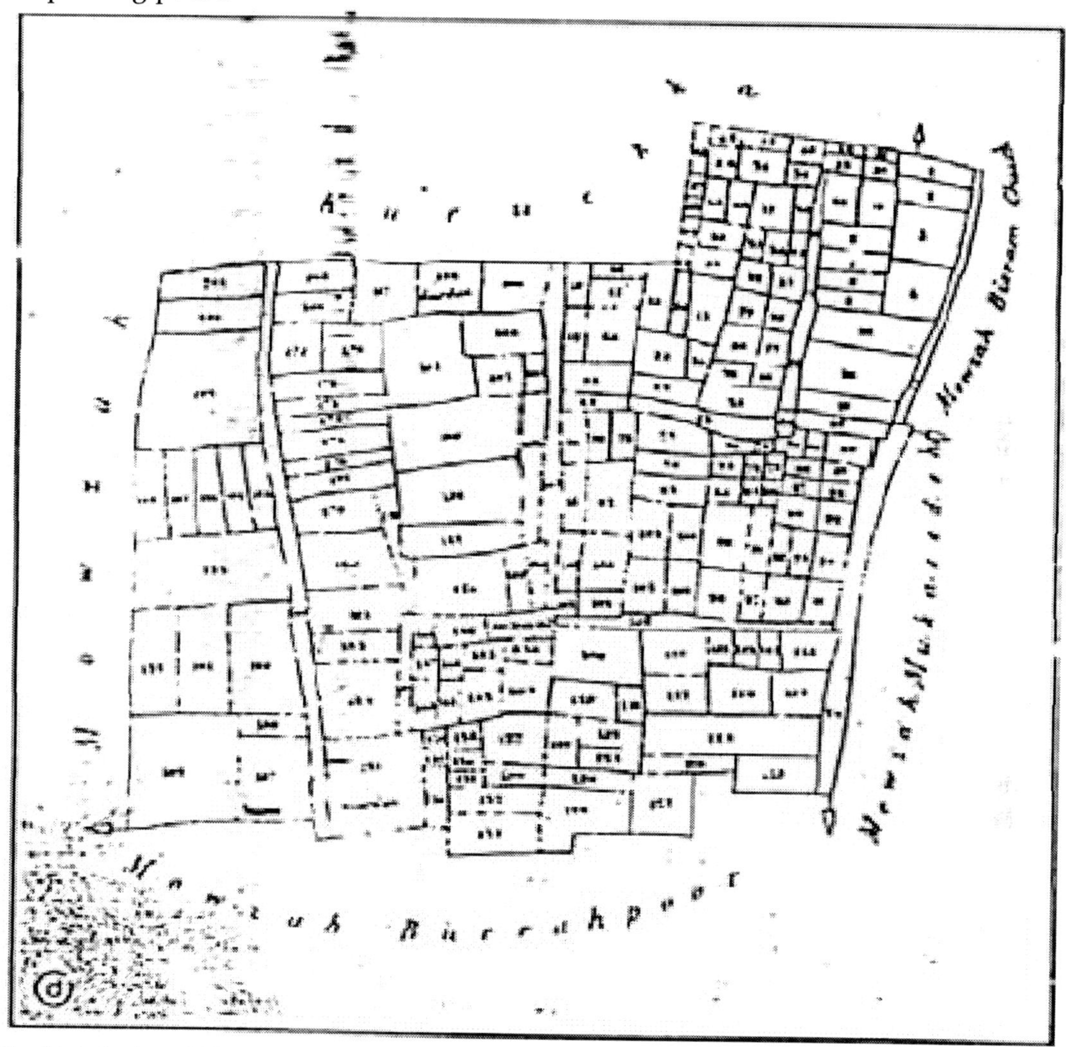

Fig. 11.1: Cadastral Map of Land Gifted to a Devasthanam in Chola Period (9th to 13th Century)

The brief analysis of the state of art with regard to cadastral mapping adequately proves that pre-British India had its own systems of cadastral and revenue mapping. What has so far come to light is only but a small fraction of the total knowledge still hidden or lost forever.

When the British launched large and medium scale surveys, they took to precise measurement of distance and directions with the help of Gunther's chain and tape and taking measures along sides and diagonals. Rennell, Dalirymple, Robert Kelly, Buchanan and others did pioneering work in this regard. These surveys were carried by trained local surveyors and collectors and the data gathered were sent to Calcutta. Bombay, and Madras. Since, the surveyors were appointed by the London office and the cartographers at central office had little control on data sent to them led to great disparities in survey standards. The East India Company management was aware of the problem but did not know how to resolve it. Many manuscript maps were lost and field officers did not get maps.

Calcutta being the national capital, work on land measurement and assessment of revenues began in Bengal first. To make the work easier, it was decided to adopt the method used by the Bengal zamindars on a permanent basis. That permanent settlement prevails even today in eastern India. The only areas excluded from it were the newly reclaimed lands in Sunderbans and 24-Parganas. In Madras Presidency, even the *jagirs* were small in area and the land holders were mainly of a *ryotwari* system. A different guideline was established and the same was extended to Bombay Presidency in 1772. In this system the settlement was a temporary one with opportunities for constant reviews from time to time. Its data base was highly comprehensive; it included the plot size, soil and its taram, slope, water access, crops seasonally, ownership, etc. The Madras Adangal system of land recording is being promoted all over the country.

Survey of India as a Department of Government of India was established in 1767. It was fully involved in the process of revenue surveying till 1904. It gave a new shape to the primary and secondary triangulation networks with minimal linear measures and precise angular measures to cover the whole country with a network of interconnected triangles. The system, however, was built from the top. It did not touch the revenue and cadastral survey system in the country. In due course of time, the Survey of India distanced itself from cadastral survey and mapping leaving it in the hands of each presidency/province of India for cadastral surveys. It evolved its own legal system of cadastral survey for revenue collection.

The process of Cadastral Survey in India has not yet taken full advantage of modern technology available in surveying and mapping. The colonial rulers in India initiated revenue surveys two century ago, which was based on the systems adopted in England and Western Europe. This survey was started with a view of collection of revenue for estates. That system continues to this day. The result is cartographic anarchy. Most of the civil court cases in India pertain to land ownership. There is hardly a parcel of land whose cadastral map is correct.

The available cadastral maps in all states and union territories of India are not necessarily based on field surveys; they are graphical representation of the land use. They do not follow any systematic scale or datum. As such they are not comparable. For the last one decade or so India has been engaged in computerization of land records of the country by storing the records in file-based database.

The rapid pace of development in cities as well as villages calls for an effective land information system. This information system is essential for optimal use of the available resources and also to protect the environment. Accurate large-scale maps in digital form for generation of GIS/LIS are in great demand. In India mapping is being performed by Survey of India. The maps on scales 1:50,000 are available with Survey of India (SoI) for entire country. Large scales maps are also being prepared by SoI for some areas on demand basis.

It is now being realized that the triangulation network should not only be linked to the topographical map grid, but also to the cadastral plans. The Survey of India too is now trying to come to terms with grassroot level realities and linkages. Problems are many: the grid, the projection for plane table level survey plan, the search for control points to merge the two, and many others. On the cadastral plans side, shrinkage due to age of the old handmade paper, subsequent plot level changes and what is worse changes in the physical landscapes by way of erosion or accretion. The job is gigantic and challenging indeed.

INSTITUTIONS RESPONSIBLE FOR CADASTRAL SURVEYS

Survey of India

As stated earlier, the management of the East India Company was aware of the problem, but it did not know how to resolve it. Many manuscript maps were lost while the field officers did not get the required maps. Calcutta being the imperial capital, work on land measurement and assessment of revenues began in Bengal first. It followed the method used by the Bengal zamindars on a permanent basis. Later, it came to be known as permanent settlement. The only areas excluded from it were the newly reclaimed lands in Sunderbans and 24-Parganas. As against this, in Madras Presidency, even the *jagirs* were small in area and the land holders were mainly ryots. The system came to be known as Adangal or *ryotwari*. It was a temporary settlement with a built-in provision for reviews from time to time. Its data base was highly comprehensive; it included the plot size, soil and its taram, slope, water access, crops seasonally, ownership, etc. It was extended to Bombay Presidency in 1772. The Adangal system of land recording is being promoted all over the country.

Survey of India as a Department of Government of India was established in 1767. It was fully involved in the process of revenue surveying till 1904. As a Central Government Organization, Survey of India was involved in cadastral surveys in India until 1904. Survey of India is one national mapping agency in the country to provide topographical maps for entire country on various scales. It gave a new shape to the primary and secondary triangulation networks with minimal linear measures and precise angular measures to cover the whole country with a network of interconnected triangles. The system, however, was built from the top. It did not touch the revenue and cadastral survey system in the country. In due course of time, the Survey of India distanced itself from cadastral survey and mapping leaving it in the hands of each presidency/province of India for cadastral surveys. It evolved its own legal system of cadastral survey for revenue collection.

Survey of India is the sole national mapping agency in the country for topographical mapping of the entire country on various scales. After 1904, the responsibility of cadastral surveys was assigned to state governments. However, Survey of India is providing technological know how to state government in the matter of cadastral surveys. Each state is adopting its own methods of cadastral surveys and mainly emphasis is given to the legal aspects, which are connected with rights, *khasara*, *khatauni*, etc.

Cadastral system in India is built upon the 'cadastre'. The cadastre is a parcel based and up-to-date land information system. The record consisting of:

1. Number of each parcel of land (Fig. 11.2),
2. Record of ownership, and
3. Map of land parcels linked to other records describing the nature of the rights or interest in the respective parcels.

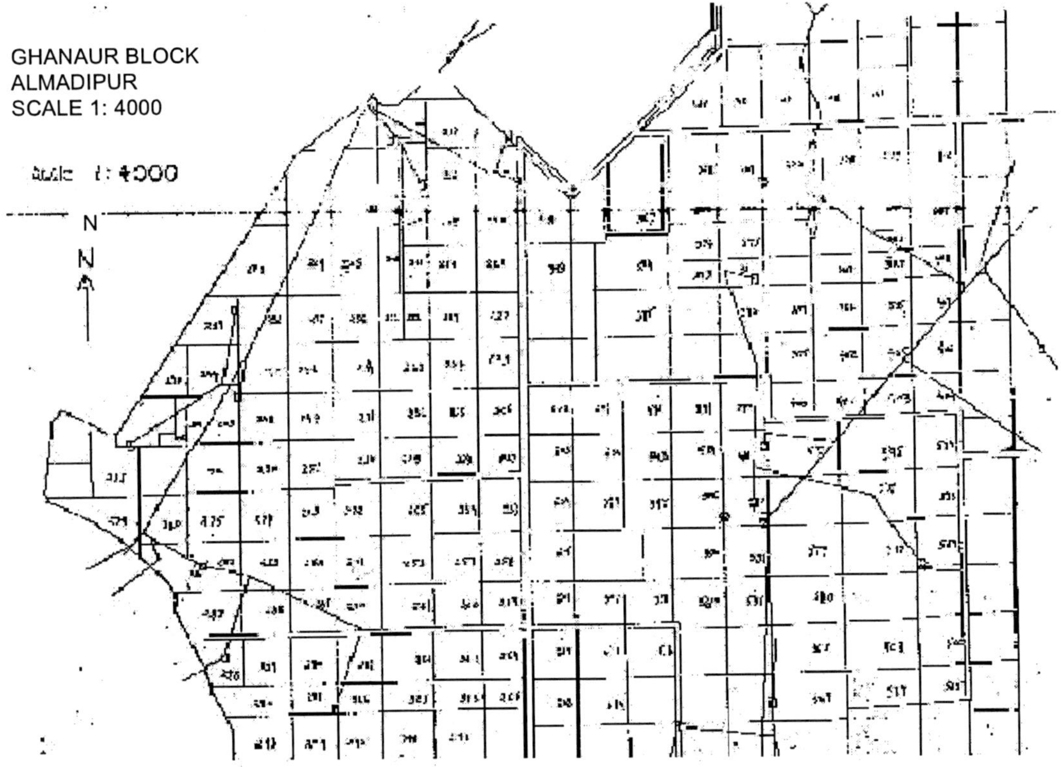

Fig. 11.2: Cadastral Map

Private and Professional Organizations

The private sector is now getting involved generating GIS and LIS for better management of land resources. There are professional organizations in India, which are involved in surveying and mapping. Survey of India is one national mapping agency in the country to provide topographical maps for entire country on various scales. The department is also generating land information system for the states on demand basis. Among the other organizations which assist the state governments in developing LIS/GIS are NRSA, NIC, SAC, IIT's, etc.

The private sector is now getting involved and generating GIS and LIS for better management of land resources. Further, a number of professional organizations too are involved in surveying and mapping. Survey of India continues generates land information system for the states on demand basis. Organizations like NRSA, NIC, SAC, IIT's, etc. too assist state governments in developing LIS/GIS.

MANAGEMENT OF LAND RECORDS IN INDIA

Cadastral system in India relating to land and land dispute comes under the charge of *Tahsildar, Kanoongo* and *Patwari* (Lekhpal). Each district is divided into tahsils and every tahsil comprises of pargana and villages. The classification of agricultural land, its measurements, the assessment and its products are properly maintained in the land record register.

The current cadastral maps in all states and union territories of the country are usually based on graphical representation. These maps are not based on any systematic scale or datum. During the last decade (2001-2010), India started computerization of land records by storing the records in file-based data storage.

The rapid pace of urbanization especially the growth of large cities, has necessitated the development of an effective land information system for land use planning with a view to better utilize it's resources while at the same time protect the environment. This has necessitated the preparation of accurate large-scale maps in digital form for generation of GIS/LIS.

In India the task of mapping is being performed by the Survey of India (SoI). Maps on the scale 1:50k are available with SoI for the entire country. Large scale maps are being prepared by SoI for some areas on demand. It is now possible to prepare vector map of digital database for cadastral application using GPS and GRAM++ software.

The cadastral graphical map of the revenue department is scanned and georeferenced using GPS control points. The GPS survey is used to pick up all the resources available on ground, and transferred to database. Tabular data i.e. names, facilities like buildings, tube-wells, schools, religious places, roads, and other facilities can then be incorporated in the data base.

Role of Cadastral Layer in Spatial Data Infrastructure (SDI)

Spatial Data Infrastructure (SDI) is one of the solutions to meet the needs of the nation because of its capability in identifying the key factors that facilitate development through better understanding the complexity of the interaction between social, economical and political issues. Main objective of SDI is to disseminate, utilize and manage spatial information for all activities related to land and its resources.

Future of Cadastral System

As the country moves forward economically and technologically, the importance of natural resources tends to increase. It is specially so in a country where nearly 60 per cent of the people depend on agriculture for their livelihood. Further, as the importance of cities as engines of economic growth increases, the importance of urban planning and development to accommodate people and various developmental activities further increases.

The cadastral surveys are invariably linked with ownership of the property. The cadastral systems support the legal ownership of the land by depicting legally created area and parcel boundaries. It becomes more important if it pertains to urban/town areas. As development activities increase, the system helps administer the land related and land based activities. Policy-makers, resource planners and administrators who make decision about the land also depend on cadastral maps. They need detailed land information than has been traditionally available. Recent land scams in Mumbai, Noida and elsewhere have resulted from tempering in the cadastral system.

The basic unit of the cadastral record is the land parcel, which is known as plot. The plots are registered in the name of the owner defining plot boundaries and number. Maintenance of land records with ownership; assessment of tax on the property and collection of tax is the responsibility of district authorities. The cadastral surveys are conducted periodically to locate and demarcate land boundaries on ground and map and then to facilitate management of land parcel information. Land records in India consist of cadastral survey and mapping, recording cadastral information, land evaluation, and land parceling. Unfortunately, the cadastral maps are fraudulently tempered with leading to property disputes and litigation.

LAND USE PLANNING

Not all the countries of the world are contemplating immediate reforms in their agrarian structure but almost every country in the world has been making strenuous efforts to improve its agriculture and rural economy. Given the current growth rate of population and the rural-urban distribution of it, the only hope for human survival and development rests in progressive improvement in the efficiency of agriculture. This is possible only if living conditions of the agriculturists improves. Food security has become a major global as well as local issue. Further, the quality of environment depends largely on land use. Thus land management has become a critical factor in development.

Most countries are naturally giving a fresh look at the use and misuse of their natural resources. Rural development has therefore got a new boost in recent years. Apart from land reforms, there are schemes for increasing agricultural production by augmenting irrigation, drainage or flood control, preventing erosion or conserving forests. No matter what the development programmes or projects, its successful completion will demand accurate understanding of the natural and human resources, actual as well as potential; and knowledge of the manner in which these resources are being used, misused or neglected.

Large-scale maps and exact knowledge of the way in which the land is held provide the basic information about natural resources. The cadastral maps belong to this category of knowledge source. Large scale maps are a must for executing irrigation works, flood control or protection works, the construction of railways, roads, canals and ports, pipelines and many urban purposes such as town planning and sewerage. Cadastral maps which show field boundaries are especially suitable for mapping land use and in classifying land on the basis of soil capabilities or productivity.

Cadastral maps are also of great use for many administrative purposes. The assessment of any tax based on area (as the most satisfactory taxes on agricultural land usually are), requires such maps, and indeed this use is the primary purpose of cadastral maps in the technical sense. Cadastral maps make possible the precise plotting of annual or seasonal variations in the area supplied with water from irrigation canals, and thus facilitate calculations of the actual amount of water supplied to any farmer or field. They also allow the effects of changes in water supply on crops to be clearly and accurately recorded, and permit a close control over water distribution to ensure the optimum available supply.

The use of maps to record annual crop statistics is only too common to repeat it here. The countries which do have accurate mapping system fail to estimate their annual crop yields. Measurement of areas is important and the cadastral surveys give precise figures for farm areas, the data will remain defective. Cadastral maps can also be used as base maps for classification of the productivity of the land where there are considerable local variations or to facilitate direct sampling of annual yields.

ENVIRONMENTAL PLANNING

Environmental issues have acquired added importance in recent decades. Global warming and climate change has become a major issue at the UN for as well as national fora. Planned development of the country has so far not taken note of the environment. Often, it is considered to be a cost the country and its people must pay for economic development. Now the Global warming knocking at our door, our attitude must change. While economic development in of country where at least 400 million people are below the poverty line but it can no longer be at the cost of the welfare of our children and grand children. Economic planning must be accompanied by or integrated in environmental planning.

Environmental planning needs data on land, water, forest and other natural resources. Each of these resources has to be preserved and further enriched. Economic development should not allowed to destroy our land, soil, water channels and bodies, forests, wild life and other natural endowments without compensatory programmes to redevelop them.

It is here that cadastral surveys and maps acquire new significance. Each village and city of the country should have an environment map setting aside open land, water bodies, play grounds, parks, public utilities, schools, hospitals, etc. While registering land in the name of any individual, the future need of the community must be kept in view.

There complete anarchy in land registration in India. Corrupt officers in connivance with village *pradhan/mukhia*, etc. have left no land even for houses and trees. Unless immediate steps are taken to check this menace, the villages will be destroyed forever.

The process of cadastral survey in India has not yet taken full advantage of modern technology available in surveying and mapping. The colonial rulers in India initiated revenue surveys two century ago, which was based on the systems adopted in England and Western Europe. This survey was started with a view of collection of revenue for estates. That system continues to this day. The result is cartographic anarchy. Most of the civil court cases in India pertain to land ownership. There is hardly a parcel of land whose cadastral map is correct.

The available cadastral maps in all States and Union Territories of India are not necessarily based on field surveys; they are graphical representation of the land use. They do not follow any systematic scale or datum. As such they are not comparable. For the last one decade or so India has been engaged in computerization of land records of the country by storing the records in file-based database.

The rapid pace of development in cities as well as villages calls for an effective land information system. This information system is essential for optimal use of the available resources and also to protect the environment. Accurate large-scale maps in digital form for generation of GIS/LIS are in great demand. In India mapping is being performed by Survey of India (SoI). The maps on scales 1:50,000 are available with SoI for entire country. Large scales maps are also being prepared by SoI for some areas on demand basis.

It is now being realized that the triangulation network should not only be linked to the topographical map grid, but also to the cadastral plans. The survey of India too is now trying to come to terms with grass root level realities and linkages. Problems are many: the grid, the projection for plane table level survey plan, the search for control points to merge the two, and many others. On the cadastral plans side, shrinkage due to age of the old handmade paper, subsequent plot level changes and what is worse changes in the physical landscapes by way of erosion or accretion. The job is gigantic and challenging indeed.

APPENDIX

CADASTRAL STATISTICS OF INDIA

1.	**Population (2001):**	1027 million: 72.5 per cent Rural; 27.5 per cent Urban
2.	**Total Number of Land Parcels:**	210 million 204,500 land parcels per 1 million population
3.	**Total No. of Registered Strata Titles or Condominium Units:**	41 million 39,900 strata titles per 1 million population

(Contd.)

4.	**Land Parcels in Urban Areas:**	Properly registered and surveyed: 99 per cent legally occupied, but not registered or surveyed: 1 per cent informally occupied without legal title: 0 per cent
5.	**Land Parcels in Rural Areas:**	Properly registered and surveyed: 99 per cent legally occupied, but not registered or surveyed: 1 per cent informally occupied without legal title: 0 per cent.
6.	**Number of Active Professional Land Surveyors:**	Total 30,000; Per cent time committed for cadastral matters: 30 per cent i.e. 9,000 full-time surveyors committed to cadastral matters i.e. 8.8 full-time surveyors per 1 million population
7.	**Number of Active Lawyers/Solicitors**	Total: 100,000 Per cent time committed for cadastral matters: 30 per cent 30,000 full-time equivalents committed to cadastral matters i.e. 29.2 per 1 million population

Note: The above is given the official figures. Ground realities are different especially with regard to informally occupied without legal title.

REFERENCES

Behl, Brig, K.G. (1995), *Use of GPS for Cadastral Surveys*, New Delhi: Indian National Cartographic Association (INCA)

Behl, Brig, K.G. (1996), *Use of GPS and New Technologies in Rural Development: An Overview*, New Delhi: INCA.

Behl, Brig, K.G. (1999), *Modern Technology in Cadastral/Revenue Mapping for Panchayatiraj Administration*, New Delhi: INCA.

Hendrix, Steven E. and Logan Ronald, A. (1996), Cadastral and Property Registry Modernization: Ideas on Public-Private Partnership, *GEOMATICA*, Vol. 50, No. 1, Canada.

Kaufman, J. and Slendler, D. (1998), *Cadastre 2014: A Vision for a Future Cadastral System*, Working Group I, FIG Commission 7, FIG.

Williamson, I. (1997), The Justification of Cadastral System in Developing Country, *GEOMATICA*, Vol 51, No. 1, Canada.

12
Aerial Surveying and Air Photo Interpretation

The history of aerial photography is less than a century old. The first reliable black and white infrared film was developed in Germany in 1903. The film emulsion was adjusted to be sensitive to wavelengths of energy just slightly longer than red light and just beyond the range of the human eye. By the 1930s, black and white infrared (IR) films were being used for landform studies, and from 1930 to 1932 the National Geographic Society sponsored a series of IR photographs shot from hot air balloons. Throughout the 1930s and 1940s, the military was hard at work developing colour infrared film for surveillance. By the early 1940s, it developed a film that was able to distinguish camouflaged equipment from surrounding vegetation. Within months, however, an IR reflecting paint was developed for use on military vehicles, effectively making IR film technology useless to the military. So, they dropped it.

The scope of air photography expanded fast and it continues to be an indispensable aid in Land-Use Planning and Mapping, Geologic Mapping, Archaeology, Vegetation Mapping, and Species Habitat Mapping, among others. Recent integration of Aerial Photography into GIS has further enhanced its value and importance. Although satellite imageries are now a better source of ground information, air photographs have not become redundant. They continued to be used in a variety of military and civil activities.

The first commercial companies to do aerial survey came into existence in the early 1920s. Since, then lot of experiments were made in image formatting, camera lens and films. However, today for all practical purposes 9x9 inch format camera with 150 mm focal length lens is preferred. For instance, lenses with focal lengths up to 300 mm are used for large scale urban mapping. There are also multi-spectral cameras in which the lenses have very inferior geometric proportions, but we get the spectral differentiation of the images. However, this is not used for topographic mapping purposes. Therefore, multi-spectral camera is mounted in the aircraft in conjunction with a survey camera.

Normally monochrome films are used in aerial photography because it is simple to process and does not need extremely rigorous laboratory facilities, which enables the film to be developed and checked under field conditions, which are often very far from ideal. This film is available in a number of types and speeds, which allow adequate photography to be taken over a wide range of weather and lighting conditions.

Colour films are expensive for buying and processing. The exposed film has to be sent to a fully equipped laboratory, which means, the photographic cover cannot be checked within hours of being taken. Of course, there are advantages also. For instance in urban areas with ordinary film it is difficult to differentiate the details, because of the prevalence of gray shade everywhere, whereas in colour it is easy to differentiate these details. There is yet another film known as "False colour infrared" film. Its main advantage is the delineation of the exact limits of water, especially in places of shallow water where it is often almost impossible to differentiate between water and wet mud or sand on monochrome photography.

IDEAL CONDITIONS FOR AERIAL PHOTOGRAPHY

Successful aerial surveying requires two favourable conditions. Clear air and bright sunshine is the first condition. Hygroscopic nuclei collect around smoke and dirt in the air to create a haze. A dust haze which is often found in relatively drier areas also affects the photographs because the camera picks up the dust particles as shades on the negatives. Clear air gives sharper prints. As will be explained later, shadows appearing on photographs give clues for the identification of features. Bright sunshine gives better shadows so necessary to photo interpretation. It must, however, be kept in view that with the help of some of the latest techniques now available, air photographs can be taken at night even.

The second condition which should be satisfactorily met is the proper functioning of the aircraft, the camera and the photographic materials used. Proper functioning of the camera obviously means that the proficient cameraman can expect to reproduce an image of high quality if the film is as good as his camera. If the air surveying is done in a war torn area and the plane is being shot at, the pilot cannot fly the plane at a constant level. The photos taken under such unstable conditions are not uniform in scale, tone and normal distortion. It is said that during the second world was 85 per cent of the photos had to be discarded.

Flight Procedure

The pilot must have a flight map which guides his flight lines. A flight map gives the location of the points and area to be covered. If a flight map is not available, reconnaissance flights have to be made to get control points on the ground. For successful air surveying, ground control points determined by geodetic surveying are essential. If such control points do not exist, survey parties may have to supply the necessary control before aerial operations begin.

As far as possible the flights should be flown along lines for which ground controls are in existence or which are marked by the presence of such prominent features like rivers, ridges or valleys. The most common pattern of flying for air photographs is called "serpentine pattern". This pattern is shown in Fig. 12.1. The purpose behind this pattern is to make the flight lines overlap. Overlap is necessary to produce tie between adjacent photos. The side overlap is normally 20-30 per cent and the forward overlap is from 60-80 per cent. The side overlap ensures better lateral matching among flight strips whereas the forward overlap compensates for the optical inaccuracies on each photo and permits multiplex or stereoscopic interpretation of the photographs.

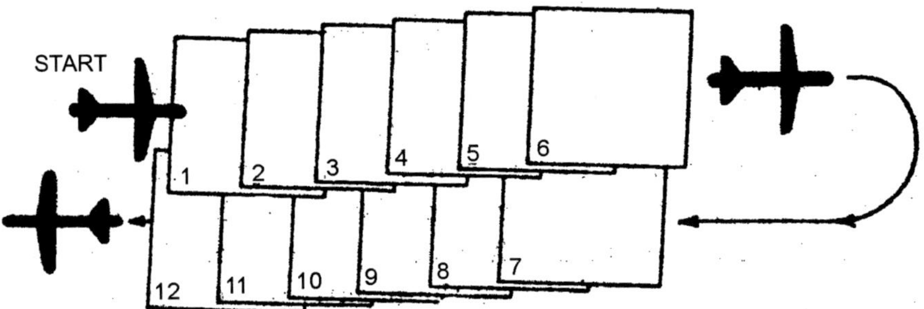

Fig. 12.1: Serpentine Pattern of Air Surveying

Aerial surveying became a major source of cartographic information since World War II. The aerial surveyors photograph the land surface and these photographs are then used by cartographers, planners, and engineers take detailed measurements from them to determine the physical and cultural geographic conditions of the part of the earth surface photographed. The final outcome of aerial surveying is a set of photographs which when interpreted become a reliable source of map information.

CHARACTERISTICS OF AIR PHOTOGRAPHS

An aerial photograph is any photograph taken from an airborne vehicle (aircraft, drones, balloons, satellites, and so forth). It is not map; it only a source material for making a map. Maps are orthogonal images of the earth's surface; they are accurate geometrically and directionally within the limitations imposed by the conversion of 3-dimensional globe into 2-dimensional plane surface. Aerial photos have another limitation. They are marked with a high degree of topographic distortions, and until these distortions are corrected, information drawn from an aerial photograph is not accurate. Now we have the tools and methods like GIS to make these corrections. And thus, aerial photographs have become a powerful tool for studying the earth's environment.

Maps and aerial photographs complement each other. A topographic map becomes obsolete with time; a recent aerial photograph shows not only what is there on the earth surface but also what changes have taken place since the map was compiled. For this reason, more and better information can be gained by using the two together than by using either alone.

Aerial photographs give the eye view of the actual area photographed from above. They can be used in conjunction with maps of the area photographed to supplement the maps with the newest details. They can also be used as source materials for preparing new maps. The greatest advantage of air photography is the amount of details it gives at such a short notice. It is for this reason that it has acquired so much importance both for military and civil planning.

The effective use of an air photograph depends upon its correct interpretation. Air photographs have a far greater range of scenery possibilities than maps. Anything that appears on the landscape is reflected on the film. Often, this means too much detail or obscuring of required details. Roads get lost under a canopy of foliage when leaves are thick upon the trees. Railroad tracks disappear into mountain sides. Houses appear reclining and not vertically erect. Heavy shadows mantle vegetation and many minor details and thus confuse the patterns made by man and nature. Such confusion can be considerably reduced by scientific interpretation of air photographs.

Advantages

An aerial photograph has the following advantages over a map:

1. It gives the latest pictorial view of the ground; map gives a generalized image of what was there earlier;
2. It can be secured within a short time; map making is time consuming;
3. Even inaccessible areas can be photographed and hence it is Used by ground soldiers;
4. It can show even the areas such defense establishments as military cantonments for which maps are not available;
5. It is of great help to armed forces and police as it gives latest visual information of even areas otherwise inaccessible; and
6. It is the best way to get a permanent and objective record of day-to-day changes in a given area.

Disadvantages

In some respects, an aerial photograph is inferior to a map:

1. It requires specialized training to interpret.
2. It does not have marginal data and, therefore, cannot be contextualized spontaneously.
3. In the absence symbols, the ground features are difficult to identify and interpret; some ground details are even obscured by trees etc., which form a canopy above the ground.
4. Exact location and scale of the areas photographed are approximate.
5. Unless the photography is overlapping and viewing is Stereoscopic, details of variations in the terrain features are not easily discernible.

6. It is difficult to do photography in poor light as the colours and tones in the photographs are not contrasting.

TYPES OF AIR PHOTOGRAPHS

Air photographs are taken in three ways. The conventional aerial photography is the main subject matter of this chapter. But there are two more types which one must know. The first is the Side Looking Radar Imagery (SLRI) which is an alternative to conventional aerial photography in those areas of the world where extensive cloud cover is persistent. It is expensive, but extremely quick to obtain and is unaffected by weather conditions. Here, the pulsed energy is beamed obliquely down towards the ground on one side of the survey aircraft. This beam scans in a vertical place over an angle of about fifty degrees. Reflections are recorded either on films or on magnetic tapes. Because the system is self-illuminating the operation is not affected by lighting conditions. Thus the imagery can be taken by day and night. Illumination of the ground from one side alone is another advantage. The shadow effect enhances the impression of land relief in the imagery and is invaluable in assisting detail interpretation. Though radar imagery can give reasonably good delineation of areas, the problem is in knowing what is contained within these areas, because of the different reflective proportion of the ground. Thus this may be a useful tool for the preparation of maps, but it cannot supply all the necessary details.

The other is the High-altitude Aerial photography which is resorted to only for military purposes and normally not available for civilian works.

Taken together, the aerial photographs may be divided into five types depending on the attitude of the camera with respect to the earth's surface when the photograph is taken.

Vertical Photographs

A vertical photograph is one which is taken with the camera pointed as straight down as possible (Fig. 12.2 and 12.3). The camera axis, should not deviate from the plumb line (perpendicular) by more than + 3°. A vertical photograph has the following characteristics:

1. It covers a relatively small area.
2. The shape of the ground area covered on a single vertical photo closely approximates a square or rectangle.
3. The axis of the lens is perpendicular to the surface of the earth.
4. Being a view from above, the view of the ground it gives is unfamiliar.
5. Distance and directions may approach the accuracy of maps if taken over flat terrain.
6. Relief is not readily apparent.

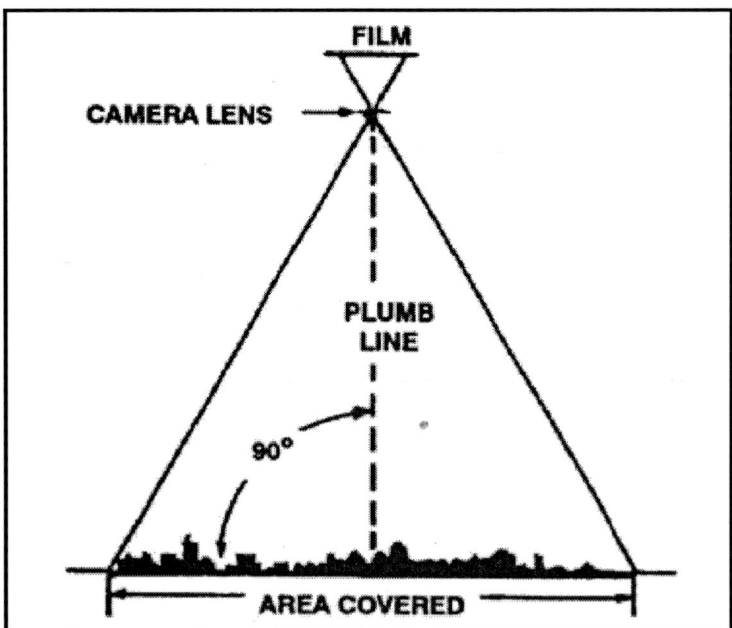

Fig. 12.2: Vertical Photograph

Oblique Photographs

These photographs have their axes purposely tilted from the vertical. The amount of till varies from 30 to 60 degrees. An oblique photograph showing the horizon is called a high oblique. One which does not have the horizon showing is called a low oblique. Oblique photographs, because of the side-view characteristic they possess, produce a more normal view of ground objects. Oblique photographs help in studying the terrain or ground features having considerable relative relief such as tall structures or slopes.

Oblique photographs are of two types: low oblique and high oblique.

Low Oblique

This photograph is taken with the camera inclined about 30° from the vertical (Fig. 12.3). It is used to supplement information available from maps. A low oblique photograph has the following characteristics:

1. It covers a small area.
2. The photo as usual is square or rectangular but the ground area covered is a trapezoid.
3. Because of the slanting images, the objects have a more familiar view.
4. Distance cannot be measures as no scale is applicable to the entire photograph. Nor is the direction (azimuth) correct as the parallel lines on the ground are not parallel on the photograph.

5. Relief is distorted.
6. The horizon is not visible.

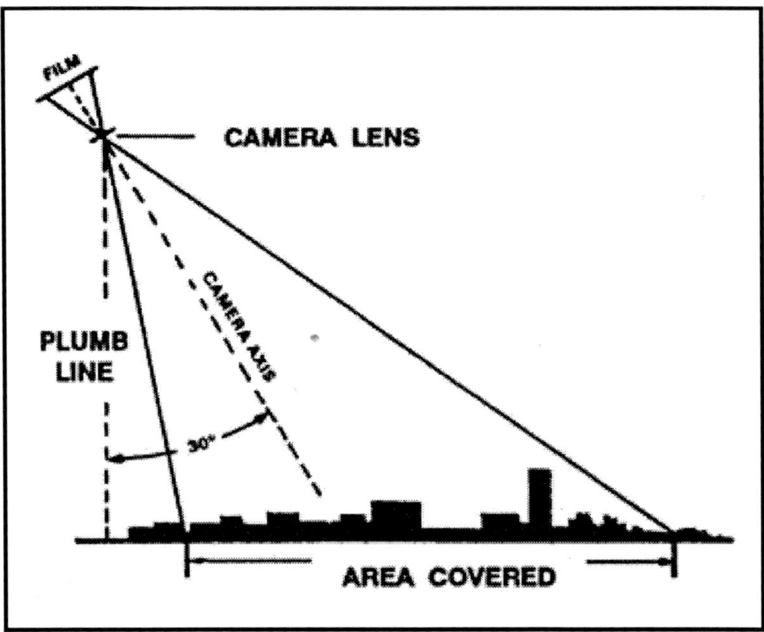

Fig. 12.3: Relationship of Low Oblique Photograph to the Ground

High Oblique

This photograph taken with the camera inclined about 60° from the vertical is called high oblique photograph (Fig. 12.4). It is used primarily in the making of aeronautical charts. It has the following characteristics:

1. It covers a very large area.
2. The ground area covered is a trapezoid but the photograph is square or rectangular.
3. The clarity in the view of the ground varies from very familiar to unfamiliar, depending on the height from which the photograph is taken.
4. Distortions in the distances and directions are similar to those in low oblique photographs.
5. Relief is highly distorted.
6. The horizon is visible.

Composite Photographs

These photographs are printed from three or more separate negatives which arc

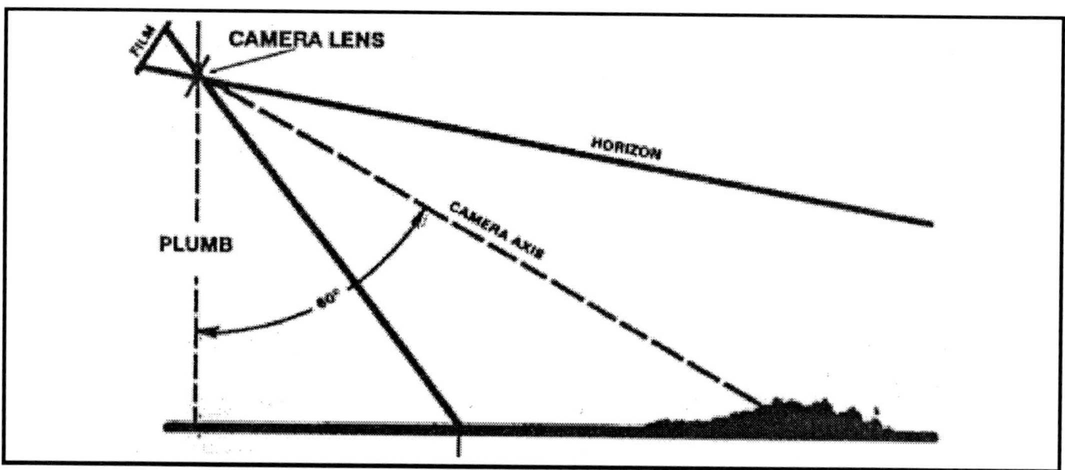

Fig. 12.4: Relationship between High Oblique Photograph and the Ground

exposed simultaneously by a multi-lens camera. The multi-lens camera has one lens aimed along a vertical line of sight and other lenses arranged in pairs tilted away from and on opposite sides of the central lens. When printed, the hit is reversed and the resulting photograph is all in one piece and m the same plane as the portion exposed by the central lens. These photographs cover large areas. With the development of wide angle cameras capable of covering equally large areas, the importance of multilens cameras has now declined.

Trimetrogon Photograph

When one vertical and two high oblique photographs are taken at the same time, in a direction at right angle to the line of flight, we get an assemblage of three photographs and composites from horizon to horizon (Fig. 12.5). Thus, when three separate photographs are exposed simultaneously by three single lens cameras mounted in the same airplane, we get trimetrogon photograph. The central camera takes vertical photographs while the other two cameras take oblique photographs to the right and left of the line of flight. The cameras are so fixed that the entire area from right horizon to the left horizon is photographed. The coverage is not printed as a composite photograph.

Multiple Lens Photographs

When we use a camera having more than one lens or more than one camera with different lenses, we get composite photographs which combine two, four, or eight oblique photographs around a vertical photograph. The oblique photographs are modified to permit their assembly.

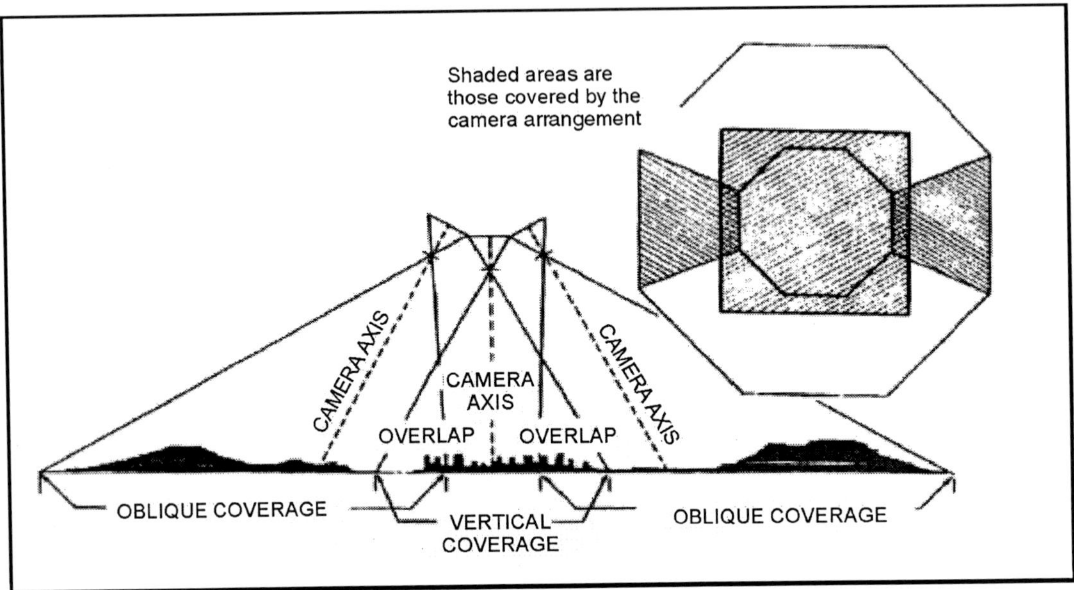

Fig. 12.5: Relationship of Cameras to Ground for Trimetrogon Photography (Three Cameras)

Sonne's Photograph

These photographs are part of a continuous strip of terrain. This is done by allowing the negative to move continuously over a fixed slit. The speed of the movement of film across the slit is adjusted to the speed and height of the aircraft.

Convergent Photograph

When one single twin-lens wide-angle camera or two single-lens wide-angle cameras are tied to the same mount and are exposed simultaneously with each camera axis converging when tilted 15 or 20° from the vertical, the resulting photographs are called convergent photographs.

Panoramic Photograph

Panoramic photography solves the usual problem of incongruence of high-resolution levels and wide-angular coverage encountered in aerial photography. Panoramic photograph covers greater details and larger areas and hence is popular all over the world. It sweeps the terrain from side to side in the direction of the flight. Continuous cover is obtained by spacing the exposures appropriately to give sufficient overlap between frames. Panoramic cameras are most advantageous when we need the resolution of small ground details from high altitudes.

DETERMINING THE SCALE OF AIR PHOTOGRAPH

An aerial photograph with scale cannot be used as a map supplement or substitute. Each photograph should have a scale and it should be printed on its margins as a Representative Fraction (RF) similar to the ones printed on maps. On a map,

$$RF = \frac{\text{MD (map distance)}}{\text{GD (ground distance)}}$$

On a photograph the RF is expressed:

$$RF = \frac{\text{PD (photo distance)}}{\text{GD (ground distance)}}$$

The RF of a vertical aerial photograph is an average determined by one of the following two methods:

1. Comparison method, or
2. Focal length-flight altitude method.

The former is derived by comparing the measured distance between two points on the photograph with the measured ground distance between the same two points. If the photo distance is 4 cm and ground distance is 1 km, the RF can be calculated as illustrated below:

$$\frac{4 \text{ cm}}{2 \text{ km} \times 100{,}000 = 200{,}000 \text{ cm}} = \frac{4}{200{,}000} = \frac{1}{50{,}000} \text{ or } 1{:}50{,}000 = RF$$

The Focal Length-Flight Altitude Method is illustrated in Fig. 12.6. It uses the marginal information of a photograph, includes the focal length (f), and the flight altitude (H) to determine the scale.

When the ground elevation is at sea level, H becomes zero, and the formula is as shown in Fig. 12.7.

IDENTIFYING EARTH FEATURES ON A PHOTOGRAPH

We can better identify the features on a photograph if the following points are kept in view:

1. Aerial photographs are pictures taken from above and hence the object look different from the ones we are familiar with.
2. The objects that are too greatly reduced in size get distorted.
3. Since most aerial photograph are printed in black and white, other colours

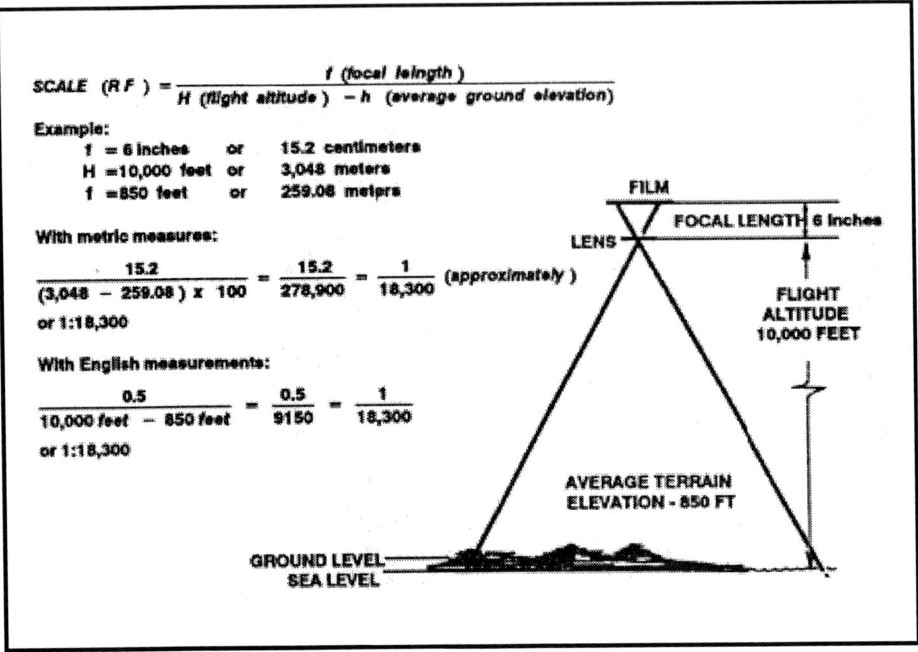

Fig. 12.6: Computation of Scale from Terrain Level

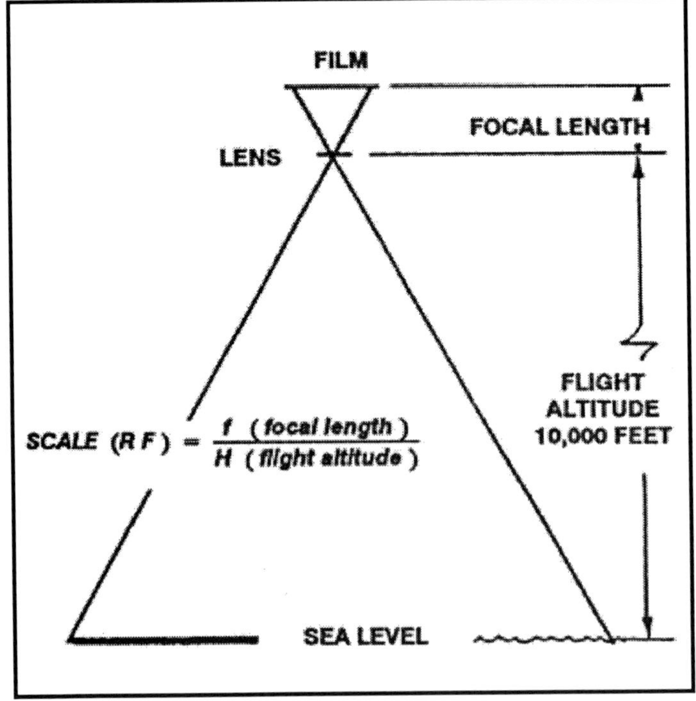

Fig. 12.7: Basic Computation of Scale from Sea Level

appear in shades of grey. And hence darker the natural colour, darker it would appear on the photograph.

4. While identifying ground features on aerial photographs, one should keep in view.

Size: To identify an unknown objects appearing in a photograph, compare its size with the size of the known objects. For example, in a densely populated area, smaller buildings are usually residential and the larger ones are commercial or administrative.

Shape (Pattern): Some features have known shape or pattern. For Example, a railway line, a canal, or a road is straight while a river is winding.

Shadows: Shadows too help in identifying objects appearing on an aerial photograph. For example, water towers or smoke chimneys, when seen from above look like a circle or dot. But their shadows show a profile that helps in identifying them. The length of the shadow gives a good idea of the height of objects.

Shade (Tone or Texture): Usually the film used for aerial photography is panchromatic which is sensitive to all the colours of the spectrum. It registers colours as shades of gray, ranging from white to black called tones. The tone is even or rough depending on the texture of the features. A smooth textured paved highway has an even tone, while a marsh has a rough tone. Reflection of the sunlight too affects the tone.

For example, a body of water would appear light if it is reflecting sunlight directly toward the camera, otherwise it would appear dark. Thus, the tone and texture are good indicators of objects appearing in aerial photographs.

Surrounding Objects: Relative position of an object is another indication of its identity. For example, a football field would be close to a school.

While studying a vertical aerial photograph, it must be oriented i.e. it should be so placed that the viewer faces the source of light that casts shadow of the object on the photograph. If it is not reoriented, the light may cause illusion of a depression appearing as a hill.

SOURCES OF DISTORTION

The vertically taken air photograph can be substituted for a map if:

1. Area photographed is flat.
2. The focal plane of the aircraft is parallel to the surface of the ground.
3. The camera is fitted with lenses which do not cause lateral distortions.

Such ideal conditions, however, rarely exist and, therefore, it is next to impossible to avoid distortions, except at and near the plumb point. Plumb is the point where a plumb-bob dropped from an airplane and through the centre of the film touches the ground (Fig. 12.2). Any feature in this alignment will be perfectly reproduced and the scale of the photograph will be true over its entire area. The distance between the centre of the film and any other point will bear the same distance relationship as distance between the plumb point and the corresponding point on the ground. Since no earth surface is perfectly horizontal this relationship between the ground and the photograph is rarely observed.

The distortion introduced by ground relief is more properly called Parallactic displacement or simply displacement. Parallax is an apparent change in the position of one object or point with respect to another when the two are viewed from different angles. When applied to aerial photography, it refers to the relative displacement of two points along the same line when viewed from an exposure point that is not directly overhead.

In Fig. 12.8, points (like Y) which are below V are displaced inward both in the negative NN′ and the positive PP′. Point Y′ will appear a t y′ and not at y. Whereas a point like X′ which is above V or datum plane K k′ is displaced outward from x to x′. The total amount of displacement is directly proportional to terrain relief and to the distance from the plumb point and is inversely proportional to flight altitude.

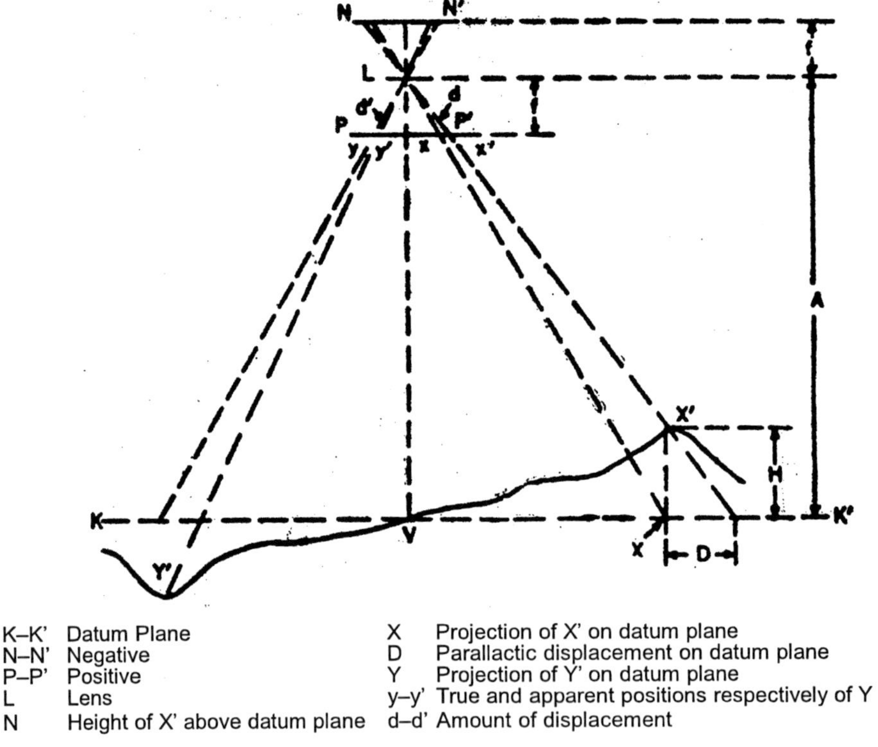

Fig. 12.8: Distortions due to Uneven Terrain

Because of the adverse weather conditions the airplanes at times have to tip up or down the horizontal alignment. If the picture is being snapped at the time of the tip, it will show distortions. In case the airplane tilts to one side, lateral displacement will take place. Such displacements are illustrated in Fig. 12.9.

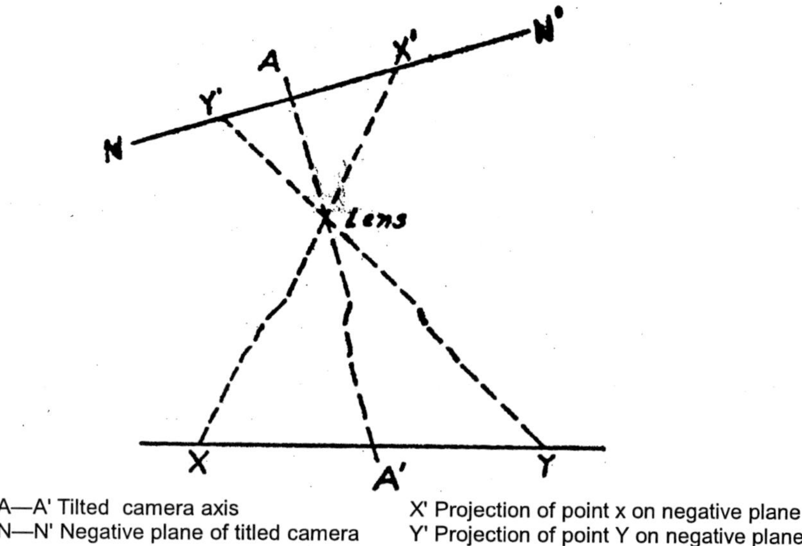

A—A' Tilted camera axis
N—N' Negative plane of titled camera
X' Projection of point x on negative plane
Y' Projection of point Y on negative plane

Fig. 12.9: Distortion Due to Tilting of Plane

Crabing i.e., nudging of a plane by wind from its flight line also creates problems because the photographs taken under these conditions cannot be perfectly aligned in strips as illustrated in Fig. 12.10.

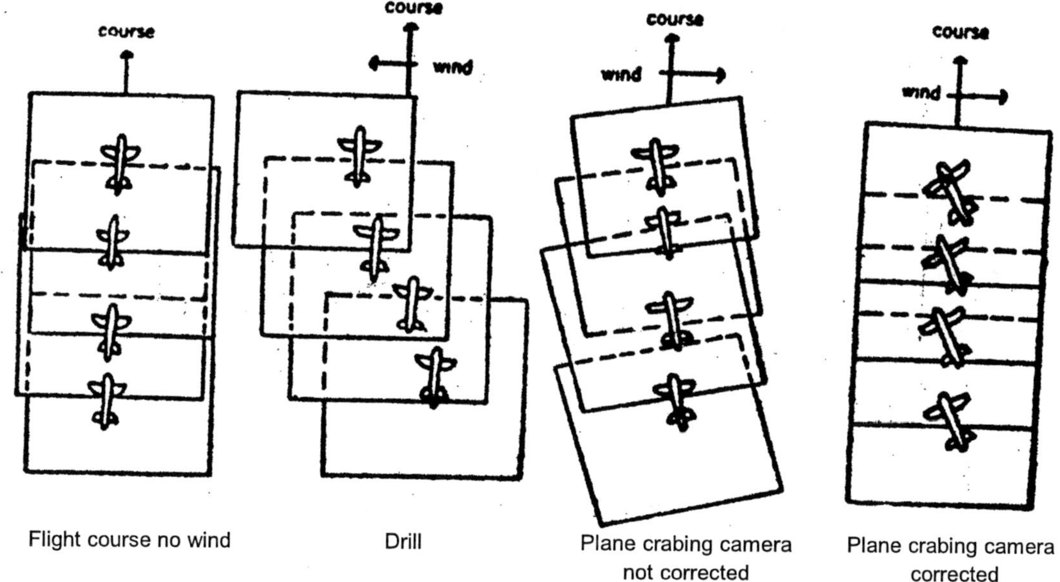

Flight course no wind
Drill
Plane crabing camera not corrected
Plane crabing camera corrected

Fig. 12.10: Distortion due to Nudging of Plane

IDENTIFICATION OF AIR PHOTOGRAPHS

Like maps, air photographs in their finished form possess certain identification marks. These marks are also called marginal information or data. They include time, date, camera type, altitude and other pertinent facts. But all these details are not given on all air photographs. Each photograph is also given an index number which gives ready access to the photograph. The indexing is done with the help of a map on which the available photographs are serially positioned. The index numbers are given either at the point where the centre of the photograph falls or on one of the sides of the area covered.

DETERMINING DIRECTIONS ON AIR PHOTOGRAPHS

To make the best and correct use of an air photograph it must be oriented like a map with north at the top. If a comparable map of the area photographed is available, the direction can be determined by comparing the location of some common features. If the time of the photograph and the latitude of the area covered arc known, the directions can be found out with the help of the shadows on the photographs. In the temperate zone of the northern hemisphere, for example, the rays of the morning sun will cast shadows to the northwest, the evening sun, to northeast and the noon sun to the true north. In the southern hemisphere the directions will be just the opposite (Fig. 12.11).

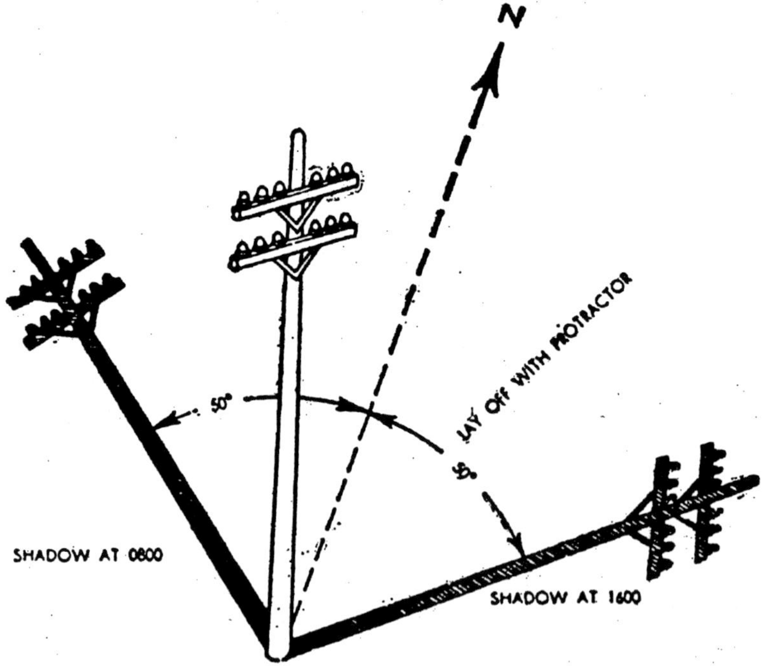

Fig. 12.11: Determine the North with the Help of Shadows Cast by the Feature Photographed

AIR PHOTO INTERPRETATION

Usually three important techniques are used in the evaluation of aerial photographs. These techniques range from the simplest (photo reading) to the more complex (Photo interpretation). But the present day practice is that they are commonly considered as parts of air photo interpretation.

Photo reading technique is mainly concerned with the identification and appearance of physical and cultural aspects of the landscape. This technique therefore treats the aerial photograph as a very detailed base map.

Photo analysis technique takes into account all aspects of photo reading, but also evaluates the interrelationships of the features that are found in the photograph. For example bridges and roads are identified and classified depending on their types and capacity. The relationships between the nature of the terrain and different land uses are ascertained. Water bodies and rock outcrops are outlined and measured. Therefore, it provides many types of basic data used for classification. Photo interpretation is the most advanced and difficult of these techniques.

Three basic conditions are essential for successful interpretation of aerial photographs:

1. Use of stereoscope,
2. Careful examination of all elements of air photo pattern, and
3. Deductive and inductive evaluation of these elements in terms of field experience and academic qualification of the interpreter.

The images shot from two different angles gives what we call sterio-view or three-dimensional view. All air photographs are therefore stereoscopic. Our eyes are so set that they have the ability to see things stereoscopically. They can see two views of a single object from two slightly different positions. The fusion or blending together of these two images in the brain gives the depth or distance. Viewing two air photographs stereoscopically gives the same view of the ground as one gets while looking down from a plane. This enables us to see the ground more realistically.

Sterio viewing requires two photographs called sterio-pair. While studying them, they are so placed that they are in the same relative position in which they were taken. The left eye sees one photograph and the right eye the other. The overlapping ground in the middle is seen by both the eyes.

Aerial photographs are generally taken in a series with 56 per cent overlap. Thus an unbroken coverage of the area is obtained. We can see the stereoscopic vision with the help of a stereoscope discussed earlier.

Aerial photographs usually give a stereoscopic view i.e. the images of the features photographed are shot from two different angles. All aerial photographs have the characteristics of stereoscopic view. Viewing two air photographs stereoscopically gives the same view of the ground as one will get while looking down from a plane. This enables the user to see the ground more realistically.

The two overlapping photographs required for stereo-viewing are called stereo pair. While studying them, we have to place them in the same relative positions in which they were taken. The left eye sees one photograph and the right eye the other. The overlapping ground in the centre is seen by both the eyes simultaneously.

STEREOSCOPES

Stereoscope is the instrument used for stereo viewing. As indicated earlier it is difficult to view air photographs without the aid of optical devices. This is mainly because:

1. Viewing photos from a close range causes eyestrain and difficulty in focusing.
2. The eyes are focused at short range (approximately 0") on the photos lying on the table, while at the same time the brain perceives parallactic angles which form to the stereo model at some depth below the table. It creates a very confusing situation.

To avoid this problem, we make use of various types of stereoscopes. Stereoscope is of two main type: Pocket Sterioscope (Fig. 12.12) and Mirror Sterioscope (Fig. 12.13). The pocket stereoscope consists of two magnifying lenses mounted in a metal frame. The mirror stereoscope is larger and heavier than the pocket stereoscope. It consists of four mirrors mounted in a metal frame.

Pocket Stereoscope

This is the simplest stereoscope and is relatively inexpensive (Fig. 12.12). This is a handy folding instrument which can be conveniently carried in the pocket. This can also be used for viewing sections of large format stereo pairs as well as stereoscopic illustrations in books. Therefore, this instrument is primarily used for training and demonstration. The photos have to be placed in such a way that the corresponding images arc slightly less than the eye base apart. This is usually 2".

Mirror Stereoscope

In this instrument binocular viewing head is either supported by four removable legs inserted into the mirror mounts (Fig. 12.13) or by supporting arm projecting from the rear. It is because of this that we get a perfect unobstructed work area for marking or for drawing and measuring on the photos.

From the point of view of viewing, here the advantage is that the two photos can be completely separated. This eliminates the problem of one photo obscuring part of the overlap of the other and clearly explains the operating principle of the mirror stereoscope. The light rays from the photos are received at the eye forming parallactic angle 0_x and 0_y. The brain automatically associates the depth point A with that parallactic angle. Thus one is above to see a stereo model.

Fig. 12.12: Pocket Stereoscope

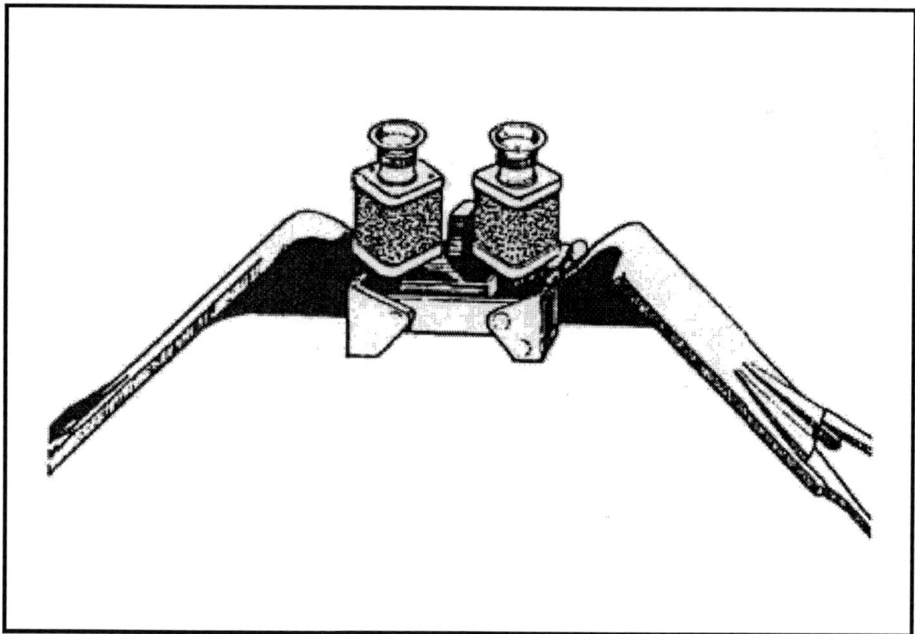

Fig. 12.13: Mirror Stereoscopes

Methods of Interpretation

The most ideal method of interpretation of aerial photographs is by examination of adjacent pairs of overlapping photographs. This pair of photographs is called a stereogram. A stereogram when seen through a stereoscope reveals to the viewer an image of the landscape—shown on the overlap of the two photographs. Photo interpretation directly depends on the analysis of the photo image. The photo image consists of the following elements: tone, texture, shadow, shape, size and pattern.

Tone or shade of grayness ranges from black to white. It is indicative of the shape of the objects and also the characteristics of the surface viewed in the photograph. For instance, in the aerial photograph the railway line appears as a dark thin line whereas the concrete highway appears in a bright tone.

Minor variations which occur in the aerial photograph give a mottled appearance to the images. This is called the texture of the objects. Texture helps in identifying areas with vegetative cover, crops, pastures, etc.

Shadow aids in the visualizing of the relief dimensions and also in the orientation of the photograph in space. Some of the features are better identified by their shadows than by their images. Tall thin objects like high tension line pylons are quickly identified only by their shadows. Besides image dimension can also be readily calculated by the dimension of their shadows.

Shape as an element in interpretation is used in its broader aspects. It is very important in the recognition of constructional landforms such as volcanic cones, sand dunes, river terraces, etc.

Size as a quantitative element in interpretation relates to the surface or volume dimension of an object. The relative sizes of images of objects generally help in their identification. The smaller buildings are more likely to be residences. The larger buildings may be commercial, industrial or cultural establishments.

The total arrangement of all cultural and natural features as they appear on the photograph is described as pattern. The individual landscape features or the collections of such features are called the elements of the pattern. Aerial photography helps in the understanding of the patterns of arrangement and areal associations. For example the pattern of drainage network gives a definite indication of the underlying rock type, its structure and also the characteristics of soil, etc.

The basic characteristics of different terrain and cultural features are of considerable help in the interpretation of aerial photographs.

Cultural Features

These are almost the first and easily recognized features in an aerial photograph. Railway lines appear as a thin and dark streak. They are commonly marked by many cuts and fills. Sometimes discontinuous railway lines indicate tunnels that are found in mountainous terrain. Certain associated features like stations and sidings make their identification easier. Roads are noticed in a lighter tone. They have a regular pattern in plain areas and less regular pattern in hilly terrain; here they are easily identified by the presence of hairpin bends. Bridges are readily recognized as they span water bodies. The shadows thrown by them help in identifying their types, whether they are suspension, truss, cantilever or arch etc., Canals appear in darker tones than roads. They form a rectilinear pattern in the plain country and follow contours in hilly terrain. When viewed through the stereoscope they appear as depressions with levees on either side. Urban features like residential buildings, offices, factories, schools, churches, airports, docks, refineries, park and race courses are easy to recognize because they have distinctive plans and silhouettes.

Agricultural patterns

Fields appear as long narrow strips with boundaries of rectangular coordinates. Their photograph appears in various shades of gray depending upon the crops grown in them at the time of aerial survey and ground conditions. Paddy fields may be identified by their rectangular pattern and by the low earthen bunds surrounding them. Grass and cereal crops photograph in light gray tones. Vineyards appear as rows of small dots and orchards in rectangular patterns of larger dots. Haystack or corn stocks show themselves as light coloured dots.

Vegetation

The distribution of natural vegetation is controlled by several factors like temperature, insulation, precipitation, altitude and soils. Therefore, a sound knowledge of these factors will aid in the interpretation of the distribution pattern of natural vegetation.

Tropical rainforest forms a near continuous canopy with few clearings. As the trees are taller than 75-80 m in height they are easily recognized under the stereoscope by their domelike tops. They appear as an unbroken sheet of dark tone on the aerial photograph. The tropical grassland-commonly known as the savanna-is marked by scattered groves. The grass appears in lighter tones of gray whereas the trees appear in darker tones. The desert plants are widely spaced and appear in small, isolated dark patches. A concentration of vegetation may be found wherever water is near the surface, or along the water courses or at the foot of the alluvial fans.

Terrain Features

Identification of various terrain features requires a thorough understanding of the types of underlying rocks and their structure and also the different processes of erosion to which they have been subjected. This will help in the easy identification of the significant elements of the terrain.

Rocks: Based on the appearance in aerial photographs, rocks have been grouped into massive and layered rocks. Many of the metamorphic and intrusive rocks belong to the massive category. The layered rocks include sedimentary, volcanic and some metamorphic rocks like schist and slate. Massive rocks appear homogeneous over large areas without any recognizable patterns except for dykes and veins appearing as dark coloured lines, whereas layered rocks produce definite patterns of parallel bands. The pattern may differ depending upon the structure and differential resistance of the rocks to erosion.

Granites appear in lighter tones of gray. In arid climates they make bold rounded outcrops with several joints. In humid climates they produce rugged mountainous landscape. Presence of large boulders is a common sight of the granite rock structure. Volcanic rocks appear dark in aerial photographs, for example basalt. Recent flow of lava is indicated by the ropy and slaggy surface of the landscape. Limestone though belonging to the sedimentary group may appear massive like granite. It is easily soluble and hence produces an intricate landscape characterized by crag like pinnacles and sinkholes. Sandstones are quite permeable and therefore support dense vegetation of shrubs and trees. Sandstone strata generally appear in light gray tone. Shale appears dark gray in tone and can be seen as grass covered patches.

Structure: The arrangement of rock masses in the earth's crust is known as structure. The deformation of rocks into folds and faults results in the alternation of strata which can be identified on the aerial photograph. Sedimentary rock structures are more readily identifiable than massive rocks. Horizontal strata appear on the valley sides as horizontal bands resembling the pattern of contour lines. Tilted strata of sedimentary rocks show as parallel bands on the surface. When subjected to stream erosion individual beds may appear on the sides of the valley. Landscape features like hogback and cuestas are characteristic of erosion resistant rock strata. In the case of folded strata the folds may appear in the photograph as anticlines and synclines. Anticline folds with outward dipping strata produce domelike landscape while the synclinal folds with inward dipping strata produce basins. Faults and linear features are easily recognized as they interrupt pattern of rock outcrops and also bring unrelated rock into contact. Rocks in the fault zones generally have shattered and distorted look and also have a tendency to develop linear stream valleys.

Erosion Processes: In humid regions streams are generally perennial and form an elaborate and interrelated system. When examined under the stereoscope they reveal the terrain type easily. Meanders and oxbow lakes are some of the features of streams flowing on plain areas. Sandbars show up as light gray areas within the river channel. Shallow waters may be recognized by their lighter tone and also by the presence of ripples on the water surface. Broken waters revealed by the stereoscope indicate the presence of waterfalls and cascades. Landscape in the arid and desert regions exhibit more varied features. Mountainous deserts appear as broad basins separated from each other by ridges. The floors of the basins are normally covered by sand and gravel. Streams will end up at the head of alluvial fans. Several of the alluvial fans coalesce and form an alluvial apron which could easily be seen on the aerial photograph. Clay or salt covered temporary lakes occupying the basins may be readily revealed by the stereoscope. Sandy deserts are characterized by long sandy ridges and barchans. Effective stream erosion, lack of vegetation, long, narrow, deep sided canyons are some of the features which help in the identification of desert landscape. Limestone regions present a landscape which is barren and sparsely covered with vegetation. These regions with horizontal beds under normal climatic conditions are marked by sinkholes, pits and coalescing sinks. Under the stereoscope a sinkhole looks mostly round or elliptical. The density of sinkholes in an area is indicative of its climatic, hydrologic and solubility conditions. Under the tropical environment sinkholes have a tendency to merge and produce 'karst' topographic features. The surface will have very few streams and even these few will disappear into the sinkholes to form subterranean drainage.

Accumulation of large quantities of snow moving down the slopes by gravity is called glaciers. They are readily identifiable from air by their light colour against the subdued background of rocks and forest cover. The valleys carved out by glaciers are generally 'U' shaped with steeper sides and flat floors. Tributaries can be seen joining the mainstream in series of cascades. They have their origin in glacial troughs which have precipitous walls of bare rock. The cirque floor may be occupied by a small lake called 'tarn'. When glaciers occupy plain areas they are called 'continental glaciers' and are liable to deposit huge quantities of glacial debris. Landscape features arising out of such deposition are known as drumlins. They are easily identified on the aerial photograph as they appear in groups and fields. They exhibit a general trend of parallelism and possess streamlined shapes. They may vary in length from a mere 50 m to several km and measure 5 m to 100 m in height. They are characterized by even dark to medium gray tone shades.

Coastlines: Four major coastlines have been recognized from the photogrammetric point of view. They are: embayed coast, plain coast, fiord coast and coral reefs. Embayed coasts are typified by the extension of sea in bays, estuaries or sounds occupying the drowned stream valleys. The adjacent landscape will be rugged or of moderate relief. They are easy to identify because they reveal a diversified pattern of headlands, beaches, islands, estuaries and marshes. Plain coasts are found next to level and plain landscape. Offshore bars and shallow lagoons between the bars and the main land are the typical features of these shorelines. The landward margin of the shoreline may be occupied by coastal marshes. On the serial photo the marsh appears dark in tone and the sand bars light gray. Fiord coasts are special features of glaciated coastal mountainous regions. The sea can be seen penetrating far into the land through the glacial valleys. Characteristic features which help identify such coastlines in the aerial photograph are the deep layers close to the shore, steep coastal cliffs, waterfalls and vast areas of barren rock. Coral reefs are found in the warm tropical waters. The corals photograph light in tone and the vegetation present on the corals appear dark. Outer margins of the reef are generally seen to be lined with white breaking surf.

COMPILING MAPS FROM AIR PHOTOGRAPHS

Making maps or map revision from air photographs is a complex process. The science of photogrammetry deals with this process. Photogrammetry may be defined as the art, science and technology of obtaining reliable information about physical objects and the environment through process of recording, measuring and interpreting photographic images and patterns of recorded radiant electromagnetic energy. We know that ground controls are essential for precise mapping. This applies to air photographs also. In an ideal case, there should be at least three geodetically determined and marked points on the ground in each air photograph. With the help of these three points the location of other points is determined. A base map, on a suitable scale and projection, is drawn for the purpose of plotting the points identified on the photographs. The points on the photograph are pinpricked or marked on transparent positives called diapositives. These are then positioned in a plotting machine and adjusted for scale, tilt, and other distortions until the points on the photograph and the map coincide. This is usually done on a stereo plotting machine. The details of the photographs can then be drawn on the map either directly or with the help of a pantograph.

REFERENCES

Alex MacLean and James Corner (2003), *Designs on the Land: Exploring America from the Air*, London: Thames and Hudson.

Arnold, R.H. (1995), *Interpretation of Airphotos and Remotely Sensed Imagery*, London: Prentice Hall;

Conway, E.D. (2000), *An Introduction to Satellite Image Interpretation*, Johns Hopkins.

Eller, Richard (1997), *Secrets of Successful Aerial Photography*, Buffalo, NY: Amherst Media Incorporated

Henbest, N. (1995), *The Planets: Portraits of New Worlds*, New York: Penguin Group.

Reynolds, Jock, *et al.* (2002), *Emmet Gowin: Changing the Earth*, Yale University Press.

Taubman, P. (2003), *Secret Empire: Eisenhower, the CIA and the Hidden Story of America's Space Espionag*, New York: Simon *and* Schuster.

Campanella, Thomas (2001), *Cities from the Sky: An Aerial Portrait of America*, Princeton Architectural Press.

Wolf, P.R. (1983), *Elements of Photogrammetry*, New York: McGraw-Hill.

Zhilin Li, Jun Chen, and Emmanuel Baltsaias (2008), *Advances in Photogrammetry, Remote Sensing and Spatial Information science*, New York: CRC Press.

13
Remote Sensing and Satellite Imageries

Remote sensing is the science and art of obtaining information about the earth, whole or part, or any object/phenomenon with the help of data acquired by a device (sensor), which is not in direct physical contact with it whether the sensor is immediately adjacent to the object or millions of kilometers away and that there is no intervening matter in between the object and the sensor. Aircrafts and satellites are common platforms for remote sensing observation. We have already discussed the process and technique of aerial photography in the previous chapter.

DEFINING REMOTE SENSING

Here are a few definitions of Remote Sensing, which give an idea of what it means and how useful it is.

"Remote Sensing is the science of deriving information about an object from measurements made at a distance from the object without actually coming in contact with it".

—*D.P. Rao* (NRSA)

"Remote Sensing is a means of acquiring information using airborne equipment and techniques to determine the characteristics of an area. Aerial photographs from aircraft and satellite are the most common form of remote sensing".

—*www.calforests.org/glossary.html*

"Remote Sensing is the acquisition of data for deriving information about an object or materials located at the earth's surface or in its atmosphere by using sensors mounted on platforms located at distance from the target".

—*P.K. Gupta* and *K.V.S. Badrinath* (NRSA)

"Remote Sensing is the science of deriving information about the earth's land and water areas from images acquired at a distance. It usually relies upon measurement of electromagnetic energy reflected or emitted from the features of interest."

—*Campbell*

"It is the art or science of telling something about an object without touching it".
—*Fisher*

"Remote sensing is the science and technology by which the characteristics of interest of objects can be identified, measured or analyzed without direct contact".
—*Japan Society of Remote Sensing*

"It is a multidisciplinary activity. It deals with the inventory, monitoring and assessment of natural resources through the analysis of data obtained by observation from a remote platform". —*The National Natural Resource Management System* (NNRMS), India

Remote sensing data basically consists of wavelength intensity information acquired by collecting the electromagnetic radiation leaving the objects at specific wavelength and measuring its intensity. The definition of remote sensing excludes electrical, magnetic and gravity surveys that measure force fields rather than electromagnetic radiation. Magnetic and radioactivity surveys are frequently made from aircraft; they are, however, treated as airborne 'Geophysical Survey', and not as 'Remote Sensing', as the term is conventionally used today.

Two technological developments of 1960s made Remote Sensing a practical proposition. First, new sensors were developed which enabled the acquisition of images at different wavelengths including thermal infrared and microwave. So photographs could be taken from very long distances from above the Earth. And second, development and deployment of manned and unmanned artificial satellites which provided an orbital vantage point for acquiring images of the earth which were far better in quality than the ones got from airborne cameras.

ADVANTAGES OF REMOTE SENSING

There are many advantages of obtaining imageries of the entire earth or a part of it by remote sensing. Important among them are listed below:

1. For sizable areas within the field of view of the sensor, spacecraft coverage is truly synoptic, because the high altitude and speed of the spacecraft permit the scientist to obtain information of large areas at a single instant of time.
2. It is possible to get real time analog and digital data.
3. One can get information on short notices, on impending diseases, forest fires, disaster areas such as earthquakes, storms etc.
4. Satellite images are permanent records providing useful information in various wave bands.
5. Repetitive coverage helps in monitoring changes which occur slowly and in remote areas such as melting of glaciers, drying up of water bodies, movement of people, etc.
6. It is less expensive considering the long-term application of information.

7. The long duration of space flights and all weather operations are highly advantageous aspects of remote sensing from outer space.
8. Technical and operational advantages result from the precise regularity of spacecraft motion, from the lack of vibration, and from the high rate of movement.
9. Remote sensing facilitates easy data acquisition over inaccessible areas like Amazon forest, Himalayan peaks, and snow covered polar areas.
10. It provides long-term record of earth surface information to be utilized later at any given time. Thus, accurate data on current inventory and rate of depletion furnishes the basic information required to anticipate forthcoming events and thus to prepare contingency plans.

HISTORY OF REMOTE SENSING

How It All Began?

The roots of remote sensing lie in five significant developments in scientific endeavours of man. The first was the invention of airplane, which aroused the desire to fly to the outer space beyond the gravity of the earth; the second was the technology to send unmanned and then manned satellites around the earth, and third was the development of sensors, which could catch the electromagnetic waves emanated from the earth; fourth was the development of cameras fitted with such sensors to convert the waves into photographs; and the fifth was the development of techniques to interpret these photographs into maps.

The story of Right brothers who are known as fathers of aeroplanes is known to all and need not be repeated here. What is important is the during the Second World War airplanes were used to bomb enemy targets by physically looking down at the targets. After the war, cameras were fitted to airplanes to take photographs of floods, forest fires, earth quakes, enemy positions and many other events and happening. As we have discussed this in the previous chapter, air photography has now become a first step in physical planning, and development.

Rocketry to go beyond the reach of the aeroplanes can be said to be the precursor of in development of artificial satellites. During the last century, three scientists in three different countries made significant contributions. They were Herman Oberth in Germany, Konstantin Tsiolkovski in Russia, and Robert Goddard in the USA. During the Second World War, London was attacked by V2 missiles, which had a range of 200 miles and travelled at 3,500 miles an hour.

After the War, the USA and the USSR proceeded further. On October 4, 1957. The USSR launched its first artificial satellite Sputnik 1 into space and on August 12, 1961 Lt. Yuri Gagarin became the first human to orbit the earth on Vostok 1. He went round the earth for 108 minutes at 327 km above the earth.

On January 28, 1958, the USA put into orbit its first satellite Explorer 1. After two years in 1961 Alan Shepard became the first American to fly in space. John Glenn became the first American to orbit the earth on February 20, 1962. On July 20, 1969,

Neil Armstrong landed on the moon. Six more missions to moon were accomplished before the end of 1972.

It must be pointed out here that before the astronauts landed on the moon its surface was well probed photographically with the help of unmanned satellites. By 1970 such orbiting communication and navigation satellites were in every day use. It was not only the earth but also the moon, mars and other planets were targets of these satellites for exploration by remote sensing photography. Mariner photographed the Mars and by end of 1970s Yoyager spacecraft sent back images of Jupiter and Saturn including their rings and moons. The highlights of the decade were the Skylab, America's first space station manned by astronauts and Apollo Soyuz Test Project. A joint US-Russian Mission, were successfully completed.

During the decade of 1980s the satellite communication carried television programmes. It was a satellite, which discovered the ozone hole over Antarctica; pinpointed forest fires and various natural and man-made disasters. Satellites were the sources of information about new planets and stars.

Launch of Space Shuttle Columbia in early 1981 made space travel less expensive. Twenty-four shuttle launches were carried out before the Challenger tragedy in 1986. The use of satellite communication and remote sensing during the Gulf War made space system more closely integrated with defence.

Now we have an International Space Station, which is being operated in collaboration with many nations. The former competitors are now working together. The Columbia disaster of 2003 has made it imperative to further improve the technology. In any case, as of today, a number of countries of the world are launching their own satellites for military and communication purposes and there is a strong competition in the commercial launch market to develop the next generation of launch systems.

The US programme is designed to ensure maximum availability of remote sensing data to educational institutions and private companies. As a matter of national policy, it—

- Supports the global preservation of the segments of the radio-frequency spectrum for use by the Earth and atmosphere observation system;
- Supports the use of remote sensing systems and to maintain these systems within NASA and NOAA;
- Supports greater use of remote Sensing data in various federal agencies;
- Encourages transfer of knowledge and technology of remote sensing to various agencies;
- Promotes the use of advanced remote sensing technologies such as hyper-spectral imaging radar and laser imaging; and
- Encourages the federal agencies to develop improved methods of sharing remote sensing equipments and services among them.

Remote Sensing in India

India, after gaining independence in 1947, paid special attention to development and application of frontier sciences like space science in nation building under the inspiring leadership of Pt Jawaharlal Nehru, the first Prime Minister of India. Dr. Vikram Sarabhai gave shape to modern Indian rocketry and space research. In the words of Dr. APJ Abdul Kalam, the President of India, "Many individuals with myopic vision questioned the relevance of space activities in a newly independent nation, which was finding it difficult to feed its population. Their vision was clear; if Indians were to play meaningful role in the community of nations, they must be second to none in the application of advanced technologies to their real-life problems. They had no intention of using it as means to display our might."

A committee known as Indian Committee for Space Research (INCOSPAR) was formed under the chairmanship of Dr. Sarabhai (then with Tata Institute of Fundamental Research (TIFR). In 1962, the Committee decided to set up Thumba Equatorial Launching Station (TERLS) at Thumba near Thiruvanthapuram, Kerala and very close to earth's magnetic equator. Dr. Abdul Kalam was part of the team assembled by Dr. Sarabhai. After six month's training in the USA, they launched the first sounding rocket (Nike-Apache) on November 20, 1963. Rohini Sounding Rocket (RSR) programme launched the first indigenously developed 32 kg Rocket (RH-75) to 7 km height in 1967. It was followed by a two-stage rocket of 100 kg which attained the height of 320 km. Sounding Rockets of other countries were also launched from Thumba as a sign of international collaboration.

Realizing the growing importance of space research, and the immense socio-economic benefits to be derived from it, it was decided to prepare satellite launch vehicles and large rockets. In 1969 Sriharikota, about 100 km north of Chennai, was chosen as the Launch Station. On August 15, the ICCOSPAR was reorganized to form Indian Space Research Organization (ISRO) under the Department of Atomic Energy of the Government of India. In 1972, the Government set-up the Space Commission and created a Department of Space (DOS) to carry forward the country's space research activities. ISRO is responsible for developing launchers and satellites two separate agencies take care of telecommunications and remote sensing. They are INSAT and IRS, respectively.

The first satellite named after famous astronomer of the ancient times, Aryabhatta was successfully launched in April 1975. Then came the launching of two experimental remote sensing satellites: Bhaskara I and Bhaskara II on June 7, 1979 and November 20, 1981 respectively. Both were by and large successful ventures. The first operational remote sensing satellite IRS-1A was launched in 1988 and the second 1B was launched in 1991. In the mean time, a data reception station was set up in Hyderabad and a series of national/regional level remote sensing application projects were started in the country.

The successful launch of second generation and indigenously built IRS-1C satellite on December 28, 1995 provided tremendous opportunities for applying space informatics in areas of disaster monitoring and natural resources management (Singh, 1991, 1994 and 1995). The IRS-1C surveys the whole earth's surface in just 24 days. With better resolution, coverage and revisit in order to provide invaluable data especially on natural resources, it proved to be a major milestone in the India's satellite remote sensing programme.

The IRS-1C satellite has three types of advanced imaging sensors. The Panchromatic

Camera (PAN) provides very high spatial resolution of 5.8 m and a ground swath of 70 km. The PAN camera can be steered to 26 degrees, which in turn increases revisit capability to 5 days. Linear Imaging and Self-Scanning (LISS-III) Sensor provides multi-spectral data collected in four bands. The Wide Field Sensor (WiFS) collects data in two spectral bands. All the three cameras are operating in real time over Indian ground station visibility circle two or three times a day (NRSA, 1996). The satellites IRS-1C and IRS-1D have the best spatial resolution of space-based remote sensing data. The demand for satellite data is going up by 15 to 20 per cent a year and with IRS-1C it would be possible to meet 90 per cent of the requirement from our own satellite. With the recent launch of IRS-1D using our own launch vehicle, PSLV, India has established capability to place 1200 kg class satellites in polar sun-synchronous orbit. India's INSAT system is one of the largest constellations of domestic communication satellites in the world.

The Organization Chart of the Indian Space Programme is given below (Fig. 13.1).

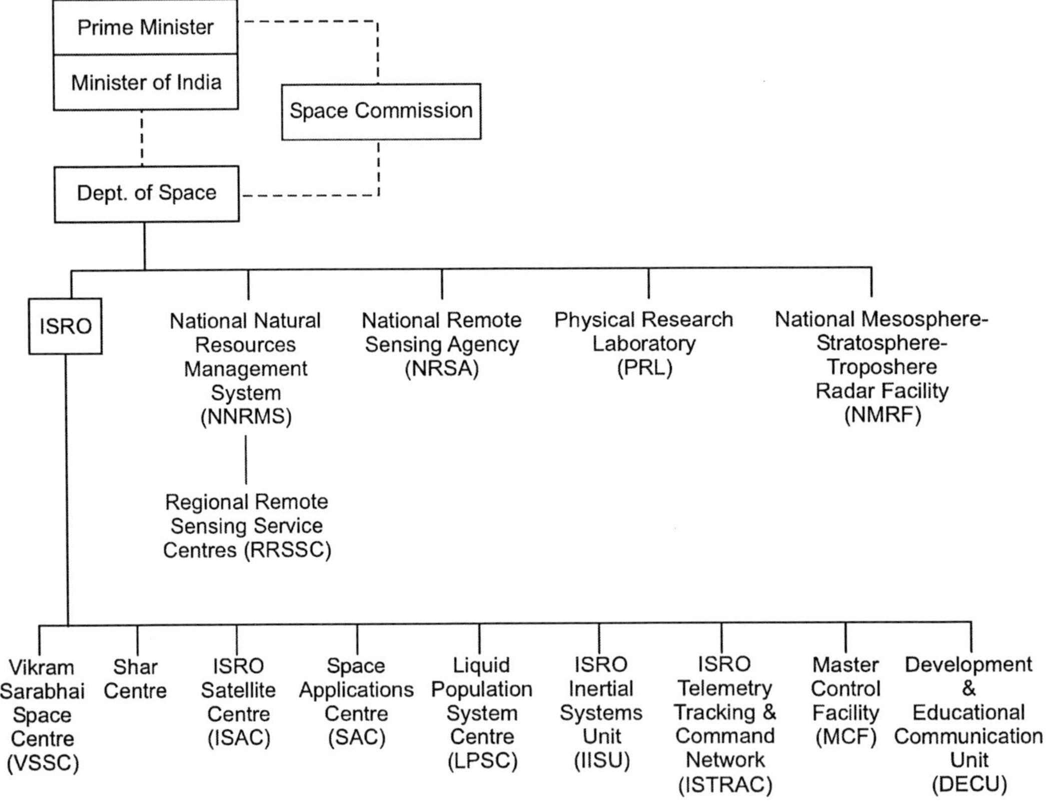

Fig. 13.1: Organizational Chart of Indian Remote Sensing Programme
Source: *www.bharat-rakshak.com/SPACE/space-history*

PROCESSES OF REMOTE SENSING

Remote Sensing involves as number of processes. The Department of Space, Government of India has identified the following steps to complete the cycle.

1. Transmission of electromagnetic energy from the source (sun) to the surface of the earth and its interaction with the intervening atmosphere.
2. Interaction of energy with the earth surface (reflection/absorption/transmission) or self-emission.
3. Transmission of the reflected/emitted energy to the remote sensor placed on suitable platform.
4. Detection of the energy by the sensor converting it into photographic image or electrical output.
5. Transmission/recording of the sensor output.
6. Pre-processing of the data to generate data products.
8. Collection of Ground Truth and other collateral information.
9. Data processing and interpretation.
10. Application of the data.

TYPOLOGY OF REMOTE SENSING

Remote sensing is of two types:

Active Remote Sensing

Active remote sensing system carries onboard its own electromagnetic radiation source. This electromagnetic radiation is directed at the surface and the energy that is scattered back from the surface is recorded. A camera is normally a passive means of recording data. If the photograph is obtained in daylight, the electromagnetic radiation within the visible waveband is reflected from the object and recorded by the film. However, if the photograph is obtained at night, an electronic flash is initially fired, which directs an electromagnetic pulse at the target in order to illuminate it and the reflected radiation is received and recorded back at the camera. It is then called Active Remote Sensing (Fig. 13.2. The commonest active remote sensing system is radar that produces electromagnetic radiation in the microwave band of the spectrum. Other active systems that have been used in remote sensing are laser and sonar devices.

Passive Remote Sensing

When the source of energy is external such as the Sun, Remote Sensing System is called Passive. It detects the solar energy reflected form the target (earth). When the energy is produced by the remote sensing system itself then it is called Active; it senses the energy reflected from the target (Fig. 13.3).

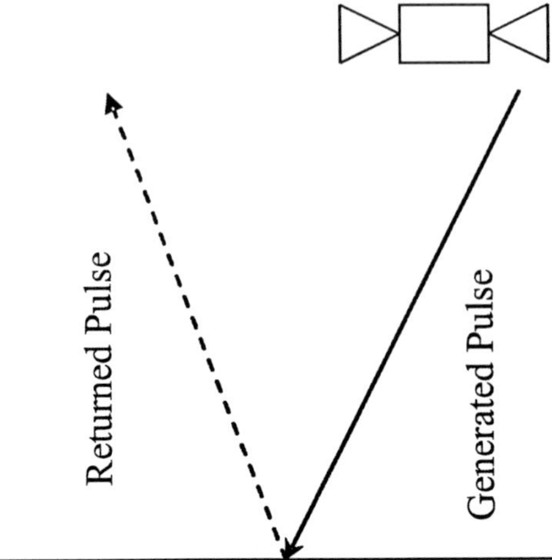

Fig. 13.2: An Active Remote Sensing System Produces its own electromegnetic Rediation

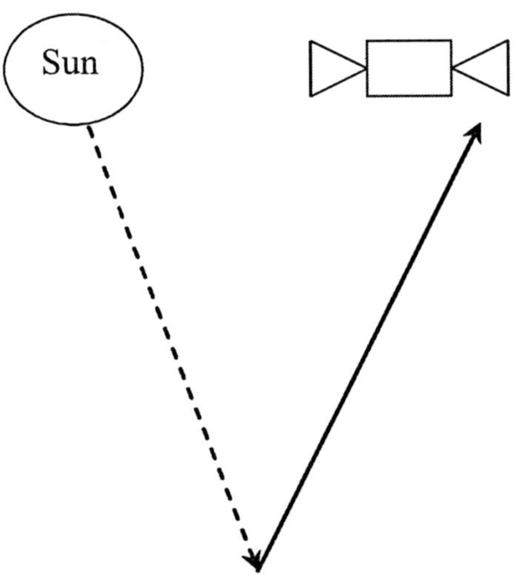

Fig. 13.3: Passive Remote Sensing system

Uses the Sun as the Source of Electromegnetic Radiation

A second strategy for passive remote sensing is to record radiation emitted from (rather than reflected from) the earth surface (Fig. 13.4). Because the emitted energy is strongest in the far infrared spectrum, this kind of remote sensing requires special instruments designed to record these wavelengths. Emitted energy from the earth surface is mainly derived from short-wave energy from the sun that has been absorbed, and then re-radiated at longer wavelengths.

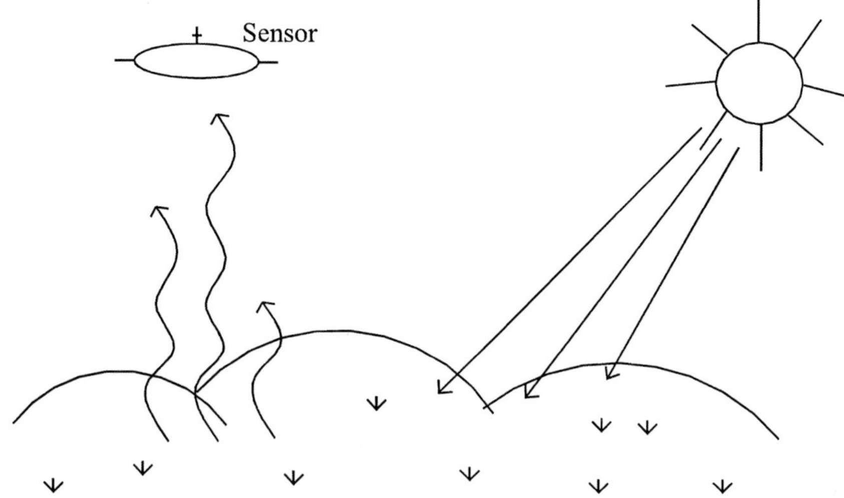

Fig. 13.4: A Passive System Using Emitted Terrestrial Radiation

COMPONENTS OF REMOTE SENSING

The components of remote sensing including the processes related thereto are:

Component	*Related Processes*
The Energy	Electromagnetic energy
Sources of Energy	The remote sensing system itself
The Target	More frequently the Sun
Transmission Path	Atmosphere
The Target	Features of the Earth surface emit and reflect energy towards the sensing system
The Remote Sensing System	Sensing Instruments

The components of remote sensing can be broadly divided into two categories:

Direct Components	*Indirect Components*
Sensors	Sun (Electromagnetic Radiation)
Spectral Signatures	Atmosphere
Platform	Earth Radiation
Data Handling System	
Target	
Multiple Users	

Direct Components

Sensor: Scientific equipment which can receive the reflected terrestrial radiation and enable the formation of spectral signature. No single sensor is sensitive to all wavelengths. All real sensors have fixed limit of spectral sensitivity. They also have a limit of how small an object on the earth surface can be recorded. The spatial resolution of a sensor is an indication of how closely and how well a sensor can record spatial details. It is generally the sensor that decides the type of platform that will be required in a remote sensing system.

Spectral Signature: Different objects have different characteristic reflectance. So it is the variation in spectral intensity and distribution of reflected radiation (which is decided by sensors), which leads to variation in spectral signature. The mechanism of energy transfer takes place through conduction, convection and radiation.

Platform: It provides a base where sensors are placed according to the payload configuration. They may vary from small satellite to large space stations. Platforms are generally controlled by a controlling facility on the ground and adjusted according to the need of sensor and of the type of data required. Platform may be of three types:

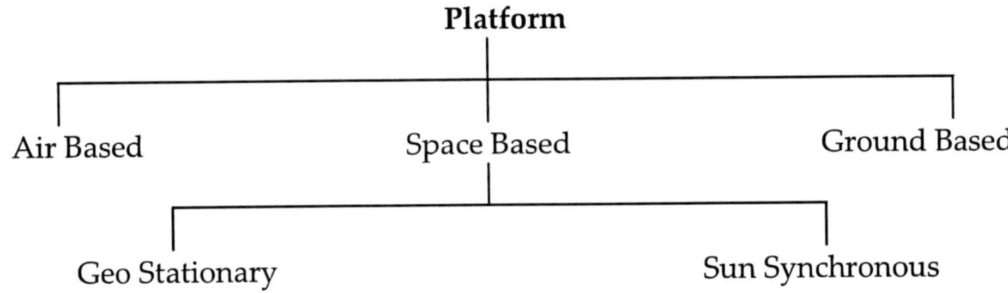

Data Handling System: The capability of remote sensing systems to generate data far exceeds the capacity for handling these data. Most of the data received from satellite is either in digital or analog format. The capacity to transfer these data into visual image depends upon the type of computer system being used and the type of instrumentation i.e., in operation. While machine and other mechanical device can do most data handling, human intervention in data processing is essential for the useful application of remote sensing data.

Target: Ultimate purpose for which data is acquired is the target. Target of remote sensing data varies from field to field. The interpretation is also done according to its use.

Multiple Data Users: The person(s) using the remotely sensed data are controlled to successful functioning of remote sensing system. The data generated by remote sensing procedure can become operational only when some one understands, interprets, analyzes and uses that data. Also no single combination of data acquisition and data analysis procedure can satisfy the need of all data users. Remotely sensed data is used by geologists, pedologists, climatologists, geomorphologists, botanists, zoologists, economists, and the most importantly for military purposes.

Indirect Components

Sun or Energy Source: Sun is the direct source of Electromagnetic Radiation (EMR). All passive remote sensing systems rely or depend on energy that is either reflected from the earth's surface or emitted from the sensor. The spectral distribution of sunlight or energy is not uniform everywhere on the surface. Also solar energy level varies with time and space. At the same time different earth surfaces emit different degree of radiation since the energy in passive remote sensing system cannot be controlled in a real system. All the reflected or emitted energy is generally non-uniform with respect to wavelength. Thus, the entire remote system has to be calibrated on the thematic basis, in order to study the specific detail of the surface we know that the EMR comprises of short wave, visible spectrum and long wave. EMR waves are propagated in the velocity of 3×10^{10} m/sec or 300,000-km/sec.

For remote sensing purpose we use only the following two bands of EMR:

1. Visible Spectrum
2. Invisible spectrum
   - Part of Microwave
   - Whole infrared Invisible

Atmosphere: It provides a medium through which EMR penetrates the earth's surface. The atmosphere modifies the strengths and spectral distance of energy received by the surface and later received by the sensor. The effect of the atmosphere also spectrally decides, "where-we-can-look". The atmosphere is characterized by various processes, which directly affect the receipt of EMR like—absorption, scattering, diffusion and reflection.

A STANDARD REMOTE SENSING SYSTEM

Data Flow Process in Satellites

As illustrated in Fig. 13.5, the data flow process in satellites has 13 elements. These are:

1. Solar Illumination.
2. Solar illumination as modified by passage through atmosphere.
3. Reflection from ground.
4. Propagation through the atmosphere of reflected energy.
5. Signal sensing at satellite.
6. Power supply from solar cells.
7. Signal processing i.e., digitization, tape recording and transmission.
8. Propagation through the atmosphere.
9. Telemetry received at ground receiving station.
10. Signal processing for geometric, radiometric and other correction.

11. Transmission to data bank.
12. Distribution and public access.
13. Analysis by users.

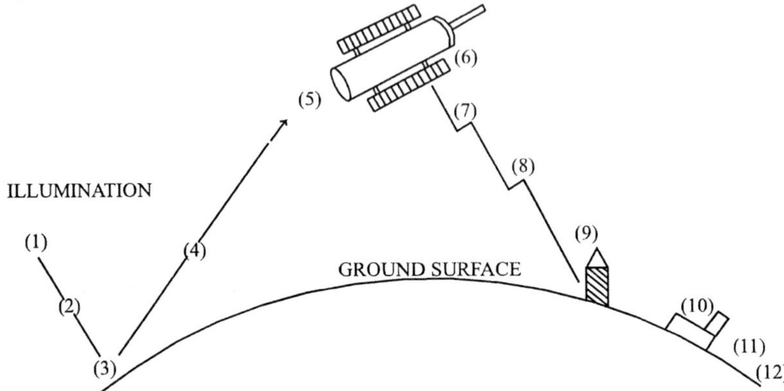

Fig. 13.5: Data Flow Process in Satellites

A sensor is a device that gathers energy and converts it into a signal and then presents it in a form that provides information about the phenomena or objects that are being secured. There are various types of sensors which can classified on the basis of source of output, mode of operation and data output.

Imaging Sensors are able to convert signals into visual format such as satellite image. An imaging sensor thus can be defined as active or passive. It measures the intensity of radiation or gives the reflectance value of any target from point to point. It provides the data, which have one to more correspondence, and shows several locations in the given area on the ground. Examples of such sensors are Multi-Spectral Scanners (MSS), thematic mapper and side looking air born radar (SLAR).

The non-Imaging sensors are those, which do not produce images. The output is in the form of digital values or analog curves. Such sensors are microwave radiometers and infrared thermograms. In such sensors the spatial distance between the pixels is not predefined. The non-imaging sensors in particular are designed to give reliable quantitative measures of the integrated intensity of electromagnetic radiation from all objects within their field of view as functions of time and wavelength (Fig. 13.6). Table 13.1.

ELECTROMAGNETIC ENERGY

Electromagnetic energy refers to all energy that moves with the velocity of light in a harmonic wave pattern. A harmonic pattern consists of waves that occur at equal interval of time. The wave concept explains how electromagnetic wave energy propagates. But this energy can only be detected as it interacts with matter. In this interaction, electromagnetic energy behaves as though it consists of many individual bodies called photons. When light bends as it propagates through media of different optical densities, it is behaving like waves. Light meter measures the intensity of light;

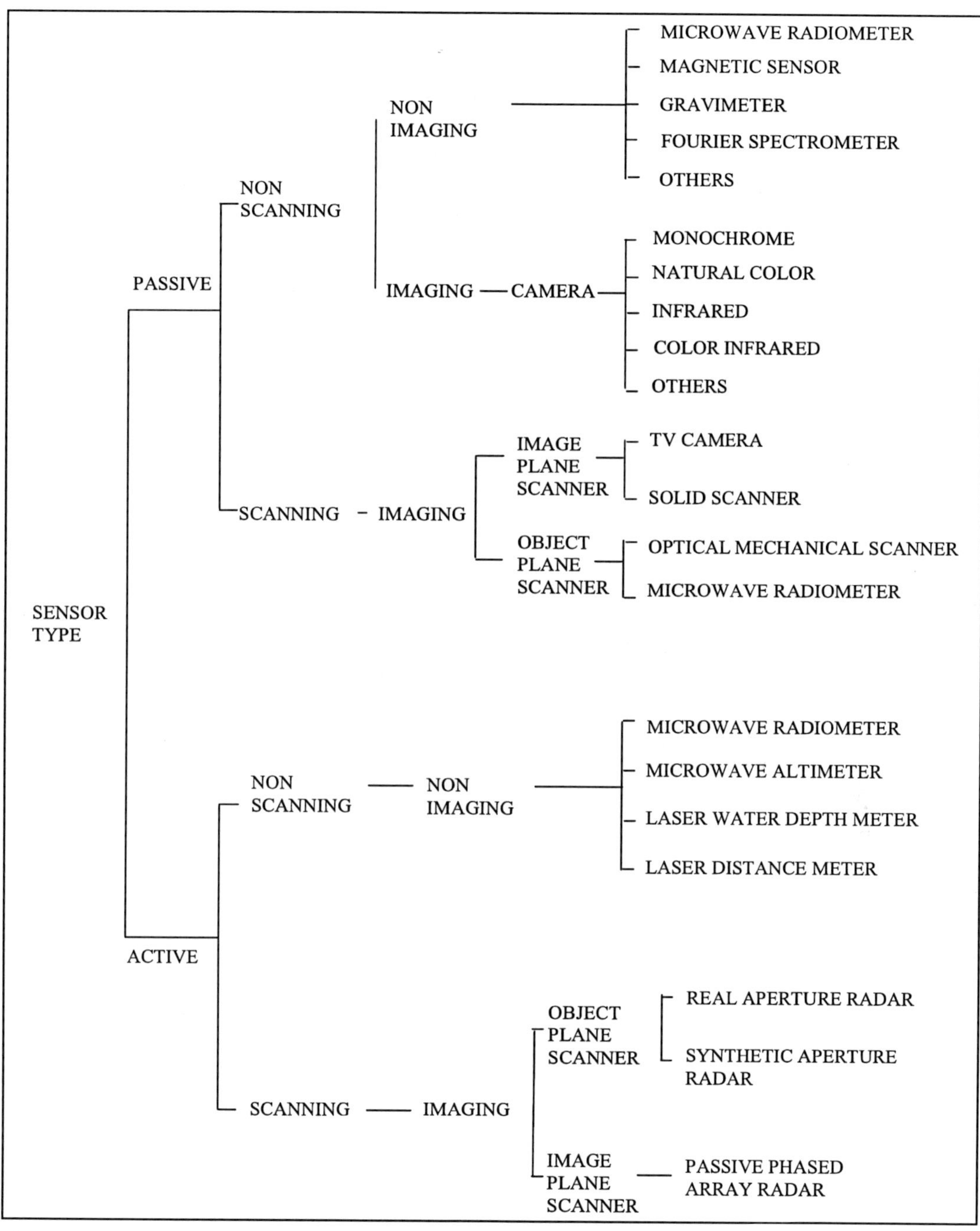

Fig. 13.6: Process of the Functioning of Sensors

however, the interaction of photons with the light sensitive photo detector produces an electrical signal that varies in strength proportional to the number of photons.

Properties

Electromagnetic waves can be described in terms of their velocity, wavelength and frequency. All electromagnetic waves travel with the same speed (c). The velocity is commonly referred to as the speed of light, which is one form of electromagnetic energy. For electromagnetic waves moving through a vacuum c= 229,773 km sec or for practical purpose $c = 3 \times 10^{10}$ m. sec.

The wavelength of electromagnetic waves is the distance from any point on one cycle or wave to the same position on the next cycle or wave. The micrometer is the convenient unit for designating wavelength of both visible and infrared radiation. Optical scientists commonly imply nanometers (nm) for measurement of visible light to avoid using decimal numbers. Unlike velocity and wavelength, which change as electromagnetic energy is propagated through media of different densities, frequency remains constant and is therefore a more fundamental property. Electronic engineers use frequency nomenclature for designing radio and radar energy regions.

Interaction Mechanism

Electromagnetic energy that encounters matter whether solid, liquid or gaseous is called incident radiation (Fig. 13.7). Interaction with matter can change the following properties of the incident radiation: intensity, direction, wavelength, polarization and phase. The science of remote sensing detects and records these changes. Scientists then interpret the resulting images and data to identify the characteristics of matter that produced those changes. During interaction between electromagnetic radiation and matter, mass and energy are conserved according to basic physical properties.

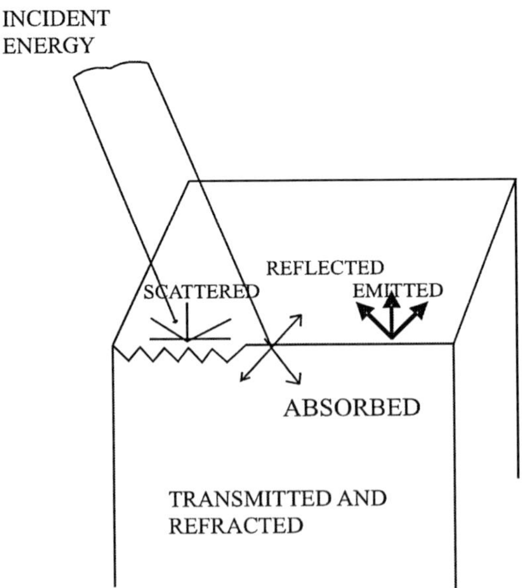

Fig. 13.7: Interaction Mechanisms between Electromegnetic Energy and Matter

The incident radiation may be:

Transmitted, that is, passed through the substance. Transmission through media of different densities, such as from air into water, causes a change in the velocity of electromagnetic radiation. The ratio of the two velocity is called index of refraction (n) and is expressed as—

$$\boxed{n = Ca/Cs}$$

where,

Ca is the velocity in vacuum, and
Cs is the velocity in the substance

Absorbed, giving up its energy largely to heating up the matter. Emitted by the substance, usually at longer wavelengths as a function of its structure and temperature.

Scattered, that is deflected in all directions. Surfaces with dimensions of relief, or roughness, comparable to the wavelength of incident energy produce scattering. Light waves are scattered by molecules and particles in atmosphere that have sizes similar to the dimension of wavelengths of light.

Reflected, that is returned from the surface of a material with the angle of reflection equal and opposite to the angle of incidence. Reflection is caused by emission; scattering and reflection are called surface phenomena because these interactions are determined primarily by properties of the surface, such as colour and roughness. Transmission and absorption are called volume phenomena because these interactions are determined by the internal characteristics of the matter such as density and conductivity. The particular combinations of surface and volume interactions with any particular material depend both on the wavelength of the electromagnetic radiation and the specific properties of that material. These interactions between matter and energy are recorded on remotely sensed images, from which one may interpret the characteristics of the matter.

Electromagnetic Spectrum

The electromagnetic spectrum is the continuum of energy that ranges from meters to nanometers in wavelengths, travel at the speed of light and propagate through a vacuum such as outer space. All matter radiates a range of electromagnetic energy with the peak intensity shifting towards progressively shorter with increasing temperature of the matter.

Electromagnetic Spectrum (Wavelength Regions)

Figure 13.8 shows the electromagnetic spectrum, which is divided on the basis of wavelength into regions that are described in Table 13.2.

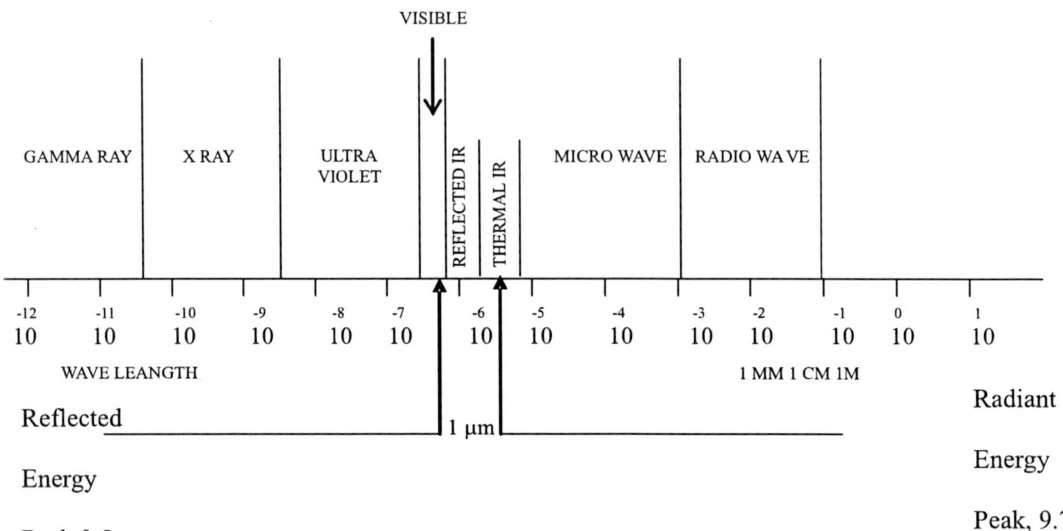

Fig. 13.8: Electromagnetic Spectrum

Table 13.2: Electromagnetic Spectral Regions

Region	*Wavelength*	*Remarks*
Gamma ray	<0.03 nm	Incoming radiation is completely absorbed by upper atmosphere and is not available for remote sensing
X-Ray	0.03 to 3.0 nm	Completely absorbed by atmosphere. Not employed in remote sensing
Ultraviolet	0.03 to 0.4 μm	Incoming wavelength less than 0.3 are completely absorbed by ozone in the upper atmosphere.
Photographic VB band	0.3 to 0.4 μm	Transmitted through atmosphere. Detectable with film and photo detectors, but atmospheric scattering is severe.
Visible	0.4 to 0.7 μm	Imaged with film and photo detectors. Includes reflected energy peak of earth at 0.5
Infrared	0.7 to 100 μm	Interaction with matter varies with wavelength. Absorption bands separate atmospheric transmission windows.
Reflected IR band	0.7 to 3.0 μm	Reflected solar radiation that contains no information about thermal properties of materials. The band from 0.7 to 0.9 is detectable with film and is called the photographic IR band.
Thermal IR band	3 to 5 8 to 14 μm	Principal atmospheric windows in the thermal regions. Images at these wavelengths are acquired by optical mechanical scanners and special vision systems but not by film.
Microwave	0.1 to 30 cm	Longer wavelengths can penetrate clouds, fog and rain. Images may be acquired in active or passive mode.
Radar	0.1 to 30 cm	Active form of microwave remote sensing. Radar images are acquired at various wavelength bands.
Radio	> 30 cms	Largest wavelength portion of electromagnetic spectrum. Some classified radars with very long wavelengths operate in this region.

The electromagnetic spectrum ranges from the very short wavelengths of the gamma ray region (measured in fraction of nanometers) to long wavelengths of the radio region (measured in meters). The visible region occupies only a small segment of the electromagnetic spectrum. Energy reflected from the earth during daytime may be recorded as a function of wavelength, which corresponds to the green band of visible region and is called the reflected energy peak. The earth also radiates energy both day and night with the maximum energy radiant at 9.7 mm wavelength. This radiant energy peak occurs in the thermal band of the IR region.

The earth's atmosphere absorbs energy in the gamma ray, x-ray and most of the ultraviolet (UV) region; therefore their regions are not used for remote sensing. Remote sensing records energy in the microwave, infrared and visible regions "Wavelengths regions of this transmission are called "atmospheric windows" the major remote sensing region are farther divided into bands, such as blue, green, red bands of the visible region.

Passive remote sensing systems record the energy that naturally radiates or reflects from an object. An active system supplies its own source of energy; flash photography is an example of active remote sensing in contrast to available light photography, which is passive. The other common form of active remote sensing is radar, which provides its own source of electromagnetic energy at microwave wavelengths.

Image Characteristics

In general, an image is any pictorial representation, irrespective of the wavelength of imaging device used to produce it. A photograph is an image that records wavelengths of 0.3 to 0.9 mm that have interacted with light sensitive chemicals in photographic film. Images can be described in forms of certain fundamental properties regardless of the wavelengths at which the images are recorded.

REMOTE SENSING OF THE FEATURES OF THE EARTH SURFACE

Vegetation

Spectral reflectance curves for healthy green vegetation almost always manifest the "peak-and-valley" configuration. The pigments in the leaves of the plants dictate the values in the visible portion of the spectrum. Chlorophyll strongly absorbs energy in the wavelength bands centred at about 0.45 and 0.65 mm. Our eyes perceive healthy vegetation as green in colour because of the very high absorption of blue and red energy by plants leaves and the very high reflectance of green energy. If a plant is subject to some form of tree that interrupts its normal growth and productivity, it may decrease or cease chlorophyll production. The result is less chlorophyll absorption in the blue and red bands. Often the red reflectance increases to the point that we see the plant turn yellow (combination of green and red).

As we go from the visible to the reflected infrared portion of the spectrum at above 0.7 mm, the reflectance of healthy vegetation increases. In the range from about 0.7 to 1.3 mm, a plant leaf reflects about 50 per cent of the energy incident upon it. Most of the remaining energy is transmitted, since absorption in this spectral region is minimal. Plant reflectance in the 0.7 to 1.3 mm range results primarily from the internal structure of plant leaves. Because, this structure is highly variable between plant species, reflectance measurements in this range often permit us to discriminate between species, even if they look the same in visible wavelengths. Likewise, many plant stresses alter the reflectance in this region and sensors operating in this range are after used for vegetation stress diction.

Beyond 1.3 mm, energy incident upon vegetation is essentially absorbed or reflected, with little to no transmittance of energy. Dips in reflectance occur at 1.4, 1.9 and 2.7 mm because water in the leaf absorbs strongly at these wavelengths. Wavelengths in these spectral regions are referred to as water absorption bands. Reflectance peaks occur at about 1.6 and 2.2 mm, between the absorption bands. Throughout the range beyond 1.3 mm, leaf reflectance is approximately inversely related to the total water present in a leaf. This total is a function of both the moisture content and the thickness of leaf.

In very simple words we can say that, the spectral characteristics of vegetation with varying wavelengths. A compound in leaves called chlorophyll strongly absorbs radiation in the red and blue wavelengths but reflects green wavelength. The internal structure of healthy leaves act as diffuse reflector of near infrared wavelengths. Leaves appears greenest to us in summer when the chlorophyll content is at its maximum. In autumn, there is less chlorophyll in the leaves, so there is less absorption and more reflection of red wavelength making the leaves appear red or yellow (combination of red and green primaries give yellow).

Soils and Rocks

The soil curve shows considerably less peak and valley variations in reflectance. The factors that influence soil reflectance act over less specific spectral bands. Some of the factors affecting soil reflectance are moisture content, soil texture (proportion of sand, silt and clay), surface roughness, the presence of iron oxide, and organic matter content. These factors are complex variable and interrelated, for example, the presence of moisture in soil will decrease its reflectance.

As with vegetation this effect is greatest in the water absorption bands at about 1.4, 1.9 and 2.7 mm. Soil moisture content is strongly related to the soil texture: coarse, sandy soils are usually well drained, resulting in low moisture content and relatively high reflectance; poorly drained fine textured soil will generally have lower reflectance. In the absence of water, the soil itself will exhibit the reverse tendency: coarse textured soils with appear darker than fine texture soils. The reflectance properties of a soil are consistent only within particular ranges of conditions. Two other factors that reduce soil reflectance are surface roughness and the content of organic matter. The presence of iron oxide in a soil will decrease reflectance, at least in visible wavelengths.

In simple words, the majority of radiation incident on a soil surface is either reflected or absorbed and little is transmitted. The characteristics of soil that determine its reflectance properties are its moisture content, organic content, texture, and structure and iron oxide content. The soil curve shows less peak and valley variations.

Water

Considering the spectral reflectance of water, probably the most distinctive characteristics are the energy absorption at reflected infrared wavelengths. In short, water absorbs energy in these wavelengths whether we are talking about water features per se (such as lakes and streams) or water contained in vegetation or soil, location and delineating water bodies with remote sensing data is done most easily in reflected infrared wavelengths because of the absorption property. Various conditions of water bodies manifest themselves in visible wavelengths. The energy/matter interaction at these wavelengths are very complex and depend on a number of interrelated factors, for example, the reflectance from a water body can stem from an interaction with the water's surface with material suspended in the water, or with the bottom of the water body. Even with deep water where bottom effects are negligible, the reflectance properties of a water body are not only a function of the water per se but also the material in the water. Clear water absorbs little energy having wavelength less than about 0.6 mm.

In short, majority of the radiation incident upon water is not reflected but is either absorbed or transmitted. Longer visible wavelengths and near infrared radiations is absorbed more by water than the visible wavelengths. Thus water looks blue or blue green due to stronger reflectance at these shorter wavelengths and darker if viewed at red or near infrared wavelengths. The factors that affect the variability in reflectance of a water body are depth of water, materials within water and surface roughness of water.

Snow

Snow in the Visible Spectrum: When clouds are not present in a visible or near-infrared image, it is a relatively straightforward task to identify the spatial distribution of snow because it is generally much brighter than the vegetation, soil or water nearby that is not covered with snow.

Snow in the Middle-Infrared and Microwave Regions**:** If one can obtain a single image of the terrain and it contains both clouds and snow, then it is possible to utilize the middle-infrared portion of the spectrum to differentiate between snow and cloud cover.

Throughout the visible and near infrared portion of the spectrum, snow reflects equal amounts of radiant flux. In the middle-infrared portion of the spectrum (especially 1.5-2.5 mm) snow reflectance and emission reduces dramatically. In the region 1.55-1.75 mm the snow has a very low reflectance and appears black.

Snow runoff can be both beneficial and disastrous. Snow provides much of the potable drinking water for mankind. Unfortunately, a tremendous build-up of snow or and untimely that can cause severe flooding and loss of life and property.

Clouds

Clouds do not reflect solar radiation equally in all directions. Therefore, a single measurement of reflectivity from a single direction makes it difficult to determine the total amount of light reflected by the cloud (its albedo) relative to the incident energy. One solution is to obtain multiple images of a cloud from different vantage points. This introduces stereoscopic parallels.

The most common thermal infrared band used in meteorological investigations is 10-12.5mm. Thermal infrared images can be obtained at night, so we have a continuous 24-hour record of cloud patterns. Clouds are usually colder than the surface of the earth and hence in visible as in images appear white. The ground or ocean surface being warmer, appear darker than clouds. Throughout the visible and near-infrared portion of the spectrum, clouds reflect equal amounts of radiant flex. In the middle-infrared portion of the spectrum (especially 1.5-2.5 mm) clouds continue to reflect or emit a substantial amount of energy. In the region 1.55-1.75 mm, the clouds have a very high reflectance and appear white in the energy

RESOLUTION

For remote sensing systems, the distance between the target and the satellite determines the detail of information obtained and the total area imaged by the sensor. Satellite sensors view a larger area, but cannot provide great detail, whereas flying onboard an airplane over a city or town; we can see fine details (individual building and cars) but only over a smaller area.

In broad terms, resolution of a system refers to its ability to record and display fine details. Resolution means firmness to temper. Resolution is the ability to distinguish between closely placed objects.

In remote sensing we deal with four types of resolution:

1. Spatial resolution
2. Spectral resolution
3. Radiometric resolution
4. Temporal resolution

Spatial Resolution

Spatial resolution is the ability to distinguish between two closely spaced objects on an image. It is the minimum distance between two objects at which the images of the objects appear distinct and separate. The visibility of details in an image depends on the spatial resolution of the sensor and refers to the size of the smallest possible feature that can be detected. It is the ground area sensed at any instant in time. It is dependent on the Instantaneous Field of View (IFOV) of the sensor.

The IFOV is the angular come of the sensor and it determines the area on the earth's surface, which is seen from a giver altitude at one particular moment in time. The size of the area viewed is determined by multiplying the IFOV by the distance from the ground to the sensor. This area on the ground is called the resolution cell and determines a sensor's maximum spatial resolution.

For a feature to be detected, its size should be equal to or larger than the resolution cell. Images where only large features are visible are said to have low resolution. Small objects can be detected in images of high-resolution. Military sensors, for example, are designed to view as much detail as possible, and therefore have very fine resolution. Commercial satellites provide imagery with resolutions varying form a few meters to several kilometers. Generally, the finer the resolution, the less total ground area can be seen.

The ratio of distance on an image or map, to actual ground distance is referred to as scale. If you had a map with a scale of 1:100000, an object of 1 cm length on map would actually be an object 100000 cm (1 km) long on the ground. Maps or images with small "map-to-ground ratio" are referred to as small scale (e.g. 1:100000), and those with larger ratios (e.g. 1:5000) are called large scale. Different satellites have different spatial resolutions; how much information can be seen is spatial resolution.

Spatial resolution is by far the most important characteristics of a remote sensing system. It determines the ability of a remote sensing system in recording spatial details.

Spatial resolution is a measure of an area or size of the smallest dimension on the earth surface over which an independent measurement can be made by the sensor.

Spectral Resolution

It describes the ability of a sensor to define fine wavelengths intervals. The finer the spectral resolution, the narrower the wavelength ranges for a particular channel or band.

Black and white film records wavelengths extending over the visible portion of the electromagnetic spectrum. Its spectral resolution is fairly low. Colour film is also sensitive to the reflected energy over the visible portion of the spectrum, but has higher spectral resolution, as it is individually sensitive to the reflected energy at the blue, green and red wavelength features of various colour based on their reflectance in each of these distinct wavelength ranges.

Many remote sensing systems record energy over several separate wavelength ranges at various spectral resolutions. These are referred to as multi-spectral sensors. Advanced multi-spectral sensors called hyper-spectral sensors, detect hundreds of vary narrow spectral bands throughout the visible, near-infrared and mid-infrared portions of the electromagnetic spectrum. Vary high spectral resolution facilitates fine discrimination between different targets based on their spectral response in each of the narrow bands.

Spectral resolution refers to the electro-magnetic radiation wavelengths to which a remote sensing system is sensitive. There are two components. The number of wavelength bands (per channels) used and the width of each wave band. A larger number of bands and a narrower bandwidth or each band will give rise to a higher spectral resolution. An aerial photograph obtained by an aerial camera with a film sensitized to the 0.4-0.7 μm (the visible spectrum) has only one wide band, which gives the name panchromatic to the film. The spectral resolution is therefore very low. In the case of Landsat MSS images, there are four spectral bans:

0.5-0.6 μm (green)
0.6-0.7 μm (red)
0.7-0.8 μm (reflected infrared)
0.8-0.9 μm (reflected infrared).

On the other hand, Landsat thematic mapper and enhanced thematic mapper plus all detect in seven bands:

0.45-0.515 μm
0.525-0.605 μm
0.63-0.69 μm
0.75-0.90 μm
1.55-1.75 μm
10.40-12.50 μm
2.09-2.35 μm.

Both these data have a much higher spectral resolution than Landsat MSS. There are three types of remote sensing with respect to wavelength:

1. Visible reflective infrared Remote Sensing.
2. Thermal infrared Remote Sensing.
3. Microwave Remote Sensing.

Radiometric Resolution

While the above two resolutions describes the spatial structure of an image, the radiometric characteristics of an image describe the actual information content in it. According to Natural Resources, Canada (www.nrcan.gc.ca), the finer the radiometric resolution of a sensor the more sensitive it is to detecting small differences in reflected or emitted energy.

"The imagery data constitute the positive digital numbers varying from 0 to a selected power of 2. This range corresponds to the number of bits used for coding numbers in binary format. Each bit records an exponent of power 2 (e.g. 1 bit = 2×1 = 2). The maximum number of brightness levels available depends on the number of bits used in representing the energy recorded. Thus, if a sensor used 8 bits to record the data, there would be 28 = 256 digital values available, ranging from 0 to 255. However, if only 4 bits were used, then only 24=16 values ranging from 0 to 15 would be available. Thus, the radiometric resolution would be much less. Image data are generally displayed in a range of grey tones, with black representing a digital number of 0 and white representing the maximum value (for example, 255 in 8-bit data).By comparing a 2-bit image with an 8-bit image, we can see that there is a large difference in the level of detail discernible depending on their radiometric resolutions.

There are tradeoffs between spatial, spectral, and radiometric resolution which must be taken into account while designing sensors for high spatial resolution, the sensor has to have a small IFOV (Instantaneous Field of View). However, this reduces the amount of energy that can be detected as the area of the ground resolution cell within the IFOV becomes smaller. This leads to reduced radiometric resolution—the ability to detect fine energy differences. To increase the amount of energy detected (and thus, the radiometric resolution) without reducing spatial resolution, we would have to broaden the wavelength range detected for a particular channel or band. Unfortunately, this would reduce the spectral resolution of the sensor. Conversely, coarser spatial resolution would allow improved radiometric and/or spectral resolution. Thus, these three types of resolution must be balanced against the desired capabilities and objectives of the sensor."

Temporal Resolution

As the name suggest, temporal resolution is linked with time. How much time takes a satellite to cover an area is known as temporal resolution. It depends upon the frequency at which the data is collected over the same scene or repeatedly.

Temporal resolution refers to the frequency of obtaining data over a giver area. It is related to revisit period, which refers to the length of time it takes for a satellite to complete one entire orbit cycle. The revisit period of a satellite sensor is usually several days, for example, LANDSAT 1, 2 and 3 had 18 days of cycle and start follows the same path after 18 days. In case of LANDSAT 4, 5 and 7 had 16 days of cycle. In case of IRS 1D has 25 days of cycle.

Therefore, the absolute temporal resolution of a remote sensing system to image the exact same area at the same viewing angle a second time is equal to this period. The actual temporal resolution of a sensor depends on a variety of factors, including the satellite/sensor capabilities, the swath overlap, and latitude.

Most important, temporal resolution of the remotely sensed data should be capable of matching the frequency of change of the landscape or phenomena to be mapped.

REMOTE SENSING SATELLITES

LANDSAT Series

The National Aeronautic and Space Administration (NASA) of United States launched the LANDSAT-1 on July 23, 1972. It was primarily designed as an experimental system to test the feasibility of collecting multispectral observation off earth from an unmanned satellite. Initially it was known as ERTS-1 (Earth Resource Technology Satellite). Until now six landsat satellites have been launched successfully while one of them LANDSAT-6 suffered a launch failure on October 5, 1993. The most recent being the LANDSAT-7 launched on April 15, 1999.

The series LANDSAT 1, 2, 3 orbited the earth in sun synchronous orbits at a height of about 900 km, every 10 minutes, 14 times a day. After 251 orbits i.e. after every 18 days, it passed over the some place on earth to provide repetitive coverage. LANDSAT 4, 5 are at an altitude of 700 km a revisit periods of 16 days.

A number of sensors have been put on board LANDSAT series Satellites, which include Return Beam Vidicon (RBV), Multispectral Scanner (MSS), Thematic Mapper (TM) and Enhanced Thematic Mapper (ETM). Each of these sensors collected data over a swath of 185 km. MSS had 4 spectral bands at spatial resolution of 80 m and a radiometric resolution of 6 bits. It is an across track electromagnetic scanner with an oscillating mirror. Six lines were collected simultaneously using six detectors for each band. TM sensor has an improvement over MSS in terms of resolutions. TM has 7 bands. For six bands the spatial resolution is 30 m and it has thermal band at 120 m resolution, ETM has 8 bands (Table 13.3 and 13.4).

Table 13.3. LANDSAT Missions

Satellite	*Launched*	*Retired*	*Sensors*	*Revisit Period*	*Altitude km*	*Equatorial crossing time*	*Orbit*
Landsat-1	23 June, 1978	6 June, 1978	MSS, RBV	18	900	0930 hours	Near Polar Sun synchronous
Landsat-2	22 June, 1975	27 July, 1983	MSS, RBV	18	900	0930 hours	Near Polar Sun synchronous
Landsat-3	5 March, 1978	7 Sept., 1983	MSS, RBV	18	900	0930 hours	Near Polar Sun synchronous
Landsat-4	16 July, 1982	—	MSS, TM	16	705	0945 hours	Near Polar Sun synchronous
Landsat-5	1 March, 1984	—	MSS, TM	16	705	0945 hours	Near Polar Sun synchronous
Landsat-6	5 October, 1993	Launch fail	ETM	16	705	0945 hours	Near Polar Sun synchronous
Landsat-7	15 April, 1999	Launch fail	ETM+	16	705	0945 hours	Near Polar Sun synchronous

MULTI-SPECTRAL SCANNER SYSTEM (MSS)

Oceansat and Resourcesat

ISROs Polar Satellite Launcher Vehicle (PSLV) from SHAR Centre, Sriharikota, launched Oceansat in the year 2002, IRS-P4_OCEANSAT is the first satellite primarily built for ocean application. The 1050 kg satellite will be placed in Polar Sunsynchronous orbit of 720 km height. IRS-P4 will carry on board an Ocean Colour Monitor (OCM) and a Multifrequency Scanning Microwave Radiometer (MSMR). OCM is a solid-state camera operating in eight narrow spectral bands. The camera is used to collect data on chlorophyll concentration, detect and monitor phytoplankton blooms and obtain data n atmospheric aerosols and suspended sediments in the water. MSMR, which operates in four microwave frequencies both in vertical and horizontal polarization, is used to collect data on sea surface temperature, wind speed, cloud

Table 13.4: Sensors on Board Landsat

Sensors	*Mission*	*Channel*	*Spectral Resolution*	*Spatial Resolution*	*Radiometric Resolution*	*Swath (km)*
RVB	Landsat 1,2,3	1	0.475-0.575	80	6 Bits	185
		2	0.580-0.687	80	6 Bits	185
		3	0.690-0.830	80	6 Bits	185
		4	0.505-0.750	80	6 Bits	185
MSS	Landsat 1,2,3,4,5	1	0.5-0.6	79	6 Bits	185
		2	0.6-0.7	79	6 Bits	185
		3	0.7-0.8	79	6 Bits	185
		4	0.8-1.1	82	6 Bits	185
		5	10.4-12.6	240	6 Bits	185
TM	Landsat 4,8,5	1	.45-.52	30	8 Bits	185
		2	.52-.60	30	8 Bits	185
		3	.6.76-.90	30	8 Bits	185
		4	.63-.69	30	8 Bits	185
		5	1.55-1.75	30	8 Bits	185
		6	2.08-2.35	30	8 Bits	185
		7	10.4-12.5	120	8 Bits	185
ETM+	Landsat 6,7	1	.42-.52	30	8 Bits	185
		2	.52-.60	30	8 Bits	185
		3	.63-.69	30	8 Bits	185
		4	.76-.90	30	8 Bits	185
		5	1.55-1.75	30	8 Bits	185
		6	10.4-12.5	30	8 Bits	185
		7	2.08-2.35	30	8 Bits	185
		8	.52-.90	15	8 Bits	185

water content and water vapour content in the atmosphere above the ocean. IRS P4 will thus vastly augment the IRS satellite system of ISRO, which at present comprises four satellites, IRS-1B, IRS-1C, IRS-P3 and IRS-1 D and extend remote sensing applications to several newer areas (Table 13.5 and 13.6).

Multi-spectral imaging was upgraded in the year 2003 with LISS-4, a multi-spectral sensor with 10-meter resolution aboard the IRS_PC satellite. It is also known as RESOURCESAT and is specially designed for agricultural application and earth resources.

SPOT Satellite

The SPOT (Satellite Pour *l'*Observation de la Terre*)* is a series of satellites launched by Centre national d'etudes spatiales (CNES) of France in conjunction with Sweden and Belgium.

The SPOT satellites are sun synchronous near polar satellite at an altitude of 832 km with a revisit period of 26 days. It has two High Resolution Visible (HRV) sensors. Each HRY is capable of sensing in either a high spatial resolution single bane Panchromatic mode (PLA) or in medium resolution three bands Multi-spectral mode (MLA). HRV is a push broom sensor using the CCD array. One 6000 element array for Panchromatic records at spatial resolution of 10 m and 3000 array of multi-spectral at

Table 13.5: Payloads Oceansat and Resourcesat Salient Feature

Ocean Colour Monitor	
Swath	1420 km
Field of View	360 m
Spectral Bands	8 (from 440+ 885 nm)
Multi-frequency scanning Microwave Radiometer	
Frequency	6.6, 10.65, 18, 201 GHz
Swath	1360 km

Table 13.6: Salient Features of Oceansat and Resourcesat

Orbit	Polar Sunsynchronous Circular
Altitude	720 km
Inclination	98.28 degree
Period	99.31 min
Local time of Equator crossing	12 noon
Receptivity cycle	2 days
Size	2.8 mX1.98mX2.57m
Mass at lift off	1050 kg
Length when fully deployed	11.57 m
Altitude and Orbit Control	3-axis body-stabilized using Reaction Wheels, Magnetic, Torques and Hydrazine Thrusters.
Power	9.6 sq m Solar Array generating 750w Two 21 Ah Ni-Cd Batteries
Mission Life	5 years.

20 m spatial resolution, both having a swath of 60 km. HRV can also view sideways as much as 27° making it possible to capture stress data (stereo is the capability to perceive three dimension (3D) for depth/height of the earth through the 2D images (Table 13.7).

Table 13.7: SPOT Satellites

Satellite	*Launch Date*	*Retired*
SPOT-1	22 February, 1986	31 December, 1990
SPOT-2	21 January, 1990	—
SPOT-3	23 September, 1993	—
SPOT-4	27 March, 1998	—

Sensors Onboard SPOT

SPOT 4 launched on March 23, 1998, reflects the continuity with improved performance design philosophy. The primary imaging systems on board this satellite are the high-resolution visible and infrared (HRIVR) sensors and the vegetation instrument.

SPOT images have better spatial resolution then landsat TM images and may be required as stereopairs. SPOT lacks the spectra range of TM. The availability and cost of landsat data are more favourable for the user than are SPOT data.

Sensors	*Mission*	*Spectral Resolution*	*Spatial Resolution (m)*	*Swath (km)*
HRV	SPOT 1,2,3			60 km
Panchromatic Mode		0.51-0.73	10	
Multispectral Mode		0.5-0.59 0.61-0.68 0.79-0.89	20	
HRVIR	SPOT 4	0.5-0.59 0.61-0.68 0.79-0.89 1.5-1.75	10	60
Vegetation	SPOT 4	0.5-0.59 0.61-0.68 0.79-0.89 1.5-1.75	1.165	2250

IRS-INDIAN REMOTE SENSING SATELLITES

The Indian Remote Sensing satellite combined the features of Landsat MSS/TM and SPOT HRV. The India Remote Sensing Satellite System became operational with the launch of IRS-1A on March 17, 1998 on board a Soviet Vostok Rocket. The second satellite IRS-1b identical to IRS-1A was launched on August 29, 1991 and it is still operational. These satellites carry two sensors LISS-I and LISS-II (Linear Imaging Self Scanning) puch broom sensors. These sensors collected the data in four spectral bands in the range of 0.45 to 0.86 micrometers. LISS-I has a spatial resolution of 72.5 m and LISS-II of 36.25 m. the third satellite IRS 1C, which was launched on December 28, 1995, has three sensors on board...

- PAN
- LISS III
- WIFS.

IRS-ID similar to IRS-IC was launched on September 29, 1997 by indigenously developed launch vehicle PSLV form own soil Sriharikota SHAR centre near Hyderabad (Table 13.8 and 13.9).

ROLE OF REMOTE SENSING DATA IN RESOURCE MANAGEMENT

We need multi-dimensional and integrated geo-informatics data base at the micro-levels to understand the complex atmosphere-geo-sphere-biosphere interaction and thereby to develop our resources on sustainable basis. This includes identification of key databases related to geographical, glacial, meteorological, anthropogenic, land uses and socio-cultural. Application of geo-informatics becomes essential for the study of complex parameters, which comes into play in the overall hydrological cycle. This

Table 13.8: IRS Missions

Mission	*Launch*	*Sensors*	*Altitude (km)*	*Revisit (days)*
IRS-IA	17 March, 1988	LISS-I LISS-II	904 904	22 22
IRS-IB	29 Aug, 1991	LISS-I LISS-II	904 904	22 22
IRS-IC	28 Dec, 1995	PAN LISS-II WIFS	817 817 817	5 221 5
IRS-ID	29 Sept, 1997	PAN LISS-II WIFS	817 817 817	5 24 5
IRS-P2	15 Oct, 1994	LISS-II	817	24
IRS-P3	21 Mar, 1996	WIFS MOS Astronomy	817 817 817	24 24 24
IRS-P4	26 May, 1999	OCM MSMR	727 727	3 3

Table 13.9: Sensors on Board IRS Mission

Sensors	*Spectral Resolution*	*Spatial Resolution (m)*	*Swath (km)*
LISS-I	0.42-0.52 0.52-0.59 0.62-0.68 0.77-0.86	72.5	148
LISS-II	0.42-0.52 0.52-0.59 0.62-0.68 0.77-0.86	36.25	74
LISS-III	0.42-0.52 0.52-0.59 0.62-0.68 1.55-1.70	23.5	141
PAN	0.5-0.75	5.8	70
WIFS	0.62-0.68 0.77-0.86	188	810

is further used for monitoring and predicting natural hazards, i.e. landslides, soil erosion and floods. The hydrological hazards like floods are the result of complex interaction between various natural systems which modify topographic gradients and anthropogenic activities in a watershed. Geoinformatics can be used as decision support system for identifying and integrating natural disaster management for sustainable development (Murai, 1996; Singh and Murai, 1998).

The Ministry of Environment and Forests has started six regional offices to facilitate environmental management plans for new projects and to improve the monitoring of their implementation. These stations are coming up at Chandigarh, Shillong, Calcutta, Lucknow, Bangalore, and Bhopal and are headed either by a chief conservator of forests or by senior scientists. These locations were selected as areas having either a concentration of industries or extensive forests. This administrative decentralization is a part of a series of new measures designed to ensure that environmental responsibility is built into all future project planning. However, the entire western industrial belt is not covered. For use of Indian Satellite IRS-IA data, the Department of Space has started Regional Remote Sensing Research Centres at Bangalore, Dehradun, Nagpur, Kharagpur, and Jodhpur.

NATURAL RESOURCE DATA MANAGEMENT SYSTEM (NRDMS)

The Department of Science and Technology launched a comprehensive programme in 1982 on the development of multi-disciplinary Natural Resources Data Management Systems (NRDMS) by setting up computerized data bases at micro-levels, taking the district as the unit. Natural resources are a great asset for any economy or region. An integrated approach to the utilization, monitoring, conservation, and management of natural resources along with protection and preservation of the environment is essential. The main objective of the NRDMS is to integrate the natural resources and socio-economic databases in order to facilitate monitoring exercises by providing information on different spatial unit's village, block, and district. Conventional land-based and remotely sensed data are integrated so as to develop comprehensive data bases in a particular area that can be used for micro-level planning. The system involves two distinct tasks: (i) development of interface technology to link satellite remote sensing to conventional district database, and (ii) adoption of a grid-based geo-coded integrated data system for macro and micro-level analysis as well as planning.

NATIONAL NATURAL RESOURCE MANAGEMENT SYSTEM (NNRMS)

The Planning Commission of the Government of India set-up the National Natural Resources Management System (NNRMS) in 1983 in order to achieve optimum utilization of natural resources through a proper and systematic inventory of resource availability. The Department of Space has been identified as the nodal agency for establishing NNRMS in the country. Taking into account the recommendations and suggestions by the task forces, six standing committees have been set-up to cover the following sectors: (i) Agriculture and soils, (ii) Bio-resources and the Environment, (iii) Geology and Mineral Resources, (iv) Ocean Resources, (v) Water Resources, and (vi) Remote Sensing Technology and Training sectors. The various ongoing projects include crop acreage and production estimation, soil moisture estimation using ERS-I, SAR, marine fisheries, coastal zone mapping, brackish water aquaculture, water-land mapping, grasslands of the Banni area in Kachchha, watershed prioritisation, environmental impact studies for the Narmada and Tehri projects, land use/land cover mapping for agro-climatic regional planning, drought monitoring, flooded area mapping and damage assessment, wasteland mapping and snow-melt run-off forecasting. Recently NNRMS has approved the projects begun with the initiative of the Ministry of the Environment and Forests for Natural Resources Audit as part of UNCED's Agenda 21 recommendations.

NATIONAL (NATURAL) RESOURCES INFORMATION SYSTEM (NRIS)

Since a comprehensive system is essential for policy-makers to ensure the optimum utilization of natural resources, development of a NRIS has been conceived as a major component under NNRMS. The information system would provide updated and systematic information on natural resources related to land, water, forest, minerals, soils, and oceans, etc., which are, in turn, being integrated with the socio-economic data. A computer system based on a Geographical Information System (GIS) is being developed with capabilities for data integration and easy retrieval. The NRIS has advanced capability and is proposed as an integrated information system with linkages with the other existing systems. Such an integrated information system of spatial (maps) and non-spatial (socio-economic) data at the districts, states and country, provide an efficient and powerful tool for resource managers and policy makers.

INTEGRATED MISSION FOR SUSTAINABLE DEVELOPMENT

The Integrated Mission for Sustainable Development is a new initiative by the Department of Space, Government of India for generating thematic maps covering arid and semi-arid regions in India, in 157 districts on a 1,50,000 scale. These districts cover 45 per cent of India's geographical area perennially affected by drought and floods and are located in hilly and tribal areas. Using data from Indian Remote Sensing Satellites, the methodology of the study involves the generation of thematic maps showing current land use/land cover, types of wastelands, forest cover/types, surface water resources, drainage patterns, potential groundwater zones, landforms (geomorphology), geology (rock types, structural features, mineral occurrence), soil types, etc.

APPLICATION OF REMOTE SENSING: CASE STUDIES

Monitoring and Mitigation of Flood Disasters

In India 40 million ha of geographical area is vulnerable to floods. According to Central Water Commission estimates, the average flood damage in India is Rs. 9,375.6 million during 1955-90. As frequency and intensity of floods have increased, the area liable to flood has also doubled from 20 million ha in 1971 to 40 million ha (59 million ha as per remote sensing estimation) at present. The major flood affected tract lies between the Gandak and the Tista. Hydrological system as a linking element between mountains and plains is very sensitive to unsustainable and undated anthropogenic activities in the highlands as well as in the lowlands. There is no doubt that the Himalaya has undergone a most dynamic change in land use due to the rapid increase in the population. Subsequently, natural disasters cause great deal of harm and damage every year and it has generally increased over the last decade (Messerli and Ives, 1997; Singh and Haigh, 1995).

The occurrence of extreme events such as landslides and other types of mass wasting are becoming a common feature in Indian Himalaya. It is very difficult to say that the hazards are mainly produced by natural fragility of the region or the biotic degradation and land use practices are more significant for the disaster process. Any change in physical configuration of a watershed directly affects its environmental sustainability downstream. The combination of type and causal processes is the basic condition for the classification of hazard risks and critical zones in a watershed. The degree of criticality is determined by the magnitude, frequency and type of hazard/risk. Most of the risk zones are confined to the slopes with less vegetative cover and densely populated regions. Few important characteristics are high steep slopes, rapid bio-climatic gradient, avalanches, lithologically weak structure, high population density, irrigated and terrace agriculture, tourism based economy and construction activities. Some snow and glacier related hazards being severe are considered moderate because they occur away from human habitation. Human induced hazards are prominent around construction activities. Construction causes erratic run-off making the area locally prone to mass wasting (Gardner, Pandey and Singh, 1997). Such processes are significant for causing flood disasters in highlands.

Satellite remote sensing is very effective in mapping flood-affected areas. Multiple satellite data from IRS, landsat and ERS provides very useful information on the flood situation. Microwave data can provide information on flood-inundated areas even under clouds, which is not possible with optical data. In India microwave data from ERS-1 SAR is operationally used along with optical data from IRS-1A/1B/1C for flood management. Frequent occurrence of flash floods is reported in the state of Sikkim due to rupture of moraine-dammed lakes using IRS data. Rapid retreat of glaciers can lead to the formation of terminal moraines which obstruct the glacial melt run-off, and form a lake. After reaching a certain level, water can spill over and break the barrier, sometimes resulting in a catastrophic flood. The later can also get triggered due to a surging glacier or avalanches. Remote sensing provides useful information about such glacial lakes and valleys prone to flash floods. Space technology is also being used for issuing of flood warning through collected data from data collection platforms.

Flood management is more appropriate than flood control. Thus for its management, the important management areas are wasteland, water storage, embankments and the construction of small scale integrated projects, to be given priority. Another effective method is the collaboration of states for Integrated Watershed Management for effective use of water. Flood control measures include channel flow regulation and intensity reduction by various measures. This problem should be considered as an integral part of National Forest and Water Policy. Flood mapping has been operational in India and management plans are drawn up for each flood risk zone, so as to arrest extensive damages consequent to floods.

Monitoring of Droughts

A major chunk of agricultural land more than sixty per cent, in India is rain fed and, therefore, is subjected to vagaries of monsoon. During the drought years, there is not only shortage in foodgrain production, but also its impact is felt due to scarcity of fodder and drinking water supply. Reduction in water in the reservoirs and rivers leads to decline in power generation and a chain of events thereafter. Drought indicates water scarcity. Drought occurs due to below-average precipitation or water available above or below ground. Frequent drought can lead to aridity and desertification. The impact of drought is governed by moisture deficiency, dry spell duration and size of affected areas. There are nine major droughts that the country suffered since 1972 (Table 13.10). The largest and severe drought of the century occurred in 1987. It is also true that the more severe a drought, the larger the areal coverage.

The drought-prone region accounts for about 20 per cent of the area and 12 per cent population of India. Over 70 million people and 18 million ha of cropped area spread over 7 states were in the grip of drought in 1992. The drought management strategy involves short and long-term plan for stabilising the economy and restoring environmental conditions. In recent years drought mitigation is closely linked with weather forecasting and groundwater management. Geographic Information System (GIS) is an automatic method to evaluate the potential of an area for groundwater recharge in crystalline hard rocks covered by granite and gneiss assemblage. The method ranks the potential of an area on the basis of recharge favourability scores by using statistical relation among factors that are assumed to be related to groundwater recharge. GIS coverage programme stores all the data needed to decide the site as to its suitability for recharge.

Table 13.10: Percentage Departures from Normal of Monsoon Rains during Recent Droughts

Year	*India and Delhi* (%)	*Haryana Chandigarh* (%)	*Neighbouring States with deficient rains*
1972	–24	–14	Rajasthan, Himachal Pradesh and Uttar Pradesh
1974	–12	–29	Entire Northwest India,
1979	–19	–46	Eastern Rajasthan, Himachal Pradesh, Punjab, Uttar Pradesh and Madhya Pradesh
1982	–15	–11	Western Rajasthan, Himachal Pradesh, Punjab, Hills of Uttar Pradesh
1986	–13	–22	Western Rajasthan and Western Uttar Pradesh
1987	–19	–67	Entire Northern and Western India
1989	+1	–29	Rajasthan, Himachal Pradesh, Western Uttar Pradesh and Western Madhya Pradesh
1991	–7	–25	Rajasthan, Himachal Pradesh, Hills of Uttar Pradesh
1992	–8	–13	—

Source: India Meteorological Department Reports.

Mapping Cropping Patterns and Land Use

Large Area Crop Acreage Estimation for several major crops (wheat, rice, groundnut, sorghum, and cotton and oil seeds) and integration with yield relationships for making production estimates has been undertaken. It is supplemented by development and demonstration of remote sensing based yield models for wheat and rice and a crop assessment system for wheat; mapping saline/alkaline soils of entire country on 1:250,000 scale and for selected districts on 1:50,000 scale; inventory and mapping of Land use/ land cover, using high resolution data, for the whole country (Level II) at 1:250,000 scale and of culturable wastelands of specific areas at 1:50,000 scale; identification, classification and monitoring of drought at district level for drought prone districts and at taluka level for Karnataka State. Projects have been undertaken at ISRO/NRSA along with Department of Agriculture and Cooperation, Ministry of Agriculture, Government of India and Central Water Commission and India Meteorological Department and in active collaboration with various State government departments on:

(i) Monitoring and early warning of droughts, and
(ii) Drought management through integrated land and water resources studies.

Under the drought monitoring and early warning project, vegetation index map, which is based on the concept that vegetation vigour is an indicator of drought, have been prepared, using NOAA-AVHRR data. The composite vegetation index maps are compared with ground parameters to establish interrelationship and the results obtained so far show a great promise. Based on the analysis of both Vegetation Index Map and Greenness Map as well vegetation index statistics, the drought assessment bulletin from National Agricultural Drought Assessment Management System (NADAMS) are regularly sent to concerned personnel at Centre/State Government Organisations/Institutions.

Drought Mitigation and Management

Drought mitigation requires input from remote sensing, meteorology, conventional data, vegetation index, and surface water index, in order to prepare short and long term integrated plan for mitigation (Rao, 1996), (Table 13.11). Drought management is a challenging task and it requires effort and suitable planning to combat it. For this purpose the desert areas are to be mapped by using NOAA/AVHRR in dry season and post rainy season. A botanical method by using mycrorrhizal fungi is being used for forestation as they are beneficial for soil micro-organisms. Three steps are necessary for drought management:

(a) short duration crops/mixed cropping
(b) water conservation/water harvesting, and
(c) improved methods of irrigations.

Table 13.11: Drought Mitigation Plan Using Space Technology

Remote Sensing Inputs	*Conventional Data*	*Integrated Drought Management Plan Short/Long-Term*
1. Geomorphological inputs for Ground Water Targeting	1. Use of Existing Bore Wells	1. Ground Water Targeting
2. Surface Water Bodies	2. Meteorological Information	2. Minor Irrigation Plans
3. Land use Patterns	3. Cropping Pattern	3. Ground Water Recharge Point
4. Drainage Patterns	4. Population Pattern	4. Watershed Development
5. Vegetation Index	5. Administrative Zones	5. Optimum Cropping Pattern
6. Waste Lands	6. Irrigation Pattern	6. Optimum Fodder Development
7. Soil Mapping	7. Evapotranspiration	7. Optimum Water Management (Surface + Ground) etc.
	8. Ground Geographical Survey etc.	

For drought monitoring and mitigation, a few innovative programmes undertaken are:

1. Cloud seeding programme,
2. Automation of weather data collection,
3. Groundwater exploration using NME based hydroscope,
4. Assessment and monitoring system,
5. Utilisation of remote sensing,
6. Improved technique of medium range weather forecasting,
7. Evapotranspiration observations,
8. Monsoon Activity Centre, and
9. Soil moisture observations.

REFERENCES

Joseph, George (2003), *Fundamentals of Remote Sensing*; University Press, Hyderabad.

Nag, Prithvish and Kudrat, M. (1998), *Digital Remote Sensing*, Concept Publishing Co.

National Remote Sensing Agency (NRSA) (1996), *IRS-1C Sensors: Interface.* 7:1-8, Hyderabad.

Murai, S. (1996), "Advanced Technologies in Geoinformatics", *Proceedings of International Seminar on Space Informatics for Sustainable Development*, UNCRD, Nagoya.

Ress, W.G., (1990), *Physical Principals of Remote Sensing*, Cambridge University, Cambridge.

Sabins, Floyd F. Jr., (1987), *Remote Sensing: Principals and Interpretation*, W.H. Freeman and Company, New York.

Singh, R.B. (ed.) (1991), *Environmental Monitoring: Application of Remote Sensing and GIS, GIC*, Hong Kong and EMS, Delhi.

Singh, R.B. (1994) *Space Technology for Disaster Monitoring and Mitigation in India*, INCEDE, University of Tokyo.

Singh, R.B. (1995), *Global Environmental Change: Perspectives of Remote Sensing and GIS*, Oxford & IBH Pub., New Delhi.

Singh, R.B. (ed.) (1996), *Research in Geography*, Vol. 1, *Disaster and Environment: Monitoring and Forecasting*, Vol. 2, *Land Use Changes and Sustainable Development*, APH Pub., New Delhi.

Singh, R.B. (ed.), (1998), *Ecological Techniques and Approaches to Vulnerable Environment: Hydrosphere-Geosphere Interaction*, Oxford & IBH Pub., New Delhi.

Singh, R.B. and Murai, S. (eds.) (1998), *Space Informatics for Sustainable Development*, Oxford & IBH Pub., New Delhi.

14
Collection and Analysis of Statistical Data

Maps are cartographic representations of ground realities as expressed by the data collected by the cartographer and/or available from various published and unpublished sources. The raw data is then statistically analyzed, and made suitable for map construction. Since map is an ideal instruments to comprehend the earth surface realities from a close distance they have been used from ancient most times in one form or the other.

Maps contain information in cartographic forms available from a variety of sources. Important among these sources are: satellite imageries, geodetic and plane surveys, socio-economic surveys and published maps, research reports, books, and publications international agencies and national, regional and local governments. The data available from the above sources are either primary or secondary. The sources of primary data are the field surveys while that of the secondary data are the already published or unpublished materials. The primary data can be assessed either by contacting every individual of the universe i.e. census survey, or by surveying only selected individuals or objects, called Sample Survey. Which of the two methods should be used, depends on the accuracy required and amount of time and resources available?

If collected by trained investigators using a well-structured questionnaire, primary data are far more reliable, representative of the population/universe and current than the secondary data. This is partly because the secondary data are not necessary focused on the research problem in hand, and partly because secondary data are usually dated. For researchers both the sources are, however, relevant and important. A good researcher would first try to lay his hands on the available secondary data and would opt for primary data collection only if secondary data are not available or are not adequate for the objectives of his research. The time and resources needed to undertake large-scale data collection using census method is, however, so stupendous that except for census of population and selected economic activities, sample surveys are the usual methods of data collection.

Both primary and secondary data are of two types: quantitative and qualitative. The qualitative data are descriptive, and based on structured or semi-structured interviews, focus groups transcripts, field notes, observation records and other sources both personal and documentary. As against this, the quantitative data are exact and expressed numerically.

SOURCES OF DATA

Primary Data

There are three main sources of primary data:

1. Geodetic and topographic surveys giving latitude, longitude, direction, elevation, location, distribution, etc. of various physical and cultural features of the area covered.
2. Socio-economic surveys conducted by the government and private agencies giving the structure, distribution and arrangement of population, agriculture, industry, etc. For example, the Census of India Organization conducts population surveys at an interval of a decade and publishes a wide-ranging data pertaining to the people of India.
3. Maps, charts, etc. including satellite imageries published by various public and private agencies depicting the distribution of both physical and cultural objects on the surface of the earth including those which depict the generalized pictures derived from statistical analysis of the primary data.

Secondary Data

The secondary sources of information are very many indeed. These sources are consulted by the researchers before as well as after the primary data are collected. Among the secondary sources of information are:

1. Encyclopaedia;
2. Professional Journals, Newspapers, Magazines, etc.;
3. Ph.D. Theses;
4. Maps, Atlases, and satellite imageries;
5. Books, Travelogues, and unpublished manuscripts;
6. Bibliographies;
7. Reports/books published by various professional organizations;
8. Reports published by various governmental organizations and departments, especially the decadal Census Reports and Plan Documents;
9. Reports of the United Nations and its specialized agencies; and
10. Internet.

Internet has now become a major source of secondary information. At the click of the mouse, one can access any information from anywhere in the world—both qualitative and quantitative. Much of this information has, however, to be cross-checked and analyzed using relevant statistical tools to derive meaningful results. Young researchers often have the tendency to reproduce materials available on the internet to prepare their research papers and doctoral dissertation without verifying their authenticity. At times the data sources are not acknowledged. This is unethical and against the academic norms. Appropriate acknowledgement of books, journals, websites consulted is obligatory. 'More the documents consulted and acknowledged better the research work' should be considered a norm.

METHODS OF PRIMARY DATA COLLECTION

There are two basic methods of primary data collection. These are:

1. Census, and
2. Sample surveys.

The census method takes each unit of 'population' or 'universe' into account. Census of India, for example, covers every Indian citizen living in India or abroad. This is the essence of Census Method. Practically all countries of the world use census method usually at the interval of ten years to find out the size, nature and distribution of their population. India conducted its first regular Census of Population in 1881. Since then, the decennial Census of India has been the primary source of information about the demographic characteristics of population of India. The 2011 Census of India is the second largest national census in human history, the first being that of People's Republic of China conducted in 2010.

Census method is used for surveying not only human population but also other vital economic resources like livestock, crops, business establishments, etc. In the case of sample studies, only representative units are studied. The samples are so drawn that they are as representative of the universe/population as possible.

WHICH METHOD TO CHOOSE AND WHEN?

Which of the two methods: census and sample be used in data collection, depends on a variety of factors? Important among them are:

1. Desired level of accuracy,
2. Purpose,
3. Urgency,
4. Resources available, and
5. Nature of the universe.

Census method, if used judiciously gives more reliable information than the sampling method. No matter how representative a sample is, the likelihood of it not being representative of the population is always there. This method can be misused too to dupe the people especially when it involves building up of confidence in certain ideas and products. The Census method, however, has its own disadvantages. It is time consuming, expensive, tedious and not without chances of some units being left out or double-counted.

The choice of method also depends on the purpose of the survey. If the area or universe is small and information on each unit thereof is necessary, one should go for census method but if the area or the universe is large, sample method is more appropriate.

If information is needed in a short time, one should opt for sampling. Census method is time consuming.

Census method is a costly affair. It needs a large number of enumerators to collect information, and a good number of assistants including the statisticians to analyze the data. If the available resources are limited, it is desirable to use sample method.

Sampling is less suited to areas where the universe is marked with irregularities in physical or cultural terms.

Sampling: Methods and Procedures

Sampling means selecting a small set of representative units from a universe and analyzing the data assuming that the results derived from the sample represent the whole universe. The chief problem in sampling is how to make the sample a true representative of the universe. This brings us to the question of how and how many samples be selected. If only typical units are to be selected the problem is what constitutes a 'typical unit?' The unit that is typical for one person may not be so for others. The samples must represent the universe.

To ensure that the samples drawn truly represent the universe, each unit must have an equal chance of being included in the sample i.e. selection is done randomly. Random sampling of the whole universe is, however, feasible only if the universe is uniform in distribution and characteristics. One can follow one of the following two procedures to do Random Sampling.

1. Arrange the units of the universe in an ascending or descending order and select every fifth or tenth, or hundredth unit, as the case may be. The ordering of the units may be alphabetic or any other basis, depending on the nature of the universe.
2. Make use of a random number table. Thus, if the data consists of 100 items listed in order of magnitude or in some other order, we can select from the table of random numbers, the number of samples needed. One can get a table of random numbers in any elementary book of statistics.

If the universe is marked with uneven distribution of a relevant characteristic, stratified sampling, which draws samples from each stratum of the universe, is used. If the characteristic is unevenly distributed, multi-level sampling is resorted to. It is a form of stratified random sampling.

Proxy sampling is method of drawing samples from a different population whose characteristics are the same as that of the real population. It is often resorted by drug manufacturers. They study the efficacy of a drug on a sample of rats presuming that if it works on rats, it would work on human beings too. Another example of proxy sampling is a study the voting behaviour of people born in 1980 and then concluding that those born in 1990 would also have the same behaviour. In both the cases, use of sampling procedure is misplaced. The rats differ genetically among themselves and hence a test on a given sample of rats need not represent all the rats. Similarly, if voting is not compulsory, it is not possible to pin point in advance of the election, who will actually vote at the forthcoming election.

In order to avoid the mistakes resulting from such imprecise populations, we should adopt a sampling frame, which can identify every single element and include anyone from it in the sample. The most straightforward frame is an electoral register or telephone directory. To ensure that all areas are included in the universe, a street map can be used for framework for door-to-door survey. Although, it does not show individual houses, we can select streets or village boundaries from the map and then visit all houses located thereon.

The sampling frames can suffer from one or more of the following drawbacks:

1. *Missing elements*: Some members of the population are not included in the frame.
2. *External elements*: The non-members of the population are included in the frame.
3. *Duplicate entries*: A member of the population is included twice.
4. *Groups or cluster entries*: The frame lists clusters instead of individuals.

The researcher has to keep the above possible shortcomings in the sample in view while choosing the method of survey.

Types of Sampling

Probability (Random) Sampling

When every unit in a population has equal or greater than zero chance of being included in the sample, it is probability sampling. For example, to estimate the annual income of the adults of a village, we randomly select one adult from each household. We interview the selected person and ask for how much earnings he has from various sources. Only one adult in a household having two adults would be included in the sample and his income would doubled to account for the income of the person not included in the sample. When every unit of the universe has the same probability of being included in the sample, it is called 'Equal Probability of Selection' (EPS) sample design. In our case, every adult has a non-zero probability of selection and hence involves random selection at some point.

There are several variants of probability sampling such as Simple Random Sampling, Systematic Sampling, Stratified Sampling, Probability Proportional to Size Sampling, and Cluster or Multistage Sampling.

Non-probability (Purposive) Sampling

In situations where some elements of the population have no chance of being selected through random sampling, one can take recourse to non-probability sampling wherein samples are selected using certain specific criteria or qualifications called *"exclusion bias"*. This is procedure is adopted when samples are meant to provide some specific information. In this case, the selection of the sample is not random and hence it is not possible to estimate the sampling error. To illustrate, let us take the example of a village survey. We visit every household in the village and interview the persons who come out to answer the doorbell. The person who comes out first need not be a family member; he may be a person who is there to answer the doorbell. He is, therefore, outside the non-probability sample. Non-probability Sampling includes Accidental Sampling, Quota Sampling and Purposive Sampling. Further, non-response can turn a probability design into a non-probability design if the reasons for the non-response are unknown. We would discuss these a bit later.

Sampling Design

One can adopt a variety of sampling designs depending on the nature of the frame; availability of auxiliary information about units of the frame; degree of accuracy and details required; and operational costs.

Simple Random Sampling

In Simple Random Sampling (SRS) each element of the universe has an equal probability of being selected: the universe is not subdivided or partitioned. Further, every *pair* or group of elements has equal chances of being selected: any variance between individual results within the sample can be taken as a proxy to the variance in the overall population. The only flaw that mars its utility is the possibility of the sample being non-representative. A sample of 10 students in a college may result in five being boys and the other five being girls. The census count may, however, record 65 boys and 35 girls or *vice versa*. Further, simple random sampling can be highly taxing in time and resources if the universe is very large.

Systematic Sampling

Systematic sampling is more rational than simple random sampling. The first task in systematic sampling is to arrange the target population in with some order and then tick mark the items at regular intervals. The ordering can be numerical (1, 2, 3....n) or alphabetical (a, b, c....z). Ticking need not begin from the item that is listed first; it can start from any randomly selected item from within the first to the *n*th element in the list. One may select every 10th name from the telephone directory but the starting point need not be the first name.

Stratified Sampling

When the population/universe in question contains distinct categories, there could be separate frames for each category to ensure adequate representation. Each category constitutes a stratum and samples are drawn from each stratum. There are several benefits of stratified sampling.

1. Dividing the population into independent sub-populations or strata helps in drawing inferences about specific sub-groups that may be lost in a more generalized random sample.
2. Stratified sampling method gives more efficient statistical results, certainly better than one can get from simple random sampling.
3. It is especially useful where data are readily available for pre-existing strata within a population along with those of the overall population.
4. Since each stratum is considered treated as a universe by itself, one can adopt different sampling approach.

Stratified sampling is most suited when:

1. Variability within the strata is minimal,
2. Variability between the strata is maximal,
3. The independent variables used in stratification are strongly correlated with dependent variable.

If the surveyor does not have adequate prior information, stratification can be done after the random survey. Known as 'post-stratification', this approach can supplement and complement the sample estimates.

Over Sampling

A sample is taken from each stratum so that the target class gets over-represented in the sample. The result built on this biased sample is not so reliable.

Cluster Sampling

Cluster sampling is 'two-stage' or 'multistage'. In the first stage, certain areas or time-periods are selected; and in the second stage, the respondents from *within* those areas or time periods are sampled.

Cluster sampling generally increases the variability of sample estimates above that of simple random sampling, depending on how the clusters differ between themselves, as compared to within-cluster variation. It is relatively less expensive and can give dependable results if the selected clusters are not biased.

Multistage Sampling

When cluster sampling with two or more levels of embedded units is attempted, it is multistage sampling. In the first stage the primary clusters are identified; in the second stage, primary units are identified in each primary cluster; in the third stage, additional samples are identified in each of the selected clusters; and in the final stage, all ultimate units are selected and surveyed. Multistage sampling is resorted to when the list of all members of the population is not final.

Quota Sampling

In the case of quota sampling, the universe is divided into a given number of mutually exclusive segments each represents a given characteristic, as in the case of stratified sampling. Samples are then drawn from each segment, in proportion to the presence of those characteristics in the universe. They constitute a quota. The interviewers get in touch with only those who want to participate in the survey and ignore those who are reluctant. This can lead to bias as not everyone gets a chance of being included in the sample. Naturally, in quota sampling the selection of the sample is non-random. This is its greatest weakness and quota *versus* probability has been a matter of controversy among the statisticians for quite some time.

Convenience or Opportunity Sampling

Convenience sampling is a typical non-probability sampling that includes in the universe only that those who can be easily approached personally or with the help of mobiles and internet. Naturally, the samples cannot represent universe. This type of sampling is, however, quite useful in pilot testing.

When new samples are selected with the help of random samples already drawn, it is known as snowball sampling. It is a variant of convenience sampling.

Panel Sampling

If a group of participants selected through random sampling technique is asked a given set of questions several times 'waves' over a given period, it is called Panel Sampling. It is used for large scale or nation-wide studies aimed at gauging the changes in the opinions of the people with regard to different issues/problems.

Event Sampling Methodology

Event Sampling Methodology (ESM) is a new form of sampling method; it allows researchers to study on going experiences and events in the ever-changing environment. It is used when a researcher has to measure the typology of activity and to capture the temporal fluctuations of work experiences. Popularity of ESM has increased in recent years because it helps researchers catch not only the inter-personal variances but also the intra-individual variances through time.

ESM has several advantages as well as disadvantages. One of the disadvantages of ESM is self-selection bias. It is particularly so when only certain types of individuals are willing to participate in the study creating a non-random sample. Some participants do not respond, or do not follow the rules.

ERRORS IN SAMPLE SURVEYS

A variety of errors both sampling and non-sampling can mar the value of data derived from sample surveys. Sampling errors are rooted in the sample design especially in sample selection. Non-sampling errors result from shortcomings in data collection and processing. They include inclusion of data from outside the population (over-coverage); exclusion of certain data from the sampling frame (undercoverage); wrong answer to a right question (measurement error); mistakes in data coding (processing error); or failure to obtain complete information from the individuals included in the sample.

The excluded spectrum can be taken care of partially the data pertaining to the strata included are weighted proportionately to the population of the excluded spectrum. For example, a simple random sample of individuals in West Bengal should include people in remote islands of Sundarbans. Interviewing them would be highly expensive. A cheaper method would be to use a stratified sample with urban and rural strata. Weighting the rural population of West Bengal in proportion to Sunderban's population would do away with this lacuna. Similarly, the households with multiple telephone lines have better chances of selection in a random digit-dialing sample. Weights can be assigned to the total population of the state to remove the lacuna.

PREPARING A QUESTIONNAIRE FOR SURVEYS

Whether it is a census or sample survey, the surveyor has to have a questionnaire to elicit responses from selected individuals. Here are a few hints for questionnaire preparation. At the top of the questionnaire should be a descriptive title of survey. Below the title, should be a short description of the purpose and objectives of the questionnaire. The body of the questionnaire should have sections with appropriate sub-titles. Items having same or similar format should go together.

Once the questionnaire is ready, indicate to the respondents what they should do next. If responses to questions are elicited by post, or *e-mail,* the forwarding letter should not appear as a *junk mail.* The respondent should feel involved in the project. If the questionnaire is administered personally the one administering it should be given clear and written instructions.

Ideal questionnaire is one, which:

- is brief enough to be covered in about 30 minutes,
- is so designed that questions fall in a logical sequence,
- contains questions, which are:
  - (a) simple, clear and unambiguous,
  - (b) not irrelevant,
  - (c) not too personal, and
  - (d) not suggestive of answers.

Questions should be brief and clear; lengthy and ambiguous questions confuse the respondent. The questions should be in logical sequence so that the respondents do not have to turn his mind forward and backward to find answers. Questions that seek detailed information about personal life and sources of income should be avoided. Lead questions like 'Don't you think that all sensible people use a particular product?' suggest the answer to be 'yes'.

Once the questionnaire is administered tabulation of the data contained therein and their analysis is equally a challenging tasks.

TABULATION OF DATA

Tabulation of data begins with the preparation of frequency tables which has columns for number of people giving each answer (N) and the percentage of the column. The following table gives the responses to a question: Do you agree with the statement that 'corruption is widespread in India today.'

Response	*N*	%
Strongly agree	284	36.18
Agree	201	25.61
Neutral	110	14.01
Disagree	121	15.41
Strongly Disagree	69	8.79
Total	785	100

Cross-tabulation becomes necessary when relationships between two or more variables are to be calculated. First, a contingency table divided into *cells* is prepared. Each cell represents the occurrence of a specific value related to each variable. The last row and column gives the total for each row and column otherwise known as 'Row Marginals' and 'Column Marginals'. Normally, the dependent variable does appear in the rows. But it may not always be so if one dependent variable is cross tabulated by a number of explanatory variables.

ANALYSIS AND INTERPRETATION OF DATA

Analysis of both primary and secondary data is influenced by the variables involved. If there is only one variable the methods of analysis would be uni-variate such as frequency table, bar chart, pie chart, etc. If the number of variables is two, the methods of analysis would be bi-variate such as cross-tabulations, scatter plots, regression, comparison of means, etc. Multi-variate analysis is resorted to when the variables number 3 or more.

The scales of measurement can differ depending on how specific one wants to be. They can range from numbers to grades, ranks, letters, and words. Numbers appear to give exact measurements of values but it is not always so. The hotel room number or registration number of an automobile differentiate one room or vehicle from the other without implying that higher the number, better it is. The values expressed by these numbers are nominal. In the same category fall the blood groups: A, B, O, etc. The blood group categorization is only nominal and hence it falls in *nominal scale* category.

When we have the *ordinal scale,* which orders or ranks certain variables keeping in view the presence or absence a given characteristics. This is how the fruits, vegetables, grains, etc. are graded. The degree of distance among the ranks in the ordinal scale is called *interval scale*. There is no absolute zero in these scales, and hence cross scale comparisons are difficult. If there is absolute zero in the interval scale (like 0° temperature), we use ratio scales. For example, 10 cm is twice as long as 5 cm. It is because equal intervals are equivalent. The unit increase in weight from 10 kg to 20 kg is the same as from 100 kg to 110 kg.

The data, no matter how collected are:

1. Qualitative data, or
2. Quantitative.

Qualitative information helps in preparing questionnaire for primary survey and also in hypothesis building. It is also important in cross-checking the quantitative data. For example, if most people in an area are considered to be illiterate but the data collected from the field show that only 50 per cent of them are illiterate then the researcher should have another look at his sample design. Either the general perception is wrong or the samples selected are unrepresentative.

There are two sources quantitative data: primary and secondary. Irrespective of the source, the authenticity of the data should be cross-checked and analyzed appropriately before use. The first step in analyzing and interpreting quantitative data is to group or classify them. Table 14.1 gives the rainfall data of a town for 21 years. The data are then converted into frequency distribution tables (see Table 14.2).

Table 14.1: Rainfall Data of Town X: 1947-1967

Year	*Rainfall in Inches*	*Year*	*Rainfall in Inches*
1990	20	2001	8
1991	36	2002	11
1992	37	2003	26
1993	15	2004	14
1994	10	2005	17
1995	5	2006	20
1996	25	2007	18
1997	23	2008	13
1998	18	2009	12
1999	35	2010	28
2000	19	2011	—

Table 14.2: Frequency of Rainfall Occurrence in Town X

Rainfall in Inches	*Frequency of Occurrence*
1-10	3
11-20	11
21-30	4
31-40	3

Once the data are classified, frequencies can be plotted to draw building blocks, linear graphs, etc. as illustrated in chapter 17.

CENTRAL TENDENCIES

We have seen how individual sets of data are discrete and give limited insight. These data have to be summarized and then analyzed to get patterns and trends. The first step in summarization and analysis is the determination of central tendencies, which are of four main types. They are:

1. Arithmetic mean/average
2. Median,
3. Mode, and
4. Geometric mean.

Arithmetic Mean

Arithmetic Mean is the most commonly used and simplest to calculate central tendency that generalizes a series of data of the same type. Here, is a simple example. Two persons working in a factory get Rs. 2000 and Rs. 8000 per month respectively as their monthly salaries. If we have to find the average monthly salary of the two, we add Rs. 2000 and Rs. 8000 and divide the total by 2. Thus the average monthly salary of the two persons = Rs. 5000. This is called Arithmetic Mean because it is based on simple arithmetical processing of data using addition, multiplication and division.

Here is another example. Suppose we have the following set of data giving the number of years ten people have lived in cities: 1, 4, 2, 3, 3, 3, 5, 2, 4, and 3. If we add these discrete numbers, and divide the total by the number of people involved, we get (30/10) = 3. This is the arithmetic mean or average of the number of years the persons surveyed lived in cities.

Arithmetic Mean can be expressed as:

$$\overline{X} = \frac{\Sigma X}{N}$$

where

X = the individual values making up the series of data

$\overline{X}$ = the arithmetic mean

N = frequencies (number of items)

Σ = (Sigma) the summation or total of all the values of $\overline{X}$

This formula is applicable only if the data are given in discrete forms. In case of grouped data, a different formula is used. Given below are the formula and its application.

Table 14.3 gives the sales checks of a Grocery Store in a grouped format. We have to calculate the arithmetic mean.

Table 14.3: Sales Checks in a Grocery Store

Class	*Mid point*	*Class frequency*	*Frequency × Mid-point*
	X	*f*	*fX*
Rs. 1 and < 3	2	3	Rs. 6
Rs. 3 and < 5	4	9	Rs. 36
Rs. 5 and < 7	6	25	Rs. 150
Rs. 7 and < 9	8	35	Rs. 280
Rs. 9 and < 11	10	17	Rs. 170
Rs. 11 and < 13	12	10	Rs. 120
Rs. 13 and < 15	14	1	Rs. 14
		100	776

Formula for the grouped data is as follows:

$$\overline{X} = \frac{\Sigma fX}{N}$$

$$N = 100\ \Sigma\ fX = \text{Rs. } 776$$

Substituting the numbers, we get

$$\overline{X} = \frac{\Sigma fX}{N} = \frac{\text{Rs. } 776}{100} = \text{Rs. } 7.76$$

Thus, the mean sale value for the store is Rs. 7.76.

Median

Median represents the mean expectation in that there are as many individual occurrences above it as there are below it. In the calculation of the median every occurrence is given the same weight, i.e. it is regarded as of equal importance no matter of what magnitude it is. This means that widely differing sets of data can return the same median value. Further, this implies that the median possesses no real mathematical qualities and cannot be used for further computations.

If we arrange the data showing the number of years the selected ten persons lived in cities in following order: 1, 2, 2, 3, **3**, 3, 4, 5, and 5 and try to find the mid-value by observation, we get 3 as there are four figures to the left and the same number to the right of 3. The mid-value of a given set of data is called Median. In this case it 3. The Arithmetic Mean of the same set of data would be 28/9 = 3.11. If we take into account the frequency of the figures in the series and select the one, which occurs most often, we get 3. The figure with highest number of occurrences in a series is called Mode. In this particular case 3 occurs three times. While arithmetic mean is represents the series best, the median and mode too represent the series to some extent at least.

We will now go into details:

Value	*Frequency*
1	1
2	2
3	3
4	1
5	2

To reiterate, median is that value of a variable which lies midway when the data are arranged in order of magnitude. Median value of n observations will be (n + 1)1/2 the observation. *Example*: to find the Median of (12, 3 and 6), first we put them in order of magnitude (3, 6, 12). The middle number is 6, so the median is 6. Here is another example: Given below are 15 numbers: 3, 13, 7, 5, 21, 23, 39, 23, 40, 23, 14, 12, 56, 23, and 29. If we put them in order, we get: 3, 5, 7, 12, 13, 14, 21, 23, 23, 23, 23, 29, 39, 40, and 56. In this set of 15 numbers, the medium number is 23. We see that "half the numbers in the list are less, and half the numbers are greater than 23." It matters little if some numbers are listed more than once in the list.

If the total observations amount to even numbers, the calculation of the median will be slightly different. In that case, we would try to find the middle pair of numbers, and then add them and divide by two. Here, is an example. Fourteen discrete numbers: 3, 13, 7, 5, 21, 23, 23, 40, 23, 14, 12, 56, 23, and 29 are then put in ascending order: 3, 5, 7, 12, 13, 14, 21, 23, 23, 23, 23, 29, 40, and 56. In this case, we do not have a middle number but we do have a pair of middle numbers: 21 and 23. We add them and divide by two to get the median value i.e. 21 + 23 = 44 ÷ 2 = 22.

And, so, the Median, in this case is 22. It is a number, which is not in the list but it does meet one criterion: half the numbers in the list are less, and the other half more than 22.

In the case of grouped data, the following formula is used:

$$\text{Median} = L_1 + \left(\frac{\frac{N}{2} - \Sigma f_1}{fm}\right).X$$

where

L_1 = lower limit of the median class.

N = total frequency

Σf_1 = the sum of all frequencies accumulated before entering the median class.

fm = the frequencies in the median class.
X = the size of the class interval of the median class.

Method of finding the median of grouped data using the above formula is exemplified by Table 14.4, which gives the number of workers in various industrial establishments.

Table 14.4: Number of Industrial Establishments and Number of Workers

Number of Workers	*No. of Establishments (= frequencies)*	*Cumulative Frequency*
0–49	46	46
50–99	59	105
100–149	45	150
150–199	37	187
200–299	31	218
300–399	20	238
400–499	13	251
500 and over	9	260

$$N = 260$$
$$\frac{N}{2} = 130$$
$$L_1 = 100$$
$$\Sigma f_1 = 105$$
$$fm = 45$$
$$X = 50$$
$$\text{Median} = 100 + \left(\frac{130 - 150}{45}\right).50$$
$$= 127.78 \text{ workers}$$

As apparent from the above example, a median divides the data into two equal parts. It is called Q_2 i.e. second quartile. Quartiles that mark the limits of lower and upper quarters are known as 1st and 3rd quartiles respectively. Similarly, particular values, which divide the total observations into 10 and 100 equal parts, are respectively known as deciles and percentiles. The quartiles are denoted by Q_1, Q_2 ... Q_n whereas deciles and percentiles by D_1, D_2 ... D_n and P_1, P_2 ... P_n etc.

The quartile observations are ascertained by the formula:

$$Q_1 = \frac{n+1}{4}\text{th observation}$$
$$Q_2 = \frac{n+1}{2}\text{th observation}$$

$$Q_3 = \frac{3(n+1)}{4}\text{th observation}$$

If the series is a discrete one, the values can be computed using the following formula:

$$Q = L_1 + \left(\frac{L_2 + L_1}{F}\right)(q - c)$$

where

L_2 and L_1 are upper and lower limits of the quartile class
F = frequency of the quartile class.
q = quartile observation.
c = cumulative frequency preceding the quartile class.

Similar there are methods to calculate deciles and percentiles.

$$D_1 = \frac{n+1}{10}\text{th observation}$$

$$D_9 = \frac{9(n+1)}{10}\text{th observation}$$

$$D = L_1 + \frac{L_2 + L_1}{F}(d - c)$$

$$P_1 = \frac{(n+1)}{100}\text{th observation}$$

$$P_{50} = 50\frac{(n+1)}{100}\text{th observation}$$

$$P = L_1 + \frac{L_2 + L_1}{F}(p - c)$$

The above derivations prove that Median = $Q_2 = D_5 = P_{50}$

Mode

Mode is the value, which occurs most often in a given distribution. In a grouped data, the class having the maximum frequency is called modal class. The mode, by its very definition, indicates what is most common or frequent. Very often, however, there is some difficulty in deciding exactly where the mode occurs. This may be for one of the two reasons: the data are multi-modal or the selection of the classes is biased. Median represents the mean expectation in that there are as many individual occurrences above it as there are below it. In the calculation of the median every occurrence is given the same weight, i.e., it is regarded as of equal importance no matter of what magnitude it is. This means that widely differing sets of data can return the same median value. Further, this implies that the median possesses no real mathematical qualities; it suffers from the same limitations as does the mode and hence it cannot be used for further computations.

Among the three types of averages, arithmetic mean alone is suitable for mathematical analysis. The first step in calculating the mean is to assign weight to each occurrence according to its magnitude and then add them together and divide them by the number of occurrences. For example, if the output of one farm is 200 quintals and that of another farm is 100 quintals, the average output of each farm would be 150 quintals. This, however, does not give a real picture. For a proper appreciation of the reality underlying a mean value, it is necessary to know how the individual variates the mean summarizes.

The formula for the grouped data is:

$$\text{Mode} = L_1 + \frac{\Delta_1}{\Delta_1 + \Delta_2}.X$$

where

L_1 = the lower limit of the modal class.
Δ_1 = differences between the frequencies in the modal class and pre-modal class.
Δ_2 = the difference between the frequencies in the modal class and post-modal class.
X = the size of the class interval of the modal class.

To find the mode we may take the example given in Table 14.5.

Table 14.5: Finding the Mode of Weekly Income of Workers

Weekly income (Rs.)	*No. of Workers*
10 and < 20	85
20 and < 30	120
30 and < 40	110
40 and < 50	67
50 and < 60	49
60 and < 70	21
70 and < 80	6
All	458

Here

$L_1 = 20$; $\Delta_1 = 35$; $\Delta_2 = 10$; $X = 10$

(*Note*: the modal class is the class with largest frequencies; in this case it is 120.

$$\text{Mode} = 20 + \left(\frac{35}{35+10}\right)10$$

$$= 20 + 7.78 = \text{Rs. } 27.78$$

Geometric Mean

The geometric mean helps in averaging the ratios and in computing average rates of increase or decrease. It is the nth root of the product of N items. If the number of items is two, we take the square root; if three, the cube root, and so on. Geometric mean can be used only if the series being analyzed does not contain zeros or negative values.

The formula for geometric mean is G = nth product of the Xs. To calculate the geometric mean of a set of data, we use the following formula:

$$\left(\prod_{i=1}^{n} a_i\right)^{1/n} = \sqrt[n]{a_1 a_2 \ldots a_n}. \quad \{a_1, a_2, \ldots, a_n\}\text{is:}$$

Thus the geometric mean of two numbers, say 2 and 8, will be square root of their product; that is $\sqrt[2]{2 \times 8} = 4$; and the geometric mean of three numbers 1, ½, ¼ will be the cube root of their product (1/8), which is 1/2; that is $\sqrt[3]{1 \times ½ \times ¼} = ½$.

Relationship among the Central Tendencies

The average, the median and the mode are the main methods of expressing the mean values of a set of data. It would be desirable to consider briefly the relationship among them and to see which one to use where, and when. Let us consider the series 1, 2, 2, 3, 3, 3, 4, 4, 5 wherein the mean, median and mode coincide. Fig. 14.1 (a) shows a histogram based on this series. The distribution shown by this histogram is called normal distribution, and the curve representing this distribution is called a normal distribution curve [Fig. 14.1(b)] The existence of normal distribution is assumed in most statistical analysis, although in practice it seldom exists. If this series is changed to 1, 1, 2, 2, 2, 3, 3, 4, 4, and 5

We get the Mean $\overline{X} = \Sigma X$

$= 27/10 = 2.7$

Median = 2.5

Mode = 2

The distribution we get is not normal. It is a skewed distribution. If the tail extends to the right, it is called positive skewness; and if it extends to the left, it is called negative skewness [Fig. 14.1(b)].

Calculating Deviation from the Mean

Deviation or variability means the distances at which the variates stand from the mean.

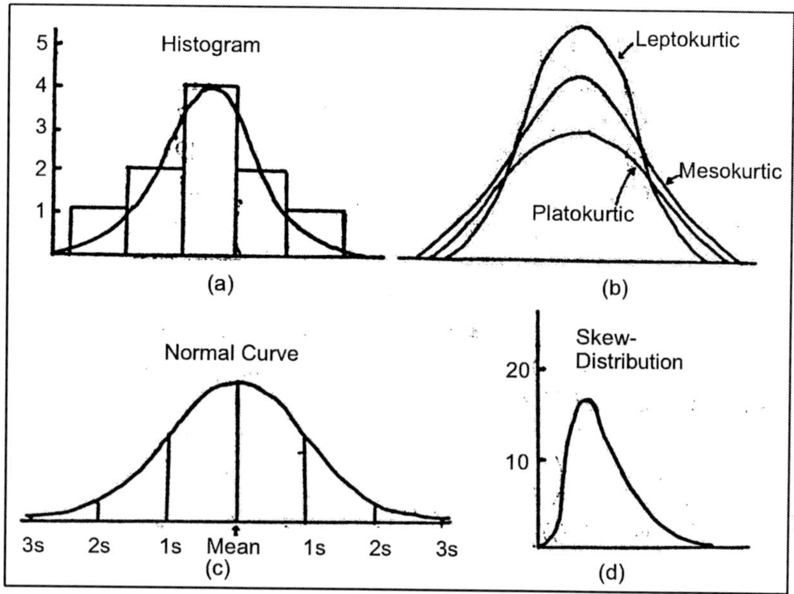

Fig. 14.1: Histogram and Various Types of Curves

In the above case, the distance at which Rs. 2000 and Rs. 1000 stand from Rs. 1500 is deviation. This can be done either by finding absolute variations i.e. Rs. 2000 – 1500 = Rs. 500, and Rs. 1000 – Rs. 1500 = – Rs. 500; or relative differences i.e. + 33.33 per cent and – 33.33 per cent. Absolute deviation is expressed in positional terms also.

Deviation is of two types:

- Range, and
- Inter-quartile

Range deviation: If we indicate the lowest and the highest values along with the mean value, then we give the range of deviation. For example, the following two sets of data have the same averages but different ranges.

Set A. 1, 3, 5, 7, 9: average = 5
Set B. 3, 4, 5, 6, 7: average = 5

In the first case, the range is: 1 to 9 while in the second it is: 3 to 7. If both average and range are given, it is easier to understand the series.

Inter-quartile Deviation: This is another form of scatter, which is used in conjunction with the median.

$$\text{I.Q.D.} = \frac{Q_1 - Q_2}{2}$$

where,

Q_1 and Q_2 are the upper and lower quartiles respectively.

Mean and Standard Deviation

Mean deviation and the standard deviation are the two most commonly used methods to find out absolute deviation. Both take into account the deviation of each item in a series from an average. The average deviation tells how much each occurrence differs from the average. To find out average deviation (D), differences are added and the sum divided by the number of items (Table 14.6).

Table 14.6: Method of Calculating Average Deviation

Items	*Values* *X*	*Deviation* *D*
1	1	– 4
2	3	– 2
3	5	0
4	7	+ 2
5	9	+ 4
Total 5	**25**	**12**

$$\overline{X} \text{ (Mean)} = 1{+}3{+}5{+}7{+}9{+}25 = 50/5 = 5$$

$$\Sigma D \text{ (Deviation)} = -4 + -2 + 0 + 2 + 4 \text{ (signs ignored)} = 12$$

$$\text{Average (mean) deviation} = \Sigma\frac{D}{X} = \frac{12}{5} = 2.4$$

In this case, 2.4 is the mean deviation. It may be noted that in this calculation, no consideration has been given to positive and negative deviations.

Standard Deviation

As mentioned earlier, we do not take into account positive (+) and negative (–) signs while calculating the mean deviation. One way of getting over this mathematically irrational approach is to square the deviations Table 14.7. This is what we do when we calculate the standard deviation.

Table 14.7: Number of persons Employed and Volume of Sales in a Store

X	*D*	D^2
1	– 4	16
3	– 2	4
5	0	0
7	+ 2	4
9	+ 4	16
25/5 = 5	12	40/5= 8

$X = 25/5 = 5, \quad D = 12$

Now $\Sigma D^2 = 40$

$$\text{the variance} = \frac{\Sigma D^2}{N} = \frac{40}{5} = 8$$

The value derived from dividing the squared deviations by the number of values is called Variance. Here 8 is the variance. If we calculate the square root of the Variance, we get a value, called Standard Deviation.

The shorthand formula for the two is given below:

$$\text{Variance } (\sigma^2) = \Sigma\left(\frac{X - X}{n}\right)^2$$

$$\text{Standard Deviation (S.D. or } \sigma) = \sqrt{\frac{\Sigma(X - X)^2}{n}}$$

Indices of Variability

The above measurements use absolute values. Dispersion can also be shown relatively by indices of variability. These are:

Index of Variability

$$\frac{\text{Quartile deviation}}{\text{Median}} \times 100\%$$

Co-efficient of Variation

$$\frac{\text{Standard devistion}}{\text{Mean}} \times 100\%$$

Relative Variability

$$\frac{\text{Mean devistion}}{\overline{X}} \times 100\%$$

Pluvio-metric Co-efficient

$$cp = \frac{r}{R\frac{m}{365}}$$

Where

cp = pluvio-metric co-efficient

r = rainfall in the month
R = annual rainfall
m = number of days in that month

NORMAL CURVE AND PROBABILITY

Thus in order to represent a set of data at least two parameters (characteristics) must be defined: first, the mean/average and second, deviation from the mean. Let us take a concrete example. The average yield of wheat per acre for a series of farms is 30 quintals and the standard deviation is 5 quintals. If the data in the series were symmetrically distributed about a central point (arithmetic mean), we will get a normal curve (Fig. 14.1a).

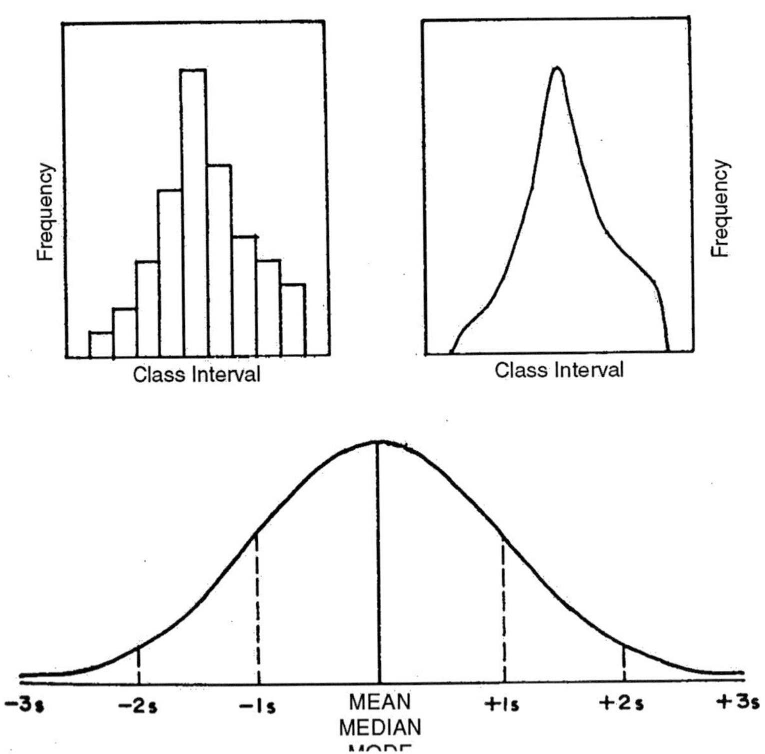

Fig.14.1(a): A Normal Curve

In this figure, 68 per cent of the farms fall standard deviation (SD) away from X (mean) i.e., 30 ± 5, and 95 per cent of the farms fall 2 SD away from the mean, and so on. More precisely:

68.30 per cent of the occurrences lie within ± 1 σ
95.45 per cent of the occurrences lie within ± 2 σ
99.7 per cent of the occurrences lie within ± 3 σ

99.99 per cent of the occurrences lie within ± 4 σ

Normal distribution of data is, however, not a common occurrence. Often they are asymmetrically distributed. In the case of symmetrical distribution, mean, median and mode are the same, but in an asymmetrical distribution, they are different. Asymmetry is marked with positive or negative skewness. In a positively skewed curve, mean exceeds the mode, and in a negatively skewed curve, mode exceeds the mean.

Degree of skewness can be expressed as co-efficient of skewness:

$$\text{Skp} = \frac{3(\overline{X} - \text{median})}{\text{SD}}$$

where

Skp = Co-efficient of skewness
X = mean
SD = Standard deviation
Higher value means more skewness.

It does not, however, mean that all asymmetrical distributions produce curves of the same shape i.e. all asymmetrical curves are not normal. Departure of an asymmetrical curve from the normal is called *kurtosis;* if it is too tapering, it is called *leptokurtic;* if too flat, it is called *plati-kurtic;* but if it is in-between and occupies a middle position, it is known as *meso-kurtic* [Fig. 14.1(b)].

LINEAR REGRESSION

It is a line describing the relationship between any two variables and may be called "the line of average relationship". In fact there are two regression lines, one for Y on X and the other for X on Y. In the first instance, it is possible to predict the approximate value of Y, given a value for X and vice versa for the regression of X on Y.

The equation for Y (unknown) on X (known) is

$$(Y - \overline{Y}) = r\frac{SDy}{SDx}(X - \overline{X})$$

For X (unknown) and Y (known) the expression is

$$(X - \overline{X}) = r\frac{SD_x}{SD_y}(Y - \overline{Y})$$

The regression equation of Y on X for the data referring to the employments and sales becomes:

$$Y - 485 = 0.85 \times \frac{265.8}{21.68}(X - 50)$$

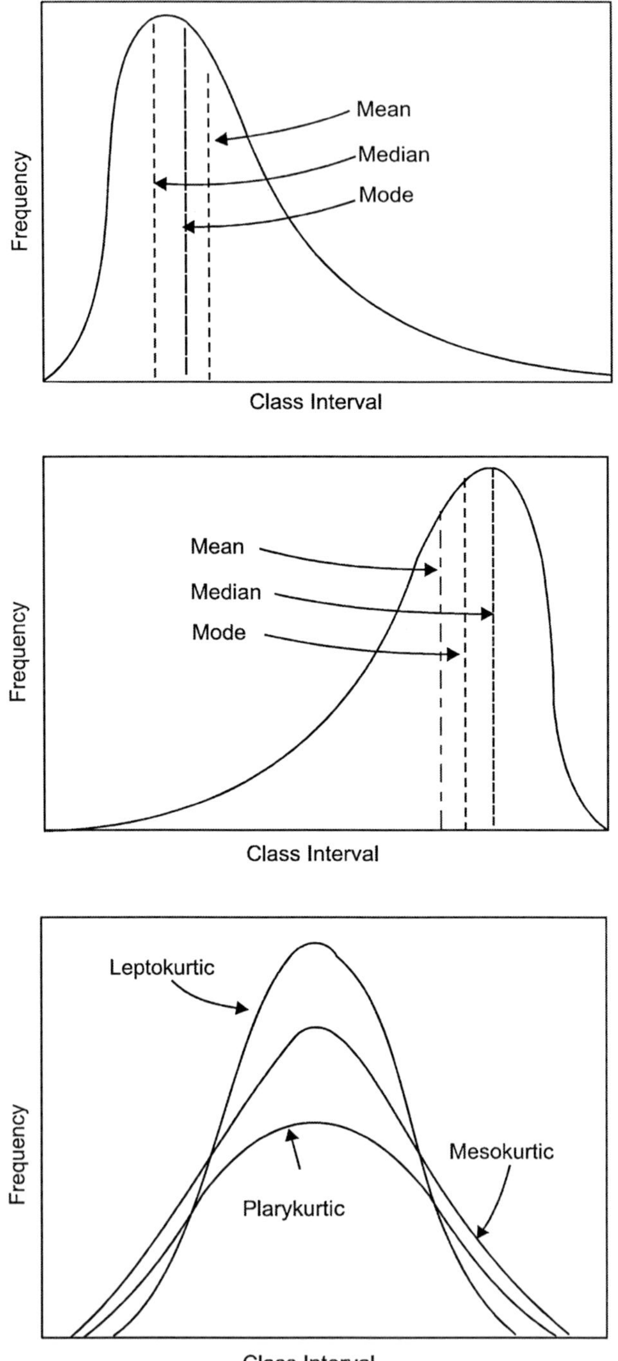

Fig. 14.1(b): Asymetrical Curves

Thus $Y = 10.17 - X\,23.50$

Taking arbitrary values,

$$X = 10,\ Y = 78.20$$
$$X = 20,\ Y = 179.90$$

The regression line for Y on X can also be found out as:

$$X - 50 = 0.83 \times \frac{21.68}{265.8}(Y - 485)$$
$$X = .08\,Y + 11.20$$

These values may be used to draw the regression lines on a scatter diagram (Fig. 14.2).

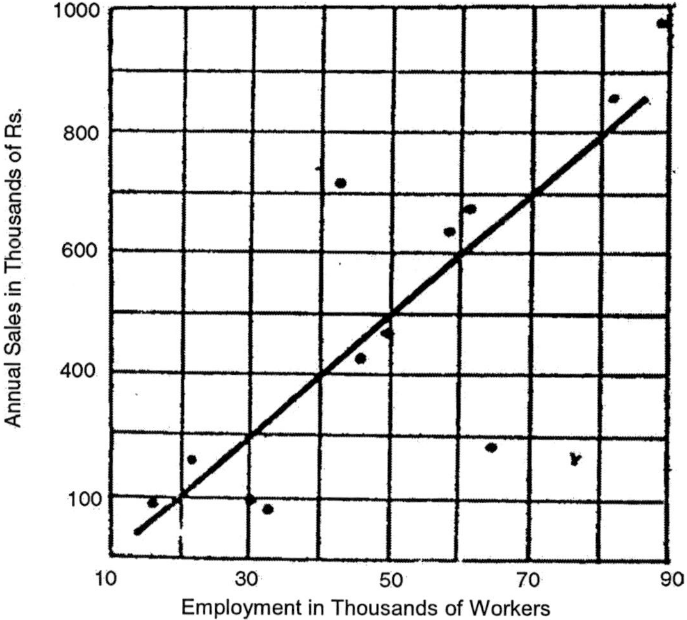

Fig. 14.2: A Regression Line

CORRELATION

So far, we discussed the characteristics of a given set of data pertaining to one variable, and the methods of analyzing and presenting it statistically. More often than not wehaveto compare two or more sets of data to find out the extent to which changes in one set of data trigger off changes in other set(s) of data. If the change in one set causes changes in other set(s), it means that individual items of the two sets of data co-exist either in time or space in such a way that the possibilities of interrelated changes cannot be ruled out. For such interrelated changes, we need indices that reflect the direction and/or magnitude in one set of data are associated with comparable changes in the other set(s). The index is Coefficient of Correlation (r).

Co-efficient of Correlation

Coefficient of correlation (r) is one of the indices that quantify this association. It varies from – 1 to + 1. If it is – 1 or nearer it, it means that the variables are negatively related to each other, i.e. when one increases, the other decreases proportionately. But if it is + 1 or nearer that, it means that they have perfect or high positive relationship. If it is 0, it means that they are not related to each other.

Let us take one specific example. Table 14.8 gives the employment figure as well as the sales in rupees in a given store.

Table 14.8: Number of People Employed and Annual Sales in a Store

Employment (workers in thousands)	*Annual Sales (Rs. in thousands)*
22	250
31	200
90	850
82	850
43	710
65	280
59	630
16	180
61	670
46	420

The basic formula for the derivation of coefficient of correlation is,

$$r = \frac{\Sigma \frac{x}{SD_x} \cdot \frac{y}{SD_y}}{N} \text{ or } \frac{\Sigma xy}{\sqrt{\Sigma x^2 \cdot y^2}}$$

where

SD_x and SD_y = Standard deviations.

N = Number of items

x and y = Deviation values of X and Y respectively.

$$\therefore \qquad r = \frac{\Sigma xy}{\sqrt{\Sigma x^2 \cdot \Sigma y^2}} = \frac{57,220}{\sqrt{5612 \times 17,500}}$$

$$= +0.83$$

The coefficient of + 0.83 implies that there is a high degree of positive correlation between employment and sales. Whenever employment increases, the sales also

Table 14.9: Computation of Co-efficient of Correlation (r) (using the data given in Table 14.8)

X	Y	x= X-X	y= Y-Y	xy	x^2	y^2
22	250	– 28	– 235	+6,580	784	55.225
31	200	– 19	– 285	+5,415	301	81.225
90	980	+ 40	+495	+19.800	1600	245,025
82	850	+32	+365	+11.680	1024	133,225
43	710	– 7	+225	–1.575	49	50.625
65	283	+ 15	– 205	–3.075	225	42.025
59	630	+ 19	+145	+10,370	1156	93.025
16	180	– 34	–305	+10,370	1156	93.025
61	670	+11	+185	+2.035	121	34.225
46	420	– 4	– 65	+260	121	4.225
35	190	– 15	+295	+4,425	225	87.025
50	460	0	– 25	0	0	625
600	5820	0	0	+57,220	5642	847,500
ΣX	ΣY	Σ_x	Σ_y	Σ_{xy}	Σ_x^2	Σ_y^2
X = 50	y = 485			SDx = 21.68	SD = 2658	

increase. In general the coefficients between + 0.5 and + 1 indicates a high degree of correlation and between –0.5 and + 0.5 indicates low level correlation. If it is zero, it means that the sets of data pertaining to X and Y change without affecting each other i.e. there is no correlation between the two. The above said analysis of correlation coefficient is, however, not complete unless we determine the significance of level of it. This can be done by conducting t test. The formula for t test is:

$$t = \frac{\sqrt{r^2 (N-2)}}{1-r^2}$$

Rank Correlation

Rank Correlation measures the degree of association between two series. Its advantages are as follows:

(a) If the data are given, such as ranks, scores etc., the degree of association can be found out by this method;
(b) It is quick, less-labourious and useful where intensive study is not called for;
(c) It is is helpful in correlating two sets of qualitative observations such as poor, fair, good, superior, etc.

Table 14.10: gives the data for calculating rank correlation between the size of labour force (X) and sales volume (Y):

Table 14.10: Labour Force and Sales Volume

Size of Labour Force	*Sales Volume*	*Rank of X*	*Rank of Y*
16	180	1	1
22	250	2	4
31	200	3	3
35	190	4	2
43	710	5	10
46	420	6	6
50	460	7	7
59	630	8	8
61	670	9	9
65	280	10	5
82	850	11	11
99	980	12	12

The formula for coefficient of rank correlation is–

$$P = I - \frac{6\,\Sigma D^2}{N\,(N^2 - 1)}$$

where

P = stands for the coefficient of rank correlation,

D = stands for the difference between each pair of ranks,

N = stands for Number of paired items, and

D stands for each pair of ranks. It can be found out by subtracting the rank of Y from the rank of its corresponding X (Table 14.11).

Table 14.11: Calculation of Difference in the Pair of Ranks

Rank of X	*Rank of Y*	*D*	*D^2*
1	1	0	0
2	4	– 2	4
3	3	0	0
4	2	2	4
5	10	– 5	25
6	6	0	0
7	7	0	0
8	8	0	0
9	9	0	0
10	5	5	25
11	11	0	0
12	12	0	0
		0	58

Now $$P = 1 - \frac{6\,(58)}{12\,(144 - 1)} = +0.80$$

Sampling Errors in Correlation

If two samples of the same size are randomly selected from a given population, they would not give the same distribution on the scatter-diagram. The *r* values of the two would differ and none of them can be treated as perfectly reliable indication of the *r* obtained from the entire population. To determine the deviation of the *r* of a sample from the true *r* we have to calculate the standard error:

$$\sigma_r = \frac{1 - r^2}{\sqrt{N}}$$

Suppose that the correlation (*r*) between X and Y is $r = .60$ for a given random sample of 100 cases, the standard error will be:

$$\frac{1 - .6^2}{\sqrt{100}} = .064$$

This means that if *r* was similarly obtained they will fall into normal distribution with a standard deviation of approximately .064.

With the help of these methods, one can analyze the data and use them for statistical mapping. Only a few of the very elementary methods have been given in this chapter. Those interested in advanced work are referred to books on general statistics.

MULTIVARIATE ANALYSIS

Multivariate statistics are employed for developing systems of classification, for group variates and to generate and test hypotheses. The concept underlying multivariate analysis that everything is linked to everything else; nothing stands alone in the world. In order to understand a phenomenon, we have to know what is there behind and underlying the phenomena we want to understand. This curiosity to know, leads us to multivariate analysis.

In multiple regression and analysis of variance, several variables are used; and a dependent variable is generally predicted or explained by means of independent variables and covariates. Then there are methods to find out interrelationships among the variable such as factor analysis, multidimensional scaling (MDS) and cluster analysis. All these statistical techniques are discussed below.

Multiple-Regression

Multiple-Regression was first used by Pearson in 1908. It helps in finding out relationship between a number of independent variables and a dependent variable. For example, an agricultural extension worker may like to know why output of wheat per acre in different farms is not the same. Could it be because of poor quality of seeds, low input of fertilizers, or size of the farm? The question is how these variables relate to output of the farms.

In the natural and social science researches multiple regression procedures are very widely used particularly when one has to seek the best predictor of a phenomenon. In the above case, an extension officer wants to know the best predictors of high yields of wheat. An educational researcher may want to know the best predictors of success in high school. A psychologists may seek to know which personality variable best predicts social adjustment. Sociologists may want to find out which of the multiple social indicators best predict whether or not a new immigrant group will adapt and be absorbed into society. A geographer may try to know the indicators of development disparities.

As an example, let us take an independent variable X (IQ of students measured by a test) and one dependent variable Y (school achievement as reflected by average grade point – GPA). Each point in the scatterplot (Fig. 14.3) represents respective student's IQ and GPA.

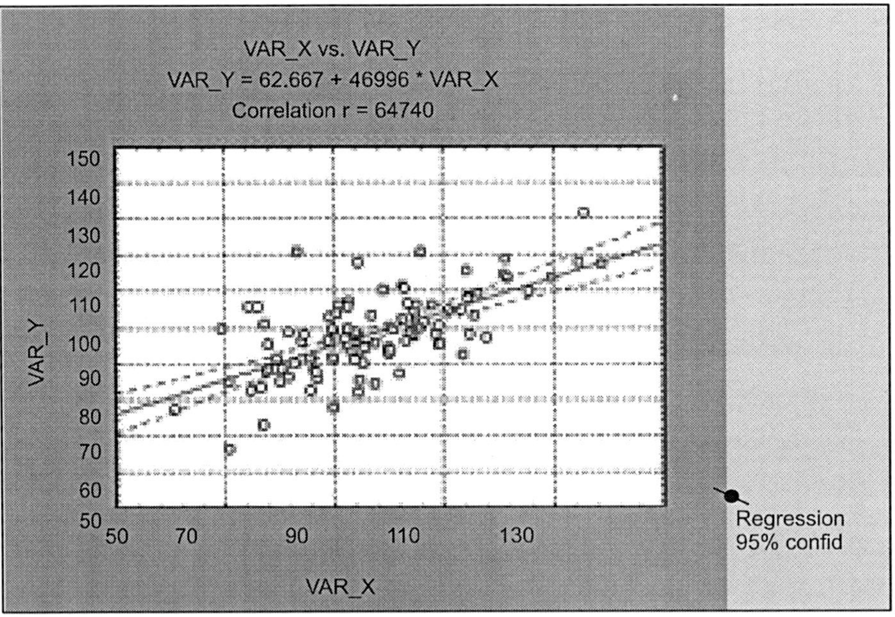

Fig. 14.3: Scatterplot of a Simple Case of Multiple Regression

The scatterplot is formed by an independent or *X* variable (IQ), and a dependent or *Y* variable (GPA). Each point in the plot represents one student's IQ and GPA. The *Y* variable can be expressed in terms of a constant (*a*) and a slope (*b*) times the *X* variable i.e. $Y = a + b/X$. The constant is also referred to as the *intercept*, and the slope as the *regression coefficient* or *B coefficient*. The GPA may be predicted as $1 + .02/IQ$. Thus, If a student has an *IQ* of 130 we can get his/her GPA to be 3.6 (since, $1 + .02/130 = 3.6$).

In the multivariate case, there is more than one independent variable hence the regression line is not visualized in the two dimensional space. But the computational process is similar and easy. Suppose, that there are three independent/predictor variables: IQ, Motivation, and Self-discipline. Now we construct a linear equation containing all those variables. It will take the following form:

$$Y = a + b_1/X_1 + b_2/X_2 + \ldots + b_p/X_p.$$

In this case the regression coefficients or *B* coefficients, represent the contributions of each of the three independent variables to the prediction of the dependent variable.

Table 14.12 gives the data on agricultural productivity and its determinants for Tamil Nadu districts. We have to calculate the coefficient of correlation between agricultural productivity measured in rupees and the factors influencing/determining it.

Table 14.12: Agricultural Productivity and Its Determinants (Tamil Nadu, 1972-75)

S. No.	*Districts*	*y*	*x1*	*x2*	*x3*	*x4*		*x5*	*x6*
1.	Chingleput	2300	135.26	68.03	60.63	5.4	1	9.14	87.50
2.	South Arcot	2682	121.82	56.64	52.15	5.0	2	6.97	584.70
3.	North Arcot	2297	128.68	53.57	36.31	5.4	3	6.19	69.47
4.	Salem and Dharmapuri	1556	118.64	27.34	22.07	3.8	2	5.10	65.68
5.	Coimbatore	2028	117.07	43.42	46.20	4.9	3	1.46	55.72
6.	Tiruchirapalli	1593	111.74	40.04	41.40	2.4	1	8.15	61.40
7.	Thanjavur	2023	161.28	71.06	53.17	1.7	1	0.92	87.04
8.	Madurai	1784	113.14	41.47	33.81	2.9	2	0.43	70.10
9.	Ramanathapuram	1226	103.16	38.76	15.98	1.2	2	5.32	59.46
10.	Tirunelveli	1794	120.18	44.40	25.59	2.0	2	9.35	80.94
11.	Nilgiris	3517	100.11	0.38	45.22	0.8	6	3.29	150.60
12.	Kanya Kumari	2364	144.70	50.57	33.44	0.4	2	3.97	188.78

Y = Agricultural Productivity per ha in rupees
X_1 = Intensity of cropping
X_2 = Proportion of GOA to GCA
X_3 = Consumption of fertilizer per 1000 ha
X_4 = Mechanisation index
X_5 = Proportion of area under non-food crops
X_6 = Mean annual rainfall in Cms

Source: Parimala, G., *Levels of Agricultural Development in Tamil Nadu: A Talukawise Study M Phil Dissertation Submitted in CSRD,* JNU.

As a first step, a multiple correlation matrix like the one given in Table 14.13 is obtained.

Table 14.13: Multiple-Correlation Matrix

	Y	X_1	X_2	X_3	X_4	X_5	X_6
Y	1.000						
X_1	0.544	1.00					
X_2	0.644	0.767	1.00				
X_3	0.690	0.527	0.793	1.00			
X_4	0.521	– 0.150	0.221	0.575	1.00		
X_5	0.110	–0.509	– 0.425	– 0.391	– 0.330	1.00	
X_6	0.574	0.725	0.449	0.207	– 0.234	– 0.260	1.00

The correlation co-efficient values in Table 14.13 show the mutual inter-correlation between the dependent variable and independent variables. Let us now interpret this computer created matrix. We notice that X_1 has high positive correlation (+ 0.767) with $X_{2,}$ and X_3 has still higher positive correlation (+ 0.793) with X_2. It happens if the independent variables are not so independent. This is known as multi-colinearity. In the present case, this has occurred because X_3 (irrigation) has high correlation with cropping intensity (X_1).

The correlation matrix provides the base for further stepwise regression. Table 14.14 presents the results of this regression exercise.

Table 14.14: Order of the Variables added in the Stepwise Regression

Steps	*Included variable*	*R*	*R^2 x 100*	*Increase in R^2 x 100*	*Steward error*	*R*	*R^2 x 100*	*Increase in R^2 × 100*
I.	X_3	0.690	47.7		325.010	0.690	47.7	
II.	X_3X_6	0.819	67.1	19.4	288.009	0.797	63.5	15.8
III.	$X_3X_6X_5$	0.965	93.2	26.1	148.652	0.956	91.3	27.8
IV.	$X_3X_6X_5X_2$	0.966	93.4	0.2	169.313	0.951	90.4	–0.9
V.	$X_3X_6X_5X_2X_4$	0.967	93.6	0.2	196.781	0.945	89.3	–0.1
VI.	$X_3X_6X_5X_2X_4X_1$	0.968	93.6	0.0	240.921	0.934	87.2	–2.1

In the above exercise, fertilizer enters in the first step. All the independent variables together explain 93.6 per cent of the variations in agriculture productivity in Tamil Nadu. The variables are more important in explaining the variations only up to the third step. R^2 also starts decreasing as X_2 enters the fourth step. Thus, as the exercise proceeds the explanatory power of independent variables increases but the significance level decreases.

Once the problem of multi-colinearity is detected, further processing of the data can be done to eliminate this problem. Much depends on the usage of this technique in relevant research areas.

In general, the degree to which two or more independent variables (*X*) are related to the dependent variable (*Y*) is expressed in the correlation coefficient *R which in the case of multiple-regression* can assume values between 0 and 1. To interpret the direction of the relationship between the variables, the plus or minus signs of the regression or *B* coefficients are kept in view. If a *B* coefficient is positive, then the relationship of this variable with the dependent variable is positive (e.g., the greater the IQ, the better the grade point average); if the *B* coefficient is negative then the relationship is negative (e.g., the lower the class size, the better the average test scores). A 0 coefficient points to the absence any relationship between the variables.

ANALYSIS OF VARIANCE (ANOVA)

An ANOVA is used to find out differences between groups on a given variable. For example, a teacher may be interested in finding out if the class performance of his students is related to final grade of students or not. First, he would divide the students according to their grades and then see if performance is different across the grades.

The simplest version of ANOVA is one-way between groups model. Here, it is only the final grade that is being used to define the groups. A two-way between groups ANOVA is used to look at complex groupings. For example, the grades by class performance analysis could be extended to find out the differences in the performance of students from two different states of India.

Procedure for calculating ANOVA, using the data pertaining to the performance of the students is as follows:

1. *Step I*: calculate the Group Mean i.e. the mean for each of the final grading groups on the class performance data.
2. *Step II*: calculate the Overall Mean i.e. the mean for all the groups combined.
3. *Step III*: calculate Within-Group Variation i.e. total deviation of each individual's score within each group from the Group Mean.
4. *Step IV*: calculate Between Group Variation i.e. the deviation of each Group Mean from the Overall Mean.
5. *Step V*: produce the F statistic which is the ratio Between Group Variation to the Within Group Variation.

If the Between Group Variation is significantly greater than the Within Group Variation, then it is likely that there is a statistically significant difference between the groups.

CHI SQUARE (χ^2)

Chi is a Greek letter pronounced as 'kye'. It is a non-parametric statistical technique expressed as χ^2. It is used to determine whether a distribution of observed frequencies is the same as theoretically expected or not. The frequencies may be expressed in nominal (categorical) or ordinal mode but they should not be expressed as percentages, ratios, means and variances.

Chi-square is the sum of the squared difference between observed (*o*) and the expected (*e*) figures (or the deviation), (*d*), divided by the expected data. To put in the form of a formula:

$$\chi^2 = (o-e)^2/^e$$

where χ^2 is the chi-square statistic, *'o'* is the observed frequency and *'e'* is the expected frequency.

Data used in a chi-square analysis should satisfy following conditions:

1. They should be randomly drawn from the population,
2. They should be reported in raw counts of frequency,
3. Their measured variables must be independent,
4. Their observed frequencies should not be too small, and
5. The values of independent and dependent variables must be mutually exclusive.

There are two types of χ^2 test:

1. Chi-square test for goodness of fit; it compares expected and observed frequencies to determine how well the predictions fit the data; and
2. Chi-square test for independence; it compares two sets of categories to determine how differently the groups are distributed among the categories.

Chi-square Test for Goodness of Fit

If we toss a coin 100 times, the outcome would generally be 50 heads and 50 tails. But if the outcome is 47 heads and 53 tails, it may be because the coin is defective or because of a chance. To test it, we first formulate our hypothesis which, may be stated in one of the following two forms:

1. *Null hypothesis*: The coin is fair
2. *Alternate hypothesis*: The coin is biased

The null hypothesis states that there is no relation between the phenomena under investigation, or at least not of the form given by the alternative hypothesis. Null hypothesis assumes that no statistical significance exists in a set of given observations and if any kind of difference or significance is seen it must be due to chance. It is assumed to be correct unless there is sufficient evidence from the sample and the given criteria (significance level) to reject it. In the above case, there is no statistically significant difference in the grades of boys and girls.

If we hypothesize that the coin is unbiased and hence the observed values are close to the predicted values, we call it null hypothesis. But if we hypothesize that observed values are not close to the predicted values because the coin is biased, we call it alternative hypothesis. These hypotheses hold for all Chi-square goodness of fit tests.

Tabulated Results of Observed and Expected Frequencies

	Heads	*Tails*
Observed (o)	47	53
Expected (e)	50	50

To calculate the χ^2 (chi-square statistic), we substitute values for *o* and *e* in the formula:

$$\chi^2 = (o-e)^2/^e$$

For Heads: $\chi^2 = (47-50)^2/50 = 0.18$

For Tails $\chi^2 = (53-50)^2/50 = 0.18$

The sum of these categories is 0.18 + 0.18 = 0.36

Chi-square Test for Independence

The chi-square test for independence determines to what extent the two variables of a sample are related to each other. In social science research, we are generally interested in finding the related factors, e.g. education and income, occupation and prestige, age and voting behaviour. If the factors are independent of each other, it means there is no relationship between the two. As an example, let us try to find out how the boys or girls fair in examinations. Table 14.15 documents the frequency of boys and girls who score grade A.

Table 14.15: Observed and Expected frequency of Boys and Girls Scoring Grade A

	Scored A (Observed)	*Did Not Score A (Observed)*	*Total*	*Scored A (Expected)*	*Did not Score A (Expected)*
Boys	46	71	117	40.97	(76.02)
Girls	37	83	120	42.3	(77.97)
Total	83	154	237		

The question is who among the boys and girls got distinction marks more often? To answer this question, let us first establish a hypothesis which is an idea which is proposed in order to explain something, which seems to have some logical or philosophical validity but which has not yet been tested. It is thus a provisional idea which has to be tested against the reality. For proper evaluation, the framer of a hypothesis has to define specifics in operational terms. Once a hypothesis is confirmed it may become a theory. Hypothesis may be stated in the form of a mathematical model or an existential or universal statement. It leads to experimentation to answer questions and explore observations.

The basic components of hypotheses are concepts which it connects by specifying the expected relationships among them. For example, a simple relational hypothesis such as 'education increases income' specifies a positive relationship between two concepts 'education' and 'income.' Both the terms are, however, abstract and hence the hypothesis cannot be tested. First, the terms must be operationalized or situated in the real world. We can improve this statement by saying 'Education increases Income.' But even here the meaning of education and income is not operationalized. To test the hypothesis the abstract meaning of education and income must be operationalized to make them measurable. For example, education could be measured by 'years of schooling' or 'highest degree completed' etc. and income could be measured by 'annual/monthly salary' or 'hourly rate of wages', etc.

If we calculate the expected value for each cell of Table 14.2, the key idea behind the chi-square test for independence is a comparison of observed and expected values. In the case of the data given in Table 14.1, we do not know what the distribution should look like (as we did with tossing the coin). Rather the expected values should be calculated on the basis of the row and column totals given in Table 14.2. We calculate the expected value for each cell of the table using the following equation:

Expected value = Row total or Column total/Overall Total. The expected values (in parentheses, italics and bold) for each cell are also presented in the table.

Let us now calculate the Chi-square statistics using the data given in Table 14.15.

The chi-square equation:

$$\chi^2 = (o-e)^2/^e$$

$$\chi^2 = (46-40.97)2/40.97 + (37-42.03)2/42.03 + (71-76.03)\ 2/76.03 + (83-77.97)2/77.97=1.87$$

In the χ^2 for independence, the degree of freedom is equal to the number of columns in the table minus one multiplied by the number of rows in the table minus one i.e. degree of freedom (*dof*) = $(r-1)(c-1) = 1$. Thus the value calculated from the formula above is compared with values in the chi-square distribution table. The value returned from the table is $p < 20$ per cent. Therefore, the null hypothesis is not rejected, i.e. boys are not significantly more likely to score distinction marks than girls. It is a statistical test used to compare observed data with the expected data.

LORENZ CURVE

The Lorenz Curve measures inequality. It tells how far let us income distribution among the people is away from perfect equality. Suppose we plot the proportion of total income of a population (*y*-axis) that is cumulatively earned by the bottom x per cent of population (Fig. 14.4), the line at 45 degrees represents perfect equality of incomes. If the income is equitably distributed among the people, 45 degree line would become the Lorenz Curve. If all the income is owned by just one person, the Lorenz Curve would be the baseline. Nearer, the Lorenz Curve is to the Line of Equality, lesser the inequality.

GINI-COEFFICIENT

Gini Coefficient was formulated by Italian Statistician Corrado Gini in 1912. It is closely related to Lorenz Curve and shows how far a given variable deviates from equal distribution. Gini coefficient is thus the ratio of the area that lies between the line of equality and the Lorenz Curve (marked 'A' in the figure) over the total area under the line of equality (marked 'A' and 'B' in the figure); i.e. G = A/(A + B).

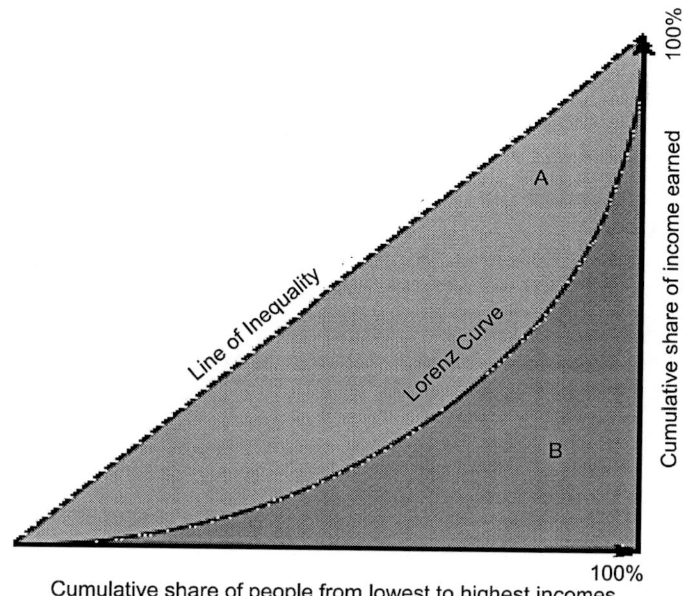

Fig. 14.4: Lorenz Curve

The Gini coefficient ranges from 0 to 1; at times it is multiplied by 100 to range between 0 and 100. Lower the Gini coefficient more equal the distribution, and higher the Gini coefficient more unequal the distribution. Thus, Gini = 0 means complete equality i.e. every person receives the same income, and Gini = 1 means complete inequality i.e. a single person receives 100 per cent of the total income and the remaining people receive none.

Given a Lorenz Curve, G can be calculated mathematically too. Since A + B = 0.5, the Gini index, G = A/(0.5) = 2A = 1–2B. If the Lorenz Curve is represented by the function Y = L(X), the value of B can be found with integration and:

$$G = 1-2\int_0^1 L(X)\,dX.$$

$$G = \frac{1}{n}\left(n+1-2\left(\frac{\sum_{i=1}^{n}(n+1-i)y_i}{\sum_{i=1}^{n} yi}\right)\right)$$

$$G = 1-\frac{\sum_{i=1}^{n} f(y_i)\,(S_{i-1}+S_i)}{S_n}$$

While Gini-coefficient is a very commonly used measure of inequality, it has certain disadvantages too. Its main advantage is that it is a simple measure of inequality and can be easily interpreted. It measures inequality by means of a ratio analysis, rather than a variable that does not represent most of the population, such as per capita income or GDP. It can be used to compare income distributions across different components of the population (like urban and rural) as well as countries. It can be used to indicate how the distribution of income has changed over a period of time, thus it is possible to see if inequality is increasing or decreasing.

Gini-coefficient suffers from certain disadvantages too. It does not tell why the inequality in income. It does not tell the level of the economy. Both rich and poor countries can have the same G. It does not express regional variations. The Lorenz curve may understate the actual amount of inequality if richer households are able to use income more efficiently than lower income households or vice versa. It measures income rather than wealth. As with other inequality coefficients, the Gini coefficient is influenced by the granularity of the measurements. It is not a measure of egalitarianism; it is only a measure of income dispersion. The Gini coefficient is a point-estimate of equality at a certain time, hence it ignores life-span changes in income. It does not take into account the benefits people get from the government in terms of subsidized food, education, health facilities etc.

PARETO DISTRIBUTION

Named after the famous Swiss economist/sociologist Wilfredo Pareto, Pareto distribution is a power law probability distribution found in a large number of real-world situations. The power-law implies that large occurrences are rare while the small occurrences are extremely common. This regularity or 'law' is sometimes also referred to as Zipf or Bradford distribution. It explains why many man-made and natural phenomena, including wealth, incomes, city sizes, word frequencies, and earthquake magnitudes follow a power-law distribution.

While studying family income distribution in Switzerland, Pareto discovered that there were more families with a low income than families with a large income, and that these distributions followed a very smooth pattern. Seeing the regularity of this phenomenon, larger portion of the wealth of any society is owned by a smaller pattern of wealth distribution, Pareto converted it into a principle which is often referred to as '80-20 rule' or Pareto principle, which says that 20 per cent of the people control 80 per cent of the wealth. It further says that the fraction of population (80%) that owns a small amount of wealth (20%) per person decreases steadily as total wealth increases. In other words, less developed an economy more unevenly distributed is the wealth of a nation. This is the ground on which economists contend that rapid economic growth would lead to better distribution of wealth.

If we draw a probability density function (PDF) graph of the population f(x), the probability, or fraction of f(x) owning a small amount of wealth per head, is high. As wealth increases, the probability or fraction of population owing small amount of wealth, decreases as wealth increases.

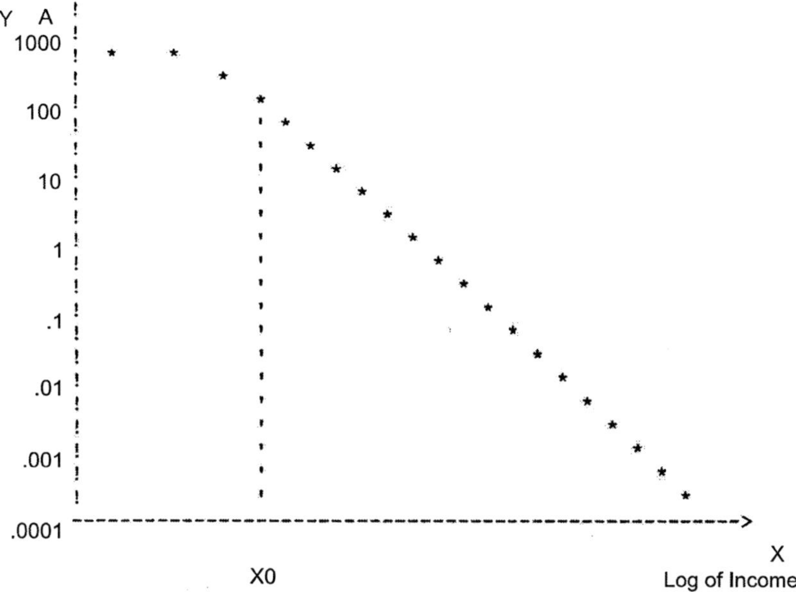

Fig. 14.5: Pareto Distrbution

Mathematically, the Straight line part of the graph can be expressed as:

$$Y = A X + B$$

where

X is the logarithm of the income level x

Y is the logarithm of the number of people having an in come x

The Pareto distribution, however, fails to give a good fit for the lower end of the size spectrum. It was admitted by Pareto himself that his principle was valid only above a minimum size X0.

REFERENCES

Abelson, R.P. (1995), *Statistics as Principled Argument,* NJ: Lawrence Erlbaum Associates.

Argyrous, G. (2000), *Statistics for Social and Health Research,* (with a guide to SPSS), London: Sage Publications.

Brown, F.L., Amos, J.R., and Mink, O.G. (1995), *Statistical Concepts: A Basic Programme,* NY: Harper Collins College Publishers.

Best, J. (2004), *More Damned Lies and Statistics: How Numbers Confuse Public Issues*. CA: University of California Press.

Craft, J.L. (1990), *Statistics and Data Analysis for Social Workers,* 2nd Edition, Itasca, IL: F.E. Peacock Publishers.

Diamond, I. and Jefferies, J. (2001), *Beginning Statistics: An Introduction for Social Scientists*, CA: Sage Publications.
Foster, J.J. (2001), *Data Analysis Using SPSS for Windows Versions 8 to 10: A Beginner's Guide*, CA: Sage Publications.
Freed, M.N., Ryan, J.M., and Kess, R.K. (1991), *Handbook of Statistical Procedures and Their Computer Applications to Education and the Behavioural Sciences*, New York: American Council on Education, Macmillan Publishing Company.
Gigerenzer, G. (2002), *Calculated Risks: How To Know When Numbers Deceive You*, New York: Simon & Shuster MacMillan Co.
Graham, A. (2003), *Teach Yourself Statistics*, New York: McGrawHill.
Jaisingh, L. (2000), *Statistics for the Utterly Confused*, New York: McGraw Hill.
Kaplan, M. and Kaplan, E. (2006). *Chances Are... Adventures in Probability*, New York: Penguin/Viking Group.
Langley, R. (1970), *Practical Statistics*, New York: Dover books.
Levitt, Steven D. and Dubner, Stephen J. (2005), *Freakonomics. A Rogue Economist Explores the Hidden Side of Everything*. New York: HarperCollins Publishers.
Mlodinow, L. (2008), *The Drunkard's Walk. How Randomness Rules Our Lives* New York: Vintage Books.
Moore, D.S. (1991), *Statistics Concepts and Controversies*, 3rd Edition. New York: WH Freeman & Co.
Morgan, S.E., Reichert, T., and Harrison, T.J. (2002), *From Numbers to Words. Reporting Statistical Results for the Social Sciences*, MA: Allyn and Bacon.
Newton, R.R., and Rudestam, K.E. (1999), *Your Statistical Consultant : Answers to Your Data Analysis Questions*, CA: Sage Publications.
Niederman, D., and Boyum, D. (2003), *What the Numbers Say: A Field Guide to Mastering our Numerical World*, New York: Broadway Books.
Phillips, Jr., J.L. (1992), *How to Think About Statistics*, New York: W.H. Freeman & Co.
Ramsey, F.L. and Schafer, D.W. (1997), *The Statistical Sleuth: A Course in Methods of Data Analysis*, CA: Duxbury Press.
Rosenthal, J.S. (2006), *Struck by Lightning: The Curious World of Probabilities*, Washington DC: John Henry Press.
Spence, J.T., Cotton, J.W., Underwood, B.J., and Duncan, C.P. (1990), *Elementary Statistics*, 5th Edition, NJ: Princeton Hall.
Tal, J. (2001), *Reading Between the Numbers: Statistical Thinking in Everyday Life*, New York: McGraw Hill.
van Belle, G. (2002), *Statistical Rules of Thumb*, MA: Wiley-Interscience.
Wallgren, A., Wallgren, B., Persson, R., Jorner, U., Haaland, J.A. (1996), *Graphing Statistics and Data: Creating Better Charts*, (originally in Swedish language), Thousand Oaks, CA: Sage.
Weinbach, R.W., and Grinnell, R.M. (1991), *Statistics for Social Workers*, New York: Longman.
Weiss, N.A., and Hassett, M.J. (1991), *Introductory Statistics*, New York: Addison-Wesley Publishing Co.
Zeisel, H. (1985), *Say it with Figures*, (6th Edition), New York: Harper & Row.

Part III

Map Design and Construction

15
Map Design and Layout

Map design and layout determines the aesthetic quality of a map. The information at the disposal of the cartographer is of great variety and complexity, but to depict them all that is available and useful with clarity, simplicity, accuracy, legibly and aesthetically is not easy. Map making is a scientific exercise but it is no less an artistic job.

Use of computers in making maps has not changed the situation. Computers only replace manual work involved in drawing symbols and placing them at appropriate places. We still need artistic bent of mind to make a good map. Like an artist, a computer operator too has to follow certain principles of visual art but he does not have complete freedom of portrayal techniques and media. A cartographer whether he prepares the maps manually or with the help of a computer shows his artistic talents within the framework of two constraints:

1. Cartographic traditions and conventions; and
2. Basic requirements of maps themselves.

A map is designed to serve certain utilitarian purposes. Its aesthetic value is, therefore, desirable but cannot supersede its utilitarian value. The artistic talent has, therefore, to function within the framework of the utilitarian requirements. But an ugly, misleading, and difficult to read is a disaster. It detracts people away from maps and thus away from cartography, the science of map making.

A cartographer uses a variety of symbols to represent selected information on a map. Each bit of information is an integral component of the map. In order to produce a good map, it is not enough to represent each of these components clearly, simply and aesthetically. This is indubitably important but more important is the presentation of these individual components in such a manner as to create a good integrated picture of the whole. In a map, the individual components get their meaning only with reference to the map as a whole. Map design involves the development of this integrated plan and style of the map and its individual components, and layout involves the arrangement of these individual components on the map.

As mentioned earlier, there are innumerable things which can be shown on maps. But everything that exists or that can be conceived to exist is shown on all maps. Each map is made to represent only a select few features, natural and human. There are certain marginal details which are given in all maps such as the title of the map, the legend, the scale, and the land and see boundaries and the graticules. Then there are features which help the reader understand locations and distributions better such as main roads, railways, etc.

Maps can broadly be classified into two types: general, and thematic. The maps, which represent a variety of things—the things about which people in general are more intimately concerned in their day-to-day life, are called general or reference maps. The maps which give specialized information only such as political map, physical map, population density map, etc. are called thematic maps. In these maps only a few details get visual prominence. The other details are either not shown at all or are shown merely to produce a background effect. The problem of making certain things visually significant without giving the impression of imbalance in the total design of a map is the most crucial problem in map design and layout. A map has to be an integrated whole, but within this holistic vision certain components get prominence over many others, not necessarily because they are more important in reality but because they are more important for the persons for whom the cartographer makes the map.

In order to prepare a balanced map and, yet, to make certain component of it visually more significant, we have to have the understanding of:

1. Theory of visual perception;
2. Techniques of making things visually significant; and
3. Limitations within which cartographers function.

THEORY OF VISUAL PERCEPTION

Perception means awareness of things and ideas we come across. This awareness comes through sensations like sight, sound, taste, touch, mental image, etc. The Epicurean Greek philosophers said that the objects which we see emit something of themselves to our eyes. After Newton's discoveries, it was generally believed that the objects, rather than emitting something of themselves, reflect light to our eyes, very much as it happens when we use a camera to take photographs. This image is supposed to be the exact copy of the objects seen. Experiments later disproved this theory also. Various optical illusions were given as the basis for discarding this theory. Then came the theory of perception which lays emphasis on the interaction between the perceiver and the objects perceived. This theory can be explained better by taking the example of sunlight falling on an object, say, grass. We see grass to be green because it absorbs all the light waves except the green. The light waves creating green are not green so long as certain cells in the retina of the eye do not interact with the light wave in certain ways to produce the experience of green. Neither the light emitted by the grass nor the sensory cells in the retina are green. It is the interaction between the two, which produces green.

The most recent theory of visual perception disapproves all the above theories. It suggests that perception is in reality a transition and that various shapes, colours etc., that we perceive exist only in terms of the situation in which they are perceived. It says that:

1. Total environment in which the object lies enters into perception as an active input; and
2. The object, if removed from the given environment, loses its whole meaning.

An example will make it clear. When we look at the sunset, we do not see certain colours, lines, shapes and shades. We see the sunset. We do see the details of the colour and shapes but only sub-consciously. What we get, however, is a total visual experience. The bright colours we see in the sunset do not bother us but they surely bother us if we have wall papers of the same colours.

The above noted shift in the theory of perception is very pertinent and relevant to map design. It, in effect, tells us that various symbols that we use to make a map, acquire their desired meanings only in the context of the map as a whole. They will convey little or no meaning if they are removed from the total situation i.e., map as a whole. Further, it also indicates, that the same symbols and their arrangements cannot be used in all maps, for each map is designed to present a different complex of things and situations. What is good in one map may prove to be disastrous in others.

This, however, does not mean that individual components of a map are also map specific and vary in importance. A line, one of the simplest elements of a map, for example, can be manipulated to produce different perceptions. A line involves relationships within itself and with its surroundings. A horizontal line produces the perception of line direction. A vertical line tends to produce more tension and excitement than a horizontal line. A diagonal line produces discomfort to the observer's eyes because it lacks balance (Fig. 15.1).

The positioning of symbols in association with various types of lines produces different perceptions. For example, if a bending man is shown standing on a diagonal, he appears to be picking up something from the ground. But if he is placed beneath the diagonal, he appears to bend under the weight of space above him (Fig. 15.2). Several lines put together in different orders produce different perceptions. Lines forming various shapes, like squares, triangles, etc., lose their existence altogether. When we see a triangle, we never perceive it to be three lines put in a certain order. We see a figure with three sides. The new theory of perception also tells us that like other symbols, colours are also seen against the background in which they appear.

When we look at a visual symbol or a group of symbols (map) we first perceive only vague shapes, colours, lines, and letterings. These symbols slowly begin to relate to one another. At first we perceive the distribution of land and water quickly if the two features are shown in a familiar way i.e., water is shown in blue colour. And we take quite much of time, if the colour scheme has been reversed and they are shown in not familiar.

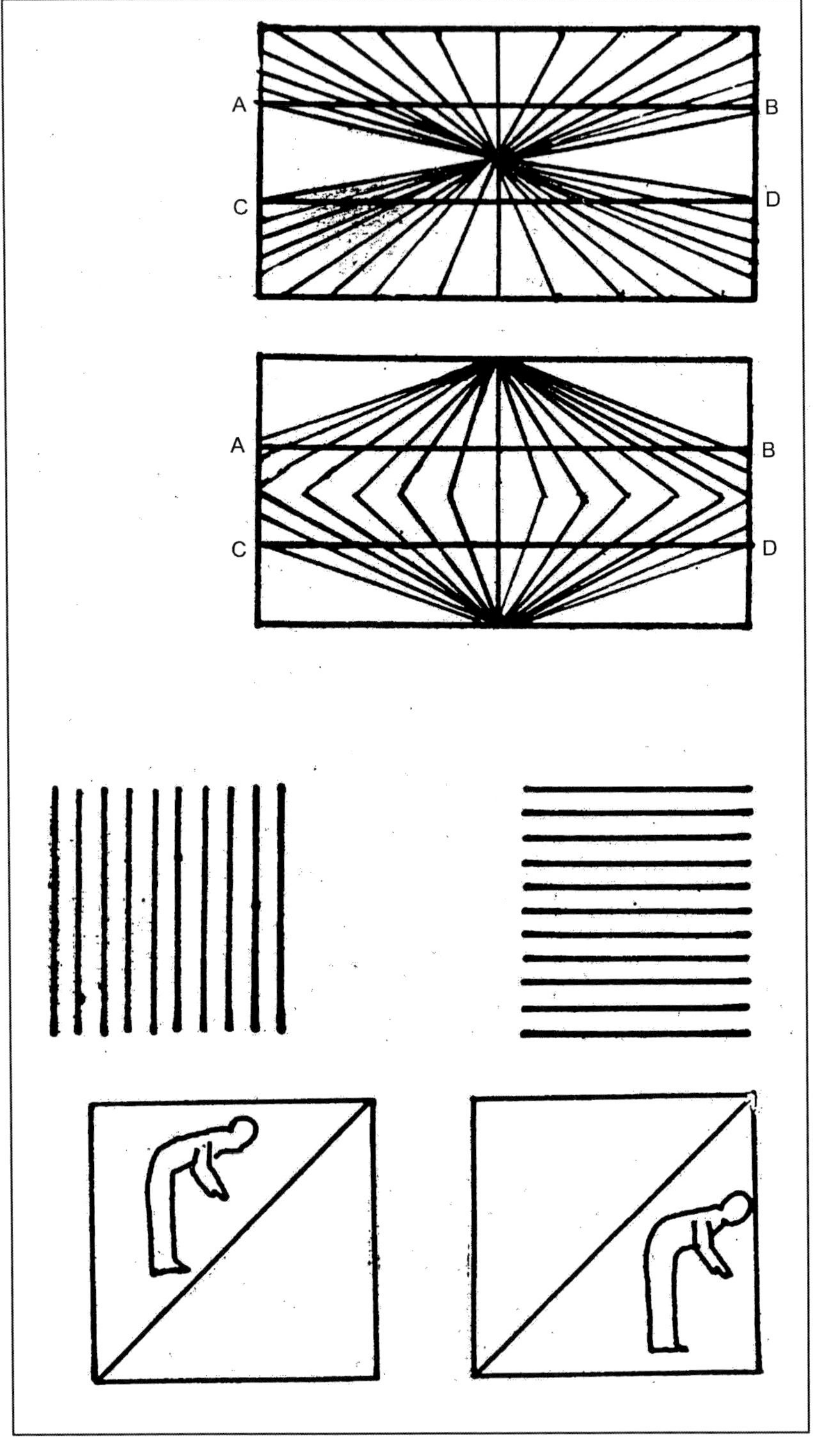

Fig. 15.1: Visual Impacts of Line Orientations on Perception

Once these broad features are recognized, the smaller internal relations begin to unfold themselves. We see the rivers, railways, roads, canals, and towns and the interrelations among them. This is, just like the situation of visiting a factory for the first time. At first everything is inarticulate, but soon we distinguish various sounds, machines and sections. Finally, and after a necessary period of time has elapsed, the various parts appear to present a meaningful whole.

Making Symbols Visually Significant

We see the above mentioned integrated picture of maps in three phases:

1. Diffusion phase
2. Differentiation phase
3. Integration phase

Diffusion Phase

During this phase, the map reader sees only the visual outline of the map. Whether it is a reference map or a thematic map, the visual outline gives to the reader an idea of what the map is trying to lay emphasis on. According to Robinson the fundamental elements of the visual outline are:

1. The place,
2. The phenomena portrayed,
3. The position of the phenomena in the area, and
4. The relative position of the symbols.

If the objective of the map is to lay emphasis on places, they should stand out distinctly in the map. This is done when we prepare a map showing location of cities. If the distribution of certain phenomenon are to be emphasized, the specific region where they are distributed should be made distinct and if certain things are to be emphasized within that area, first the area should be made distinct from its surroundings and then the region where the phenomenon is distributed should be made distinct from the rest of the area. Finally, if the distribution of two different phenomena is to be given prominence, they should be made quite distinct by using contrasting colours or shades.

An arrangement of the type suggested above will make the prominent aspects of the map stand out from the rest of the aspects. Such an arrangement enables the map reader to instantly know the main thrust of the map as soon as he looks at it. The visual outline serves the same purpose as the chapter headings in a book. By making the outline prominent, the map maker is able to transform his products into effective instruments of visual communication.

Differentiation Phase

In this phase of visual communication, the eyes of the map reader look for further details of the phenomenon portrayed in the map. A good map is marked with two characteristics of the data represented on it:

1. They are authentic, and
2. They are represented clearly, legibly and effectively.

Both the qualities must be present simultaneously. A clear, legible and effective representation is no substitute to correct representation, and *vice versa* is also true. A good balance has to be struck between accuracy and effective, clear and legible representation.

Presentation of Symbols: To make a map clear and legible, symbols used must be adequately differentiated from each other. For example, all lines must be clear, sharp and uniform. To differentiate one line symbol from the other, we can either use varying thickness or varying designs of lines.

Size and shape: The size or width of symbols used in a map should be large and distinct enough to enable the map reader to see them. In this connection it should be kept in view that the unfamiliar symbols can be seen by a normally sighted person only if it subtends an angle of at least one minute at the eye. It means that farther away a symbol is from the observer's eyes, larger it should be in size in order to be legible. As all the people who have normal sight are not really so, it is better to keep this limit a bit higher. Experiments in this regard indicate that the angle subtended by the symbol with the eye should be between 1.75 minutes to 2.5 minutes.

The Table 15.1 gives the average viewing conditions:

Table 15.1: Relationship between Viewing Distance and Size of Symbols

Viewing Distance (in meters)	*Size of Symbol (width in mm)*
0.30 m	0.02 mm
0.45 m	0.205 mm
0.75 m	0.04 mm
1.50 m	0.08 mm
4.00 m	0.16 mm
8.00 m	0.32 mm
16.00 m	0.75 mm
32.00 m	01.5 mm
40.00 m	01.75 mm

In addition to being sufficiently large in size, symbols used on a map in conjunction with each other should be such that they stand out individually on the map. They should present a good contrast with their surroundings. The contrast can be achieved by varying the shape, shade and size of symbols. How various line symbols can be contrasted with each other has already been discussed earlier. We may vary the thickness or the design of the lines for this purpose.

Colour and Shade: Colour is by far the most important single medium in map design. It enables us to create a better contrast in symbols. Used in conjunction with other graphic symbols, it makes the portrayal of data visually most interesting. It also enables more information to be fed in a map without congesting and cluttering it.

In order to make an effective use of colours, we must know something about its relevant characteristics. Colour is an ingredient of sun light. It can be separated in the form of familiar rainbow or spectrum by passing light through a prism (Fig. 15.2).

The wavelengths of the seven colours which constitute sun light vary from 1/79,375 cm for the red to 1/157,813 cm for the violet. Wave-lengths of less than 1/157,813 cm or more than 1/79,375 cm cannot be perceived by human eyes as colour. This is, however, not the case with all life forms. Some insects can perceive ultra-violet rays in the· form of a colour because their sensory cells are made to respond to them as a colour experience.

The spectrum of the seven colours which we can see is derived from three primary colours. These are red, yellow and blue. These are called primary colours because all other colours can be produced by a combination of these three.

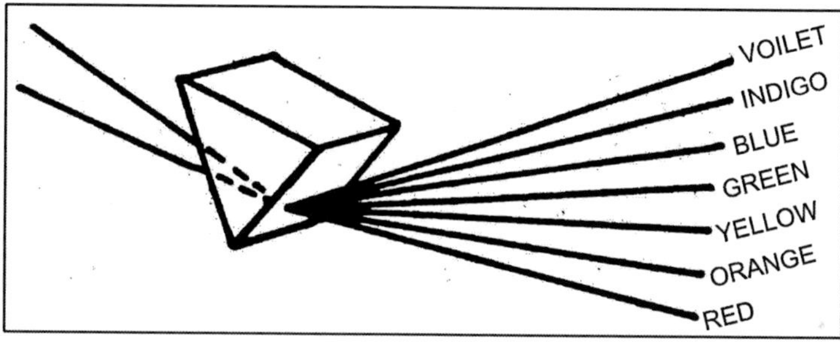

Fig. 15.2: Rainbow colours

Despite the fact that our eyes perceive only a limited wavelengths as colour we can perceive a wide ranging combinations of colours. The colours that we perceive have three characteristics. These are:

1. Hue,
2. Value, and
3. Intensity.

Hue is the quality which differentiates one colour from the other. In all, there are ten hues:

1. Yellow-red (orange),
2. Yellow,
3. Yellow-green,
4. Green,
5. Blue-green,
6. Blue,
7. Purple-blue,
8. Violet,
9. Purple, and
10. Red-purple (magenta).

The shades between any two hues can be further subdivided into ten steps, to make 100 hues in all. Each of the hundred hues is further divided vertically and horizontally. The vertical divisions are called values. Each hue has ten shades of values ranging from white at the top to black at the bottom. Each hue is again divided into 10 shades to give the varying degrees of intensity or brightness or relative saturation of the colour area. These 10 shades vary from gray to pure colour. Thus, we get combinations to create over 5000 colours. These colours are designated to give all the three characteristics. For example, Y.R. 4/3 means (yellow-red) of fourth gradation in darkness (value) between black and white and third gradation in brilliance (intensity) from gray to full yellow-red.

Human eye is most sensitive to red, followed by green, yellow, blue and purple colours in that order. Certain colours like yellow, blue, green, red, white and black appear as distinct colours while the others appear as mixed ones. While choosing colours to depict different features on maps, these facts should be kept in view. Another thing to be borne in mind is that colours maintain their original hue, value and intensity only when shown individually. In association with other colours they tend to lose their specific characteristics. For example, a dark area shown next to a light area will make the dark appear darker and the light lighter. It means that by rearranging the location of various symbols with respect to each other, we can create different apparent values. The symbols, including colours, used in a map have a value rating. Whether they maintain the rated values or not depends on their arrangement with respect to each other and within the frame of the map.

Patterns: Patterns also help us differentiate the phenomena represented on a map. Patterns can be and are used in place of colours where use of the latter is not possible, for reasons explained later. Patterns are made by varying arrangements of lines and dots, separately or together (Fig. 15.3). The possibilities of getting patterns of lines or dots or both are surely unlimited. But not more than 100 patterns are generally used in designing maps. The most common among them do not exceed 25.

From a visual viewpoint, dot patterns are better than line patterns. As lines have directions, viewer's eyes tend to give an image of something unstable. At times they join the boundary lines and make letterings merge in patterns. Unless line patterns are formed by closely set fine lines, they are irritating to the eye. As against this, dot patterns give an indication of stability, and are pleasing to the eye.

Integration Phase

The final phase of observation of a map or perceptual development occurs when various articulated elements of a map are composed into a coherent whole. A well integrated map will give sufficient material or information to bring to light the purpose of the map. It will be simple without ornate artistry so that the attention of the reader is not diffused. Such a map will evoke similar responses among many map readers.

One of the ways of achieving better integration is the positioning of various

Fig. 15.3: Dot and Line Patterns

components of a map in such a way that their relationships appear logical. In a well balanced design nothing is too small, too large, too bright, or too light. The importance of each component is directly related to its position and visual significance.

To determine whether a map is balanced or not we have to view it with respect to its visual centre which is little above the centre of the area enclosed within the neat line of the map (Fig. 15.4). If an item is out of balance it may be above or below the visual plane. Cartographers aim at balancing the various components of a map so that they appear natural and realistic.

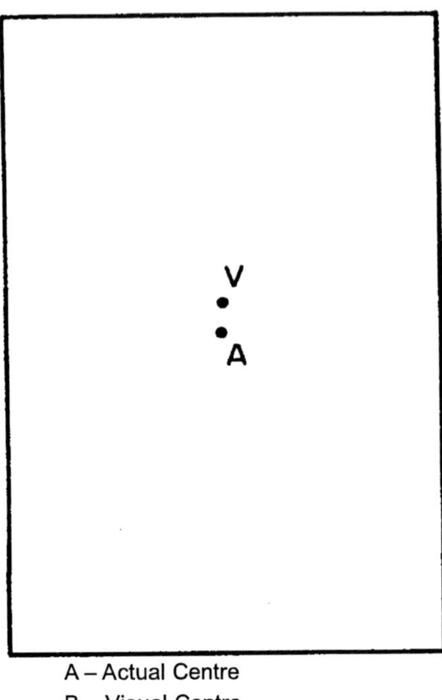

Fig. 15.4: Visual Centre of a Map

CONSTRAINTS IN MAP DESIGN

As discussed earlier, a cartographer is not as free as an artist in representing the objects he selects for the map. The constraints within which a cartographer functions can be grouped under three heads:

1. Cartographic
2. Technical
3. Resource

Cartographic Constraints

A number of cartographic conventions have been evolved which are usually followed while designing a map. Many of these conventions have now been internationalized, so that any departure from them appears to be unnatural and incorrect to the map reader. Take, for example, the question of representing water features by blue and plains by green. We know that all water bodies do not necessarily appear blue nor are all the plains green. On our physical maps the evergreen Sunderban forests are shown as green as are parts of Rajasthan desert. Supposing that we try to reverse the colour scheme and show the water features by green and the plains by blue, the map users would consider it wrong. The only logic behind the existing method of representation is conventional usage.

If we compare the topographic maps published by various government agencies the world over, we will find that most of the symbols used therein are common. These symbols are often referred to as conventional signs because they have been conventionalized. Whether these symbols are logical and aesthetic or not, is of secondary importance. The map users have become used to such symbols and hence their replacement will create confusion.

It does not, however, mean that a cartographer has no freedom of innovation. First, there are only few symbols which have been conventionalized and secondly, even the conventional symbols to a certain extent can be modified in size and style to present a different picture. For example, it is a convention to represent water in blue but there is no restriction on using different tones of blue: flat, half-tone, etc. We can change the hue, value as well as the intensity of blue to see that the water features do not give an unpleasant look.

Technical Constraints

While designing a map, a cartographer has to work within a number of technical constraints such as publisher's specifications, data availability, scale and projection of the map, and method of reproduction.

Publishers Specifications: The publishers have certain set ideas about the kind of map they want to publish. There are people who want are great variety of information to be fed in a single map. They prepare a long list of facts to be portrayed and also suggest the size of the printed map.

Maps serve a number of purposes. Some of them serve as wall maps, others as atlas maps, but a vast majority of them go as illustrations in various publications. For each purpose, the design has to be different, although the information given might be the same. Wall maps are to be seen from a distance of more than five feet. In these maps only outstanding features are depicted clearly, legibly and boldly. Atlas maps are to be seen from a distance of about 1.5 feet. Hence, the visual significance of various objects is not based on bold representation; it is based on contrasts created by appropriate positioning of symbols, and proper selection of the colour scheme. The book illustrations are used in conjunction with the text material. Although each illustration should be self-contained, it is clear that it can have no independent existence. Removed from the book, many of the illustrations lose their significance. All book illustrations have to be fitted within a given space. Not too often, is the size of a book determined by the size of its illustrations.

Data, Scales and Projections: The nature of the data, including the size and the shape of the area to be represented, also influences the design of a map. For example, a map of Chile has to be an elongated one. A map of the USSR has to be longitudinally extensive and relatively narrow latitudinally. The nature of the data also influences the process of symbolization. Some facts can be shown by line symbols, certain others by point symbols, and still others by area symbols.

Scale also influences the design of a map. On a large scale map objects can be represented in greater detail without creating cluttering. If the same data have to be shown on a small scale map, many of the details may have to be dropped in order to keep the map legible.

Projections are also influence the design of a map. Projections are of different types but there is none which can represent area, shape and direction truly at the same time. Most of the distributional data have to be shown on equal-area projections. To show correct direction, one has to select the Mercator's or one of the zenithal projections. A cartographer has very little choice in this regard.

Reproduction Processes: The reproduction technique to be used has a great influence on the design of a map. If direct contact prints are to be made, using the original as a positive, as in the case of zerox or deskjet printers the reproduced map will be of the design as the original one even if it enlarged or reduced in size. The scale of the map to be reproduced should be linear so that the reduction and enlargement would not affect the scale of the map.

On the contrary, if the original has to be reduced photographically to get a negative or positive for final printing, it will have to be designed differently. A map to be printed in multi-colour, will have as many originals as the colours used in the map. Far more data can be shown on a multi-colour map than on a black and white map. Different 'printing processes require different kinds and numbers of originals, which in turn require different map designs.

In most of the printing processes maps are drafted at a scale larger than the reproduction scale. This is done with a view to get a more refined picture of the fair drawing. It is, however, often forgotten that a well designed original does not necessarily give a well-designed print. In fact the design of such maps should have the scale of the reproduced map in view.

Resource Constraints

The third set of constraints under which cartographers function is related to cartographic materials, instruments, time and finances. Design of a map and its quality is often determined by the training of the cartographer if computer mapping is attempted, and cartographic materials and instruments used in drawing if the map is drawn manually. Availability of time and finances also influences the design. A multi-colour map is a costly proposition. If finances are limited, one will be forced to think of a black and white map. Similarly, the time at the disposal of the cartographer influences the design.

SYMBOLIZATION

Map itself is a symbol; it is a symbol of symbols. Symbols are like words. As the words giving same meanings differ from language to language, so also the symbols from map to map. Except for a few conventionalized symbols, a cartographer has far greater freedom to develop symbols than a linguist has to create words. Words take the meaning given by its users. Symbols take their meaning given by the cartographers.

When several words are put together in a definite order, we get a sentence. Similarly when several symbols are put together in a definite order, we get a map. Similarly, many sentences put together in a definite order make an essay. These orderly arranged symbols give a meaning which individual symbols fail to give in the same way as orderly arranged sentences convey a definite message which individual sentences cannot. Symbolization and the arrangement of the symbols in a map are, therefore, crucial processes in map design. No book can be popular if the choice of words and sentences is bad. So also no map can be popular if the choice of symbols is bad.

Symbols are of three types:

1. Point symbols,
2. Line symbols, and
3. Area symbols.

Point Symbols

Point symbols are those which give the location of an object or the quantitative value represented by it exactly at the point of its location. Point symbols are of two types:

1. Qualitative, and
2. Quantitative.

Qualitative symbols are used to suggest the existence of an object. For example, a dot is put for a town and a cross for a hospital. Such symbols do not represent any quantitative data such as the size of the town or the hospital.

The quantitative point symbols are used to indicate:

1. Presence of an object,
2. Length, width and height of the object
3. Size and area of the object, and
4. Volume of the object.

Uniform dot symbols can be used to represent the existence of a certain phenomenon in partially quantitative terms. The amount by which an object or idea is characterized can be represented either by bars or circles or spheres depending upon the type of data to be represented. Representation by bars indicates the length or height; by circle or squares, the size and by cubes or spheres, the volume. In this connection it may be noted that the cubes· and spheres are three dimensional and, hence, they take less space than squares and circles.

Line Symbols

Like point symbols, the line symbols too are used to represent qualitative as well as quantitative data. In the first category fall the latitudes, longitudes, boundaries, lines of transport and communication, streams, coastlines, etc. The thickness of these line symbols is not dependent on quantitative values of the objects represented. In fact certain objects like geographic coordinates and coastlines do not exist in reality. The width of the transport and communication lines as well as of the streams and boundaries are highly exaggerated. They are not drawn to scale.

Line symbols represent quantitative values too. The iso-lines such as contours and isarithm represent quantitative values. Similarly, the flow lines show the amount of the object moving from one place to another.

Area Symbols

Both point and line symbols are used to give a combined effect of spatial distribution of selected phenomena. Area symbols also are of two types:

1. Qualitative, and
2. Quantitative.

Qualitative symbols indicate the spatial spread of a given phenomenon but not its density such as the spread of swamps, forests, deserts, political units or soil types. When symbols are used to give the relative density of a phenomenon whether by administrative units or by isarithm lines, they acquire quantitative values.

MAP FORMAT

All maps must show a few common features. These are title, legend, direction, scale, and source of information and in some cases insets. The title of a map may be placed anywhere within the neat line that makes the frame of the map. Most appropriate place is the top right of the frame. It can also be placed at the top left or bottom left or bottom right depending on the space available. The title should include the name of the area represented, and the nature of the data shown. If the data pertain to a given year this should also be given. The title should always be given in bold and simple letters.

The legend of a map is usually placed in a corner within the neat line. The position of the legend is so selected that it does not interfere with other details. Every symbol and abbreviation used in a map should be explained in the legend.

Direction is shown in one of the corners by an arrow pointing to the north. Scales can be expressed in one or more of the several ways explained in Chapter 6. In an original drawn for reproduction, the scale should conform to the requirements of the printed map. If the scale is given in R.F. it should be the scale of the printed map and not of the original drawing. The best thing to do is to give a bar scale because the bar is also reduced with the original drawing. The scale of a map should be placed at a prominent place. It can be placed just below the title or somewhere at the bottom. Every map must give the source of the data used. Most of the maps we use, do not mention the source, thus they keep the map reader in suspense about their accuracy. The source should normally be given outside the frame of the map on the bottom right. On the bottom left should be given the name of the author, publisher, etc.

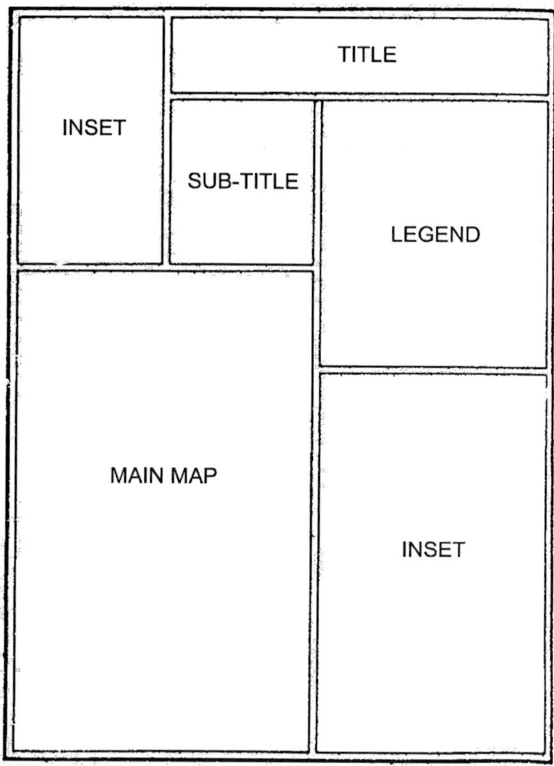

Fig. 15.5: Over all design and lay out of a map

In many maps it becomes necessary to give an inset map (Fig. 15.5). An inset map can be used to show the location of an unfamiliar area within a relatively familiar area. For example, if a district is represented on a map, an inset map may be given to show the location of the district in the state or the country. It can also be used to give a portion of the area represented in the map in greater detail. For example, while showing the population distribution in a state, an inset can be given to show the details of the population distribution in a metropolitan region. There are many other details which can be fruitfully shown in an inset map.

The border of a map usually consists of two lines 1/4" to 1/2" apart. In between the two lines are given the degrees of latitudes and longitudes. The outer line is thicker than the inner line. The inner line is also called the neat line. In some maps only the neat line forms the border. The degrees are written inside the frame. The margin outside the outer line should not normally be less than 1/2". If marginal information is also given, as in the case of the topographical maps, margins will have to be wider.

Lettering forms an important component of map design. It is discussed in the following discussion.

Lettering and Toponomy

Letters are verbal symbols. They form words which give us certain meanings. These letters and words have to be incorporated in the body of the map along with other symbols. Letters are conventional symbols of the linguists. They have to be used as they are given to us whether we like them or not. What we can do, however, is to change the style, form, size, colour, etc. of these symbols to suit our specific needs.

Lettering has always been an important aspect of map design. In the past ornate letters were common. They were partly used as a device to fill up vacant spaces in the map. In those days all lettering was done freehand. Subsequently, it came to be engraved. In all these processes the artistic aspect of the work was considered to be more important than its usefulness. Ornate letterings are difficult to read.

The style of lettering has changed with the change in the printing technology and the taste of the people. Now-a-days one can get any style of lettering with the help of internet and computers. At present the best lettering is considered to be one which can be read easily.

While lettering on a map, one has to decide the following:

1. Style
2. Form
3. Size
4. Colour
5. Method
6. Position
7. Relation to reproduction
8. Standardization

Style of Lettering

There are three main styles of letterings. But now the styles of letters available on computers are so vast, that cartographers have far more choices than ever before. Here we are discussing only three types of letterings. They are:

1. Classical,
2. Modern, and
3. Sanserif.

The origin of the classical style is Roman. In this style the proportion of thick to thin lines making the letters is not great. The strokes of the letters have long and curved serifs. It is an ornate style and difficult to read. The modern style was developed around A.D. 1800. It has precise geometrical shapes, and the difference between the thick and thin lines making the letters is often excessive to give an unbalanced picture.

The lines are marked by small horizontal strokes. The san serif style is the most modern and up-to-date one. It has no serifs at all. It gives a clear cut, new and nontraditional appearance. It is the best style from the point of view of legibility (Fig. 15.6).

Form of Lettering

Within these styles one can develop several sub-styles or forms by changing the slant, thickness, and complexity. The style which can be considered to be good is one which is easy to read. Ornate and fancy designs are good to look at but difficult to read.

Baskerville BE Regular
Baskerville Italic
Baskerville Bold
Baskerville Bold Italic

GoudyHandtooled BT
GoudyoldSt BT
GoudyoldSt BT Italic
GoudyoldSt BT Bold
GoudyoldSt BT Bold Italic

Some classic or old style letter styles

Caslon
Caslon Open Face

Bodoni
Bodoni italic
Bodoni Bold
Bodoni Bold Italic

Some Modern Letter Style

COPPER PLATE
COPPER PLATE GOTHIC BOLD
COPPER PLATE GOTHIC LIGHT
COPPER PLATE GOTHIC BD BT

Some sans serif Letter Style

FUTURA LT BT
Futura Md BT
Arial Normal
Arial Italic
Arial Bold
Arial Bold Italic

Fette Frakur ABC abc 123

Text type Style

Fig. 15.6: Styles of Letterings

Size of Lettering

In view of the complex and varied nature of the data represented on maps, it is often desirable to use several lettering styles to create contrast. But this should not be overdone. Within the modern style, we can have several combinations by using the capital and small letters and by varying the size and thickness of letters.

Size of lettering plays an important role in map design. It is true that certain styles are difficult to read but even the most modern and legible style will be of no use if the size is not properly selected. The size of letters is designated by points. Points 1 is equal to one twelfth of an inch. Lettering that is one fourths of an inch in height is equivalent to 18 points. Perhaps point 3 is the smallest type which can be read from a distance of 1 foot. It is safer to use 4 or 5 point types (Fig. 15.7).

BY the ideal book I suppose we are to understand a book not limine
BY the ideal book I suppose we are to understand a book no
BY the ideal book I suppose we are to understand a b
BY the ideal book I suppose we are to understan
BY the ideal book I suppose we are to unders
BY the ideal book I suppose we are to u
BY the ideal book I suppose we ar
BY the ideal book I suppos
BY the ideal book I
BY the ideal bo
BY the ideal
BY the idea
BY the id
BY the i

Fig. 15.7: Size of Letterings

Table 15.2: Relative Visibility of Size of Letters

Size in points	*Relative visibility from 18 inches/45 cm*
3	1.10
4	1.60
5	2.11
6	2.64
8	3.64
10	4.65
12	5.66
14	6.67
18	8.67
24	11.68

Table 15.2 can be used to select the size of letters while preparing a map. For example, if we want to increase the relative visibility of a letter by 5 times (point 3 being the base), we will have to select point 12 and not point 15.

Colour and Background

Another way of creating contrasts and making letterings more legible and easily perceptible is to put them in varying colours and against contrasting backgrounds. The greater is the contrast between letters and background, the more is legibility and perceptibility. Black lettering against white background stands out most prominently (Fig. 15.8); on a gray background it looks faint and blurred.

Fig. 15.8: Background and Lettering

Positioning of Letters

Positioning letters means placing them in relation to other symbols in the map. The layout of letters should normally be parallel to the top and bottom of the neat line. This, however, creates some imbalances if the parallels and meridians are also shown in the map, for in many cases the latitudinal lines do not run parallel to the neat line. It is, therefore, desirable to eliminate the graticules from the land areas of the map.

They can be shown by strokes along the neat line. They need not be eliminated from the water bodies especially the oceans. This should not, however, be considered to be a rule, for in many cases the graticules may have to be shown even on the land to serve certain specific purposes.

In cases where the features to be named have spatial spread such as countries, mountains, etc., letters should be spread to include the entire feature. They should be equally spaced and easily distinguishable. Names of the rivers should be positioned along their courses and the letters should be slanting. The alignment of lettering used for railways, roads, canals, telegraph lines, air-routes, sea routes, etc., should be the same as that of the objects. Place names should be so positioned that they do not get mixed up with symbols. They should be placed a little above or below on the right or left of the symbol, to avoid confusion. The titles and legends if put in more than one line, should be positioned around a central line to give a sense of balance.

Mechanics of Lettering

With the introduction of computers, the mechanics of lettering has changed drastically. Now it is easy to experiment with various sizes and styles and to select one that is most pleasing and clear. Until very recently, lettering on maps was a very tedious and time consuming process. For the benefit of the computer and digital cartographers, the methods used in lettering are discussed below although only briefly.

There are four methods of letterings:

1. Freehand,
2. Stick up,
3. Mechanical, and
4. Computer

Freehand Letterings

It is done with the help of a pen. It follows the guidelines drawn with the help of a ruler, curve or lettering angle. For all capital lettering only two guide lines are needed but for mixed letterings, three lines are more appropriate. First, the letters are drawn with a lead pencil. Freehand lettering requires considerable practice. It is quick and more suited for maps in which letters have to follow certain crooked lines. To ensure that letters of the same size, parallel lines can be drawn as shown in Fig. 15.9.

Stick up Letterings

Under this system, letters were written by hand or mechanically and cut and pasted at appropriate places on the map (Fig. 15.10). To make the task easier, beautifully printed alphabets and figures of various sizes and styles are available in the market which can be transferred on the map as per requirement. Quick and precise lettering can be done with the help of computers.

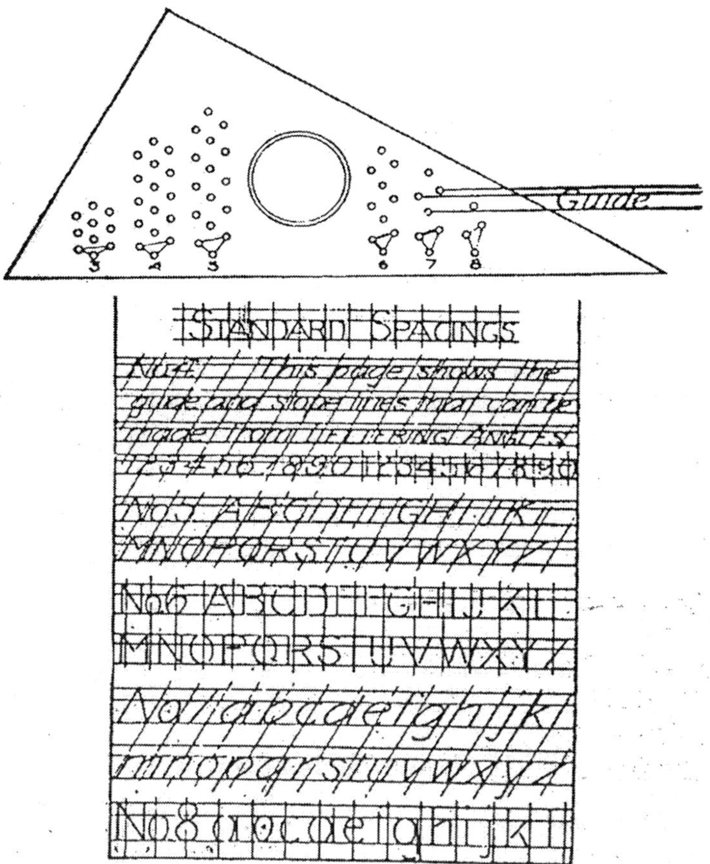

Fig. 15.9: Freehand Letterings

Mechanical Letterings

In this process, the size, style and thickness of letters is controlled mechanically. The latest instrument used is the computer. With the coming of computers, the old mechanical devices are no longer used but one should know how the maps were prepared earlier (Fig. 15.11). Among the instruments used until very recently, were:

1. *UNO Pen*: It was a pen attached to a small tube which contained ink. The size and style of letters are determined by the templates in which the letters are stenciled.
2. *Leroy Set*: It contains a scriber to which a pin and a pen are attached. The pin moves along a groove in the template and the pen gives the required letters. The templates contain a variety of letter styles and symbols.
3. *Varigraph*: It is the most mechanized of the lettering devices of recent past. It also consists of a template with engraved letters and a stylus. It's functioning is based on the principles of a pantograph. Adjustments to make the letters large, small, elongated, etc. are possible.

Fig. 15.10: Stick-up Lettering

4. *Wrico*: It uses stencils as in the case of the UNO pen but the pen is not the same.
5. *Hope graph*: It was the latest in the mechanical lettering devices computers came into the picture. This equipment is more functional and precise than those discussed earlier.

Computer Letterings

Introduction of computers has made lettering on maps easier, better and faster. It is now possible to print letters in:

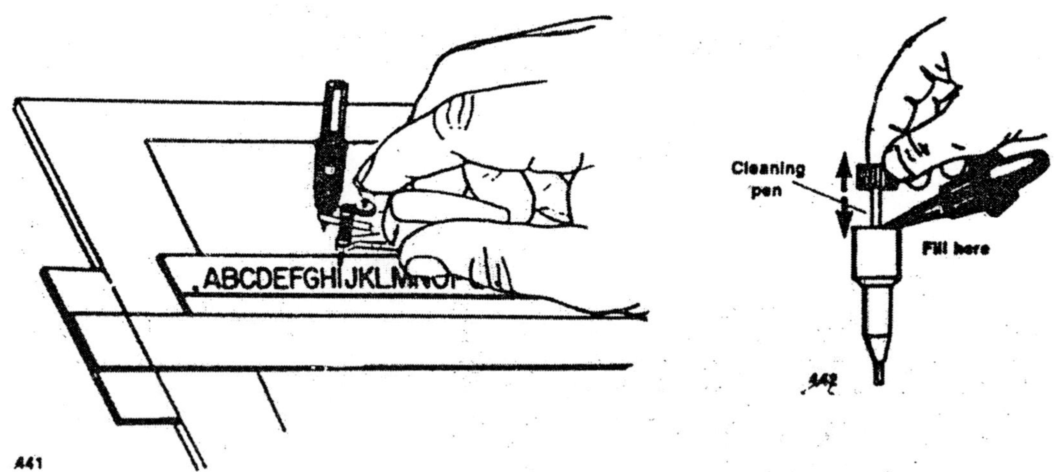

Fig. 15.11: Leroy Set

1. Any style,
2. Any size (point),
3. Any language,
4. Any colour, and
5. Any design.

One can see the letters on a computer screen, change them to the desired size, style and format and then lift and place them at appropriate places on the map. While selecting the size and style of lettering, it must be kept in view that the letters fit the scale of the map and the size and shape of the other symbols included in the map. They should be so selected that they are legible once the map is printed.

GEOGRAPHICAL NAMES

Attempts to rationalize and standardize geographical have been on for the last 125 years. The International Geographical Union (IGU) took this matter up as early as 1872. It has not, however, proved to be an easy task and the recommendations of the Union have been disregarded by almost all the member countries.

The problem involved in the standardization of geographical names can be gauged from the following example. In the 1930s a river was discovered in the western half of Papua New Guinea. It was named after its discoverer as Father Le Coq D'Armandville River. The Popuans of New Guinea could not make either head or tail of it and continued to call it 'The Broad River' (in their language). Up to 1957 the capital of Dutch New Guinea was called Hollandia. After the Dutch withdrawal in 1957, it was renamed as Kota Bam. Subsequently, it was renamed as Sukarna Pura. Now it is called Ayapura. The principle of giving local names (advocated by the IGU) had been disregarded in this case.

The irrational changes in place names are only too common. In the USSR almost all the names given by the Czar regime were changed to commemorate Lenin or Stalin. And now after the fall of communism, all those names have been changed again. In India, Calcutta has been changed into Kolkata, Bombay into Mumbai, Bangalore into Bangaluru, Madras into Chennai. There is also a problem of disparity between the official names and the popular names. For example, Banaras in U.P. is now officially called Varanasi but the people continue to call it either Banaras or Kashi. Duplication of place names is another problem. Many of the names are too often repeated in the same country. There are so many Washingtons in the USA.

It thus appears that it is difficult to standardize geographical names unless all the member countries adhere to the principles laid down by the I.G.U. in this regard. A cartographer has to use his judgment in determining the authenticity of place names. It may involve considerable amount of library and at times even field research.

16
Mechanics of Map Construction

A good map should not only be well designed, it should be well drawn also. Drawing or drafting of maps whether manually, mechanically or with the help of computers requires some artistic skills. And artistic skills can be acquired only by regular and patient practice. Those who do not have patience cannot become good cartographers. Apart from patience and practice they must be conversant with different computer programmes which help in creating maps. New programmes are being developed ultimately leading virtual maps. Satellite imageries are now available for countries, state/provinces, districts, cities, villages and even streets and houses. All one needs to do is to apply computer programmes to list the relevant details available on Google, etc. and transform them into good maps.

This chapter gives the details of the traditional approach to map construction. This is still being used in places where the use of computers is not yet prevalent. Irrespective of how a map is made, the paper on which it is being made or printed should be of good quality. Poor quality papers shrink with change in weather. This leads to change in scale too. If the map is drawn on a scale of 1 cm to 4 km, its reduction into half will change the scale to 1 cm to 8 km. But if the original drawing of the map has shrunk or expanded, the reduction of the size of the map to half will not give the scale of 1 cm = 8 km.

While drawing a map by hand, one should know that the quality of the paper on which a map is drawn determines its capacity to take ink. The soft and porous papers soak ink and diffuse it through fibers. Good quality drawing sheets hold on the ink to give a sharp image. If the surface of the paper is rough the drawing on it would not be smooth. This applied to drawing by computers too. The surface of the paper should be smooth and strong enough to take erasing.

The Indian ink used in original drawings by hand is a permanent suspension of fine carbon in a liquid medium. It dries dense black which is very important for reproduction and is waterproof. While using this ink, it should be kept in mind that drawing surfaces easily pick up oily substances from the hand. The ink will not hold on to the oily surface. The ink used should be absolutely opaque black. If the drawing is to be reproduced photographically, white opaque paint or ink can be used for corrections. This should not, however, be done on drawings which are to be used for direct contact printing.

DRAWING INSTRUMENTS

Some of the equipments used in manual drawing of maps are as follows:

1. Drawing board or table (Fig. 16.1)
2. Drafting machine (Fig. 16.2)
3. T-square (Fig. 16.3)
4. Magnifying glass (Fig. 16.3)
5. Straight edge-12" scale and a slide rule
6. Curves of various types (Fig. 16.3)

Fig. 16.1: Drawing Board or Table

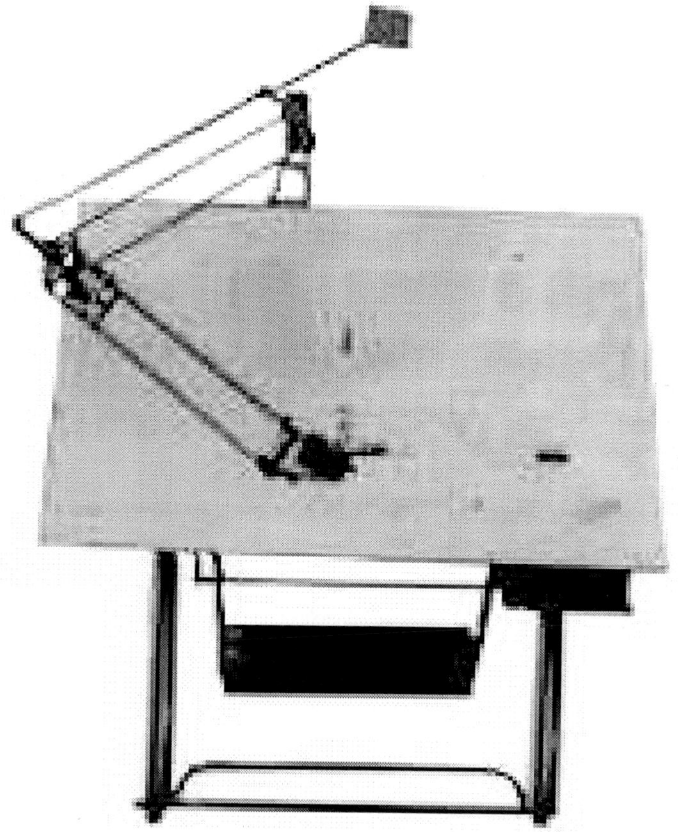

Fig. 16.2: Drafting Machine

Maps are usually drawn on drawing tables manufactured specially for drafting purposes. A drawing table consists of a drawing board measuring at least 24″ × 18″ made of soft wood. A drawing table with a board is preferable. Drawing tables have mechanisms for adjusting the height as well as the tilt. The drawing paper should be affixed to the board either with a tape or thumb tacks. The latter leave holes and hence should be fixed only on the extreme corners. They also obstruct the movement of the T-square. Tapes are therefore preferable.

A special kind of table called tracing or light table is a must in a cartographic laboratory. It is used in tracing on an opaque surface. It differs from an ordinary drawing table in that its surface is made of glass and not of soft wood. The glass is illuminated from beneath.

A table can also be custom made which combines the functions of a drawing and a tracing table. In the board of the table on opening of, say, 12″ × 9″, can be made to fix a glass.

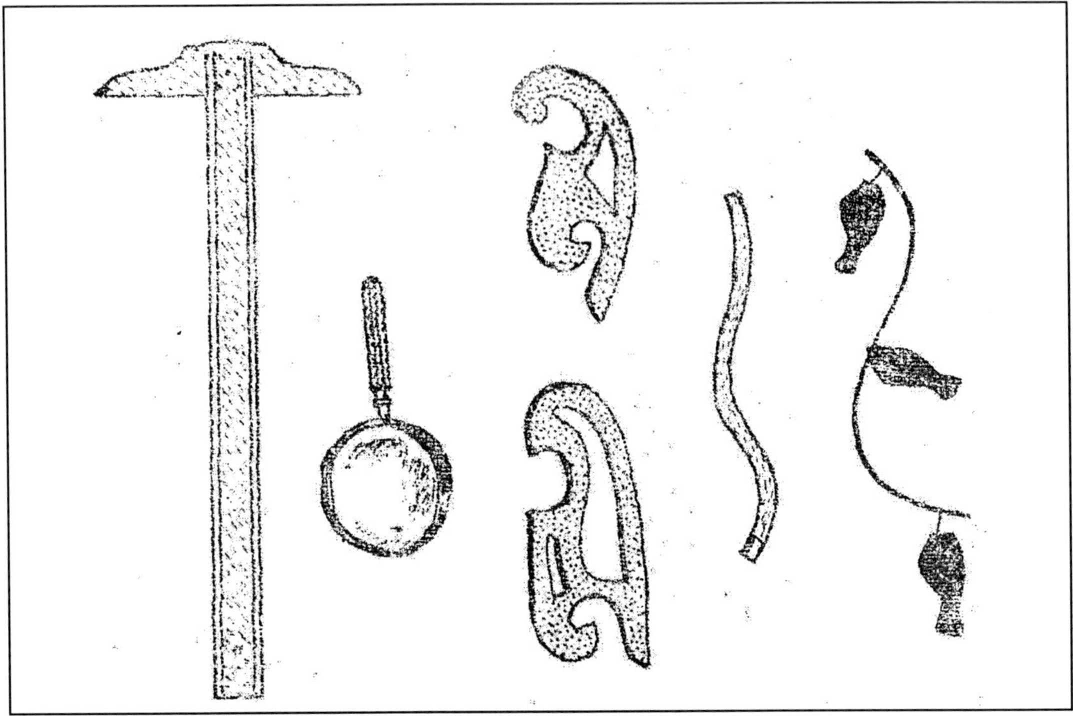

Fig. 16.3: T-square, magnifying glass, curves, adjustable curve rule, and spline

All tables and drafting machines must be properly lighted. Cartographic work puts lot of strain on the eyes. The lights should be such that they do not leave shadows and are adjustable. It is desirable to use fluorescent lamps.

A T-square and a set of triangles are used in line drawing work. T-squares are made of metal or solid wood. One with transparent plastic edges is better, for it enables the draugftsman to see the drawings below. It slides up and down the side of the table or board. The triangles help in drawing vertical lines with varying slants.

The curves of various types help in the drawing of curved lines. For smaller and sharp curves, a set of 'French Curves' is enough, bur for large curves one has to take recourse to railway curves or spine with weights. Recently flexible curves of different make have also been introduced for drawing curves of various shape and sizes. For example, Japanese flexible curve is made of rubber with small steel balls inside the rectangular tube with leveled edge. It is of different lengths which could be moulded and be given the required shape. Magnifying and reducing glasses are useful in judging the look of the drawings at various scales of reduction and enlargement. Stencils of symbols of various types, if at hand, can help in quick drawing of symbols.

There are a number of drawing instruments one can get in a drawing box. It is better to have complete set, because the case in which the instruments are kept provides a convenient storage place. The basic instruments consist of: 1. pens, 2. compasses, 3. dividers, 4. computer; 5. printers.

Pens

The most common pen used by a cartographer is a 'crow-quill' pen. It is a sharp pointed pen and is used for fine drawings. The use of a crow-quill pen requires a lot of training.

Ruling pen also known as lining pen happened to be used for drawing lines of various thicknesses. It has two blades which could be adjusted as per requirement of the thickness of line, with the help of a side screw. Thus the lines of different thicknesses can be made with the same pen. It requires frequent cleaning as the ink clogs in between the blades. Some ruling pens have their blades assembly fixed in a swivel. They are used in drawing curved lines such as contours. Pens with two sets of blades make the so-called road pens which are used in drawing parallel and swinging lines. (Fig. 16.4).

However, in ruling pens adjustment of blades for drawing lines of various thicknesses is done manually which is difficult to maintain at precise level. Therefore, the precise variant pens have been introduced. Among these pens the most commonly used ones are: (i) U.N.O. Pen (ii) Leroy Pen, (iii) Stedlars Mars, (iv) Ratring, (v) Payzant Pens, (vi) Graphics, and (vii) Keo-Aristo.

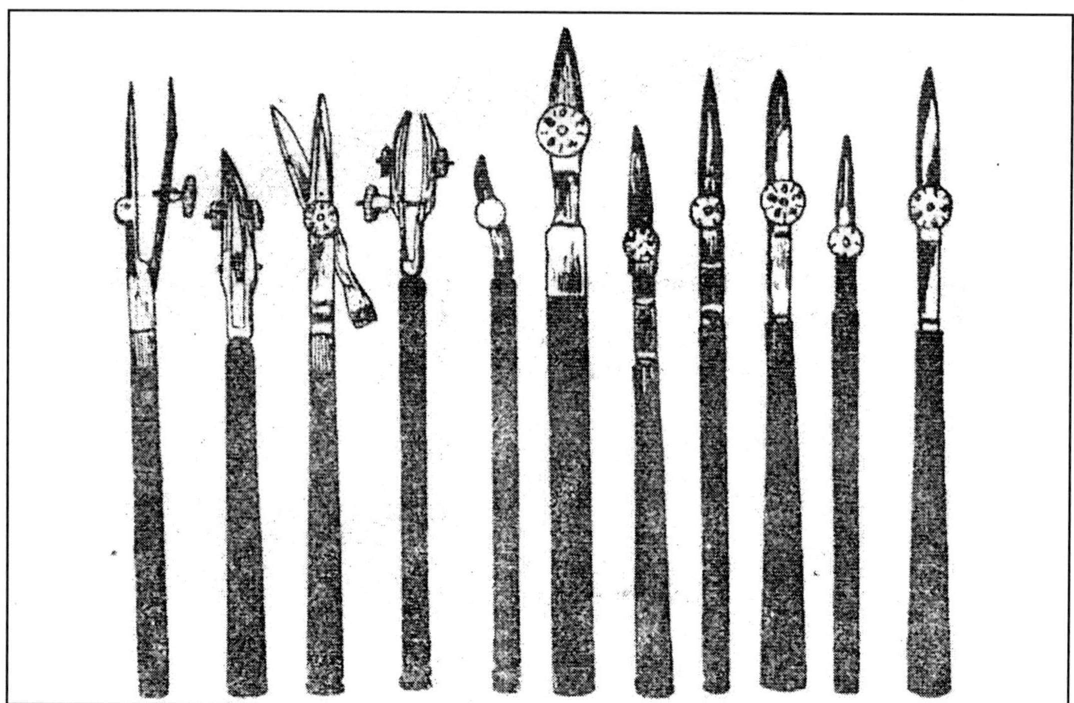

Fig. 16.4: Ruling Pens

A leroy pen comes with leroy lettering set. It can be bought also separately. It has a cylinder with a point and ink is fed through an opening in the cylinder. The cylinder is attached to a holder. Leroy pens are numbered-from 0000 to 8. Each number gives a constant width of line or dot, and variation pressure does not affect the thickness of line. It is especially useful for those who do not have steady hand while drawing.

The Barch-Payzant pen is also designed for lettering. It is useful in drawing uniform lines and dots. The speed ball pens have different shapes of nibs to give different kinds of lines. These are inexpensive and easy to operate; the graphos are like fountain pens.

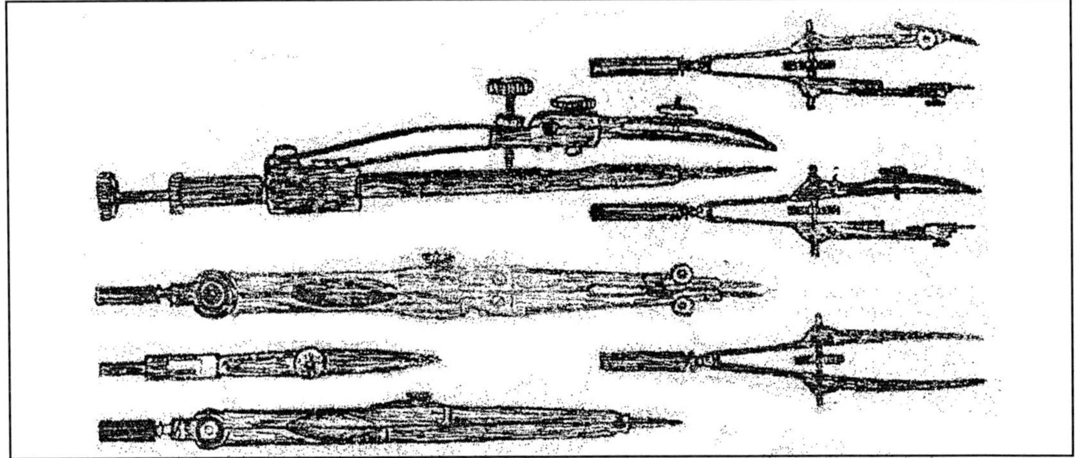

Fig. 16.5: Bow Compasses, Dividers, and Drop Bow Compass

Compasses and Dividers

Compass is used in drawing arcs and circles and dividers are used to layout distances. A compass is a divider with one leg fitted with a holder for pencil or ruling pen. Proportional dividers have two sets of needle points one at each end. Its mechanism and use has been explained earlier.

There are several other types of compasses. For making big circles, we have the beam compass and for small circles or round dots we have drop-bow compasses (Fig. 16.5).

The other important instruments are: (1) rotameter, (2) planimeter, and (3) slide rule. A rotameter is used for measuring the length of a curved line; the planimeter is used to measure area; and a slide rule is used for quick calculation (Fig. 16.6).

DRAWING THE ORIGINALS

Patience and perseverance are the two qualities that a cartographer must possess in order to be successful. There are very few short-cuts in cartographic drawing. It requires patience to spend hours but visibly producing very little. This aspect is scarcely appreciated by a map user. He is unable to realize the amount of thinking and effort

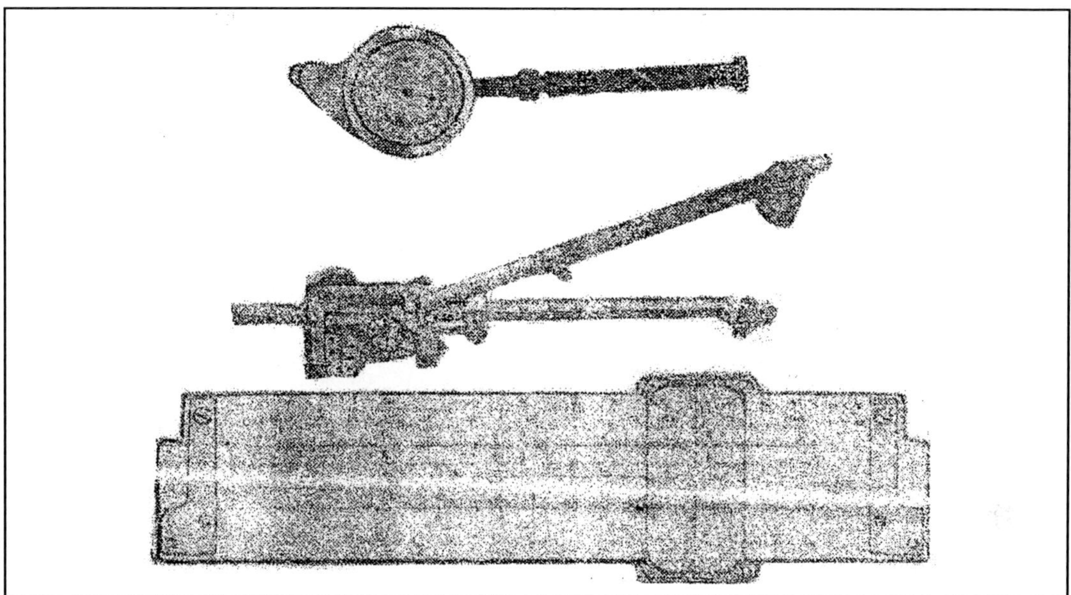

Fig. 16.6: (a) Rotameter (b) Planimeter and (c) Slide rule

that goes in making a map. This applies not only to manual drawings but also computer mapping.

Much of the cartographic work consists of line work. A regular line is drawn with a line pen held against a drawing edge such as a T-square or a curve; conversely most irregular lines have to be drawn free hand. While drawing a line the pen or the pencil should be held against the drawing edge in such a way that it remains perpendicular to the plane of drawing. When kept in this position, it will keep a wedge between the bottom of the edge and itself. This is necessary to stop the ink from running under the edge of the guide. This also allows the cartographer to see the line as he draws it.

17
Mapping the Terrain

Since the early days of cartography—the depiction of terrain on maps has posed special problems. As against many other visible features of the earth's surface, terrain has three dimensions i.e., length, breadth and elevation. Because of the difficulties in depicting the third dimension on a two-dimensional surface, early cartographers took note of only two dimensions. They depicted mountains as piles of crags and the mountain ranges as arrays of sugar loaves (Fig. 17.1).

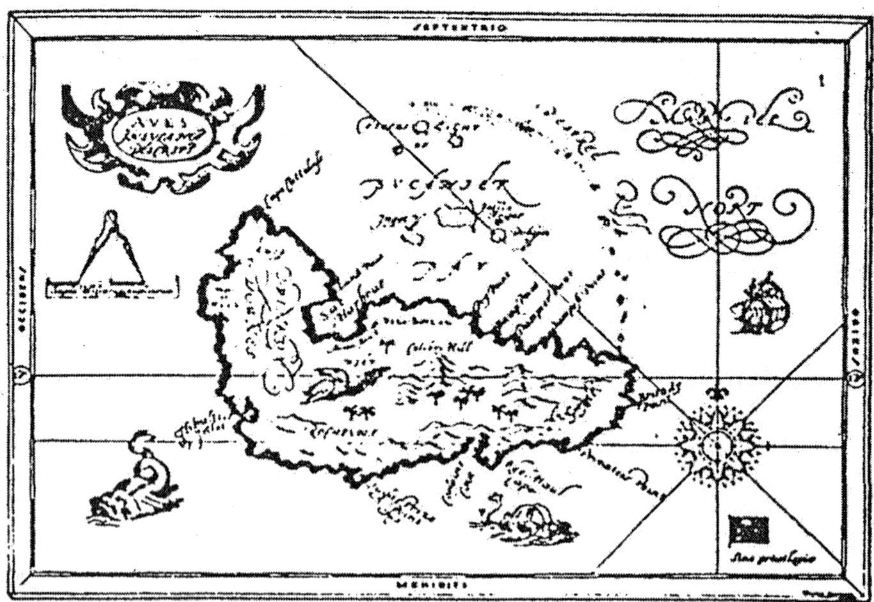

Fig. 17.1: Depiction of Terrain in Early Times

The pictorial depiction of terrain was resorted to due to two reasons: firstly the elevations of the mountains were not known as accurately as their lateral spread (length and breadth) and secondly, symbols to depict elevations were just not there. It has to wait for modern surveying to get correct measurement of the third dimension (elevation of land).

There are some unique characteristics of the third dimension (elevation) which call for special cartographic techniques to represent them. Firstly, height (positive, zero, or negative) is a continuous and universal phenomenon. Secondly, it is more apparent than length and breadth. The sea level has come to be known as the zero level above which the third dimension is called elevation or height, and below which it is known as depth.

Representation of relief on large scale topographical sheets is relatively less problematic than its representation on small scale maps. On large scale maps the symbols selected appear natural and are measurable. But what is measurable is not always effective visually. The use of colour makes symbolization a bit easier but the problem remains unresolved on small scale maps. The representation of terrain on these maps involves the generalization of landform data. It also involves the creation of a balance between terrain and other details so that one does not over-shadow the other. The representation of landforms on wall maps requires techniques which keep the slope, relief, elevation and direction clearly visible from a distance.

METHODS OF REPRESENTATION

Spot-heights

When heights are marked on the map by numerical values (feet, meters, etc.) at appropriate locations, they are called spot heights. The chief merit of this technique is that it provides definite and precise information. Spot heights, however, fail to give any visual impression of the relief pattern. They can, therefore, be fruitfully used only in conjunction with other cartographic techniques.

Hachuring

Lehman, an Austrian army officer in 1799 suggested a kind of flow-line symbol called hachure for the depiction of terrain. It consists of a set of finely drawn disconnected lines. Each line follows the direction of the slope and the ups and downs of the terrain. To highlight the relief, the hachure lines are drawn thick and close together on steep slopes. When so many lines are drawn close to each other, they collectively indicate slopes; thin and widely spaced lines indicate gentle slopes, and thick and closely spaced lines indicate steep slopes. The level areas whether plains or plateaus are left blank (Fig. 17.2).

Hachuring was a very popular device among the eighteenth century cartographers. The drawings were made artistically, but they did not prove effective in portraying the terrain on small scale maps. They looked like 'hairy caterpillars'. But on a large scale maps they turned out to be tolerably good.

Contouring

Contour is another line symbol invented during the first half of the eighteenth century. A Dutchman named Cruquius appears to be the first to use this technique in 1730, a

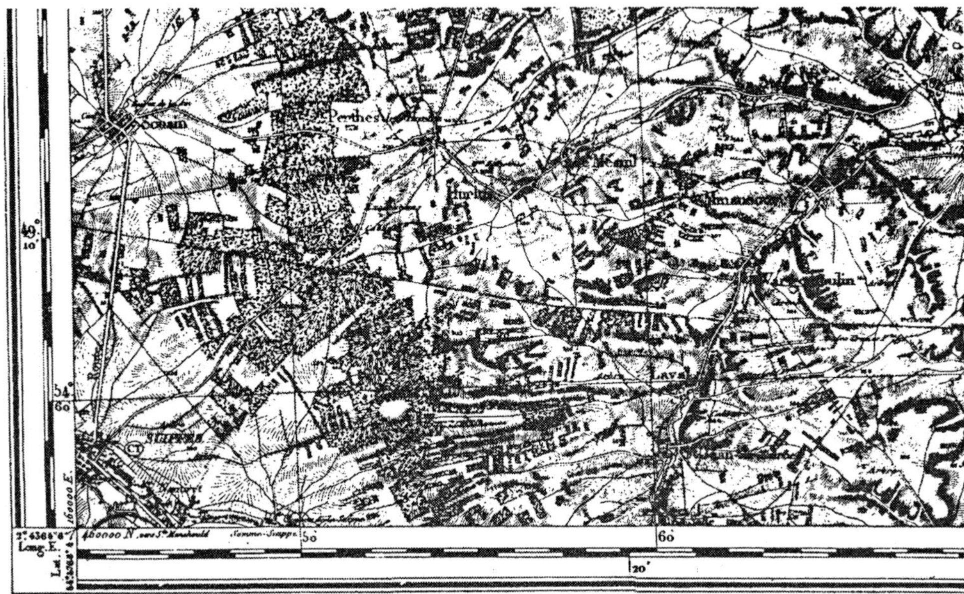

Fig. 17.2: Hachuring to Show Slopes

little before the hachure was first used. It was, however, not before later half of the nineteenth century that it became a common method of depicting relief on topographic maps.

Contours are isarithm, which connect points of same elevation. The elevation of points used in drawing the contour lines is measured with respect to a given datum level, which is the zero elevation level. Moving up or downwards, we can derive contours of various elevations in the same way. When we draw successive lines of equal elevation at a given interval and view these lines vertically, we get a contour map of the area. The principles underlying contouring are best illustrated if we imagine an island amidst an ocean whose level fluctuates to submerge and then uncover the island. If we draw successive lines along the water levels at different times, we will get contour lines of different heights (Fig. 17.3a). In Fig. 17.3b, contours are drawn to show the relief of the body of a cow.

Principles of Contouring

Looking at the contour map of an area, we conclude that all points falling on a particular contour have the same elevation. This is, however, not so. Contour connects the spot heights of sample points, not of all points on a given contour. Contour is thus an average line connecting selected points of a given elevation not all points that make a contour when joined together. The direction of a contour connecting any two points whose elevations have been measured is estimated with reference to local relief. The

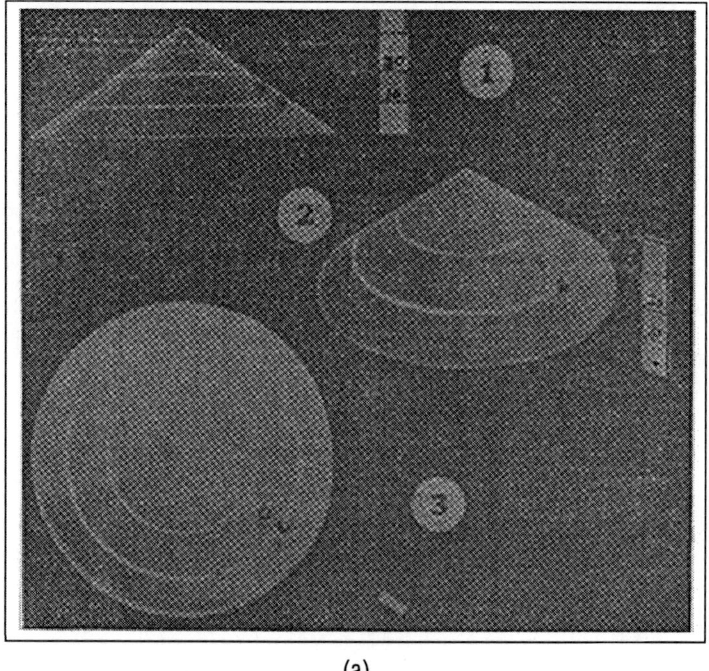

(a)

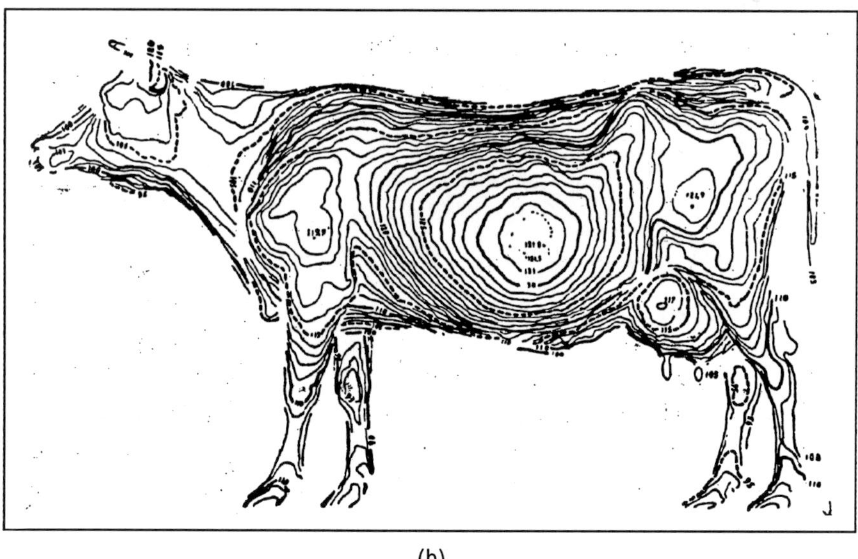

(b)

Fig. 17.3: Concept of Contouring; (a) A Cone Viewed Horizontally from 45 Degrees, and Vertically; and (b) A Cow Represented by Contours

selected points through which the line joining the two measured points passes are presumed to be at the same elevation at which are the measured points. In other words, we assume the existence of a statistical surface. The points which can be selected for actual measurement constitute only a sample of the total 'universe'.

The selection of the sample points for the measurement of elevation complicates the drawing of contour lines. Elevations of these points do not coincide with the values of contour lines everywhere. The contours are placed at proportionate distances from the known points as shown in Fig. 17.4. This amounts to saying that they do not always represent real elevations.

There is another type of error that enters into contouring. The selection of points for which measurements are made, is not done on any scientific basis. Statistical geography suggests that for a given size of the 'universe', there is some lowest limit of the number of samples which must be taken to make the samples representative of the universe. This principle of statistics is not yet applied to topographical surveying and contouring. We do not know the ideal number of points to be selected for measurement, given the degree of slope and the amount of relief.

The arbitrarily selected points for which we have elevation data are called control points. Each control point has its x, y and z axes values. Error in any of these three values undermines the accuracy of a contour map. The desired accuracy is further undermined by the fact that contour values do not always conform to the values derived for the control points as illustrated in Fig. 17.4.

Contours are only indicative of slopes and some other terrain details (Fig. 17.5). They are marked with five characteristics:

1. They are horizontal to the dip of the land and perpendicular to the direction in which the surface water runs.
2. They are closed lines.
3. Steeper the slope, closer the contour.
4. They bend up-stream while crossing a river.
5. They do not show minor relief features.

Interpolation of Contours

Drawing of new contour line by estimating the points in between various spot heights is called interpolation of contours. A river course and number of spot heights are given in Fig. 17.4. We have to draw contours at an interval of 10 feet. If the minimum height given is 21, we have to find out spots with 30, 40, 60, 70 ft. and so on elevations. If there are several points having 30 ft elevation, they should be joined to give 30 ft contour. But in case, for example, there is no point having 30 feet elevation, we have to draw the contour by interpolating 30 feet contour between points having higher than 30 ft and lower than 30 ft values. If two such points have 28 and 43 ft elevations, the 30 ft contour has to pass through a point lying in between these two points. Actual location of the point will be two-fifteenth of the distance between 28 and 43 feet point from the first point. We can determine the other points also in the same way.

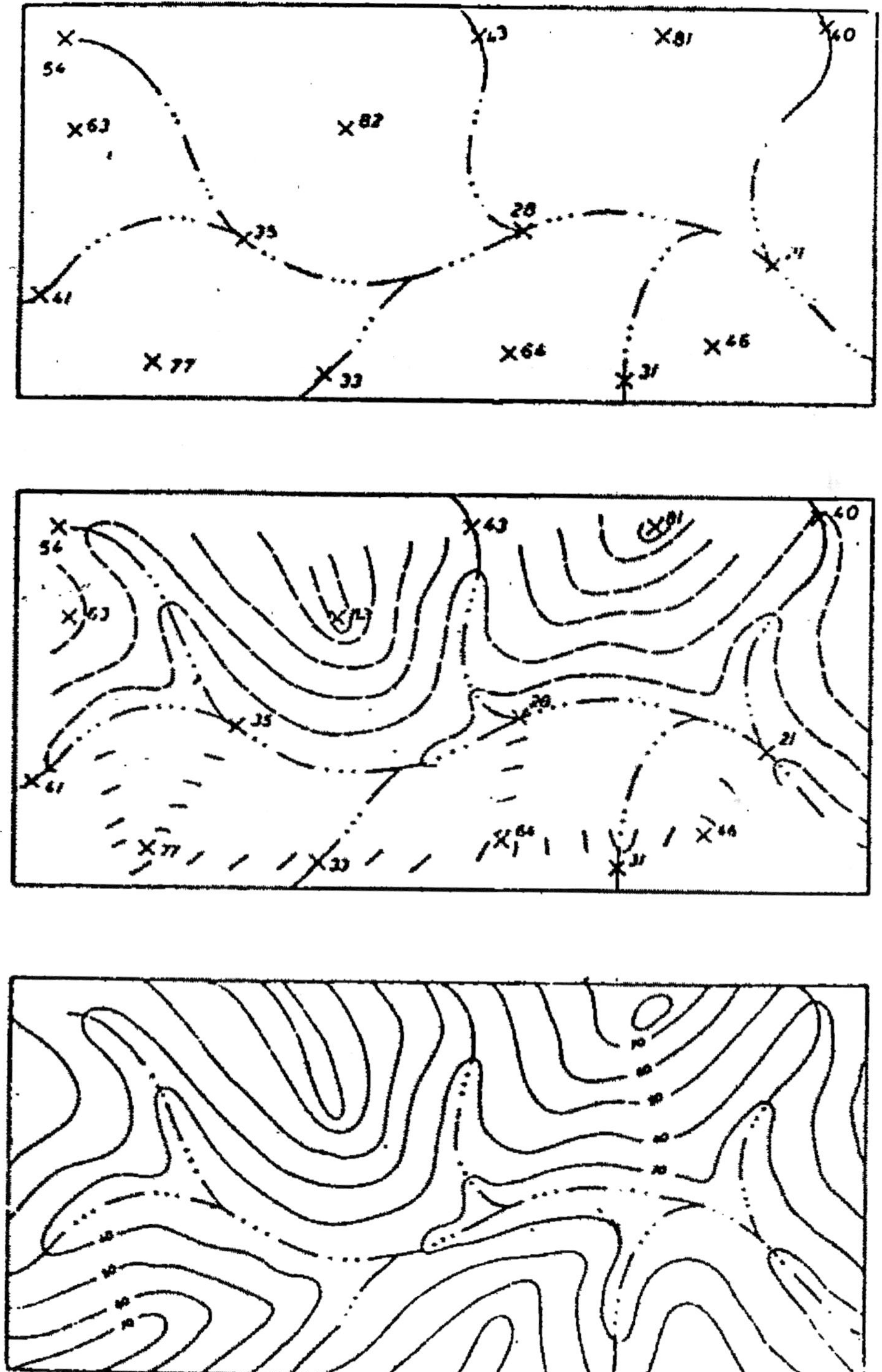

Fig. 17.4: The Stages of Contour Interpolation

Representation of Slopes by Contours

The distances between successive contours as measured on planimetric maps indicate the nature of the slope. Contours are made at regular intervals but they do not appear equi-distant when viewed vertically. They come close together when slopes are steeper. As the slope gradients flatten, the distances between the contours increase. Thus tile spacing of contours at equal intervals enable us to get a generalized impression of the nature of terrain (Fig. 17.5).

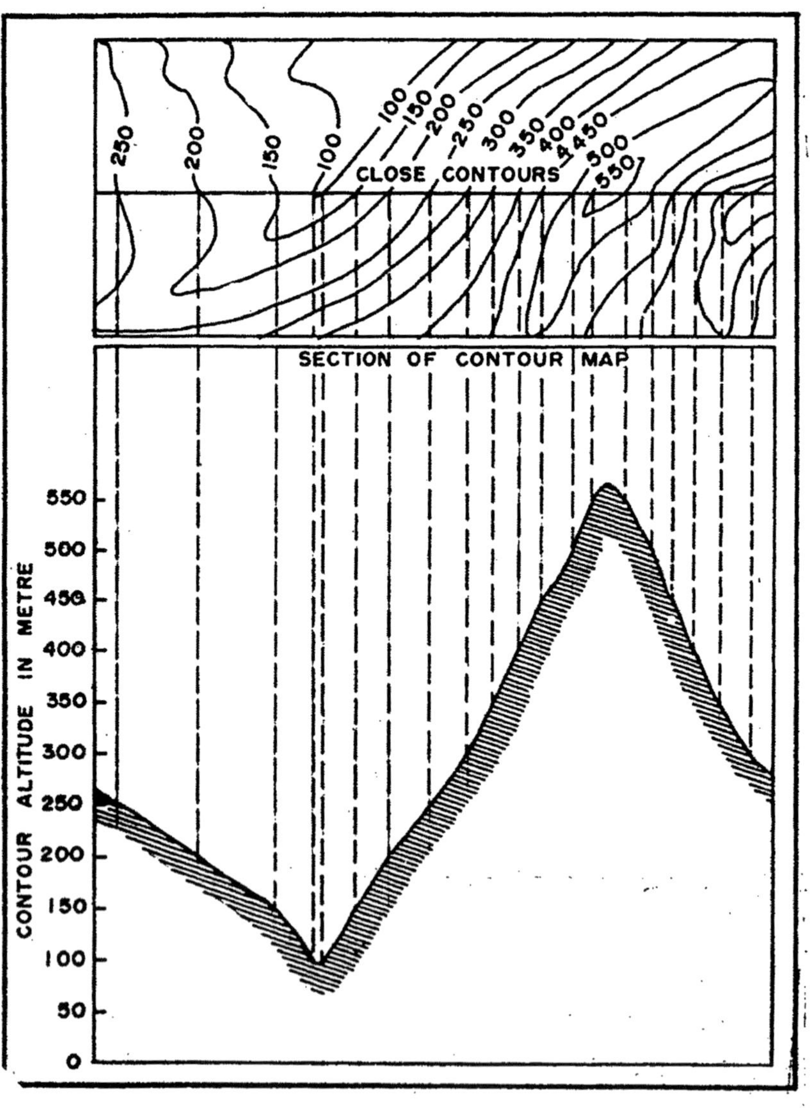

Fig. 17.5: Relation between Slope and Contour Interval

Slopes are of three types:

1. Uniform,
2. Concave, and
3. Convex

These are shown by contours and slope profiles in Figs. 17.6 (a) and 17.6 (b), and 17.7(a), (b), and (c).

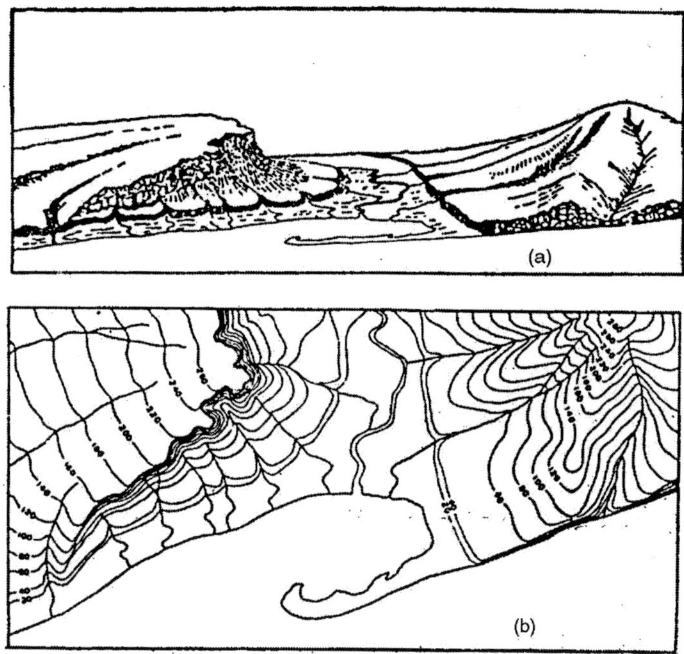

Fig. 17.6: (a) A Sketch of Landscape and (b) A Contour Map of Same Area

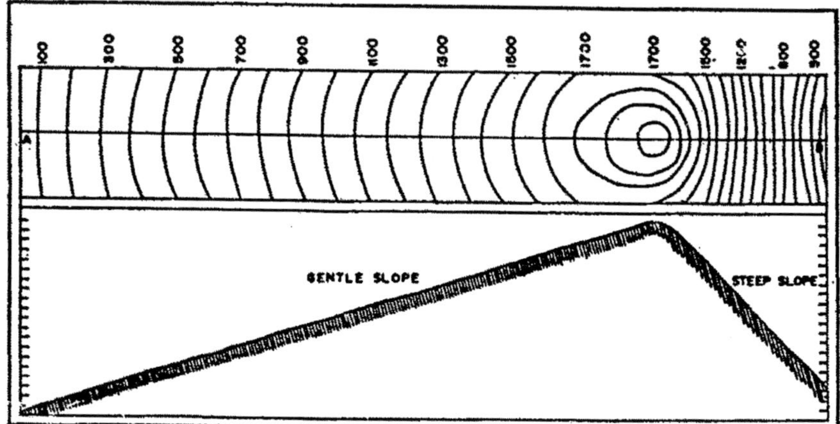

Fig. 17.7(a): Slope Representation by Contour: Uniform Slope

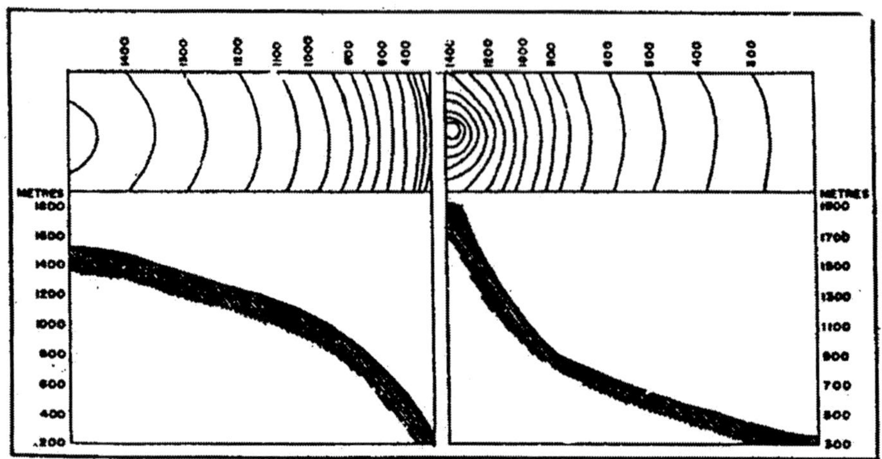

Fig. 17.7(b): (a) Concave Slope, and (b) Convex Slope

Representation of Landforms by Contours

Some of the landforms represented by contours and slope profiles are given in Fig. 17.8. For drawing a slope profiles, see section under profiles in this chapter.

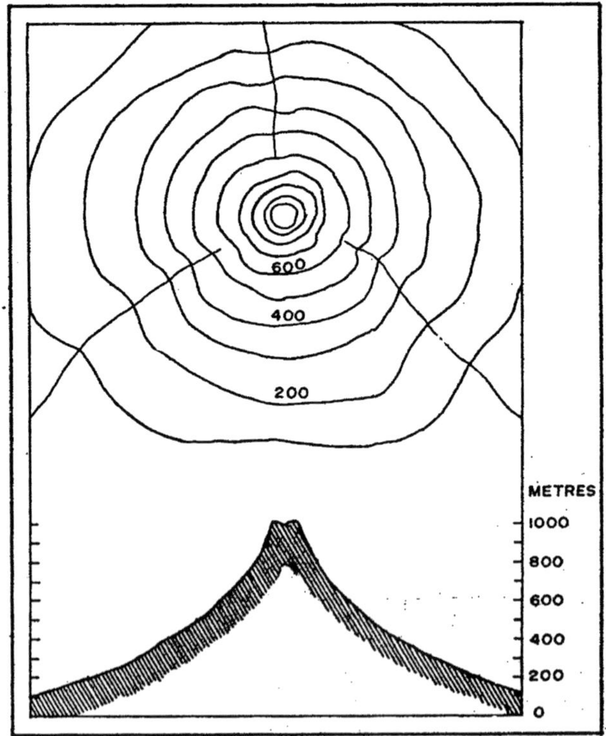

Fig. 17.8: Landforms Represented by Contours and Slope Profiles (*Contd.*)

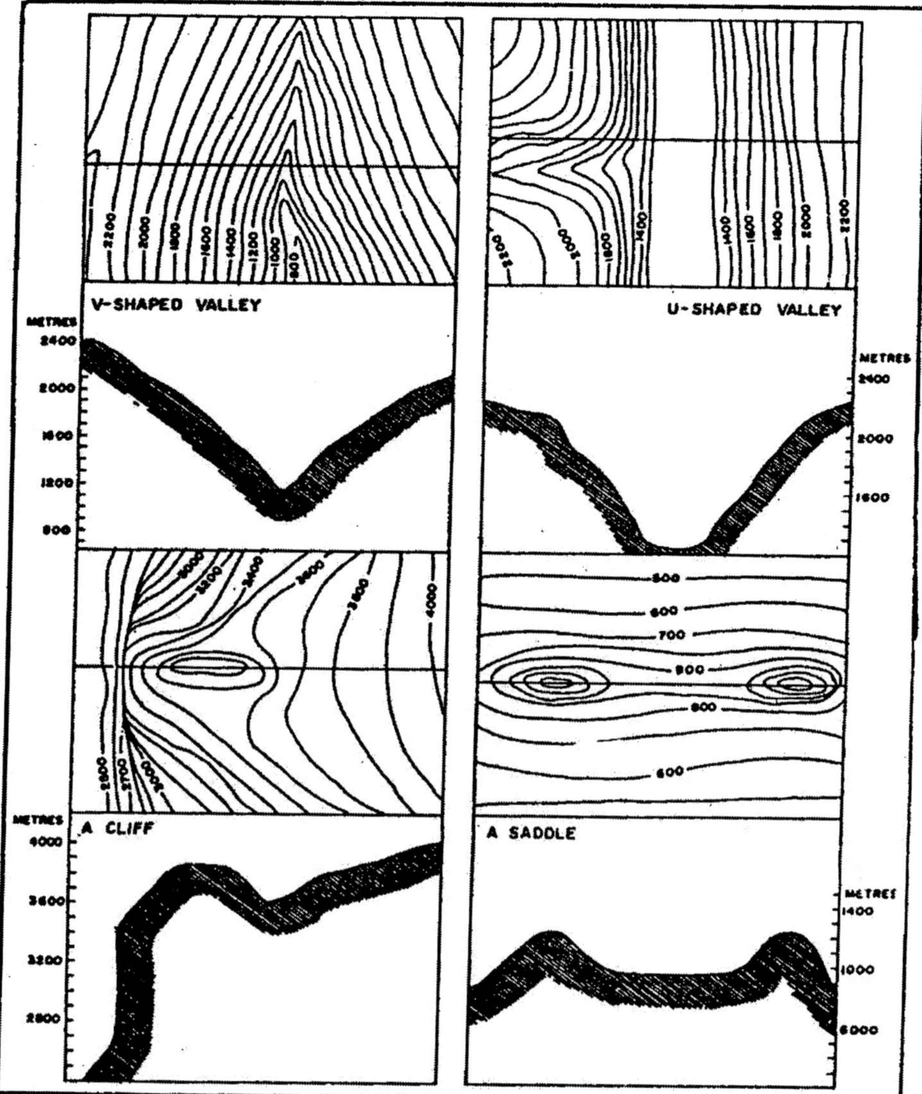

Fig. 17.8 (*Contd.*): Landforms Represented by Contours and Slope Profiles

Inter-visibility

In military operations, and explorations, it is often necessary to determine inter-visibility between two given points on a contour map. There are two methods available to us. The first method consists of drawing of a slope profile. If a straight line joining the two points on the slope profile is not cut by an intervening relief feature, the points will be inter-visible (Fig. 17.9). This is called the graphical method. The other method is based on the calculation of gradient. To determine the inter-visibility select the highest point C in between A and B. Calculate the gradient of B from A, and then of C from A. If the gradient of AC is higher than that of AB, A and B are not intervisible; but if the gradient of AB is more than that of AC, the two points are intervisible.

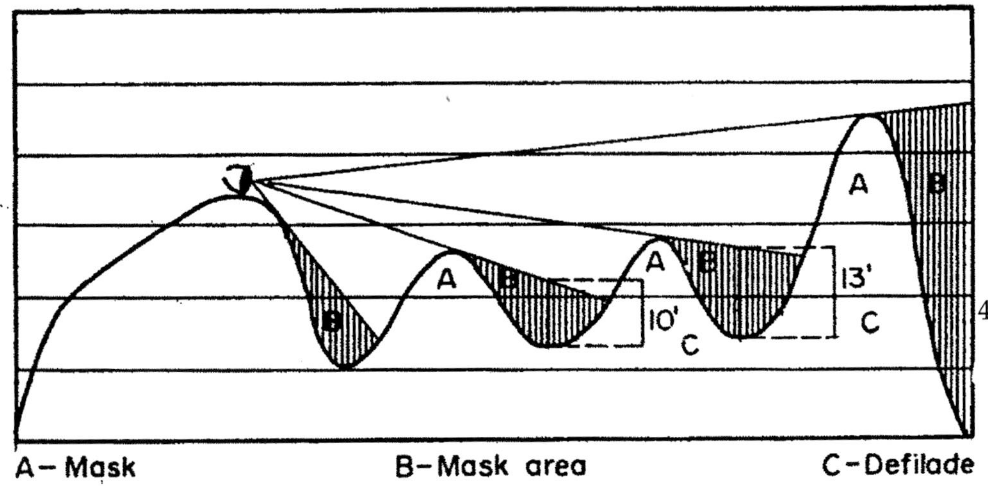

Fig. 17.9: Determination of Inter-visibility (Reproduced from Map Intelligence)

Plastic Shading

Contours are an effective means of portraying terrain but they suffer from a very serious perceptual limitation. They fail to give the third dimension visually. Plastic shading is a technique which, when used in conjunction with contours, goes a long way in removing this deficiency.

In its simplest form, shading is a technique of representing relief as viewed from a horizontally lighted relief model. If we photograph a relief model illuminated from one side, we will get a picture which will be similar to that of a shaded picture. Usually, light is supposed to be focused from the north-west. The south-eastern slopes appear to be relatively darker. For the same reason, the south-western and north-eastern slopes appear gray (between white and black). The principles underlying plastic shading are illustrated in Fig. 16.10. The shading of a contour map is done with the help of an air-brush or carbon pencil or charcoal or crayon, the slopes are shaded according to the principle stated above. When completed, a shaded map, unlike a contour map, gives a continuous gradation from white to black. This means, that while printing the map, the terrain plate will need half toning (see chapter on map reproduction).

The usefulness of shading as a technique of representing relief is beyond any doubt. It gives a very useful relief background for many of the thematic maps showing cultural data. The population maps of the *National Atlas of India* (English edition) have effectively used the shaded relief to indicate the relationship between distributions. This cannot be done, if contours or hachure had to be used. Contours and hachure clutter the map and make its reading difficult.

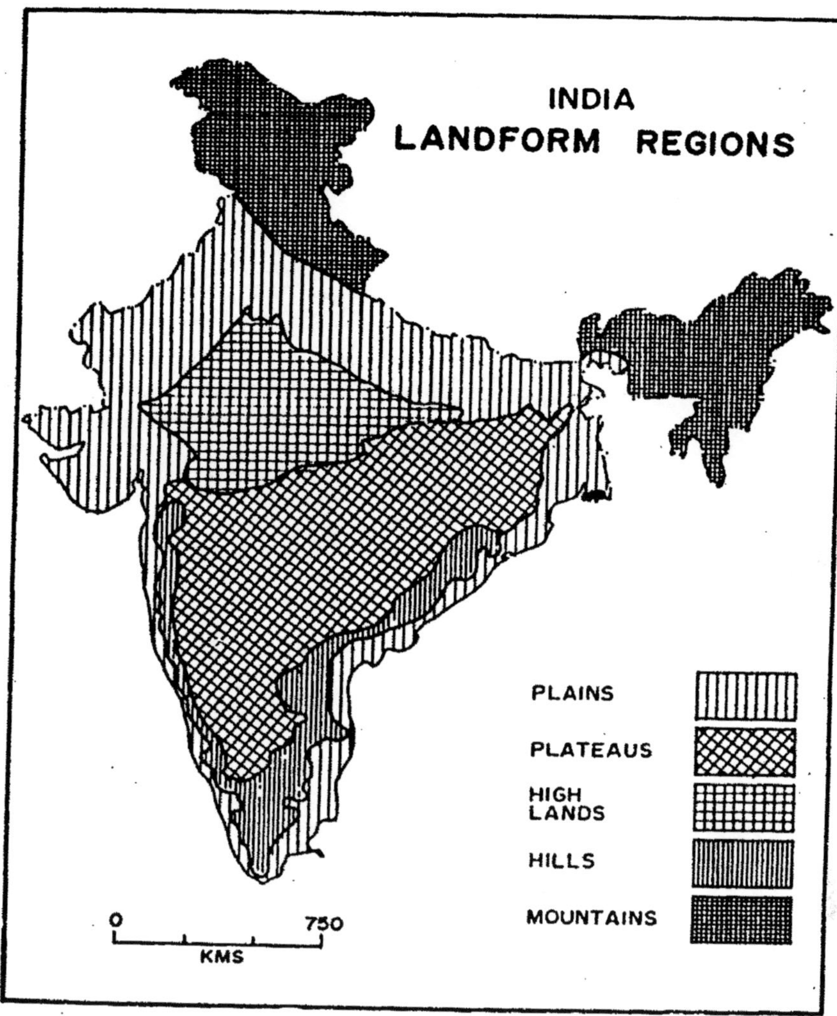

Fig. 17.10: Plastic Shading

Tanaka Kitiro's Method

This technique was developed by Prof. Tanaka Kitiro of Japan in 1931. It transforms a contour map into a kind of profile shading. First of all a contour map is ruled with closely set horizontal lines, as in Fig. 17.11.

Beginning with the intersection of a horizontal line with the lowest contour successive intersections with next higher horizontal line and higher contour line are joined by a line. This process is continued until one gets to the highest contour. Thereafter, the intersections of the successive horizontal lines and contours downwards are joined together. The resulting wavy lines appear to be profiles produced by an inclined plane (Fig. 17.12).

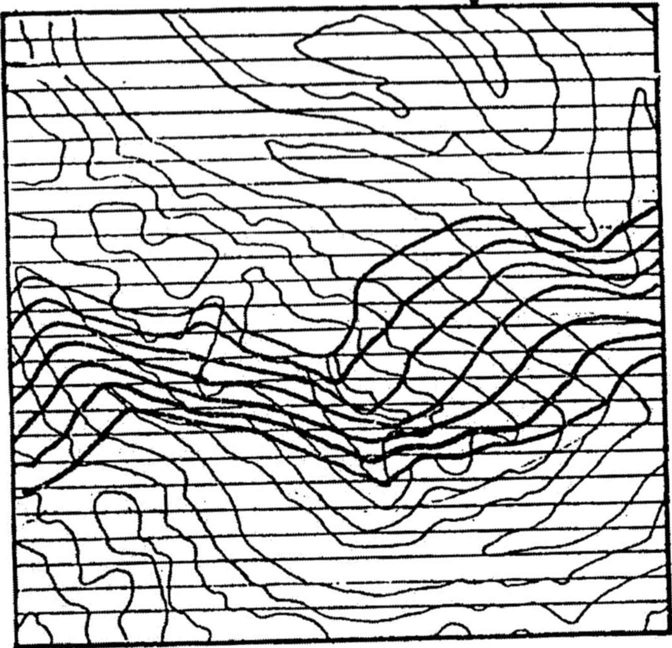

Fig. 17.11: Principles of Relief Representation by Tanaka Method

Fig. 17.12: An Example of Relief Representation by Tanaka Method

In 1950, Tanaka devised the technique of illuminated contours. On a gray background, the contours on the north-western side of the slope are shown white whereas on the south-eastern slope black. The lines are thick but thin out at and near the junctions of the white and black contours (Fig. 16.12). Robinson and Thrower, later in 1957, devised another version of this method which according to them is simpler to use and more effective visually.

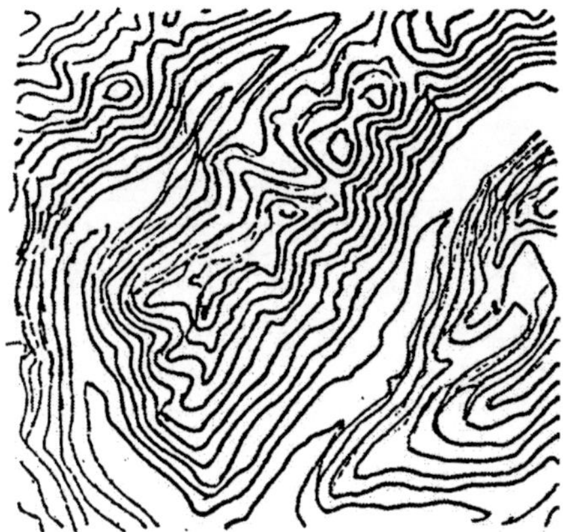

Fig. 17.13: Illuminated Contours Drawn against White Background (A Modified Version of Tanaka Method)

Layer Shading or Tinting

To show the distribution of relatively high and low areas, areas lying in between successive contours are shaded. This technique is very useful in depicting landforms like plains, plateau, etc. It is, therefore, useful for areas having varied relief. This technique, however, misleads the amateur map readers because a single colour between any two contours gives the impression of evenness of surface relief.

There are two methods of layer shading: Line shading and colour shading. In colour shading either different shades of a single colour or different colours are used. Contour interval at which the colours change has to be selected very carefully. Layer tinting on small scale maps portrays little about the land surface except the elevation zones. But on a larger scale maps, it represents the terrain more effectively (Fig. 17.14).

Morphographic Method of Relief Representation

This is pictorial method of representing relief. It involves a systematic application of a standardized set of conventional pictorial symbols for various relief features. The features so represented are viewed obliquely from the air at an angle of about 45°. Late Prof. E. Raisz has been pioneer in the development of this method. The advantages of this method are quite apparent. It brings out the outstanding relief features and yet is pictorial so that even a beginner can appreciate the relief depicted on a map. It is an effective method of teaching landform geography (Fig. 17.15).

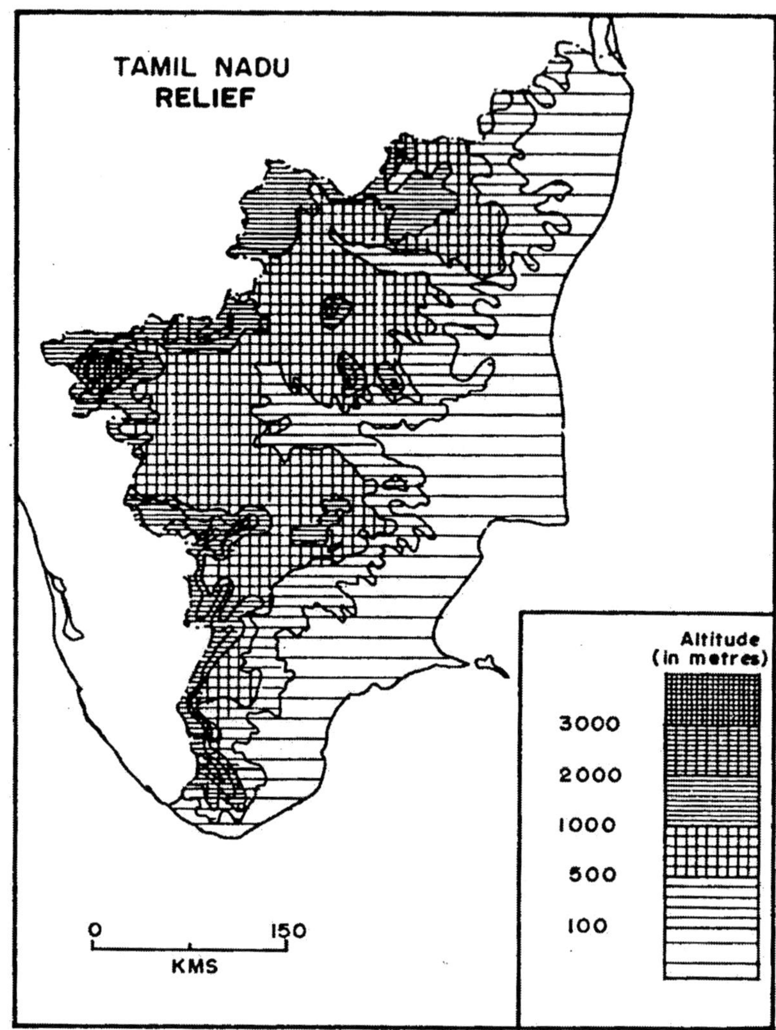

Fig. 17.14: Layer Shading

TERRAIN TYPE MAP

If the objective is to indicate broad and generalized landform types, the land can be divided into various categories such as:

1. High Mountains,
2. Mountains,
3. Upland and plateau,
4. Plains,
5. Ice covered areas

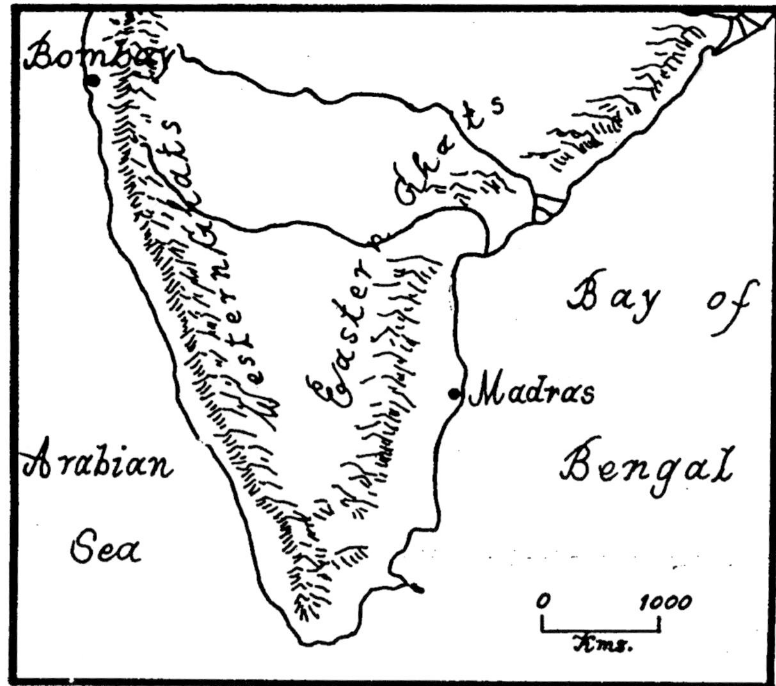

Fig. 17.15: Morphographic Method of Relief Representation

Each of these categories can be line shaded or stippled to produce a terrain types map. We often encounter a simple four-fold classification of landforms into plains, plateaus, hills, and mountains. This can be 'fruitfully shown on small scale maps. It is similar to layer shading except that it gives importance to terrain type rather than to contour lines (Fig. 17.16).

RELATIVE RELIEF MAPS

While preparing a relative relief map, the differences between the highest and the lowest elevations within a limited map area have to be determined first. To start with, a topographical sheet is divided into squares of say 5 minutes of longitude and latitude. Differences between the highest and the lowest points within each of the squares are marked and plotted on a small-scale base map. Isarithmic lines are then drawn by joining places of same difference to get a choropleth map showing relative relief (Fig. 17.17).

This technique is quite effective in depicting relief of maturely dissected plateaus, and plains horizontal rock structures. It is less suited to geologically complex regions. It is also not very suitable for differentiating minute yet important terrain details as they are often too small to extend beyond the confines of the squares chosen for determining the relative relief.

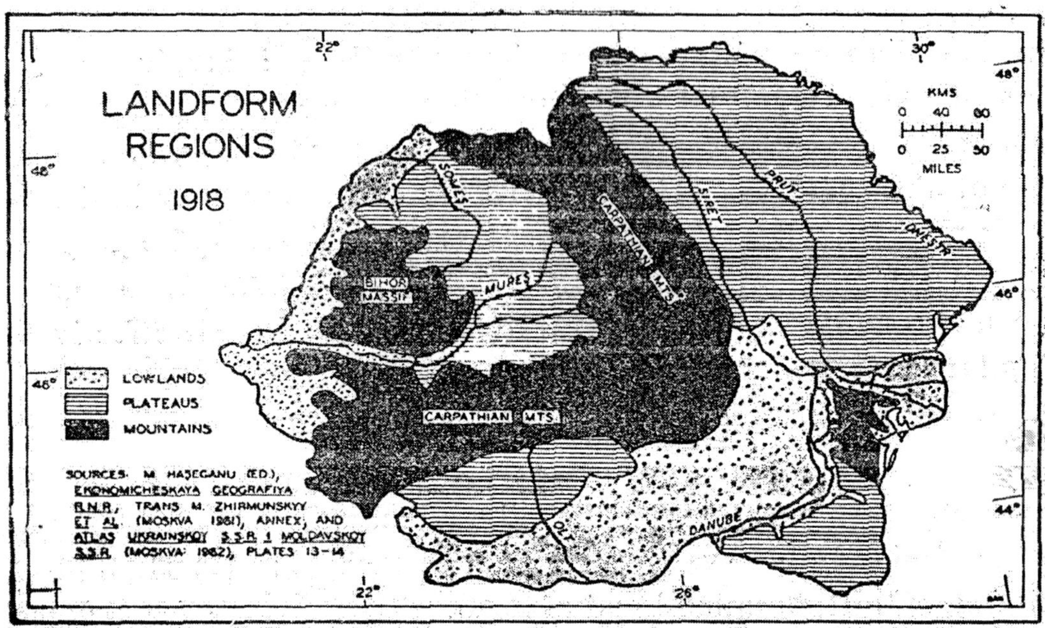

Fig. 17.16: Terrain Type Map of Romania

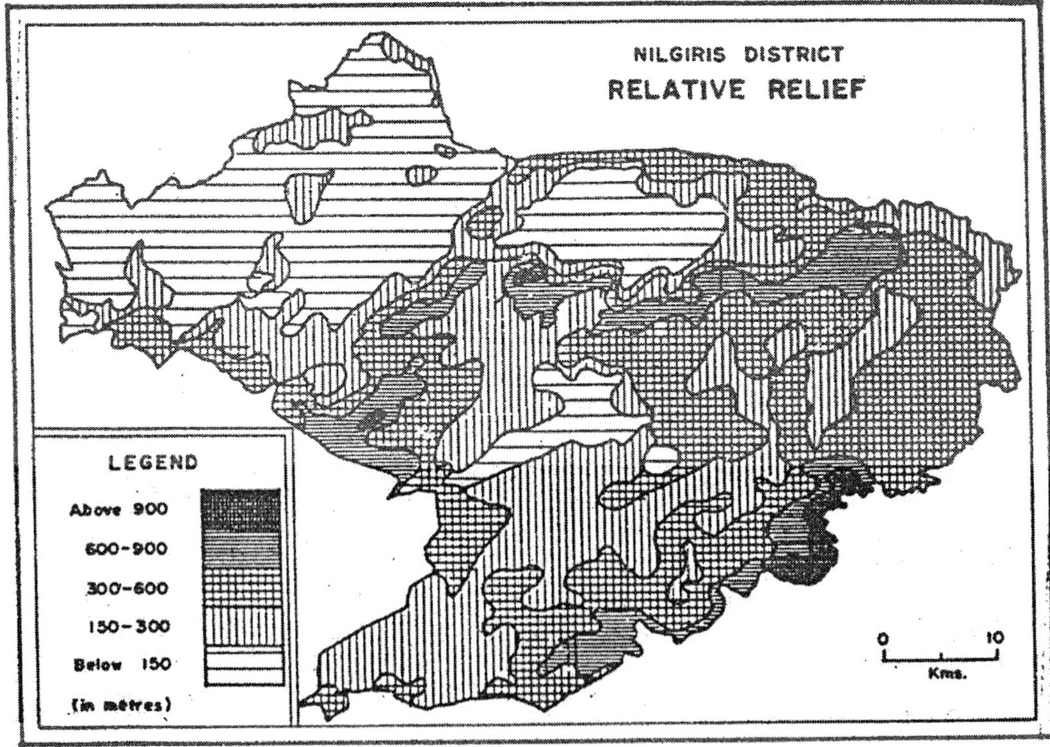

Fig. 17.17: Relative Relief Map of Nilgiris

SLOPE VALUE MAPS

Hachuring and shading provide graphic representation of slope on medium and large-scale maps. The question of representing actual slope values on small-scale maps is not easily resolved. The technique suggested by Raisz and Henry solves this problem at least partly. It outlines and depicts slope by area symbols in a dasymetric way. The relationships between slope and important minor topographic features are well presented; the representation of major topographic features is, however, poor. In this respect it is somewhat opposite to the relative relief map.

Operational procedure for this technique is as follows. A topographical map is divided into small areas, within each of which the contours have the same standard spacing i.e., the same number of contour-lines per mile of horizontal equivalent. A horizontal scale showing standard contour spacing is drawn from which the slope categories are ascertained by careful inspection. The chosen slope categories may be, under 50 ft., 50 to 100 ft. per mile, and so on. These categories are demarcated on large-scale maps. Later the boundaries are transferred on a small-scale map (Fig. 17.18).

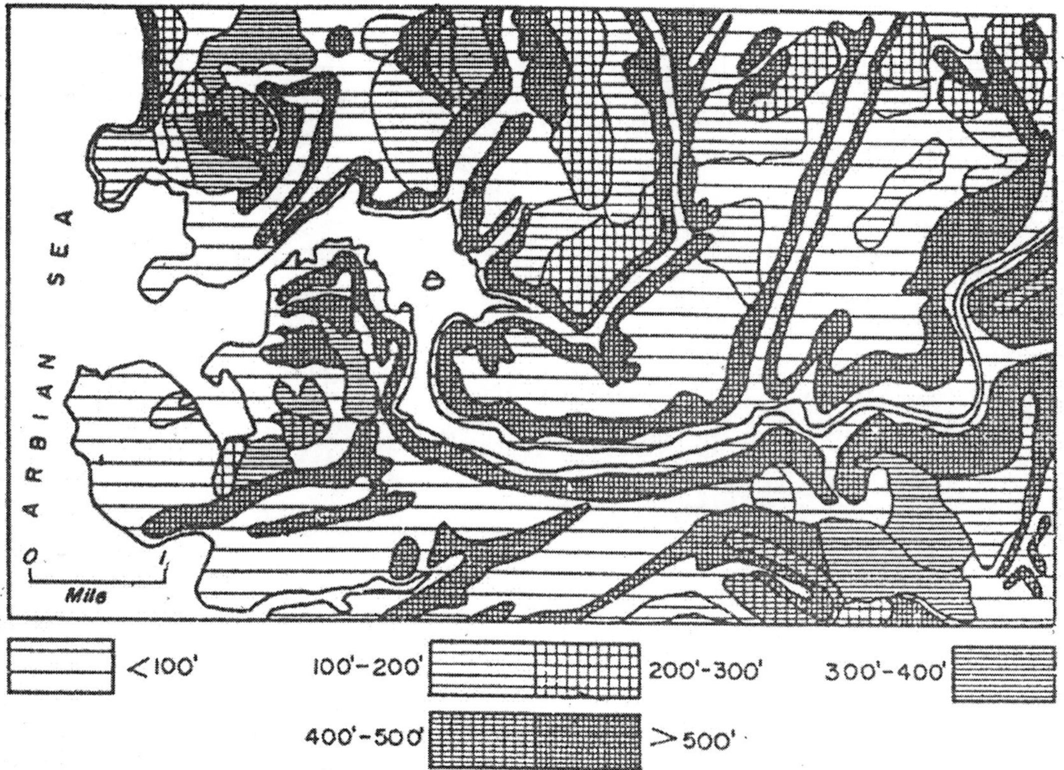

Fig. 17.18: Slope Value Map of a Part of West Coast of India

Dot Method

Attempts have been made to represent slope by dots. The approach is: steeper the slope, denser the dots. For the construction of such maps, a contour map is divided into squares. Average slope is determined for each of the squares. Dots are then put in each square in proportion to the degree of slope.

PROFILES

Cross-profiles drawn with the help of a contour map often prove to be of great help in visualizing the relief. Drawing of profiles is relatively easy. The contour maps offer the base material for cross-profile drawing.

The first step in the drawing a cross-profile is to lay a strip of paper on the line along which the profile is to be drawn. All the contour inter-sections, spot-heights, rivers, summits and other defined points are marked on the paper. A base line equal to the line along which profile is to be made is drawn separately. Two vertical lines are drawn at both ends of the base-line. Taking into account the relative relief of the area, a vertical scale is marked off and lines parallel to the base line are drawn at selected intervals. The vertical scale is usually larger than the horizontal scale to make undulations along the profile stand out. The relation between horizontal and vertical scales is known as vertical exaggeration. The ratio of vertical exaggeration should invariably be noted below the profile. Starting from the base line, transfer the contour intersections marked on the edge of the paper on respective lines. The lowest contour should be marked on the base line. When all the points are plotted, join them by a smooth line [Fig. 17.19(a) and 17.19(b)].

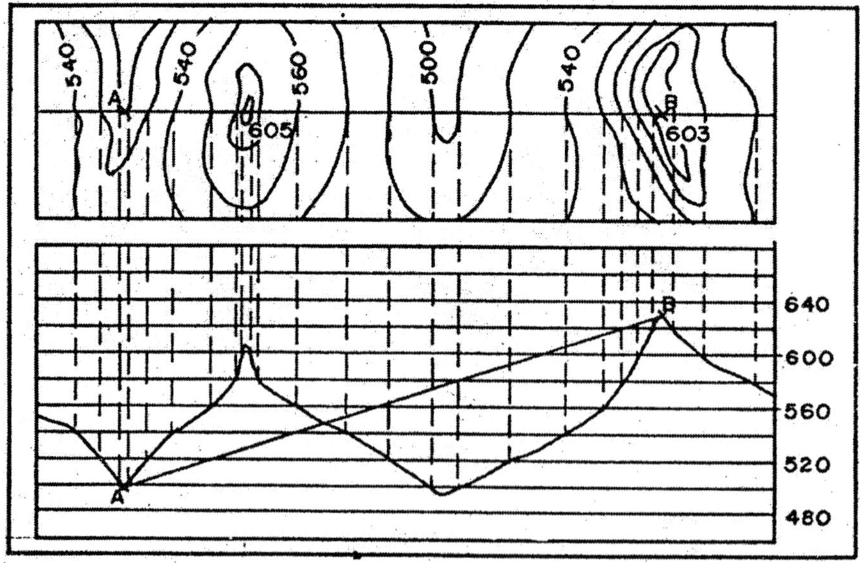

Fig. 17.19(a): Cross Section Profile

When a profile is drawn along the stretch of a road, or a river, it is called longitudinal Profile [Fig. 16.19(b)].

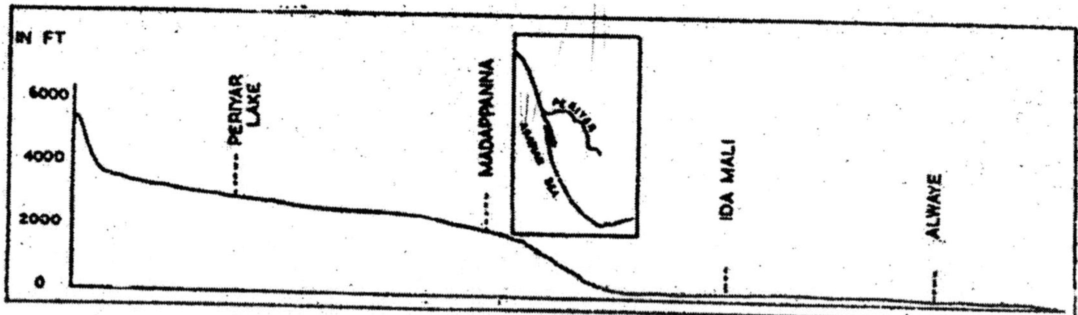

Fig. 17.19 (b): Longitudinal Profile

When cross profiles are first drawn along a series of lines and then are put together, we call them serial profiles [Fig. 17.19(c)].

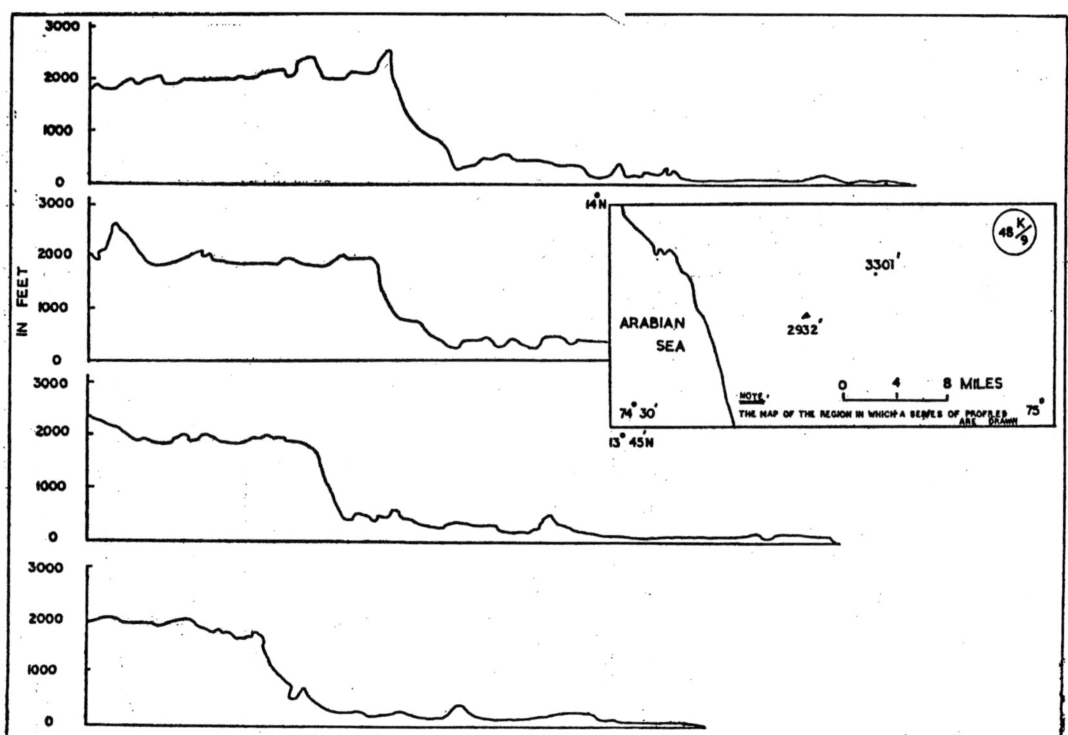

Fig. 17.19(c): Serial Profiles

When several profiles drawn along parallel lines are put within one frame, we get a superimposed profile. When the highest parts of a series of profiles are shown to represent the relief as viewed in, horizontal plane of the summit-levels from an infinite distance, we call it a composite profile. If we remove those parts of the profiles from a superimposed profile which show the relief hidden behind higher elevations, we get a, projected profile. A projected profile gives a panoramic view with a distant sky-line, a middle-ground and a fore-ground [Fig. 17.19(d)].

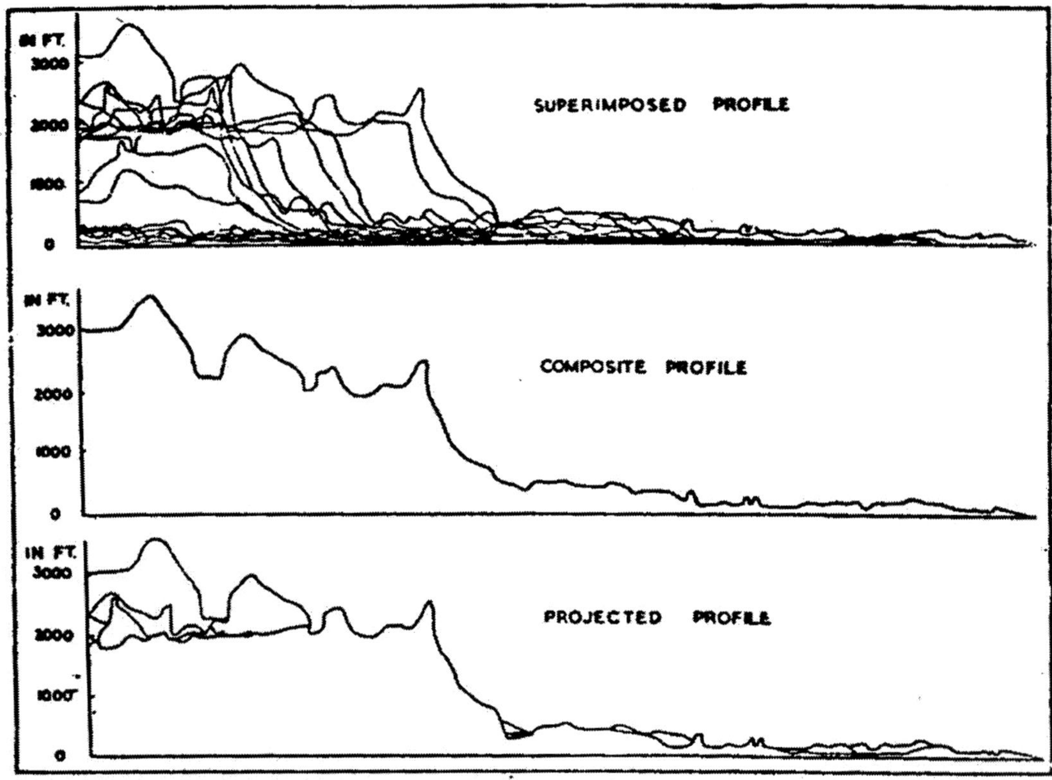

Fig. 17.19(d): Superimposed Profiles; Composite Profile and Projected Profile

REPRESENTATION OF RELIEF BY CURVES

Hypsometric Curve

Hypsometric curve indicates the proportion of surface area at various elevations. The procedure for constructing a hypsometric curve is as follows. Select a convenient horizontal scale to represent the area in square miles. Putting zero on the left, construct a vertical scale on the left side of the base line to represent elevation above sea level at suitable intervals. Calculate the area enclosed within different contours (0′ and 1000′; 1000′ and 2000′, etc.). Plot each area against its corresponding contour interval with origin indicating the total area. Join the points so marked with a smooth curve [Fig. 17.20(a)]. Hypsometric curve is useful for indicating relief of an island. Percentage, in stead of absolute, area can also be used in making this curve.

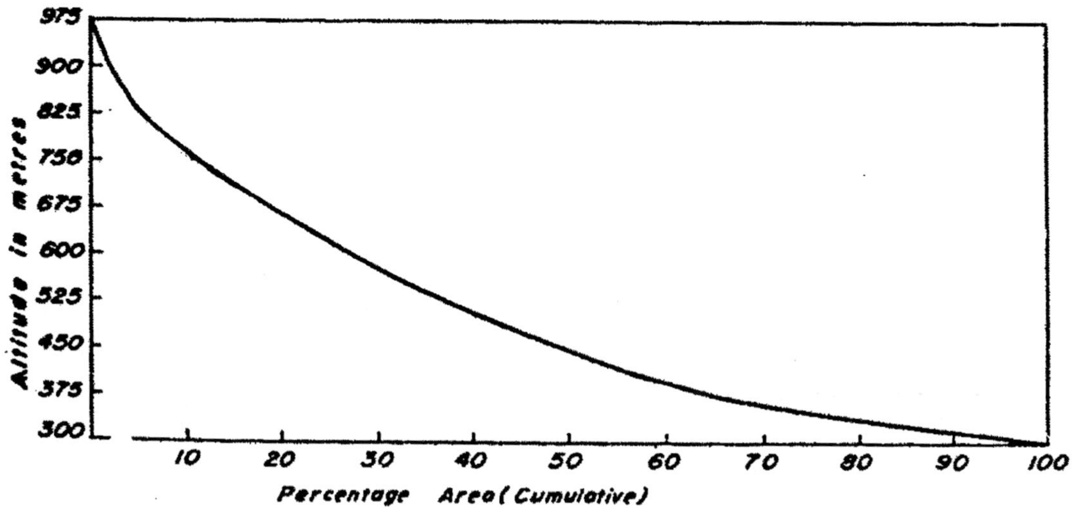

Fig. 17.20(a): Hypsometric Curve

Clinographic Curve

It illustrates the average gradient between successive contours and shows up prominently any abrupt change in the general relief depicted on a map. To construct this curve, first of all the area of land enclosed by sea level contour should be calculated. Then the area between sea level and the next higher (say 100′) contour should be calculated. This process is repeated until areas between all successive pairs of contours are found out. Presuming that the areas so calculated are enclosed within circles, we determine the radii of these circles by using the formula 2n r^2 (where r = radius). The radii can be converted by scale factor into inches. If a triangle ABC is drawn with < B = 90°, CB = radius of the circle and AB= height (contour interval), the angle ACB which is the angle of slope can be found out using the tangent formula i.e., tan O = AB/BC = h/r. We can thus calculate the angles of the slopes between each succeeding pairs of contours. These data can then be used to prepare a clinographic curve [Fig. 17.20 (b)].

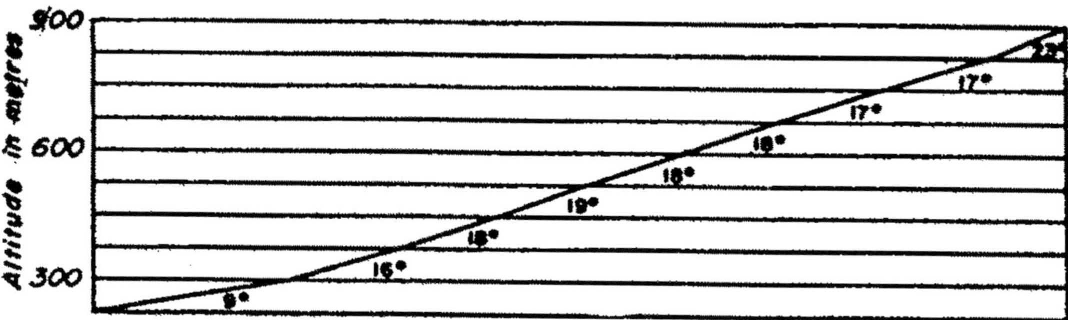

Fig. 17.20(b): Clinographic Curve

Another device which is useful in geomorphological analysis is *altimetric frequency graph*. It shows the frequency of occurrence of heights above sea level (Fig. 17.21).

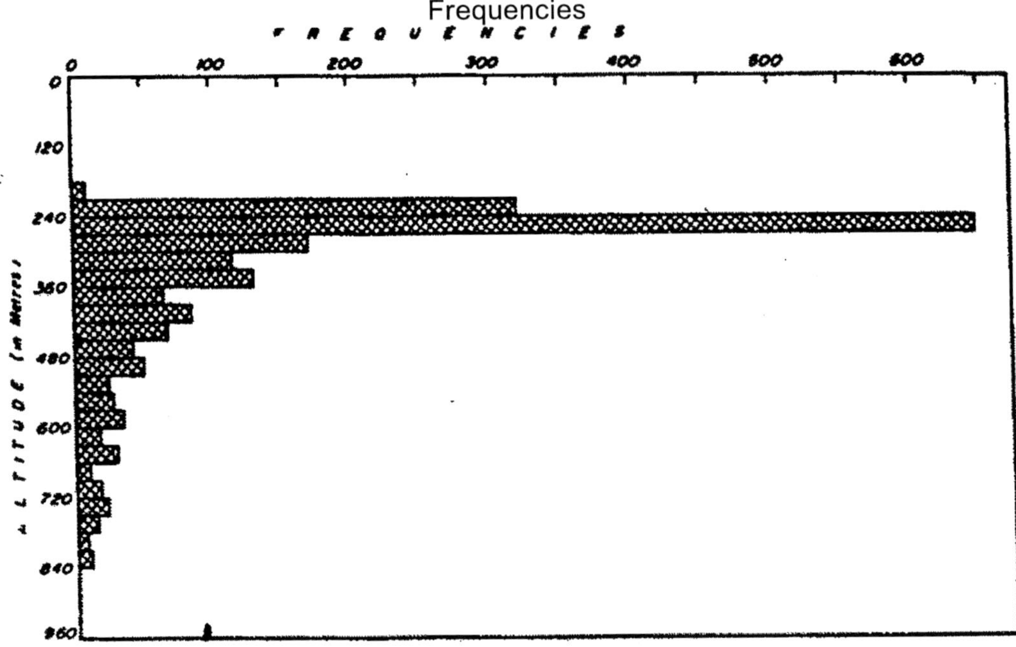

Fig. 17.21: Altimetric Frequency Graph

FIELD SKETCHING

Field sketching is a technique of making perspective sketches of prominent features of landscape as observed from a specific point. A pencil, an eraser, a scale, a compass and a drawing sheet or sketch book are the only necessary tools and materials for field sketching. It requires relatively little artistic ability, for all it needs is strict adherence to a few 'rules of thumb' to do field sketching.

The first step in field sketching is to choose the landscape to be included in it. Once the area to be sketched is selected, a central point is selected and a vertical line passing through this point is drawn. Usually, part of the horizon is also included in the area. Once this is done, a line formed by the horizon or level of a distant lake or a communication line is drawn. The third step consists of selection of some prominent features and the measurement of their distances from the centre with an inch/cm scale held at an arms length. These points are located at their respective places on the paper. For greater exactness, cut a hole in a piece of cardboard of the size of the desired sketch. Cover this hole with a piece of celluloid with grid rules. Hold it at a fixed distance from the eye and transfer the details seen, square by square on the paper. To see that the landscape does not give an impression of being flat, the vertical scale should be exaggerated by 1.5 to 2.5 times depending on the relative relief of the area sketched.

After the framework of the sketch is ready, the minor details can be filled by detailed observation. Some of these details can be shown in their true shape but many others may have to be symbolized. Names of the prominent features like hills, peaks, villages, etc. can also be added later (Fig. 17.22).

Fig. 17.22: A Field Sketch

BLOCK DIAGRAMS

Block diagrams are only too often used in geological and geographical literature to need introduction. They are sketches of relief models with the added advantage of having geological sections appended to their sides. This enables the map user to correlate the landscape with the underlying geological structure. The concept behind a block diagram is that a block of the earth crust is cut out and viewed obliquely from above.

Block diagrams can be of three types:

1. Isometric block diagrams,
2. Perspective block diagrams, and
3. Panoramic sections.

Isometric Block Diagrams

Isometric block diagrams are of two types:

1. Those drawn without the help of contours, and
2. Those drawn with the help of contours.

Isometric block diagrams are a sort of pseudo-perspective block diagrams. Although they appear to be perspective, they are only isometric based on simple geometrical figures. Isometric graph paper provides ready guide lines for drawing geometrical figures upon which isometric block diagrams can be built. When drawn with the help of geometrical figures like cones, pyramids, spheroids, cylinders etc., we get the isometric block-diagrams of the first type (Fig. 17.23).

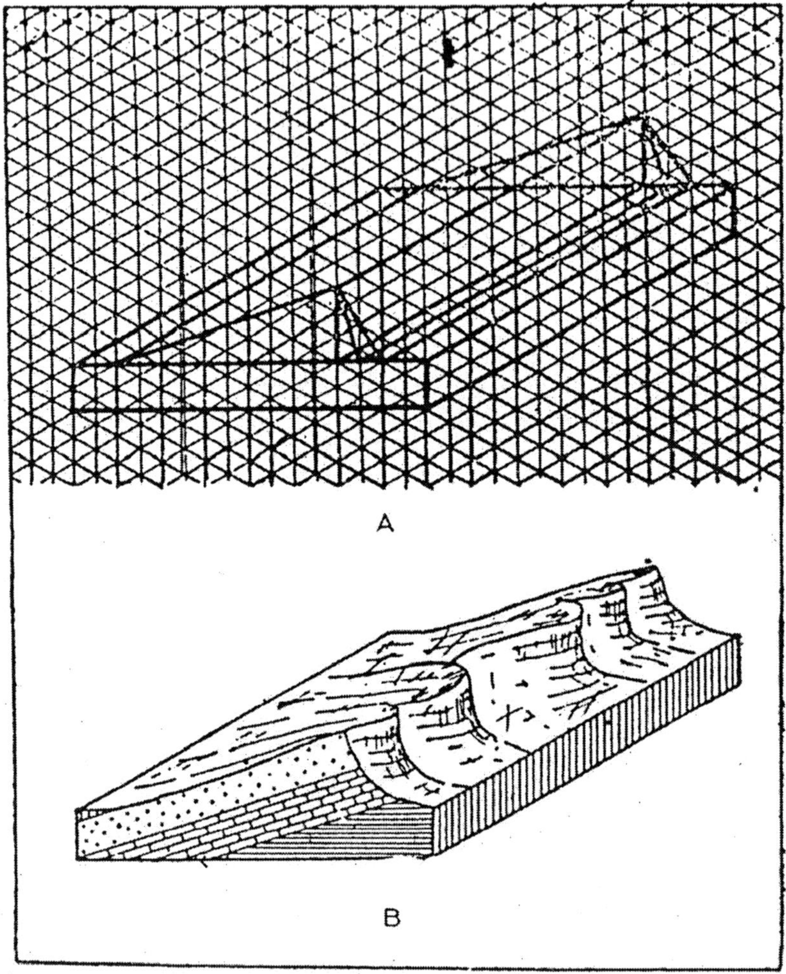

Fig. 17.23: Isometric Block Diagram Drawn with the Help of Geometrical Figures (Robinson)

The second type of isometric block-diagram is the one which is drawn with the help of contours. To draw such a diagram, trace a plan view square from a contour map. Draw horizontal and vertical lines to divide the plan into several squares. Determine the position from which the observer is looking at the plane. After deciding the orientation, project the plan with its squares into a rhombus as shown in Fig. 17.24. The obliquity is determined by the angle between the base and the visible sides. This angle should be between thirty and forty-five degrees. Transfer the relief-details from the squares to the rhombus using one of the techniques explained below.

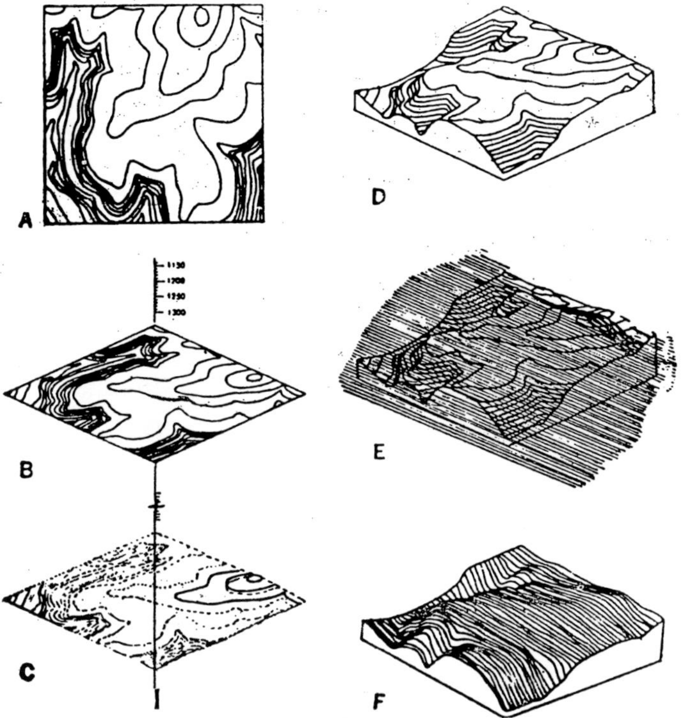

Fig. 17.24: Method of Representing Relief on Isometric Block Diagrams with the Help of Contour Maps

There are two methods of transferring relief from a square to a rhombus.

1. The multiple section method, and
2. The layer method.

The first method involves the drawing of section profiles along all the horizontal lines of the grid and then drawing of relief by form lines. The method as shown in Fig. 17.25 is self explanatory.

In the second method the highest contour is placed on an overlay. To trace the next lower contour, move the overlay to the next contour interval. Repeat this process with respect to each contour. Other details are filled subsequently (Fig. 17.26).

Perspective Block Diagrams

In the isometric block diagrams, the vertical as well as the horizontal scales are kept uniform as the sides of the square and the rhombus are kept identical. But such an arrangement does not give a natural appearance to the landscape. The natural appearance is the perspective appearance in which features at a distance appear shorter than their real size. The perspective effect can be given to a block diagram by making the rear-side of the block narrower than the front side and by allowing the side lines

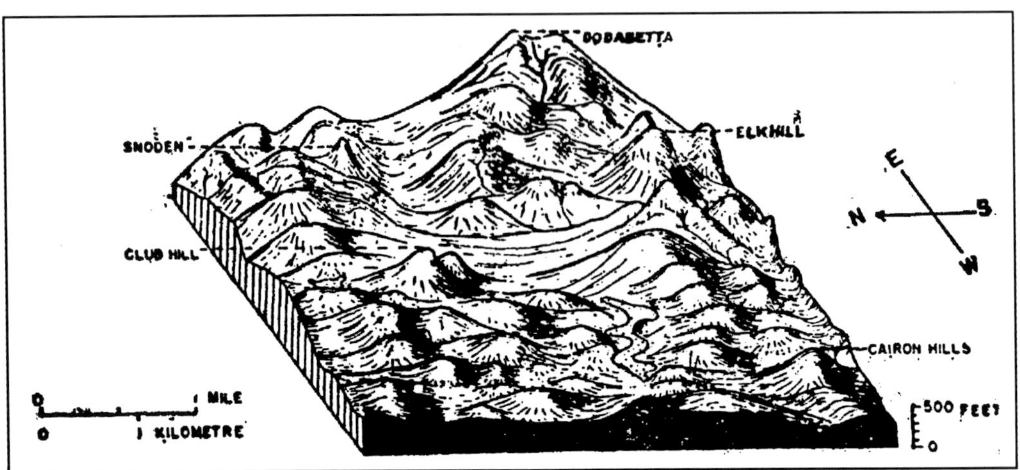

Fig. 17.25: Methods of Representing Relief on Isometric Block Diagrams: Multiple Section Profile Method

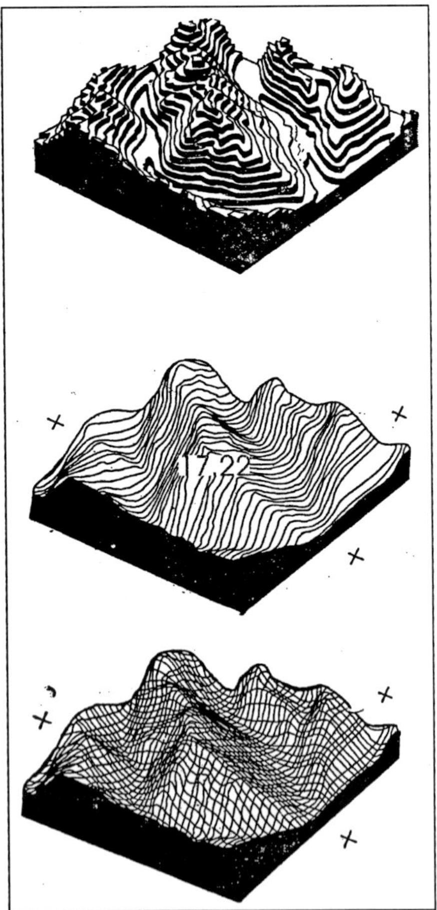

Fig. 17.26: Method of Representing Relief on Isometric Block Diagram: Layer Method

to converge somewhere on the observers horizon. The point of convergence is called the vanishing point. If only two sides of the block are made to converge at one point on the horizon, the diagram is called one-point perspective diagram. If the remaining two sides also converge at a second point on the horizon, it becomes a two-point perspective block-diagram. The principles underlying the two types of perspective block diagrams are illustrated in Figs. 17.27, and 17.28.

In one point perspective block diagram, the front of the diagram is a horizontal line drawn parallel to the back edge. The sides converge towards a common point on the observer's horizon. Thus both the horizontal and vertical scales change when a base plan is given a perspective appearance. The technique of drawing section profile along the sides is given in Fig. 17.29. If structural details are not desired, the sides can be made solid black.

The two-point perspective block-diagram is considered to have two vanishing points as shown in Fig. 18.28. By changing the location of the vanishing points or that of the block with respect to the vanishing points, one can develop as many perspective blocks as one desires. The method of transferring the topographical details from the map to the diagram is the same as in isometric block diagrams.

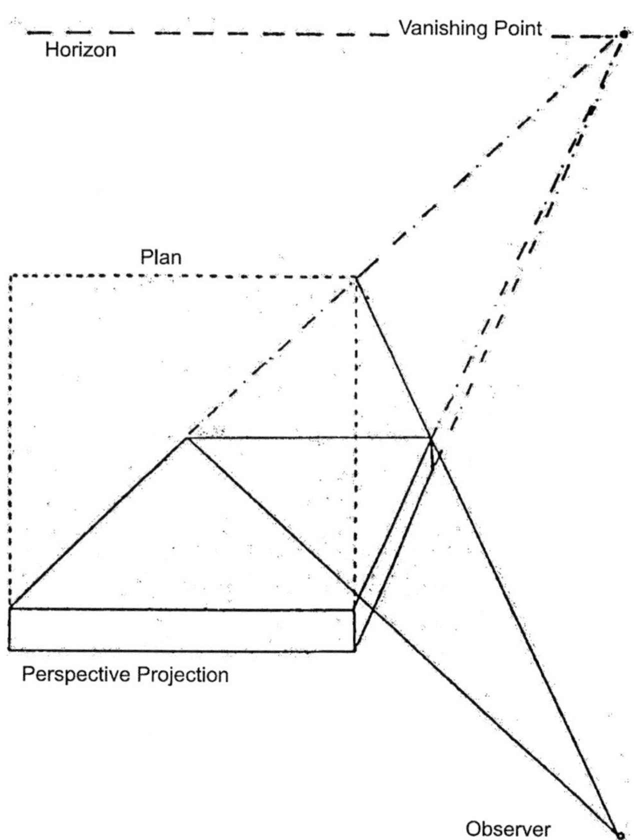

Fig. 17.27: Method of Drawing One-Point Perspective Block Diagram

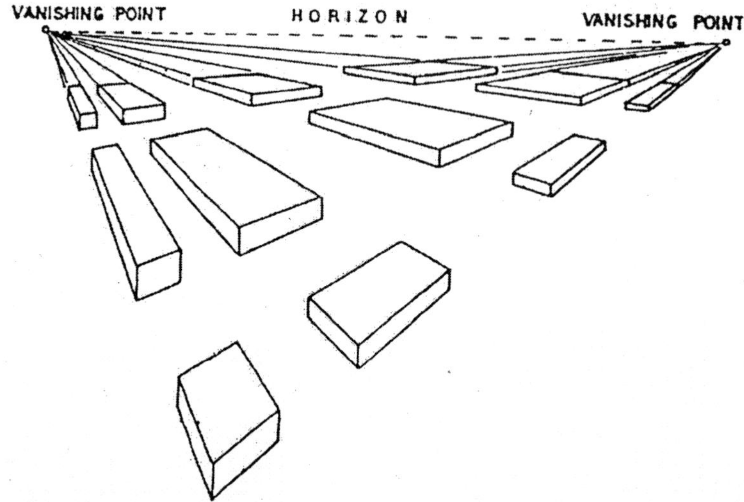

Fig. 17.28: Method of Drawing Two-Point Perspective Block Diagram

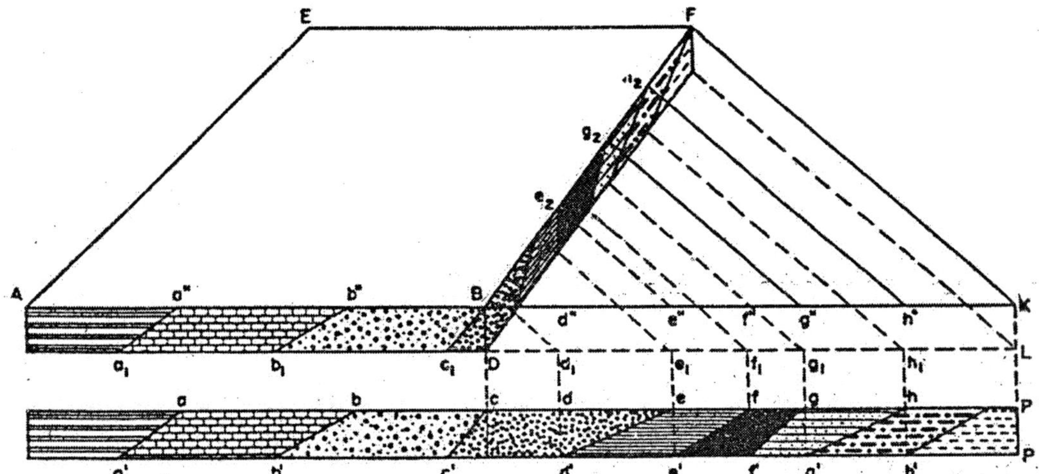

Fig. 17.29: Method of Drawing Section Profiles along the Sides of a Block Diagram

Panoramic Section

These are elongated block diagrams showing geology in the frontal section prominently and giving the surface details in a very narrow strip at the top. These are very effective tools for explaining relationship between geology and surface topography (Fig. 17.30).

MECHANICAL METHODS OF MAKING BLOCK DIAGRAMS

There are two simple mechanical devices which can be used in transforming a contour map into a block-diagram. These are:

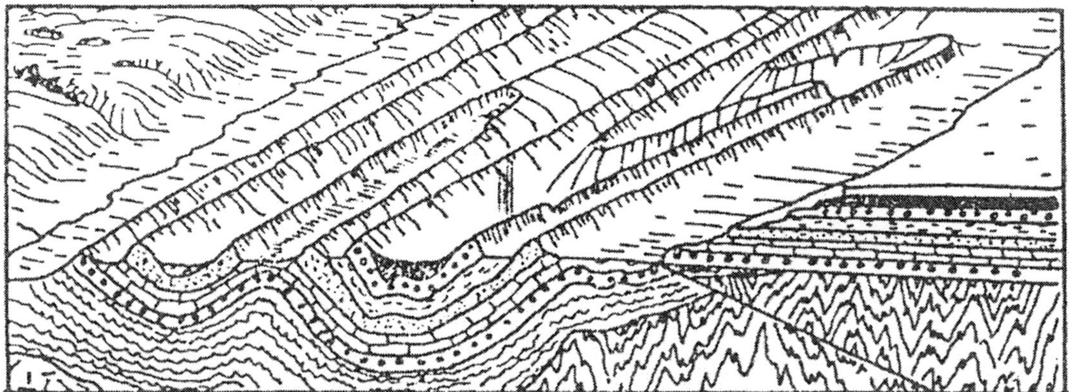

Fig. 17.30: A Panoramic Section

1. Dufour, and
2. Anamorphoser.

A dufour consists of a long rod which slides along a groove. One end of the rod has a snail which fits in the groove. At the other end of the rod is fixed a nail (pointer) or a pencil. Similarly, in the middle of the rod there is an arrangement for fixing a nail or pencil (Fig. 17.32). If the diagram is to be made on a scale larger than that of the contour map, the pointer should be fixed in the middle but if it is to be made on a reduced scale, the pointer should be fixed on the end of the rod. The contour map is placed edgewise under the pointer and a sheet of paper is placed under the pencil. As the pointer is moved along the borders of the map, the pencil draws a rhomboid. To draw the contours, move the pointer along the lowest contour. Move the paper down one contour interval and draw the next higher contour. Each time, the pointer is placed on a higher contour, the paper is moved downwards. After the contours are ready, other details are filled up manually.

The other mechanism used in making block diagram is what Prof. Jenks calls anamorphoser. It is a simple optical device which transforms images passing through two slits located on parallel planes. The two slits are oriented at right angles. The rays passing through a negative enters the first slit and converge along its linear axis. Then they passed through the rear slit and tend to converge along the opposite linear axis. The amount of transformation and change in scale in controlled by three critical dimensions:

1. Distance separating the slits,
2. Distance between the original copy and the front slit, and
3. Distance between the rear slit and the photographic negative.

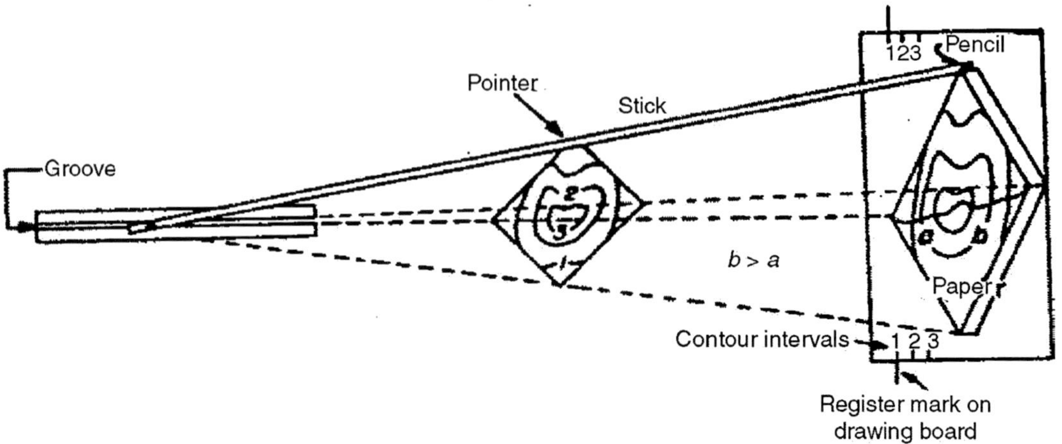

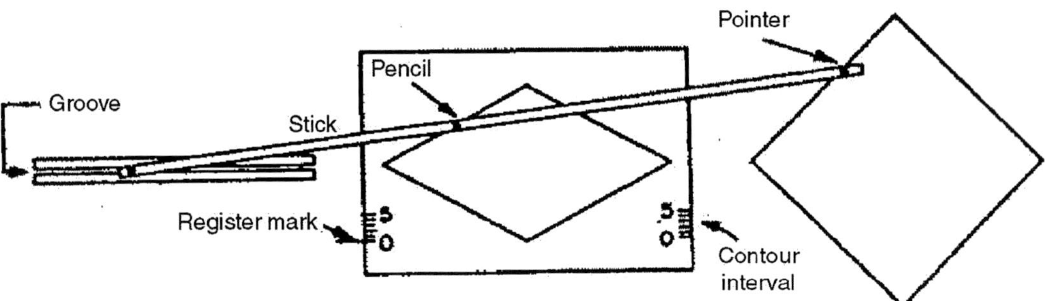

Fig. 17.31: Dufour (after Raisz)

Transformation Formulae

$$F = \frac{(a+c)}{b-c} \qquad R = \frac{c\,(F+A)}{a}$$

$$b = \frac{(a+c)\,A + cF}{F} \qquad c = \frac{a\,R}{A+F}$$

Note: The geometry of anamorphoser transformation is controlled by the relative of A, R, and F. If one of these dimensions is altered, the scale or the shape of the transformation changes. For example, 1:1 ratio of a to b could be altered to 1:2 by changing the dimension R. Similarly, the ration of b to c changes when either A or F is altered.

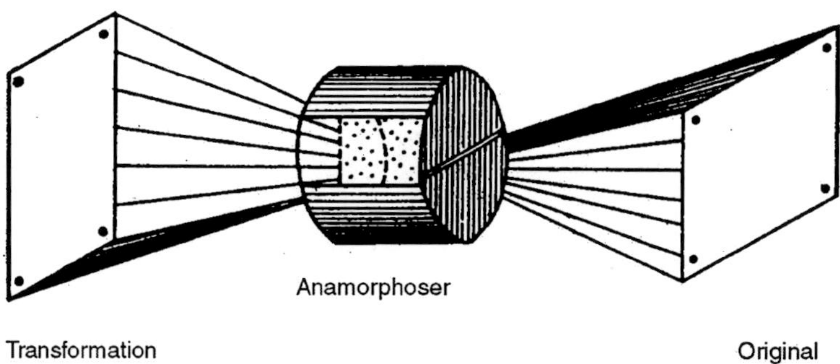

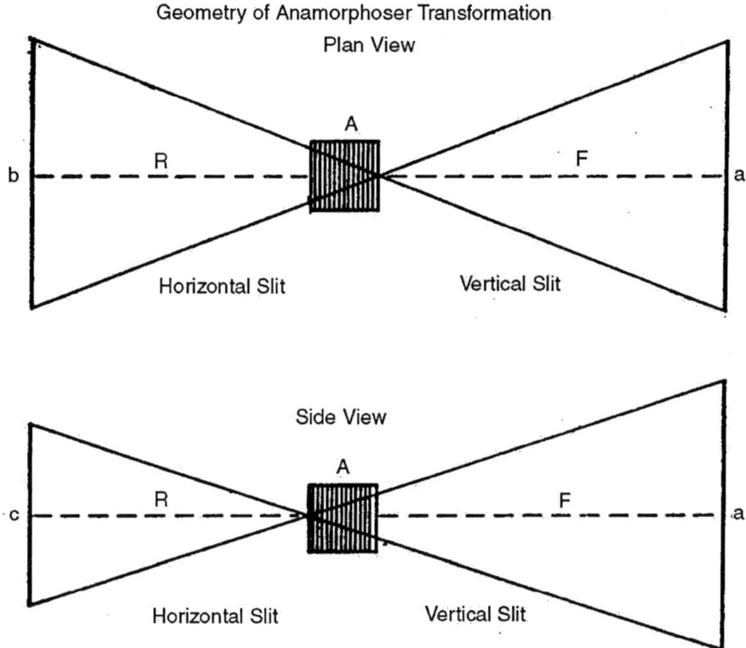

Fig. 17.32: An Anamorphoser and Geometry of Transformation (after Jenk)

a	=	A known dimension on the original map.
b	=	The long dimension of the transformed map; dimension "a" multiplied by the reduction or enlargement ratio.
c	=	Short dimension of the transformation; dimension "b" multiplied by the foreshortening ratio which equals the sine of the angle of view.
A	=	Distance between the slits of the anamorphoser.
F	=	Distance from the front slit of the anamorphoser to the original map.
R	=	Distance from the rear slit of the anamorphoser to the transformed copy.

TERRAIN REPRESENTATION AND MAP FUNCTION

The techniques that we have discussed in this chapter can be used effectively only if they are selected with reference to the function of the map. Every map is a communication device. In certain cases it contains discretely selected limited information. In certain others, it contains a variety of information designed to serve the widest range of potential users. Whatever, the nature and amount of data represented, the selection of symbols should be so judicious that it effectively meets the desired communication needs.

Selection of a particular terrain representation technique should be guided by following considerations:

1. Type of terrain information that the map is designed to give
2. Aspect of the terrain to be represented
3. Relation between terrain and other symbols
4. Perceptual problems of the map users
5. Limitations of cost and reproduction processes

If all these factors are considered carefully, the portrayal of relief will be effective and in consonance with the purpose and objective of the map.

18
Mapping the Weather and Climate Data

Atmospheric pressure, temperature, wind direction and velocity, humidity, evaporation, cloudiness, precipitation, snow cover and visibility are the important elements of weather and climate. One can sub-divide each element further to get 'sub-elements'. For instance, precipitation can be classified as hail, snow and rain. In the process of observation, the above said elements can also be combined to form 'compound elements'. Equivalent temperature (temperature and water-vapour pressure), cooling power (temperature, wind velocity and humidity) and continentality (solar radiation and temperature) are some of the examples. In contrast to this, we have 'derived elements' which include ranges, frequencies, variabilities, etc.

Since a geographer is more interested in Climate, the mean values for the above said elements are of great significance to him. Mean values give the average weather i.e., climate. It does not, however, mean that we can dispense with actual measurements of daily weather conditions. The methods of representing weather and climatic data given in this chapter, therefore, deal with both the daily and mean values.

MAPPING THE WEATHER DATA

The day-to-day weather observations are used to prepare synoptic charts and a variety of other diagrams. Synoptic charts give a composite picture of the weather conditions as observed at a particular time. Synoptic means 'seen together'. Various diagrams used to explain the dynamics of weather conditions from important elements of a book on climatology.

Synoptic Charts

There are hundreds of meteorological stations spread all over the world which observe and record weather data. Such stations are located both on land and water. In the high seas, ships function as meteorological stations. With the use of airplanes and satellites for recording and photographing weather conditions, we can now say that there are meteorological stations not only on the land and the high seas but also in the air.

These weather stations prepare special reports for farmers, masters of ships and pilots of aircrafts. They also function as laboratories where the state of the atmosphere over an entire region, country, continent or ocean is put together from reports sent by hundreds of weather stations.

The data received from various weather stations are so much and detailed that they all cannot be incorporated in a single chart unless they resort to coding designed specially to give economy of expression. These codes are called meteorological symbols. When these symbols are put at their appropriate position on a base map, we call such maps, synoptic charts.

Synoptic charts provide the primary tools for weather forecasting. They permit one to locate and identify different air masses, pressure systems, fronts and areas of precipitation. When a consecutive series of charts showing weather conditions at every three hours or so are examined, it becomes possible to predict the future weather patterns. A synoptic station model is given in Fig. 18.1.

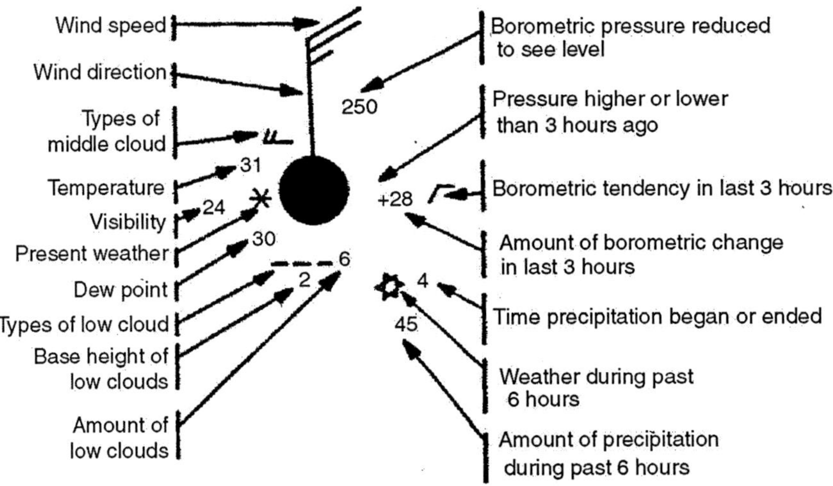

Fig. 18.1: Synoptic Station Model

A weather map or chart is essentially a representation of the frontal systems and the pressure systems that overlie the areas outlined in the map. Before attempting to interpret a weather map it is necessary to familiarize oneself with the meteorological shorthand or symbols. All the elements of a weather map are represented by symbols, majority of which are now used all over the world. These symbols are given in Fig. 18.2.

INDIAN WEATHER MAP

The meteorological observation service was first started in India in 1864 with its headquarters at Shimla. After the First World War (1914-18) the department was

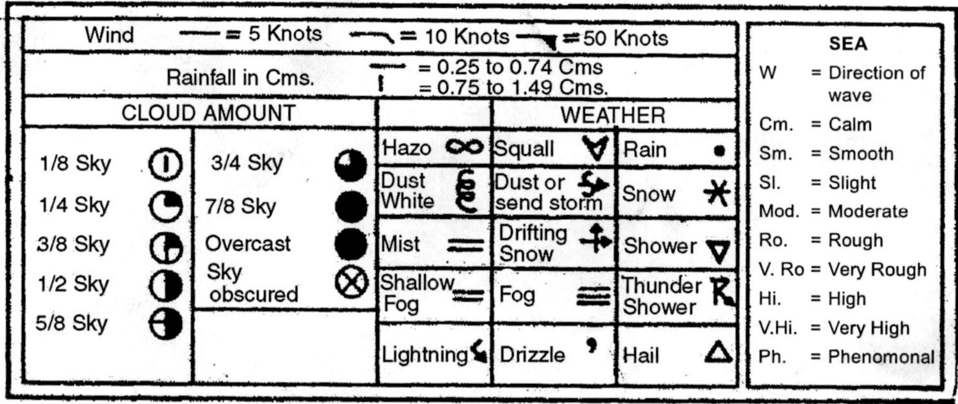

Fig. 18.2: Weather Symbols

expanded and the office was shifted to Poona. At present there are some 350 observing stations in India. These are grouped into five classes. The first class observatories

have, along with eye-reading instrument, self-recording instruments also. They transmit observations twice daily to the forecasting stations. The second class stations have generally only eye-reading instruments and they also transmit telegraphic weather messages twice daily; the third class observatories differ from the second class in that they transmit observations to the forecasting centres only once a day. The fourth class stations record only temperature and rainfall conditions and they do not telegraph messages daily. The fifth class stations only record rainfall and they telegraph at 8 hours local time the amount of rainfall during the past 24 hours.

At fixed hours in the first, second and third class observatories, the trained observers read the barometer or barograph, note the temperature, the direction and speed of winds, the amount and type of cloud, the amount of precipitation, if any, and visibility. All these results are transmitted by code to the forecasting stations. These are supplemented by observations from the fourth and fifth class stations and also by the weather messages, received from ships approaching the coast. Since, the messages are sent by telegraph or wireless they are received in a short time at the central office where the observations are immediately entered on a weather map. Upper air observations which help in forecasting are procured from hill stations, aeroplanes, pilots, balloons, etc., and are plotted separately. (Examples of Indian and foreign weather maps are given in Figs. 18.3, and 18.4.

SCHEMATIC REPRESENTATION

Schematic representation of weather is as common as synoptic mapping. Such diagrams are used to explain the dynamic nature of various weather elements. For example see Fig. 18.5.

MAPPING THE CLIMATIC DATA

Iso-lines

Much of the climatic data are represented by line symbols, the most common among them being isometric lines which can be used for any climatic data. When put on

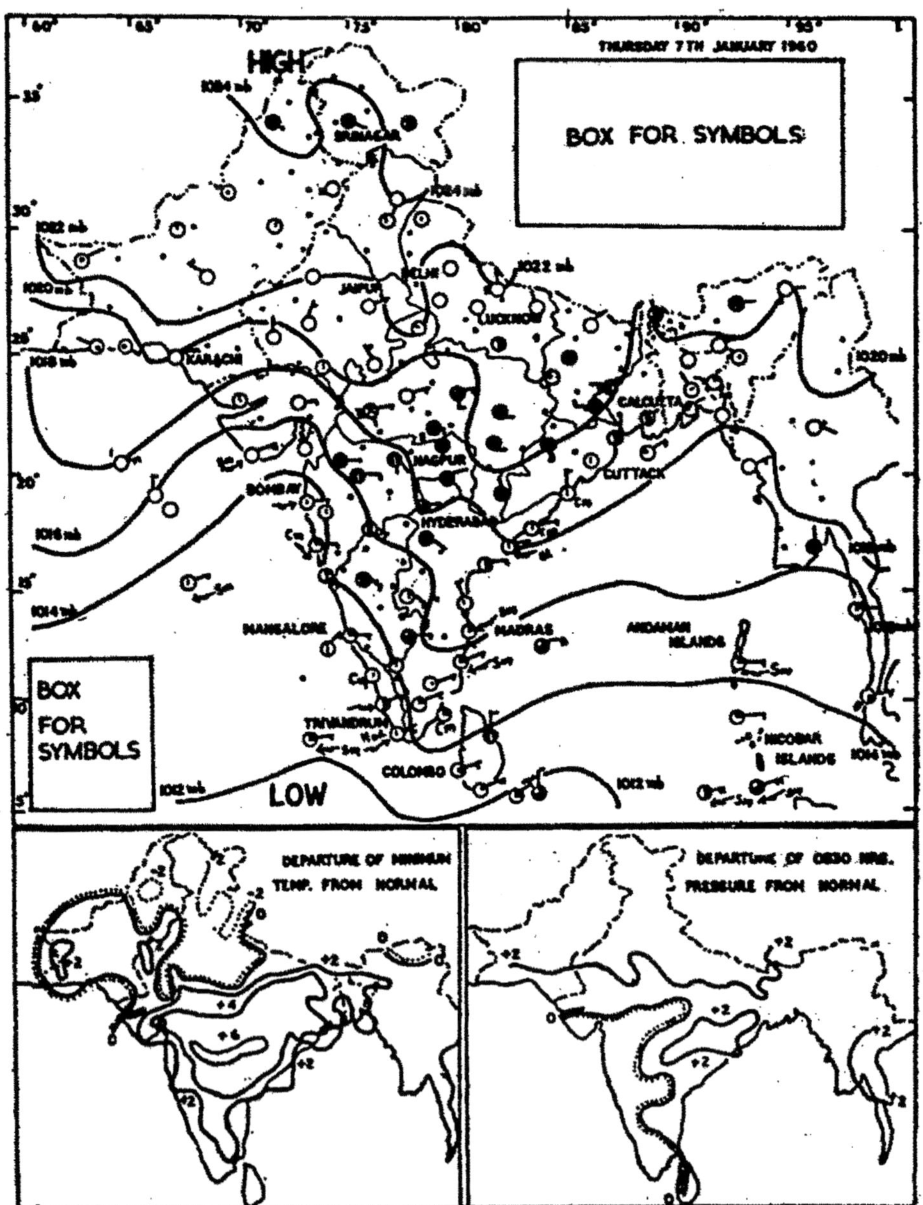

Fig. 18.3: An Indian Weather Map

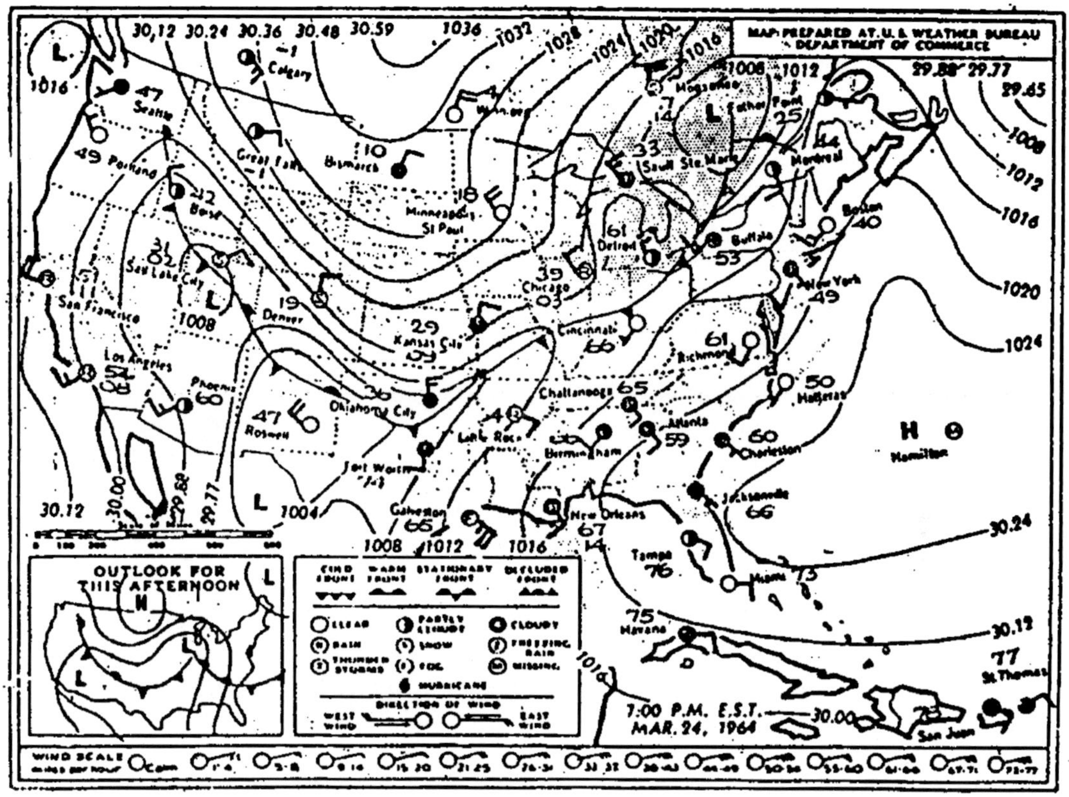

Fig. 18.4: An American Weather Map

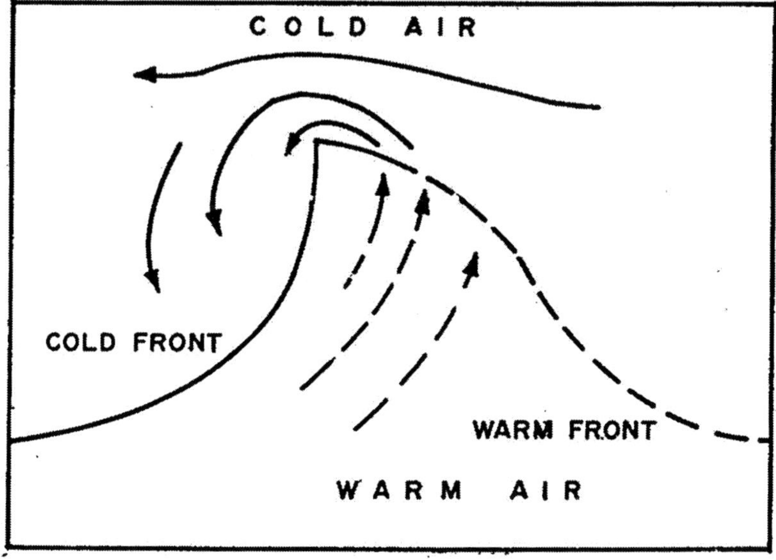

Fig. 18.5: Example of Schematic Representation of Weather Data

maps, they produce the isometric maps. Isometric lines link places having same mean values of temperature, rainfall, pressure, sunshine, frost, clouds, relative humidity, etc. Some of these lines are given below.

Isobars : Lines connecting places of same pressure (Fig. 18.6).

Isotherms : Lines connecting places having same temperature.

Isohyets : Lines connecting places having same rainfall.

Isomers : Lines connecting places having same monthly rainfall average shown as a percentage of annual total.

Isohels : Lines connecting places having same mean daily duration of sunshine.

Isonomals : Lines connecting points having same temperature deviation from the standard temperature of the points. The standard temperature for a latitude can be calculated by the formula:

$$t = -17.8 + 44.9 \cos^2 (\theta - 6° - 30') \text{ °C}$$

where θ = latitude.

Iso-frequencies: Connect places having the same mean frequency of a climatic phenomenon, expressed as number of days or number of occasions per year in which they occur.

Isorymes : Lines connecting places having same mean value of frost.

Isonephs : Lines connecting places having same mean value of cloud cover.

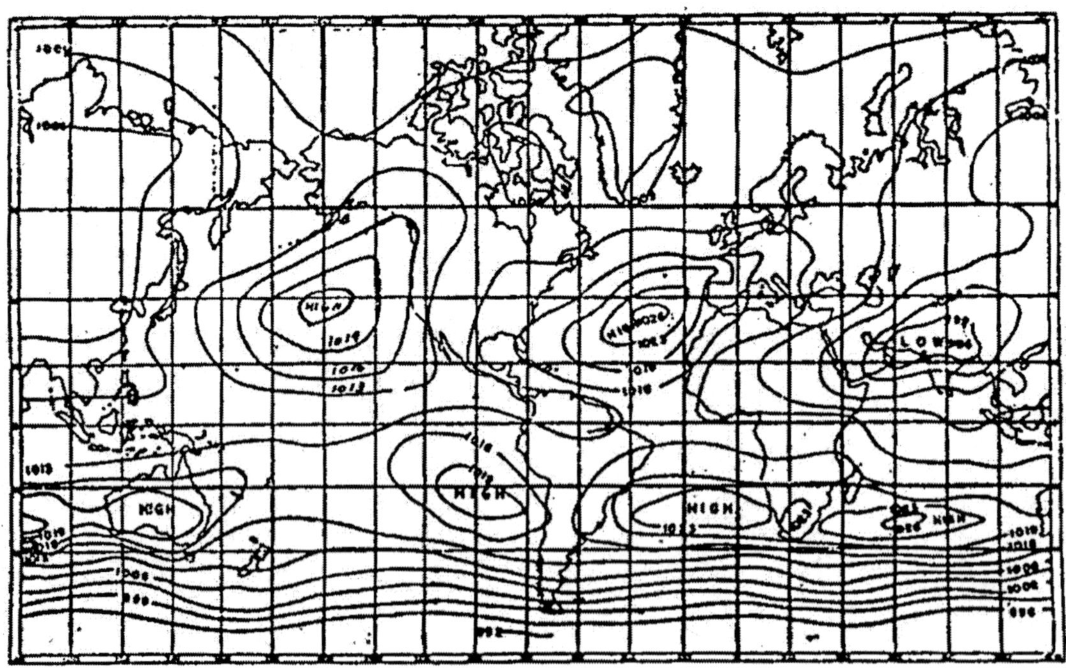

Fig. 18.6: World Map Showing Pressure Gradients of July

Equipluves

Lines joining places with same pluvimetric coefficients are called equipluves. The pluvimetric coefficient for any month can be derived with the help of the following formula:

$$\text{Pluvimetric Coefficient} = \frac{\dfrac{t}{Dm}}{\dfrac{T}{N}}$$

t = total rainfall of a particular month.
T = total rainfall of the station for the whole year.
Dm = number of days in that particular month.
N = number of days in the year (365).

Given the data in Table 18.1 for Jodhpur city, we have to calculate the pluvimetric coefficient for the month of August.

Table 18.1: Monthly rainfall in Jodhpur.

Months	J	F	M	A	M	J	J	A	S	O	N	D	Total
Rainfall in cms.	0.8	1.5	0.3	0.3	0.1	3.6	10.0	12.0	6.0	0.8	0.3	0.3	36.07

The pluvimetric coefficient for August

$$= \frac{12.0}{36.0\left(\dfrac{31}{365}\right)} = 3.9$$

Equivariables

The distribution of places with similar deviations from their average weather conditions can be shown with the aid of isopleths. This can be done by finding out the coefficient of variability for each station. The following formula can be used for calculation:

$$CV = \frac{\sigma\,(100)}{\overline{X}}$$

where
CV = co-efficient of variability

σ = standard deviation

$\overline{X}$ = Mean value

Equi-correlatives

We can calculate coefficients of correlation between the rainfall data pertaining to pairs of stations, and draw isometric lines showing same correlations. The formula for calculating the coefficient of correlation is–

$$r = \frac{\frac{1}{n}\Sigma (x - \overline{x})(y - \overline{y})}{\sigma x \ \sigma y}$$

r = co-efficient of correlation

n = number of observation

x = annual precipitation for the first station

y = annual precipitation for the second station

$\overline{x}$ = mean annual precipitation for the first station

$\overline{y}$ = mean annual precipitation for the second station

σx = standard deviation for x

σy = standard deviation for y

Note: Further details regarding the co-efficient of correlation are given in chapter 14.

ISOPLETHS FOR ARIDITY AND MOISTURE

Aridity is an important factor in the delimitation of climatic regions and in agricultural planning. It can be determined with the help of the following formula and used to prepare isopleth maps.

R. Lang's Formula

$$\text{Rain factor} = \frac{\text{Annual precipitation in millimeters}}{\text{mean annual temperature in degree centigrade}}$$

Miller's Formula

$$\text{Index of humidity} = \frac{T}{R}$$

T = mean annual temperature in degree fahrenheit

R = mean annual rainfall in inches

De Martonne's Formula

$$I = \frac{P}{(T+10)}$$

I = index of aridity
T = mean and annual temperature in degree centigrade
P = mean annual rainfall in millimetres

Thornthewaite's Formula

$$P/E = \text{Precipitation} - \text{evaporation ratio}$$

$$I = 11.5\left(\frac{P}{t-10}\right)\frac{10}{9}$$

The summation of the twelve monthly values, multiplied by 10 to avoid decimals, gives the values for the year. Index values range from 0 to 150, giving five categories.

Thermal efficiency factor:

$$I = \frac{t-32}{4}$$

t = mean monthly temperature

LINE GRAPHS

Line graphs to show a variety of data are and can be used advantageously. We can show the monthly variation in temperature by putting months on the X axis and temperature on the Y axis. Graphs to represent the temperatures of more than one town can also be drawn to give a comparative picture. (Fig. 18.7). Graphs can also be used to show trends in average precipitation or to show percentage departure from mean rainfall as shown in Figs. 18.8, and 18.9.

A graph showing the accumulated percentage deviations also show the cyclic fluctuations and trends in temperature and rainfall. For this type of graph, annual deviations from the mean and annual rainfall or from any other base are obtained by subtraction and the accumulated deviations are obtained by additions (Table 18.2 and Fig. 18.10).

FREQUENCY GRAPHS

It is possible to show the frequency of extreme temperature, heavy rainfall etc. by graphs. A histogram is typical example of frequency graphs. We can plot a histogram showing the frequency of a given amount of rain during a given period. The amount of rain is plotted on X axis and the percentage or frequency on the Y axis. Fig. 18.11 illustrates a histogram of this type.

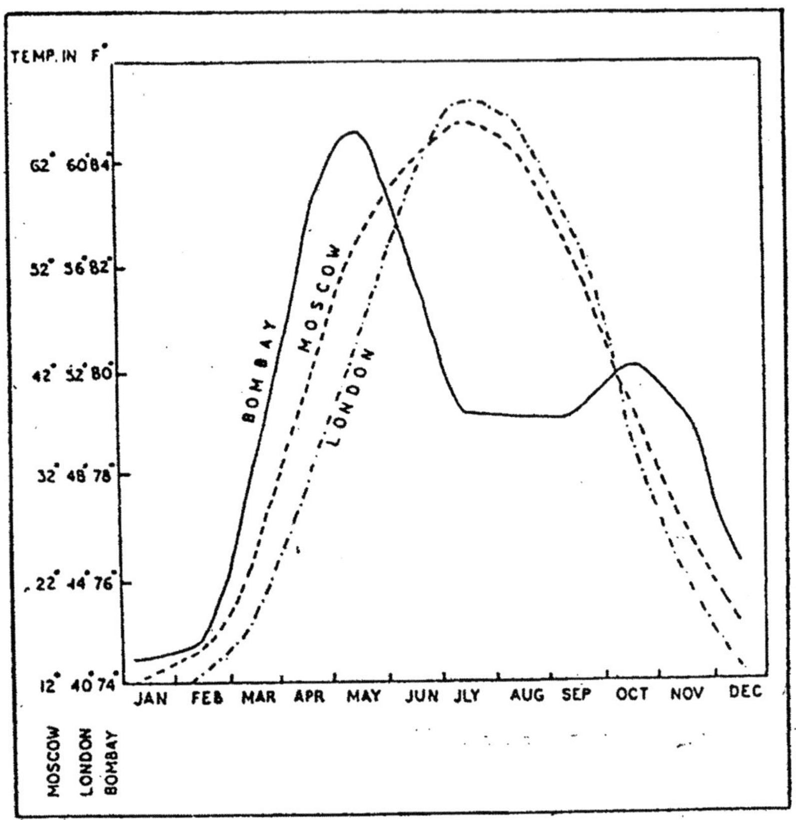

Fig. 18.7: Line Graph Showing Monthly Temperature at Moscow, London, and Mumbai

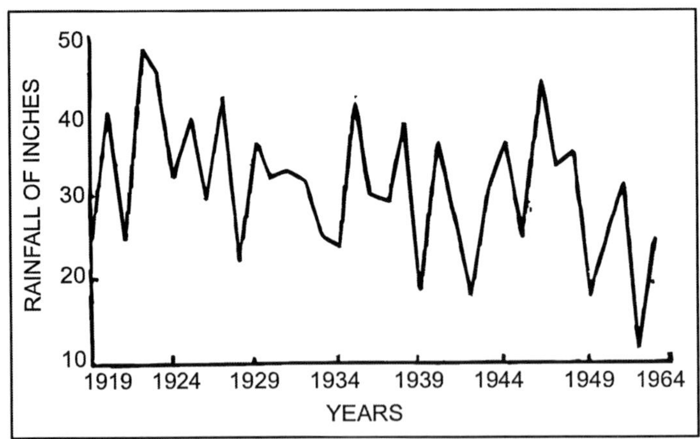

Fig. 18.8: Line Graph Showing Average Precipitation at Tirunelveli

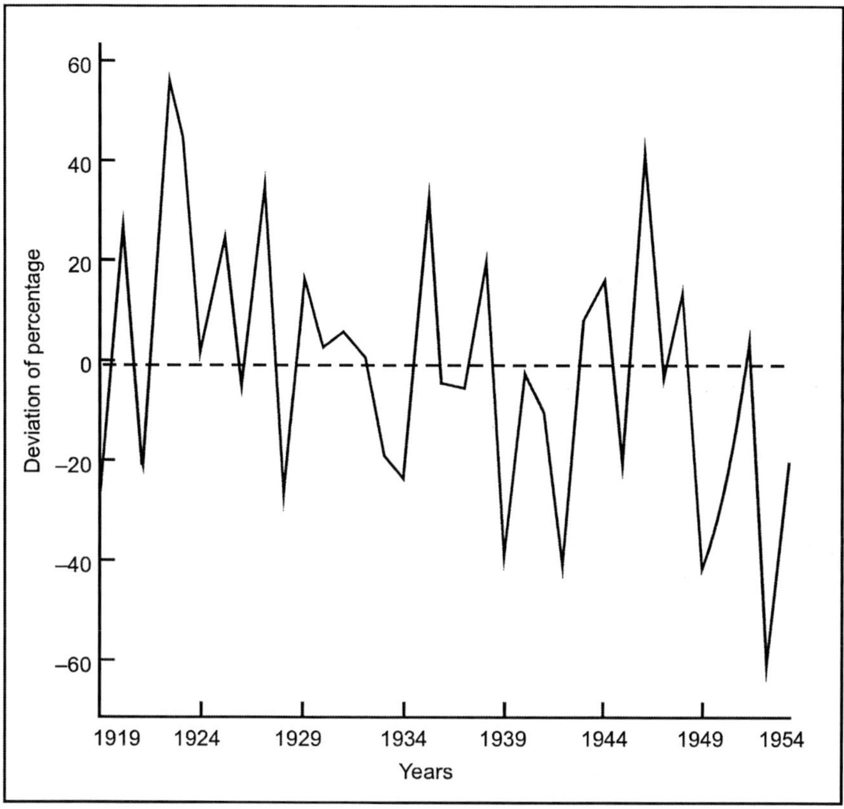

Fig. 18.9: Line Graph Showing Percentage Deviation of Rainfall at Tiruneveli

Table 18.2: Rainfall data for Tirunelveli

Year	*Annual rainfall*	*Running Mean*	*Deviation*	*Deviation %*	*Accumulated Deviation %*
1919	23.34	—	– 8.12	– 25.8	– 25.8
1920	40.03	29.41	+ 857.00	+ 27.37	+ 1.57
1921	34.87	37.64	– 659.00	– 21.05	– 19.48
1922	48.02	39.55	+ 16.56	+52.89	+ 33.41
1923	45.76	41.97	+ 14.30	+ 45.66	+ 79.07
1924	32.13	33.06	+ 0.67	+ 2.14	+ 81.21
1925	39.13	33.72	+ 7.84	+ 25.03	+ 106.24

WATER SURPLUS GRAPHS

These graphs can be grouped under line graphs. In this graph, the mean monthly potential evapo-transpiration and the mean monthly rainfall are superimposed upon each other in order to analyze the local climate in terms of seasonal moisture surplus and rainfall efficiency. Figure 18.12 shows the water surplus graph for Rome, Italy. A similar graph can be made for Mumbai using the data given in Table 18.3.

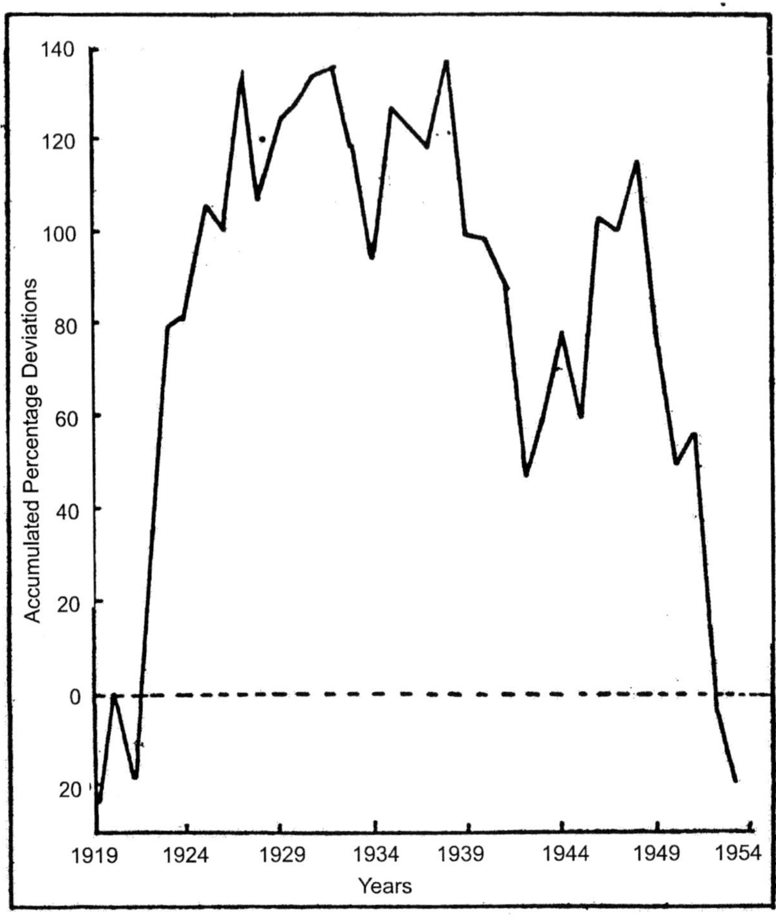

Fig. 18.10: Line Graph Showing Accumulated Percentage Deviation of Rainfall at Tirunelveli

CLIMATOGRAPHS

Several climatic phenomena can be depicted by climatographs (Fig. 18.12). It was devised by E.N. Munns and later elaborated by Hartshorne. Here, the plotting is done from the centre of the circle outward with the help of a graduated table which is based on the following formula.

$$r = \frac{X\,(\text{colog } t)}{100}$$

where

r = required distance from the centre for marking the temperature point.
x = distance from the centre of the circle to zero degree fahrenheit.
t = the particular monthly temperature.

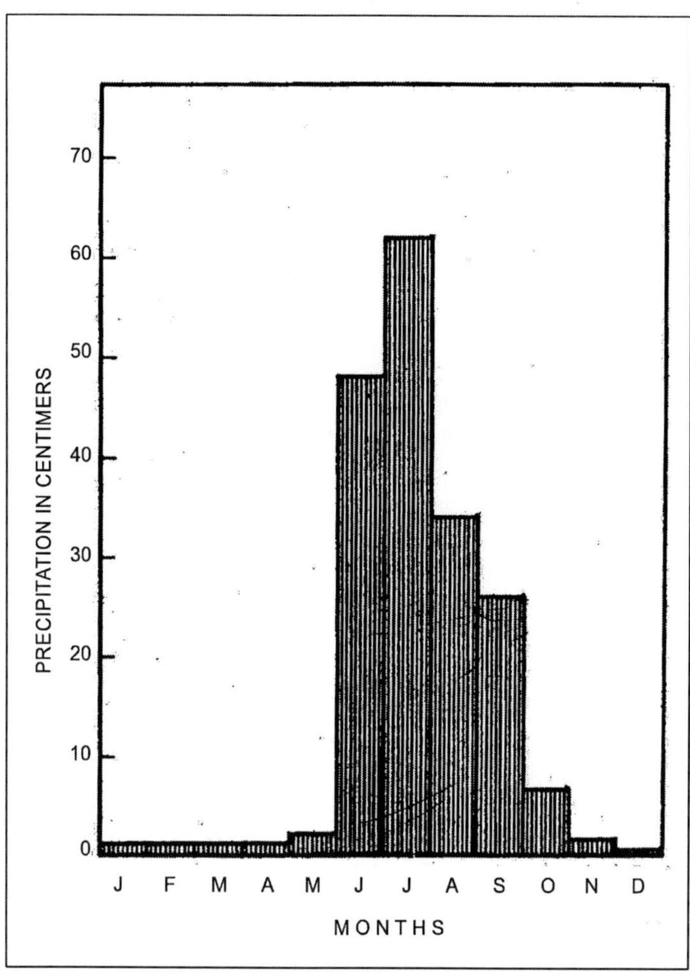

Fig. 18.11: Histogram Showing Monthly Rainfall in New Delhi

Table 18.3: Monthly Water (Rainfall) Surplus in Mumbai

Month	*Temperature °C*	*Rainfall (cms)*	*Potential Evapo-transpiration (in cms.)*
January	23.8	0.3	6.8
February	23.8	0.3	6.8
March	26.1	0.3	9.9
April	27.8	0.3	12.9
May	29.4	1.8	16.4
June	28.9	48.1	15.2
July	27.2	61.7	11.7
August	26.7	34.0	10.9
September	26.7	26.4	10.9
October	27.8	6.4	12.9
November	27.2	1.3	11.7
December	25.6	0.3	9.1

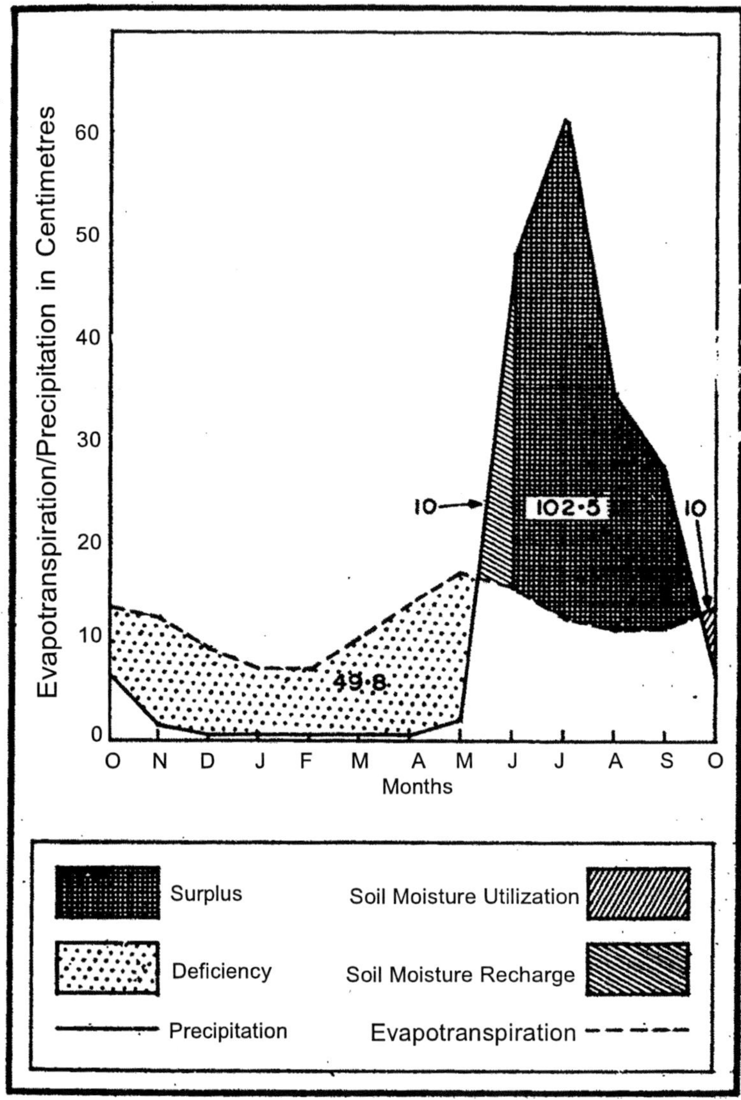

Fig. 18.12: Water Surplus Graph for Rome

Table 18.4: Monthly Temperatures, and Calculations of x to Make Climographs for Chennai, Tokyo, and Beijing

	J	*F*	*M*	*A*	*M*	*J*	*J*	*A*	*S*	*O*	*N*	*D*
Chennai	75.3	76.6	79.5	84.1	88.7	88.4	85.7	84.5	83.9	80.8	77.9	75.7
x	0.85	0.88	0.93	1.04	1.16	1.15	1.08	1.05	1.04	0.96	0.90	0.86
Tokyo	37.2	38.3	44.1	54.3	61.5	68.9	75.0	77.7	71.6	60.6	50.4	41.4
x	0.35	0.36	0.42	0.52	0.69	0.73	0.84	0.85	0.78	0.61	0.48	0.49
Beijing	23.5	29.3	41.0	56.7	67.8	76.1	78.8	76.5	67.6	54.5	38.5	27.3
x	0.26	0.29	0.39	0.55	0.71	0.87	0.92	0.87	0.71	0.52	0.36	0.28

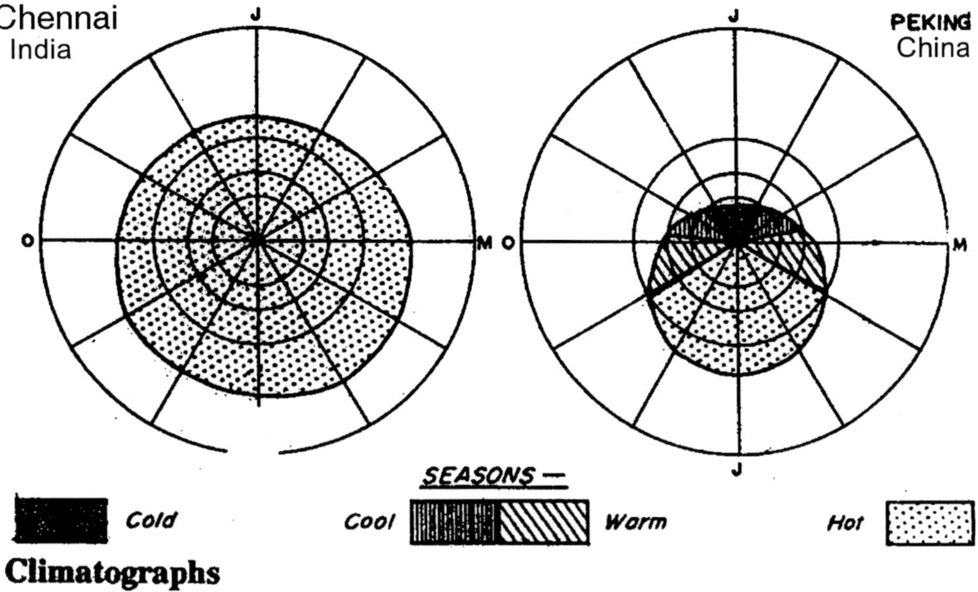

Fig. 18.13: Climatograph for Chennai (India) and Beijing (China)

WIND ROSES

The average frequency and direction of wind at a place can be shown by a star-diagram also known as wind rose. There are a variety of wind roses among which superimposed wind roses and octagonal wind roses are more important.

Superimposed Wind Roses

The Table 18.5 shows the percentage frequency of wind direction at various heights above the ground in Bangalore. With the aid of these data superimposed wind roses are drawn each representing readings at different ground levels and seasons. In this case, for the sake of clarity, frequencies are plotted from a central point and joined to form irregular octagons (Fig. 18.14).

Table 18.5: Data for Superimposed Wind Roses

Season	*Altitude*	*N*	*N.E.*	*E*	*S.E.*	*S*	*S.W.*	*W*	*N.W.*
	0'	6	7	15	7	4	14	39	8
Hot season	13,400'	4	16	26	14	11	11	10	8
	26,600'	8	13	15	13	5	23	13	10
	0'	1	1	3	1	6	19	56	13
Monsoon season	13,400'	10	14	19	16	9	10	19	3
	26,600'	0	0	75	10	15	0	0	0
	0'	7	17	46	6	3	4	14	3
Cold season	13,400'	12	28	31	10	6	3	6	4
	26,600'	9	4	20	16	11	23	7	12

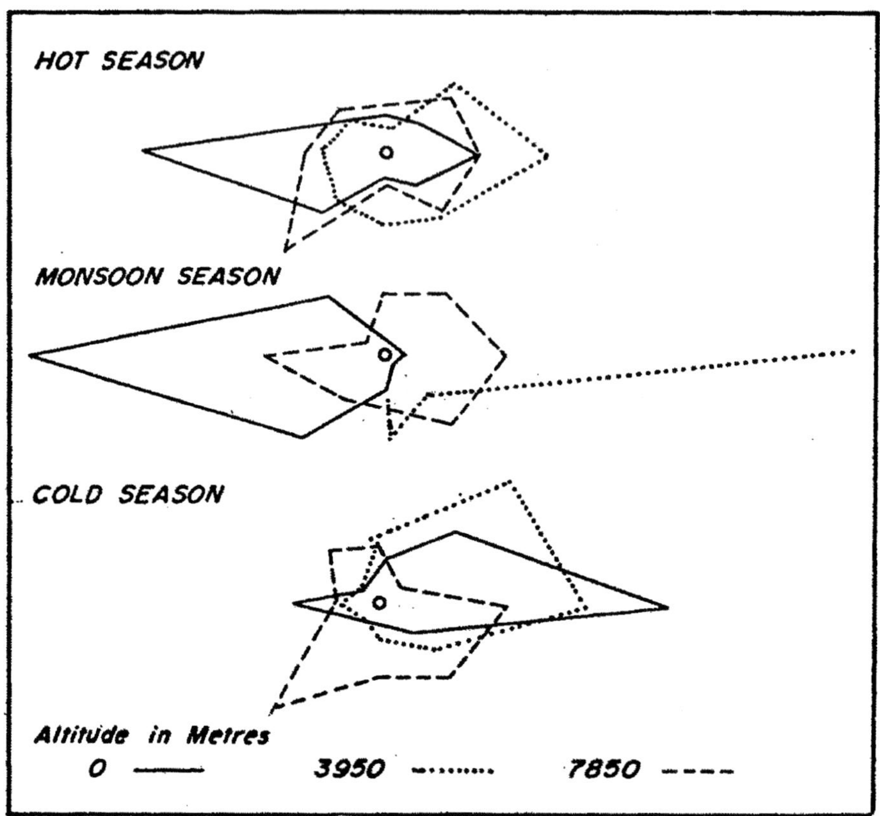

Fig. 18.14: Superimposed Wind Roses

Octagonal Wind Roses

Here, the distance between the corresponding sides of each octagon represent a frequency of 12-1/2 per cent. The eight cardinal wind directions are represented, one on each side of the octagon. The twelve mean monthly frequencies of wind from each of these directions are plotted as columns. If winds were equally frequent from each direction, eight sets of twelve equal columns would result (Fig. 18.15).

CLIMOGRAPH OR CLIMOGRAM

Here two climatic variables (monthly means) viz., wet bulb temperature and relative humidity are represented on vertical and horizontal axes respectively (Fig, 18.16). The four corners, N.W., N.E., S.W. and S.E. are marked as Scorching, Muggy, Keen and Raw respectively.

Another method is to use wet and dry bulb temperatures on the two axes. The wet bulb temperature is used on the y axis and the dry bulb on the x axis. Further details are given in Fig. 18.17.

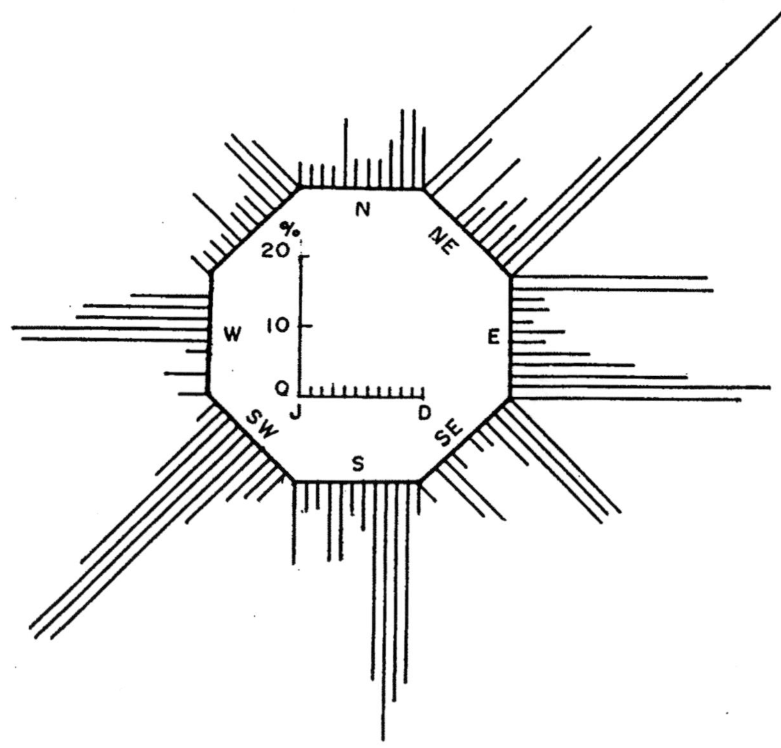

Fig. 18.15: Octagonal Wind Rose for Chennai

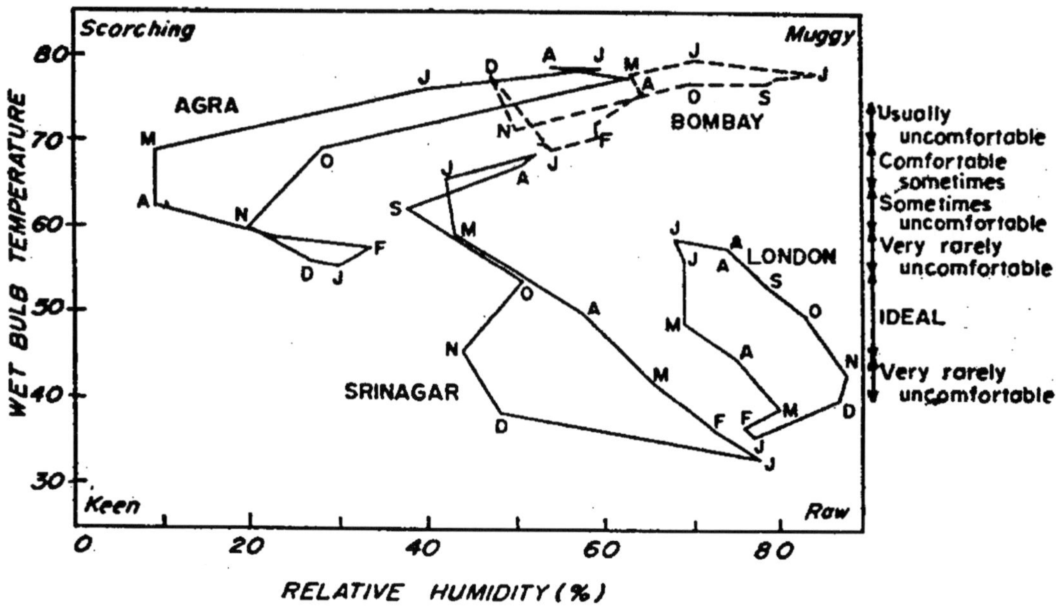

Fig. 18.16: Simple Climographs for Agra, Mumbai, Srinagar and Landon

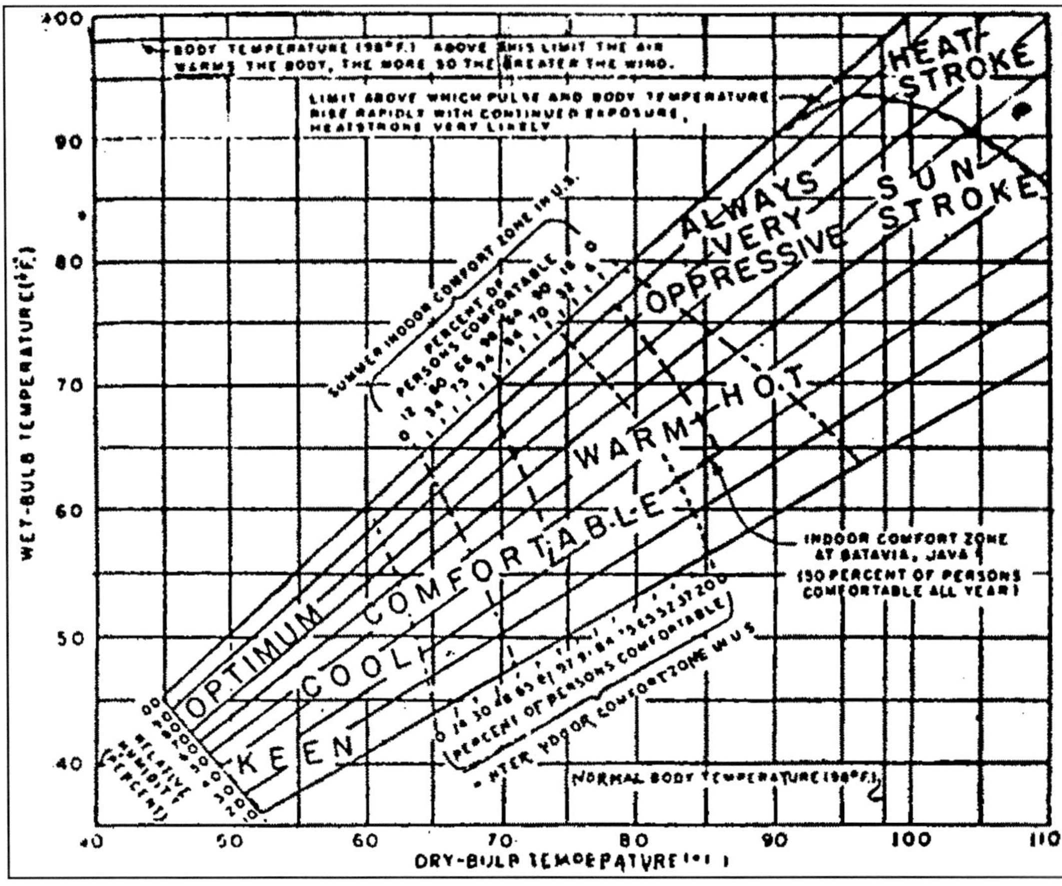

Fig. 18.17: A Complex Climograph

HYTHERGRAPH

This is a special type of climograph. Here, the mean monthly temperature is plotted on the ordinate and the mean monthly rainfall on the abscissa. (Fig. 18.18). With the aid of this graph one can analyze the climatic differences in relation to human activity.

RAINFALL DISPERSION DIAGRAM

It is an important tool in the analysis of rainfall distribution. The rainfall is represented here on the vertical axis and the months on the horizontal axis. The amount of rainfall can be plotted by means of dots of suitable sizes against each month for each year.

It may be pointed out in this connection that at least a thirty-five year record of rainfall should, be available for the above analysis. A rainfall dispersion diagram for Tirunelveli Dist. is shown in Fig. 18.19.

The median or middle value of rainfall can be found out by selecting the value which lies between the two extremes of rainfall at either end of the series. For instance,

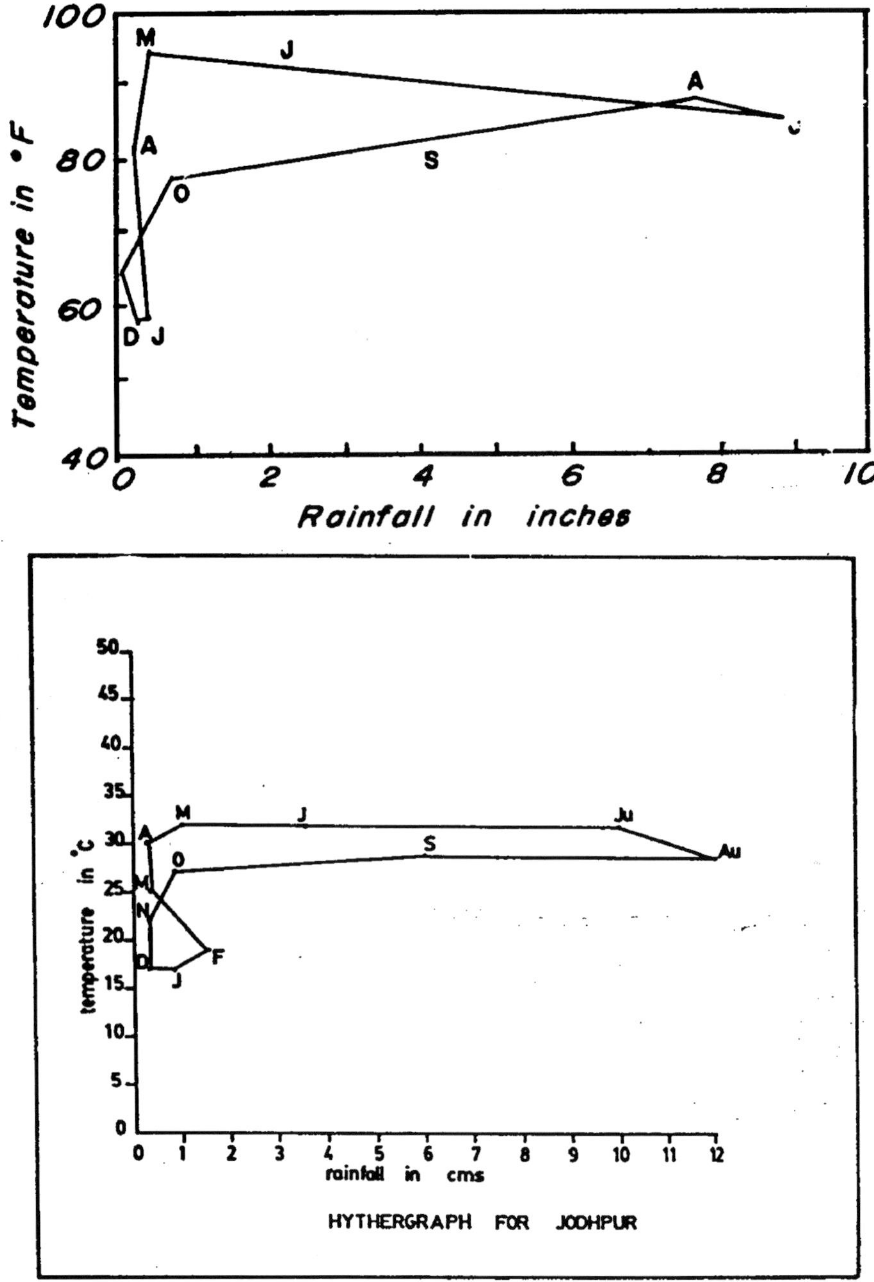

Fig. 18.18: Hythergraph for Kolkata

in a thirty-five year series, the eighteenth value from the minimum is recognized as the median value. The lower quartile value in this case lies between the ninth and the tenth figures.

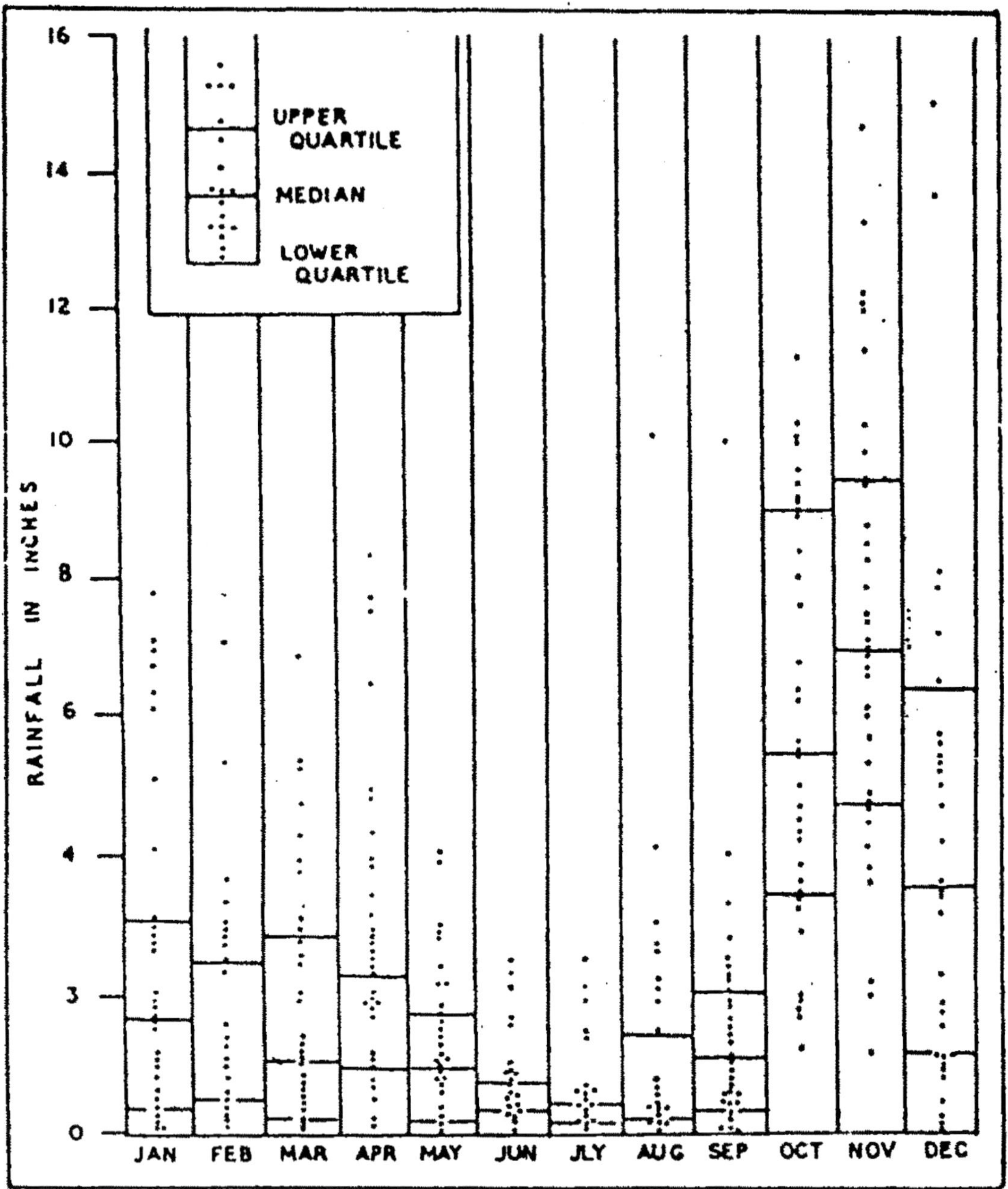

Fig. 18.19: Rainfall Dispersion Diagram for Tirunelveli

ERGOGRAPH

The relationship between season, climate and crops can also be represented graphically. The monthly average of temperature, rainfall and relative humidity here, along the vertical axis are given. The twelve months are marked on the horizontal axis. The acreage of various crops is shown below the horizontal axis on a selected scale. The net acreage of each crop is given in separate rectangular blocks. The length of each block should correspond to the growing season and its breadth is calculated according to a selected scale (Fig. 18.20). A Circular ergograph showing the amount and nature of work done during each month or season of the year has been suggested by A. Geddes and A.G. Ogilive. It is illustrated in Figs. 18.20, and 18.21.

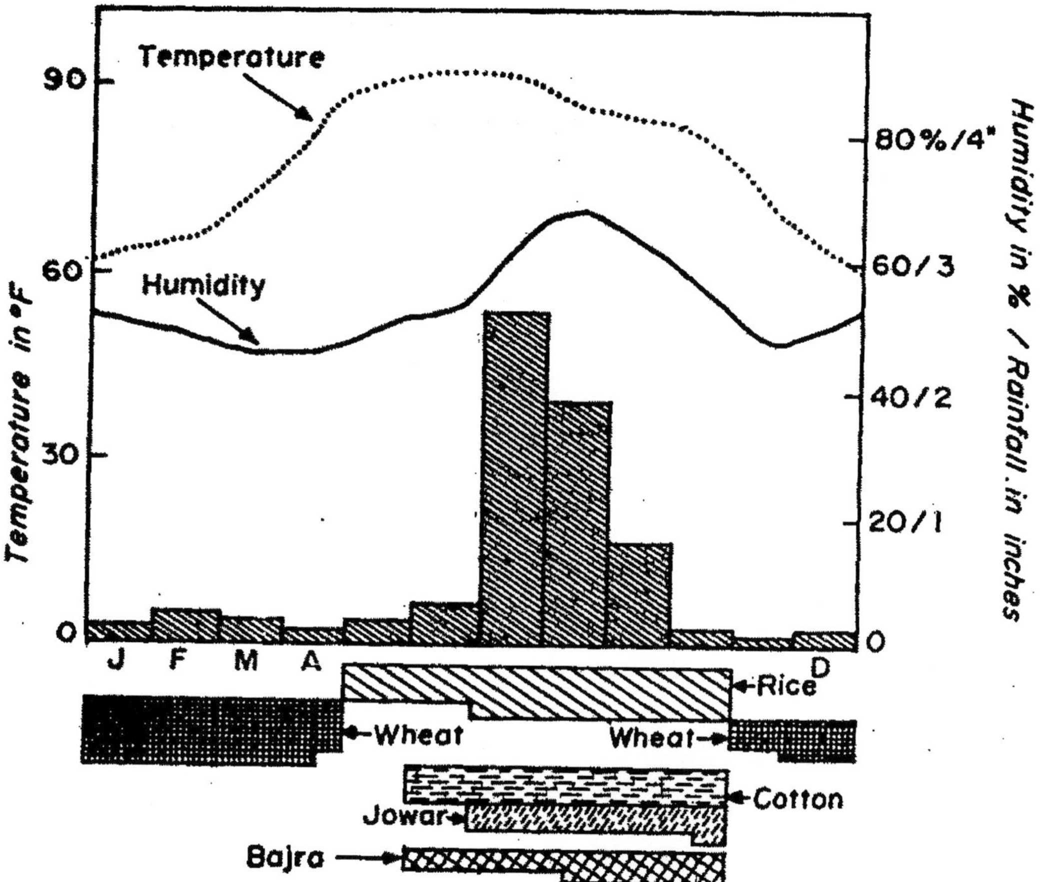

Fig. 18.20: A Simple Ergograph

COLUMNAR GRAPHS

Because of their clarity and ease of construction, they are very effective in the depiction of certain aspects of climatic data. They can be used to show rhythm of diurnal and seasonal ranges and the range and variability of various climatic elements.

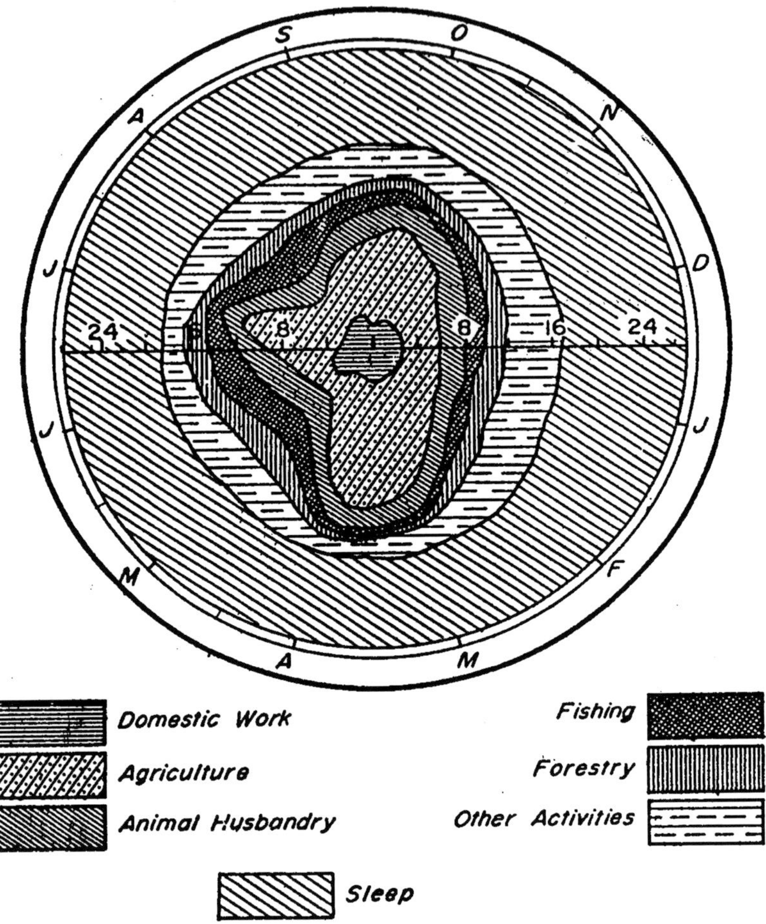

Fig. 18.21: A Circular Ergograph

In a simple columnar diagram each vertical column represents a certain value of rainfall, sunshine, etc., for a particular period of time viz., monthly (to show seasonal variation) or daily (to show daily variation). Such diagrams can also be constructed to show total amount of rainfall. For comparative purposes percentage columnar diagrams can be made. In this case, each of the monthly columns will represent percent of annual rainfall. For further details columns can be subdivided into two, one showing rainfall and the other for showing snowfall. Similarly, the columns may be sub-divided to show more details.

Deviations from mean conditions may also be effectively depicted by columnar diagrams. For example, columns may be constructed to show the maximum, mean and the minimum rainfall in each month. Further elaborations may be introduced by indicating the standard deviation and the probable deviation of rainfall from the accepted mean. Such diagrams are known as hyetographs. We can also show temperature variations in more detail to give not only the mean but also the mean

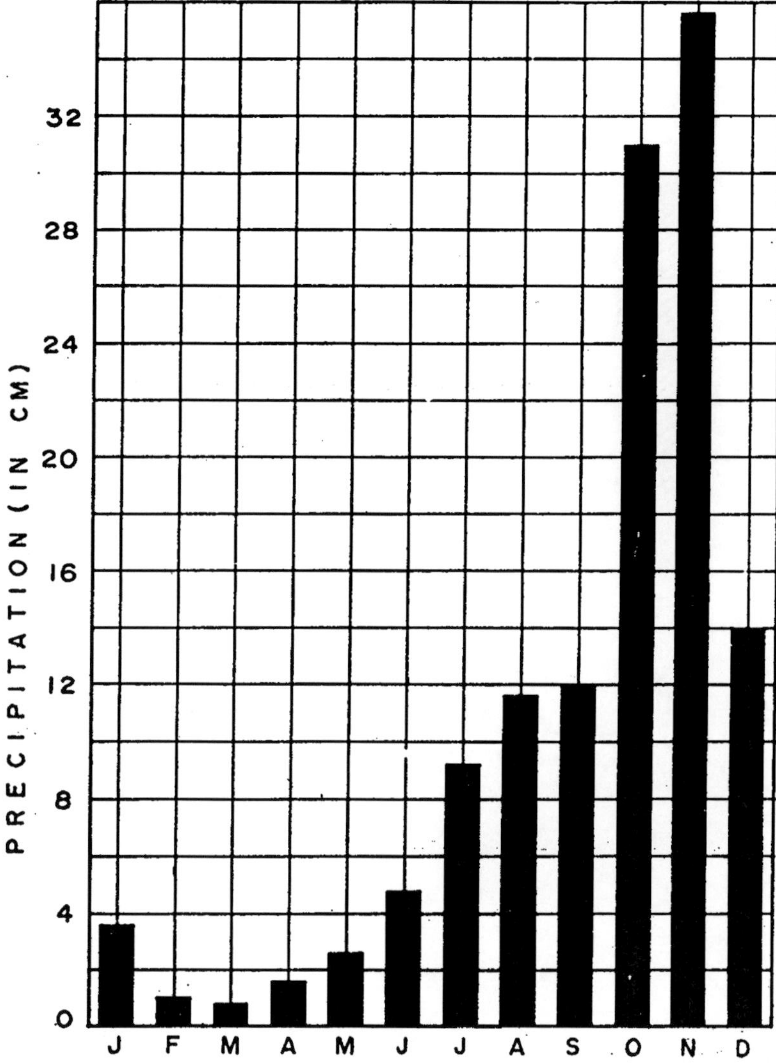

Fig. 18.22: A Simple Columnar Diagram for Brazil

daily minimum and maximum, mean absolute monthly minimum and extreme maximum and minimum records. Similarly, we can show variations in pressure.

ARROWS

There are a variety of data which can be represented by arrows. For instance the horizontal movement of air over the surface of the earth, trajectories of air, wind direction, etc., can be shown by changing the direction and width of arrows (Fig. 18.23).

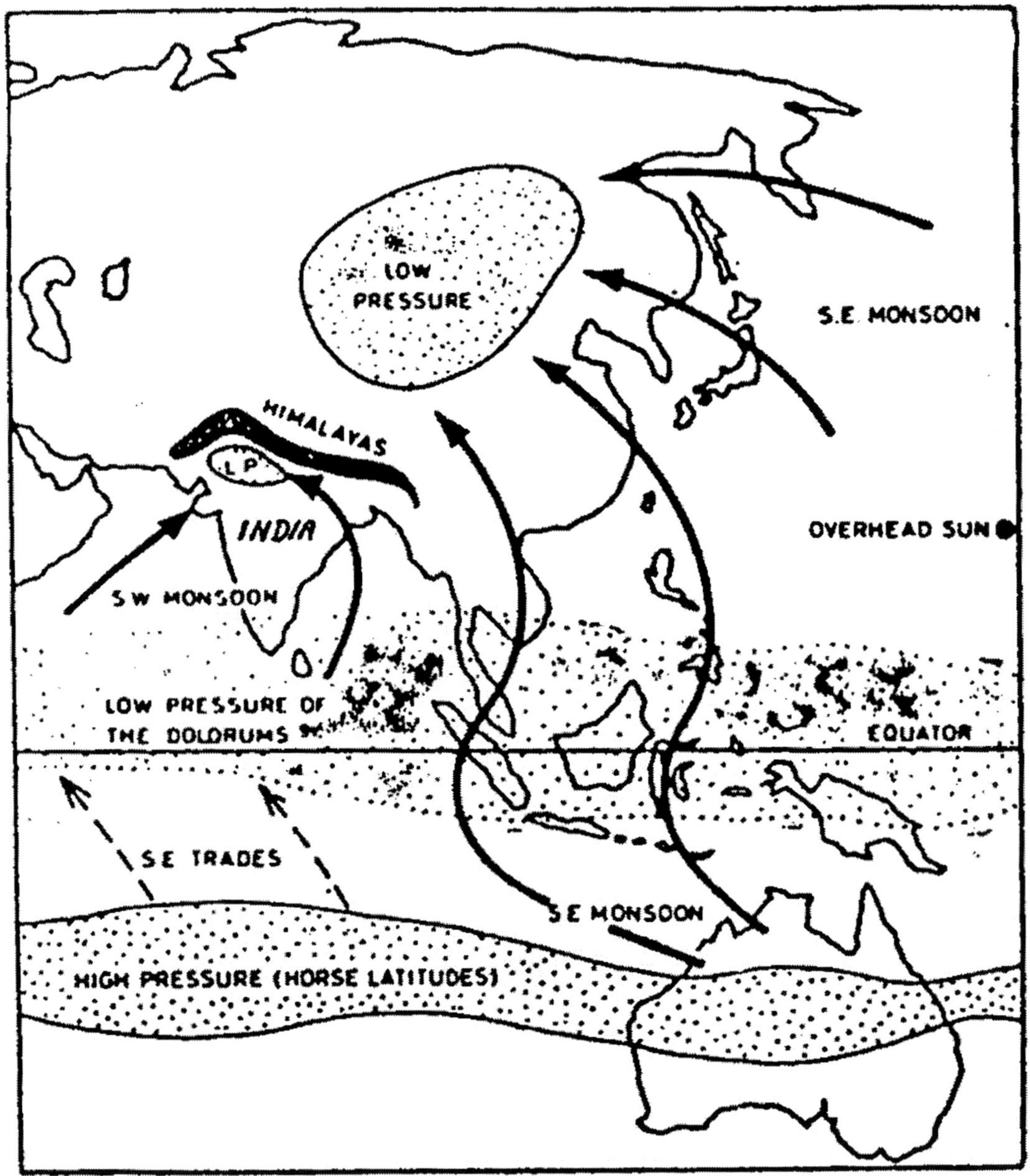

Fig. 18.23: Arrows Showing the Direction of Winds

THREE-DIMENSIONAL DIAGRAMS

Lot of climatological data can be represented by three dimensional diagrams in the same way as we represent the terrain. Jenks has attempted to represent a number of climatic features in this way. The technique is not difficult to follow because all that we have to do is to treat various types of iso-lines as contour lines. Block diagrams also can be made. Examples of some of these diagrams are given in Fig. 18.24 (a) and (b).

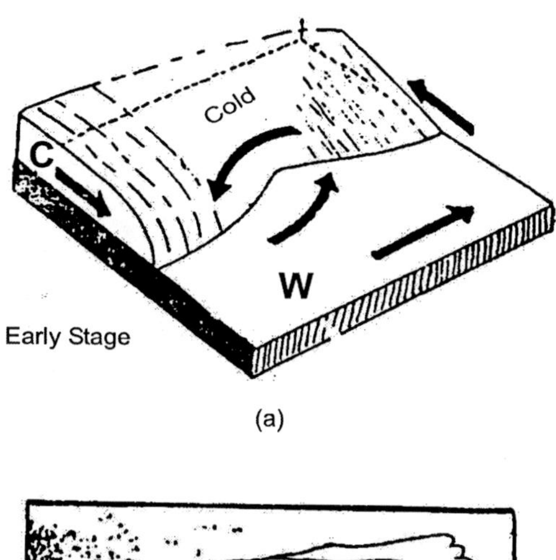

(a)

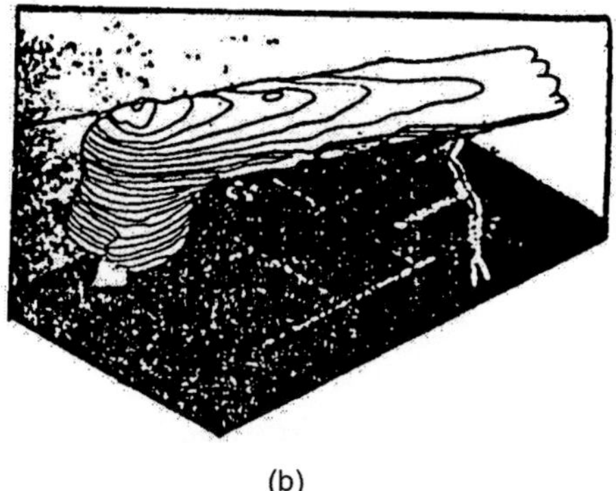

(b)

Fig. 18.24(a) and **(b)**: Three-Dimensional Climatic Diagrams: (a) Early Stage of a Moist Latitude Cyclone and (b) A Thunderstorm with Hail

MAPS SHOWING DISTRIBUTION OF CLIMATIC PHENOMENA

Climatic maps showing the distribution of a variety of climatic phenomena are only too common to need introduction. Isopleth maps can be drawn to show, for example, the distribution of rainfall. Choropleth maps are usually not made to depict such data. Cartograms showing the amount of rain received at various stations by means of bars, or the direction of wind by arrow are quite common.

Maps showing the association between two or more climatic elements can also be made. For example one can show the absolute rainfall as well as its variability in the same map.

19
Mapping the Socio-Economic Data

In the previous two chapters, we discussed the methods of representing topographic and climatic data both qualitative and quantitative. In this chapter, we propose to discuss the methods of representing of socio-economic data including those related to planning and development.

MAPPING THE QUALITATIVE DATA

In general there are three types of qualitative data:

1. Those which can be represented by point symbols,
2. Those which can be represented by line symbols, and
3. Those which can be represented by area symbols.

It should be borne in mind that symbols used to represent qualitative data too vary with variations in the scale of the map. The data pertaining to towns, capitals, cemeteries, churches, temples, light houses, bench marks, shafts, mines, hospitals etc. can be represented by area symbols on a large-scale map but they may have to be represented by point symbols on small scale maps. Line symbols used in qualitative maps are those for roads, railways, telegraph lines, streams, boundaries, parallels and meridians to name a few. And the area symbols are used to show political units, deserts, vegetation, swamps, etc.

Symbols for qualitative data are not always drawn to scale. When we put a small dot to show the location of a town, the size of the dot need not point to the size of the town. Consequently, all towns, big or small are represented by the same symbol. Similarly a symbol showing the location of mines has no relation to their size.

Line symbols used for qualitative data have similar characteristics. The widths of streams, roads, railways or boundaries are not drawn to scale and their thickness is not necessarily based on their importance measured in quantitative terms. A river drawn on a map may have a width of 1/15th of an inch. The same river drawn on another map of the same scale may appear only 1/20th of an inch wide. Although political boundaries may be shown with lines having varying widths to distinguish them as international, national, state and local boundaries, their widths is not in proportion (quantitative) to their importance. Boundaries can also be shown by different lines having the same width.

Area symbols showing qualitative values have the same characteristics. A forest area may be shown by any symbol (may be tree symbol) but this representation does not necessarily show the real spatial extent, nor does it point to the density or nature of trees.

Some text books mention population density of a given area to be thin or thick. The same idea is expressed on maps too. But we know that what is thick or thin for one person or nation is not necessarily the same for others. Terms like much, less, high, low, good, bad, hot, cold, rainy, dry, etc. give only qualitative and relative values. Unless these terms are qualified by quantitative measures, they do not mean the same to everyone.

Types of Qualitative Maps

The qualitative maps are of three types:

1. General reference maps,
2. Pictorial maps, and
3. Thematic maps.

General Reference Maps

Most of the atlas maps fall in this category. These maps usually give location of various places, political boundaries, transport networks, and prominent physical features. They are thus multiple-use maps; they give information that people in general look for.

The purpose of these maps is to give as much information as possible without cluttering it and reducing its legibility. In such maps all items of information have equal weight. In fact each one is important not only in itself but also for other items of information, for without putting them all together it is not possible to show the relative position of each.

The general reference maps are of two types:

1. Atlas maps, and
2. Wall maps.

Atlas maps give more information than wall maps. Wall maps are made to be seen from a distance, whereas the atlas maps are to be read like a book. Because of this difference, wall maps give less of details, but whatever is given is given very prominently.

Thematic Maps

In this category are included most of the historical, political and cultural maps. Some of the maps portraying socio-economic data are also qualitative in nature. Maps showing the networks of transportation lines and of irrigation canals are good examples of this type map. At times some maps in books and magazines show the crops or minerals produced in different areas by symbols. These maps are not based on any quantitative value of area or production; they are meant to give only location and distribution of select phenomenon (Fig. 19.1). Thematic maps can also be pictorial.

Pictorial maps are meant to show the distribution of various phenomena by pictorial symbols. Children's books and encyclopaedia often give such maps partly to attract the attention of the children and partly to convey certain information (Fig. 19.2).

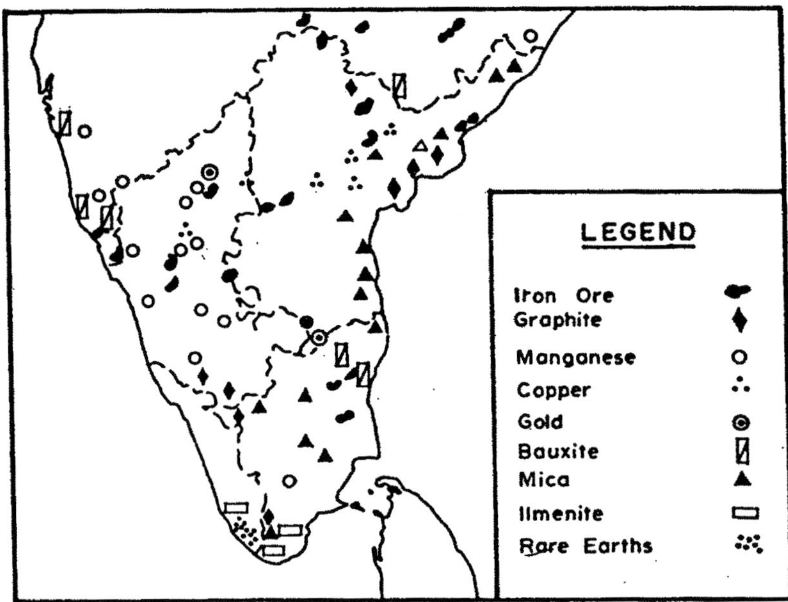

Fig. 19.1: General Thematic Map

MAPPING THE QUANTITATIVE DATA

In the foregoing section we discussed how qualitative data are represented on maps. In this section we are going to deal with the methods of representing the quantitative data. It may be recalled here that while discussing the representation of terrain and climatic data we had discussed at length how quantitative measurements can be depicted on maps. Some of the methods discussed there are going to be repeated in this chapter for they are also useful in the representation of the socio-economic data.

Mono-dot Method

Uniform size dots are used to show the distribution of a variety of socio-economic

Fig. 19.2: A Qualitative Pictorial Map Showing Distribution of Industries in Japan

data. Dot maps provide a visual impression of relative density. But one cannot get the absolute figures from them unless one counts all the thousands of dots. Compared with other map, most of the dot maps are relatively easy to prepare because they do not involve complicated computations or manual skill.

Most of the dot maps show the distribution of a single phenomenon such as the people, livestock or acres of land. On a multi-coloured dot map, dots of different colours can be used to show more than one phenomena. Thus in a population map, urban and rural population can be shown in contrasting colours.

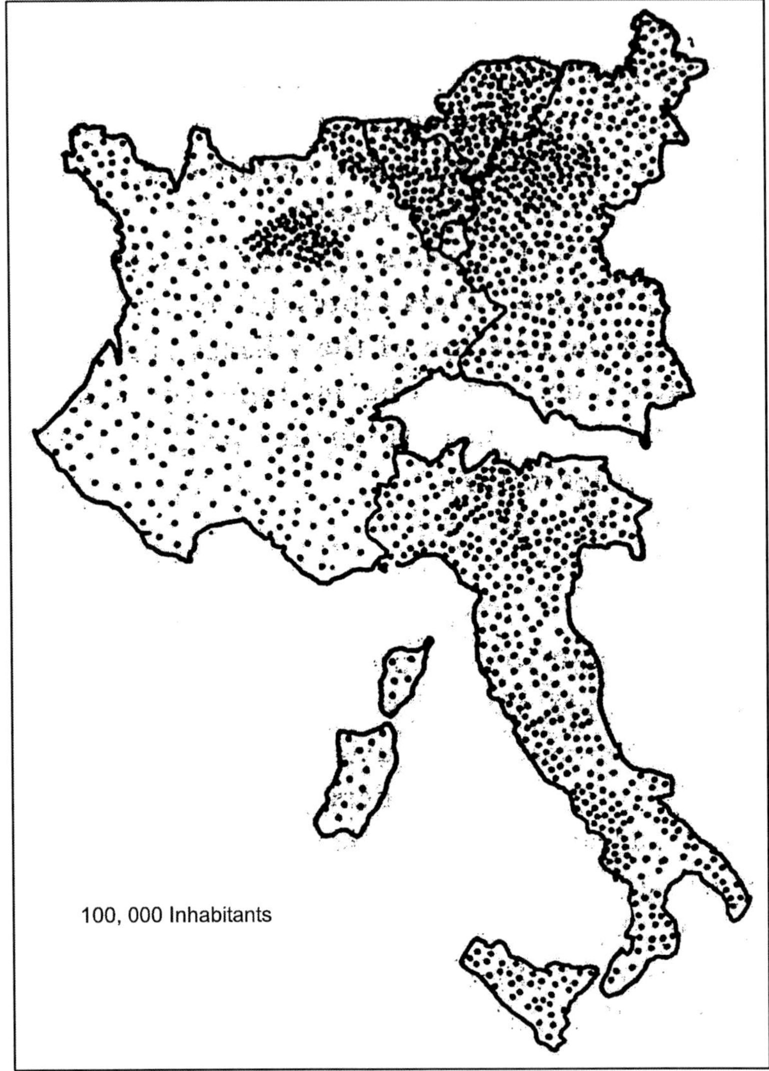

Fig. 19.3: Mono-dot Method Used to Represent Population Density in the Common Market Countries

The size of dots selected for a dot map should be such that the map remains legible. At the same time, it should not be too large to produce a 'heavy' map. The size depends upon the density of the element to be represented. This should, however, be kept in view, that too few dots give an impression of sparse distribution while too many small dots give an impression of high density. In a dot map, the chief problem is the selection of a suitable size and value of the dot.

Once the size and value of the dots have been selected judiciously, the general pattern of distribution will depend upon the areal unit of representation. If the units are large the probabilities of areal differences within these large units will be high.

The dots have to be almost equally spaced within each areal unit. A civil division within a state may have no acreage under rice, but since other divisions included in the unit of representation have rice acreage, this particular division will also show rice acreage. Great care should, therefore, be taken in selecting the areal unit of representation. For a country like India, districts should normally be the units of representation. But if the scale is 1:1,000,000 or so, one may have to go even to the Tahsil or Taluk level. Much will, however, depend upon the availability of data.

Placing of the dots also poses a problem. Dots should be equally distributed within the unit area for which data have been used? If the unit area is so located that on two sides of it there are high density units while on the other two sides are low density units, it will be desirable to put more dots on the side of high density areas and less on the other sides. Dots should not be placed over negative areas. Various accessory maps like physical, relief, etc., should be consulted to demarcate the negative areas. The most desirable state is one in which the dots are placed exactly where the elements to be mapped are located. For example, in population maps, the dots should be placed exactly at the location of the towns and villages (Fig. 19.3).

If this is done the problem of distributing the dots will not arise.

Graduated Mapping Using Multiple Dot Method

At times the density of the phenomenon to be represented in a map is so high mono-dot method does not give good results. It is particularly so if the selected dot represent large numbers/quantities. If in a population distribution map a dot represents 100,000 persons, an area having 1,080,200 persons will be represented by just 10 or 11 dots. This type of map would not be of much use. If the value represented by each dot is lowered, so that one dot represents 1,000 persons, the number of dot would go up to 1075. This may, however, be possible if the map area wherein the dots are to be placed can legibly accommodate them.

There is another and more rational way of handling this problem. That is the multiple dot method. In this system we have two or three sizes of dots standing for different values. In the above example, we can have a large dot representing 50,000 persons, a medium size dot representing 25,000 persons, and a small size dot representing 5000 persons. So, instead of putting ten large dots, we can have twenty large dots, three medium size dots and one small size dot. This method is especially useful in showing population in a country like India. In certain areas the density is too high and in certain others too low and, hence, the mono-dot method does not give good results.

The National Atlas of India has used the multiple-dot method to show the distribution of population. Dots representing 100, 200, and 500 persons have been put exactly at the location of the villages and the towns where people lived. If a village has 800 persons, it gets one large, one medium and one small dot. Urban centres are represented by them (Fig. 19.4).

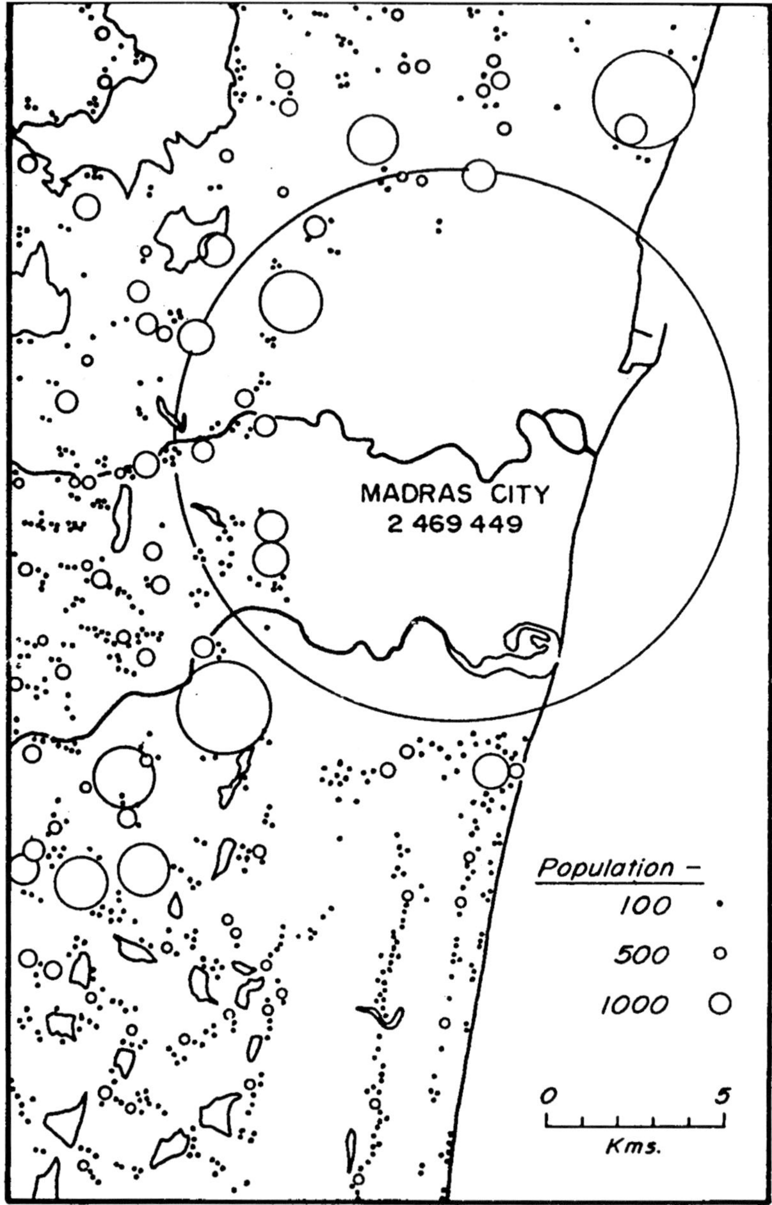

Fig. 19.4: Map of Chennai and Its Environs Showing Distribution of Population by Multiple-Dot Method

Use of Dots and Circles

If a phenomenon to be mapped is distributed too unevenly both dots and circles are used together. For example, multiple size dots are used to represent low population density rural areas, while circles are used to represent large urban centres. While using the circles, one should know that size of the circle is not proportionate to the population it represents. A circle representing 100,000 persons does not appear twice as large as a circle representing 50,000 people. It is the radius of a circle that determines its size not the circumference. The easiest way to calculate radius is to select a round number falling midway of the given set of data and select a suitable radius for it. The radii of the other circles representing the data can be calculated with the following formula:

$$\text{Required Radius} = \text{Presumed Radius} \times \sqrt{\frac{\text{Required Number}}{\text{Supposed Number}}}$$

Flannery has suggested the following alternative method of calculating the radius of a circle. The conventional method is to take the square root of the data and then to construct circles with radii proportional to the square root. According to Flannery, the following method should be used:

1. Determine the logarithm of the data;
2. Multiply the logarithm by 0.57;
3. Determine the antilogarithm of the product; and
4. Divide the antilogarithm by the chosen unit value to get the radius of the circle.

This method of calculating the radii of circles is considered better. The circles should neither be too large to appear too fullness nor too small to indicate emptiness. To ensure that two or more circles remain distinct and do not overlap, the following two methods should be used:

1. Keep the circle transparent so that their boundary lines are visible; or
2. Fill up the circles with patterns or solid colours but allow a narrow opening between them. The larger circles should be cut to accommodate the smaller ones (Fig. 19.5).

Circles can be used only if the range of data used is not too large.

Use of Spheres

If the range of data is too large, circles cannot be used effectively to show both ends of the range. In such situations spheres, in place of circles can be used. The radius of a sphere is equal to the cube root of the data to be represented.

The method of representing quantitative data by spheres is illustrated in Fig. 19.6.

Pie-Charts

Pie or circle charts are made to show fractions of a total. For *Example*: Prepare a pie chart to represent the following data:

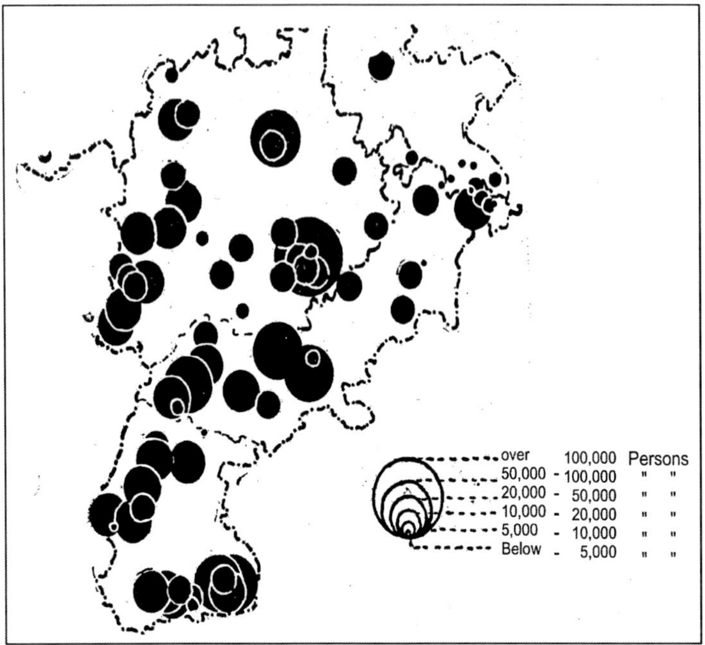

Fig. 19.5: Representation of population of urban centres in Southern Tamil Nadu by Circles

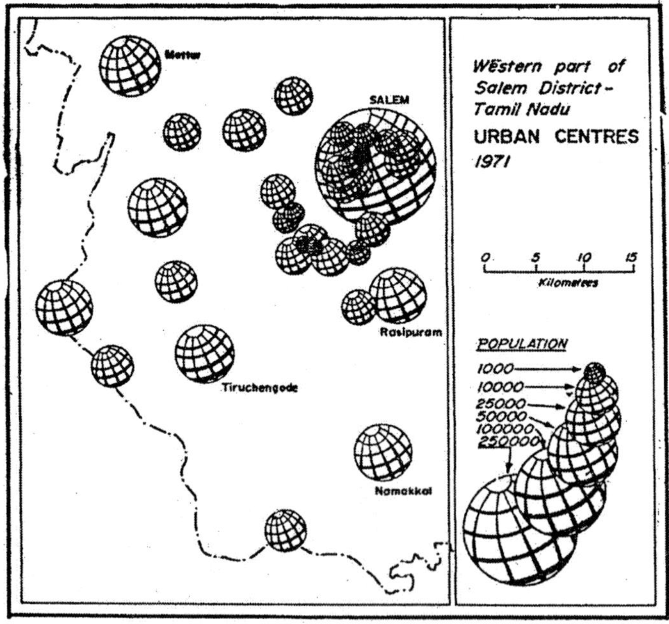

Fig. 19.6: Representation of Quantitative Data by Spheres

Total land area	1000,000 acres
Area cultivated	600,000 acres
Area under forest	300,000 acres
Wasteland	100,000 acres

Select the area to be represented by a circle of any convenient radius. The total land area i.e. 1,000,000 acres is represented by a circle which has 360°. The share of the cultivated area (600,000 acres) in 360° will be 360° × 3/5 = 216°. The segment of the circle formed by 216° will represent the cultivated area. We can determine the segments of the circle for other uses, the same way. Once, the share of 360° of the circle area is earmarked for each type of land use, we divide the circle appropriately to get a pie-chart. This pie-chart as shown in Fig. 19.7 can be made perspective (Fig. 19.8); it can also be used in maps (Fig. 19.9). In such cases the circles drawn will have to be proportional for data for each state province or district.

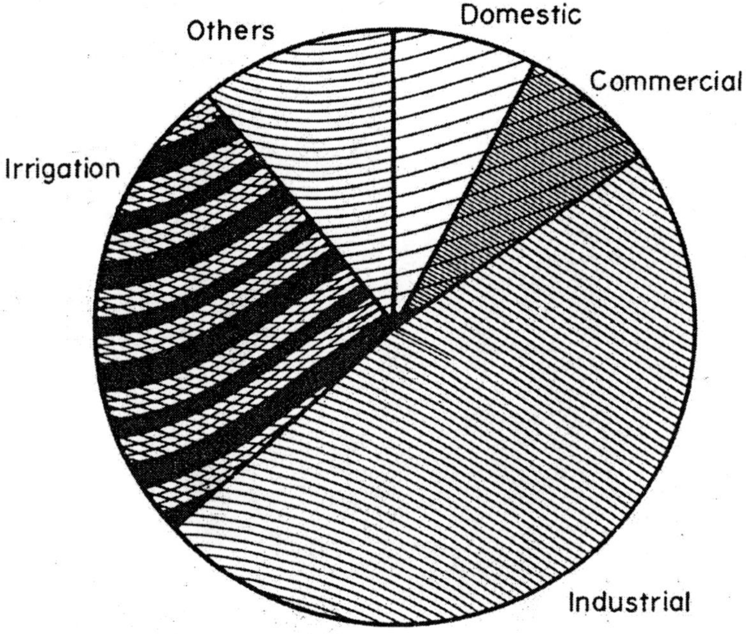

Fig. 19.7: A Pie-Chart Showing Land Development

Curve Plotting on Arithmetic Scale

Curve plotting is one of the important ways of representing statistical data graphically. Although line graphs are rarely used in maps their independent uses are quite frequent. They can be drawn to show temporal changes in the production or growth of one or several variables. To construct these graphs, time is given on the horizontal axis and the production or growth on the vertical axis. To give a comparative picture of change

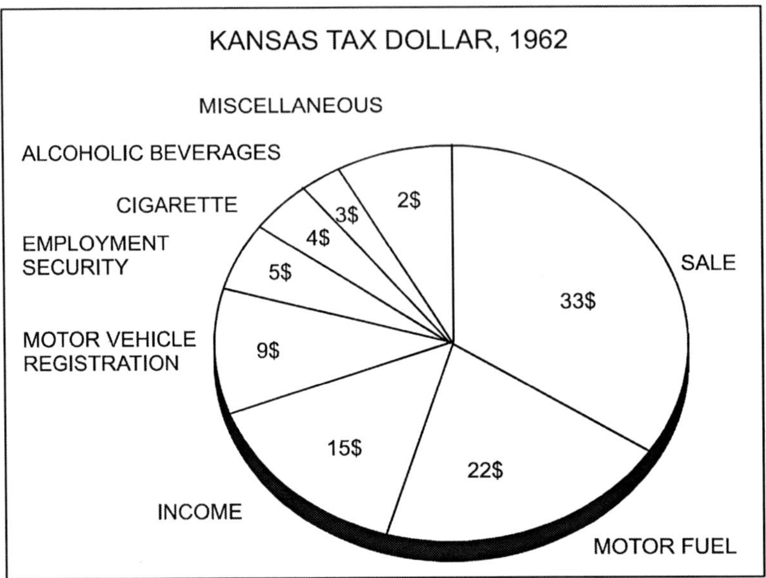

Fig. 19.8: A Perspective pie-chart

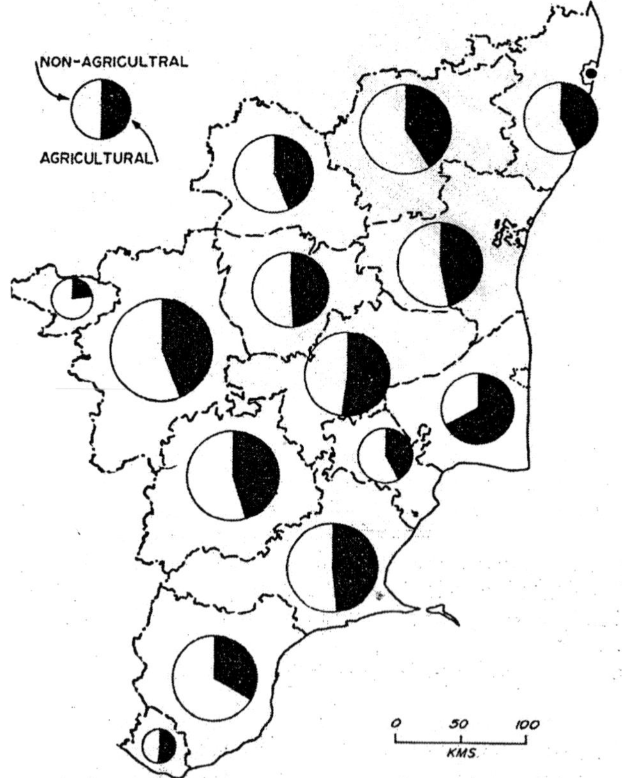

Fig. 19.9: Map of Tamil Nadu showing Land Use Data for various districts by Pie-Chart

in the production or growth of more than one commodity one can draw curves for each individual item with a common origin as shown in Fig. 19.10.

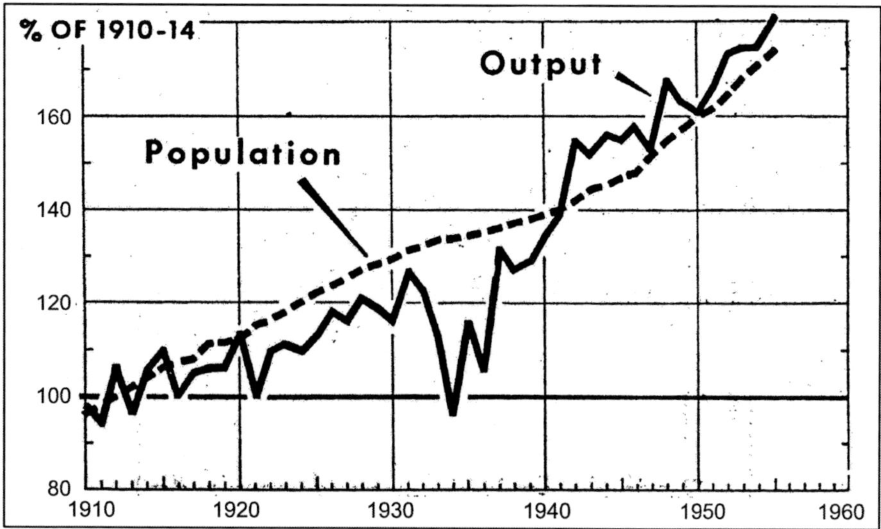

Fig. 19.10: A Simple Graph Showing US Population and Farm Output

Moving Average Curve

The curves of the type illustrated in Fig. 19.10 are marked by sharp fluctuations. To get a more generalized picture, they can be smoothened by the use of moving or cumulative averages.

To get moving average we add the figure for more than one year and divide the total by the number of years. This procedure smoothens fluctuations resulting from some abnormal conditions like epidemics, famine, recession, etc. Larger the number of years clubbed together, smoother is the curve; and smoother the curve more generalized a picture one gets Table 19.1 and Fig. 19.11.

Table 19.1: Calculating a Three-Year Moving Average

Years	*Average Yield Per Year*	*3-Years Moving Average*
1945	30	30
1946	70	44
1947	40	46
1948	38	50
1949	20	33
1950	100	55
1951	80	66
1952	10	63
1953	20	40
1954	70	33
1955	60	50
1956	30	53

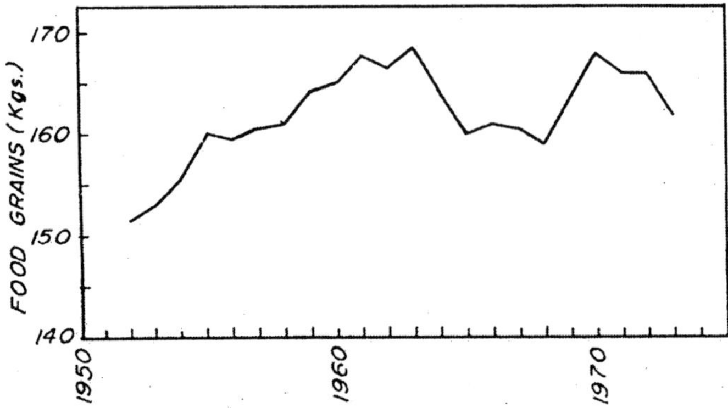

Fig. 19.11: Moving Average Curve

CUMULATIVE AVERAGE CURVE

A cumulative graph is a useful device for analyzing annual or monthly production or growth. It is designed to represent at each point the total of all the preceding separate items. The cumulative total and the total at the end of the period are the same. A cumulative graph can be drawn on either arithmetic or semi-logarithmic scale.

Fig. 19.12 showing an average cumulative curve is based on the data given in Table 19.2.

Table 19.2: Method of Calculating Cumulative Average Population

Years	Population Increase	Cumulative Average
1947	500	500
1948	300	800
1949	400	1200
1950	250	1450
1951	600	2050
1952	550	2600
1953	400	3000

COMPONENT PART CURVE

By showing total production of, say, cereals on the Y axis and the year of production on the X axis, we can depict the relative production of several commodities at different temporal stages. We may select only one commodity but different areas for comparison purposes. Such charts can also be made by using percentage figures. An example of such a graph is given in Fig. 19.13.

CURVE PLOTTING ON SEMI-LOG SCALE

The line graphs discussed and illustrated so far are usually drawn on arithmetic scale. Such graphs are very useful for showing absolute changes. Quite frequently we are,

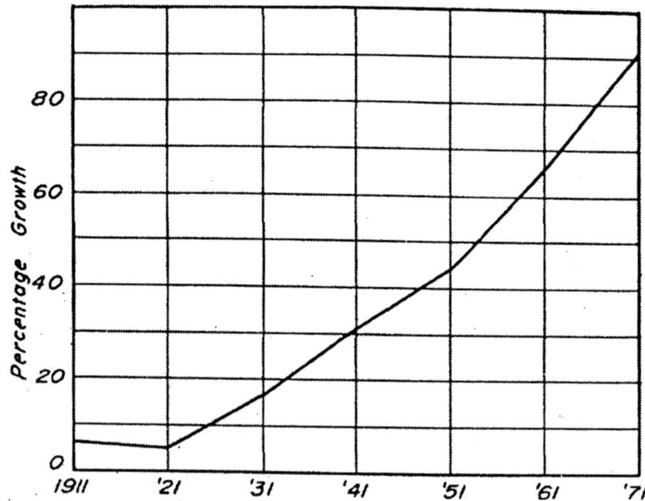

Fig. 19.12: Cumulative Average Curve

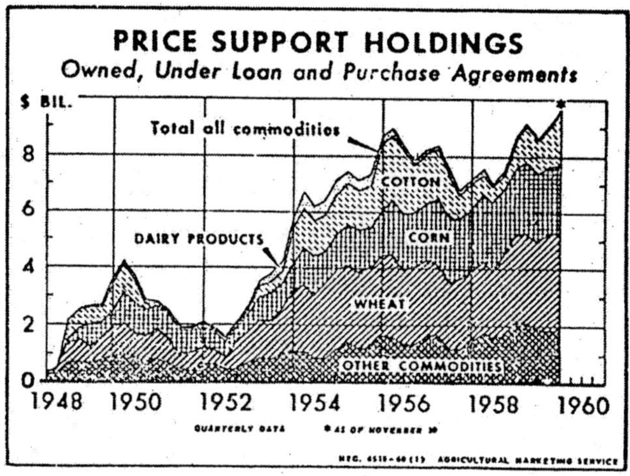

Fig. 19.13: Component Part Curve

however, interested in depicting percentage rather than absolute changes. To do this we need either to convert the data into percentages or to devise a scale which will show the rate of change even when absolute figures are used. To convert the data into percentage is a time consuming process. Hence, we often use the semi-log scale which gives the same results with absolute figures. This scale shows geometrical progression.

The vertical scale of an arithmetic chart starts with a whereas that of the semi-logarithmic chart starts with 1 or any other convenient number having 1, such as .0001, .01, 100, 1000, etc. As semi-log chart can be used in comparing per cent changes in two series, it can also be used in comparing the relative severity of fluctuations in different series (Fig. 19.14).

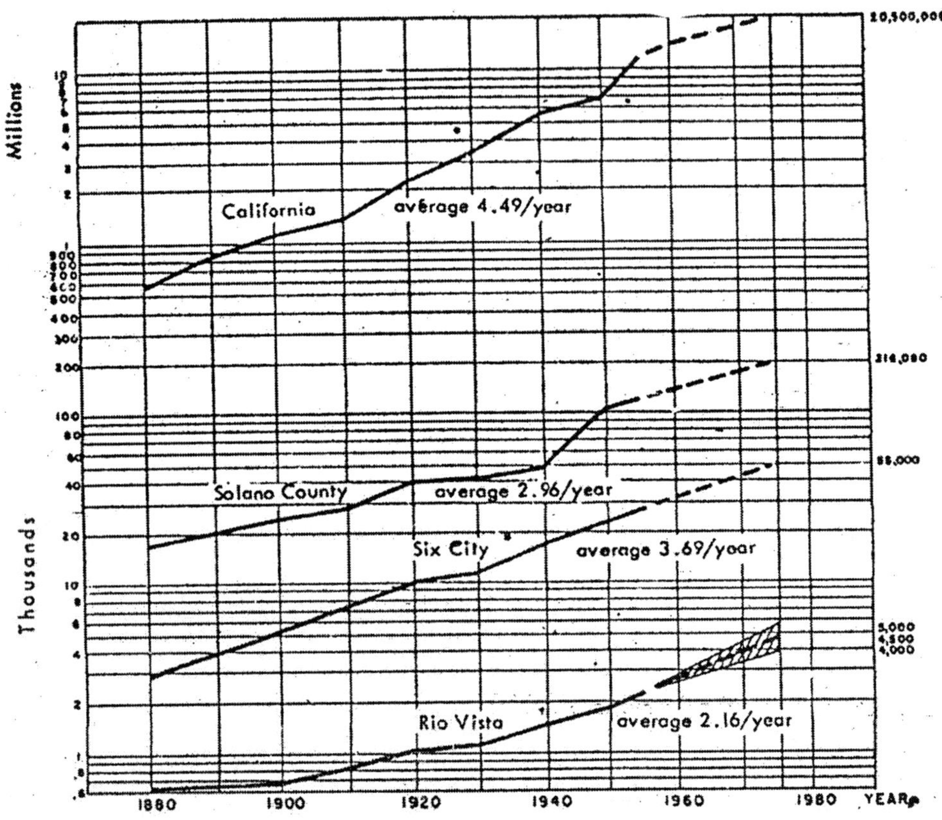

Fig. 19.14: Curve Plotting on Semi-log scale. Here the Curves Show the Population Trends and Projections (1880-1975) for Rio Vista Planning Area of Solano County, California, USA

When the percentage figures are used on the arithmetic scale, we may get a straight line. The same line can be derived by using the absolute figures on a semi-logarithmic scale.

FLOW CHARTS

The flow of goods or people can be shown by lines with o. without arrows. Arrows point the direction to which the movement takes place. The thickness of the lines is determined by the volume or quantity of flow. Flow lines are often used to represent international, national or regional flow of various commodities.

Depending upon the number of directions and destinations to which the commodity moves or people move, a flow line branches off to a number of directions from the origin. As the width of the line is determined by the quantity or volume of commodity moving or people moving, when part of the Commodity or people is diverted to a different direction, the main flow line also becomes thinner proportionately. When two flow lines cut each other, some space is left along the one to keep the two distinct from each other. Examples of flow charts are given in Figs. 19.15; 19.16; and 19.17).

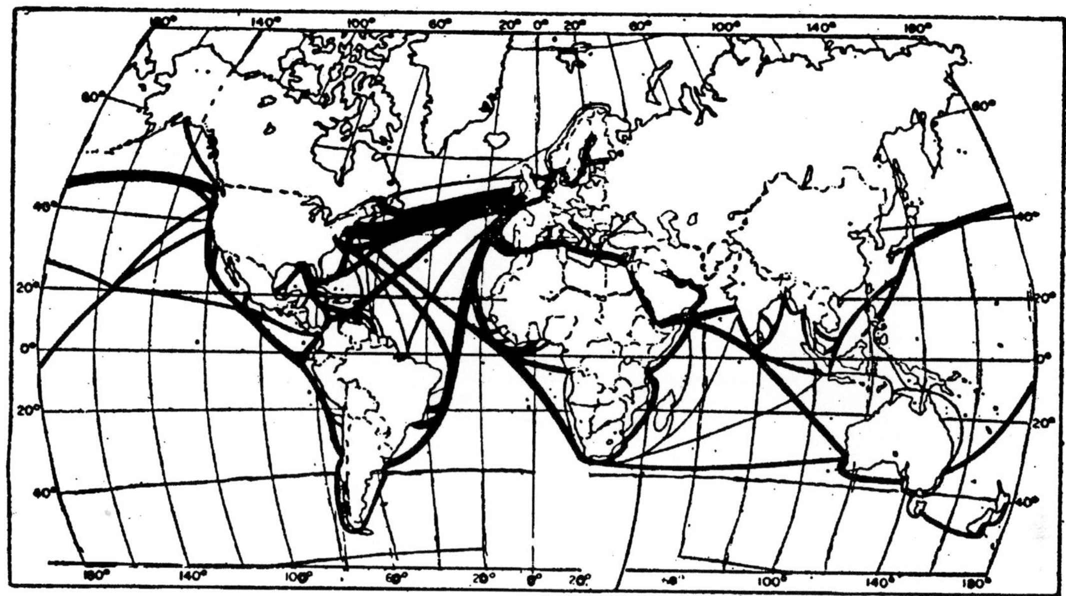

Fig. 19.15: Flow Line Charts: World Consumers and Commodity Demands

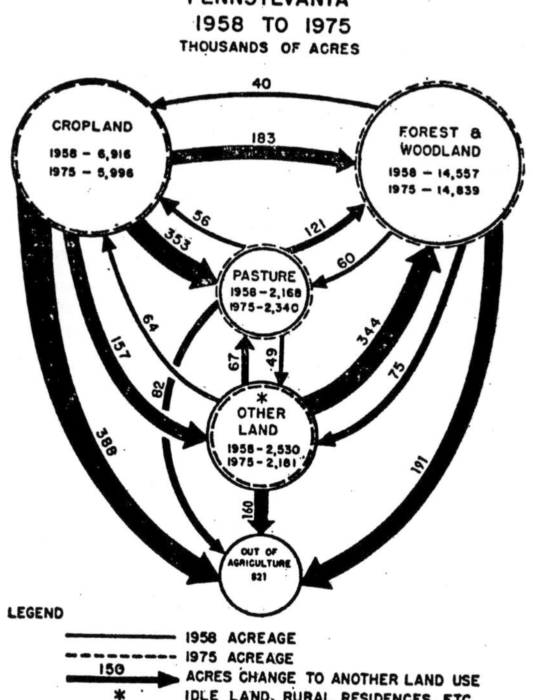

Fig. 19.16: Flow Line Charts: Land Use Conversion in Pennsylvania, USA (1958-1975)

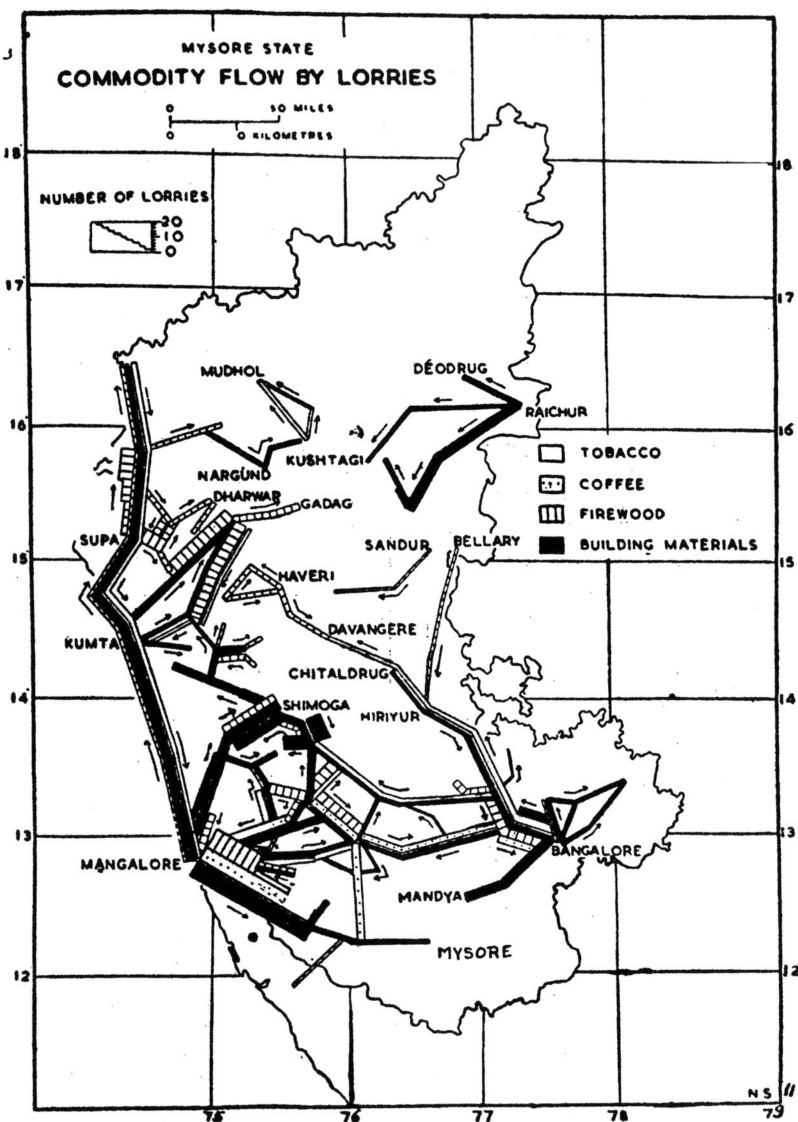

Fig. 19.17: Flow Line Charts: Commodity Flow by Lorries in Karnataka State

BAR GRAPHS/DIAGRAMS

The representation of volumes or quantities by bars is easy and simple. Bars can be placed vertically or horizontally. The length varies in proportion to the quantity represented. There are two types of bar graphs:

1. Simple, and
2. Complex.

The simple bars have no sub-divisions, so that each bar represents one commodity for a given period or place. In this type of graph, only the length of the bar changes proportionately but the width remains constant (Fig. 19.18). The data can also be converted into percentages to prepare bar diagrams. Bar diagrams showing temporal variation can also be made.

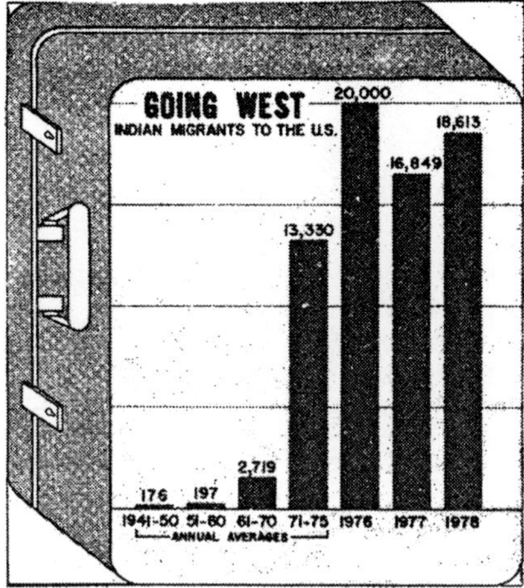

Fig. 19.18: A Simple Bar Diagram

Further complexities can be brought in by showing more than one commodity production at different places (Fig. 19.19). Another version of the bar graph is the divided rectangle or horizontal bar. The full length of the bar represents the total value. The bar is then divided into segments to show further details. For example, total land under crops can be shown by divisions.

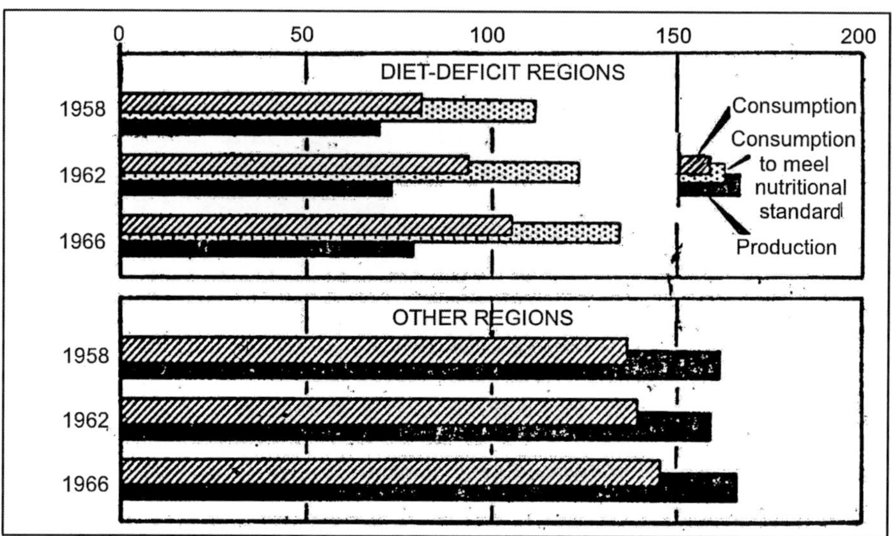

Fig. 19.19: Another Simple Bar Diagram

A number of rectangles can be drawn to provide comparisons through time. They are known as compound pyramids. Such graphs can also be used to show sex and age ratios, etc. Here they are known as age and sex pyramids (Fig. 19.20).

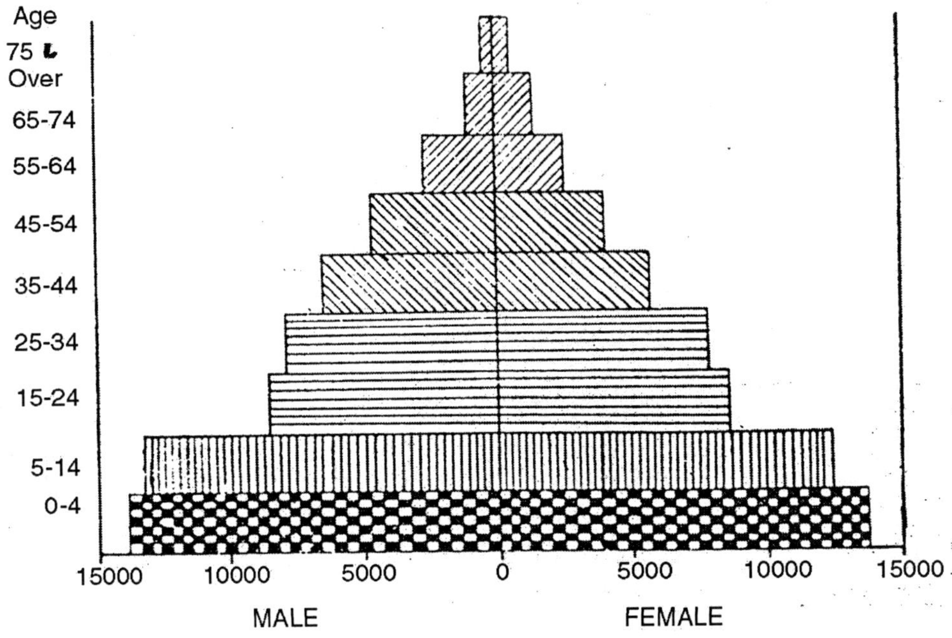

Fig. 19.20: An Example of Age and Sex Pyramid

All these bar graphs can very well be superimposed on maps to give suitable cartograms.

ISOCHRONES

Isochrones connect places which takes the same amount of time to reach from a given place. We know that geographic distance is different from time-distance. When places are connected with each other by modern means of transportation, it takes less time to go from one place to another. But in areas where the means of transportation and communication are not yet developed, it takes far more time to go from one place to the other. Isochrones are something like contour lines which connect places lying at the same distances from a given place. Fig. 19.21 shows isochrones showing number of hours it took to reach different places in India from Delhi by railways in 1958.

RAY DIAGRAMS

Ray diagrams are used to illustrate the spheres of influence of towns, etc. The affinities between the urban centres and the villages in surrounding countryside can be indicated

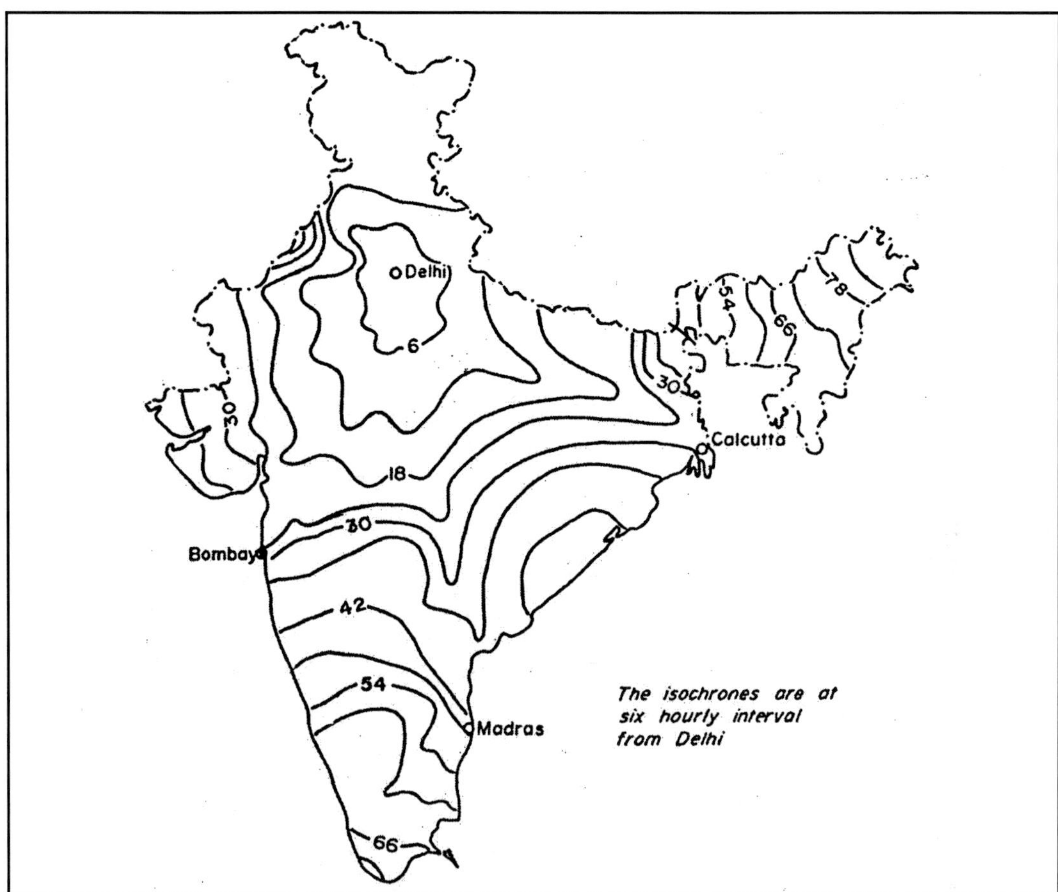

Fig. 19.21: Map of India Showing Time-Distance Isochrones

by simple lines radiating from centres. Such diagrams can also be used to show the relative importance of the functions various towns are known for. Often, these are called wind rose diagrams (Fig. 19.22).

BLOCK PILE DIAGRAM

This method involves the piling up of a number of cubicles one above the other in such a way that each one of them can be easily counted. Being three dimensional, one of its sides is shaded so that it gives a perspective view. It is relatively easy to make. All that is needed is to assume a small unit cube to represent a given quantity. Such unit cubes are then piled one above the other to produce a block. One of the sides of the block is sub-divided into 10 equal parts to represent sub-divisions (Fig. 19.23).

This diagram gives an impressive visual result. The values represented by a block pile diagram can be easily counted and understood. It occupies less space than a circle but more than a sphere.

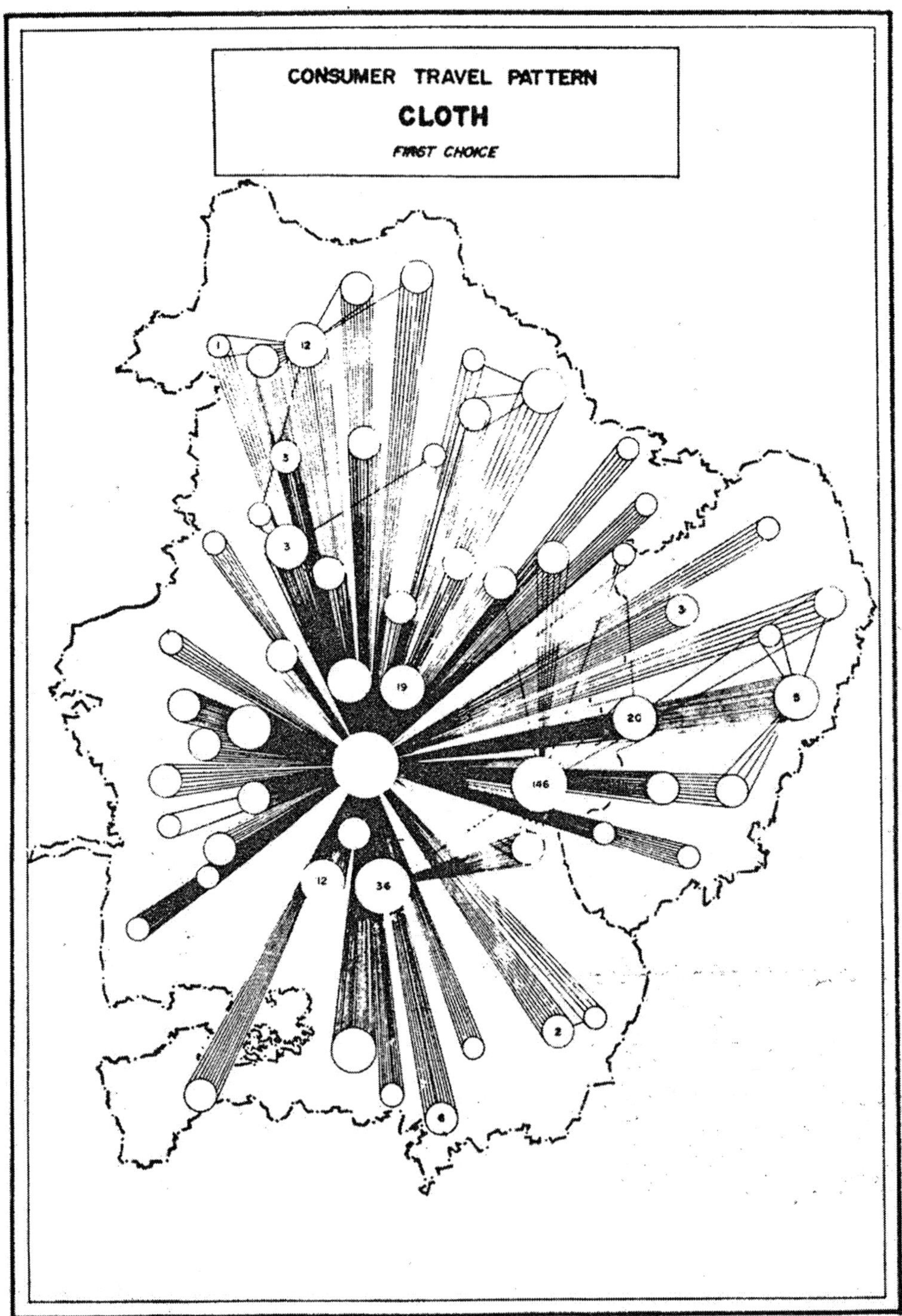

Fig. 19.22: Ray Diagram Showing Urban Services and Amenities in Karnataka State in 1960s

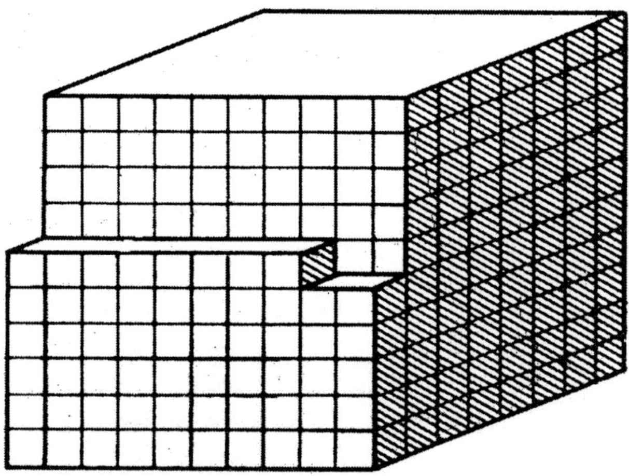

Fig. 19.23: Block-pile Diagram

CHOROPLETH, DASYMETRIC AND ISARITHMIC MAPS

"Geographical quantities which exist over an area exist in vast array and range from the simple to complex, including many very useful abstractions". It is due to this fact, that a cartographer faces a serious problem in their representation. Geographic variables are of two types:

1. Discrete, and
2. Continuous.

The real difference between the two is more theoretical than practical because one can easily be converted into the other. Let us take the example of population distribution. We know that people live in clusters called villages and towns. Population distribution thus cannot be called a continuous phenomenon. But when we make a population density map, add the population of various clusters and then add divide the total population by the area to get the density per sq. km. We know that density is a continuous phenomenon, but not the distribution of population which provides the data for calculating density. Here, we have converted discrete data into continuous data. The total population of an area can be shown by a column built at a central position to show the total population of each administrative unit. In doing so we have to convert the spatially dispersed data into discrete data, for we assume that the total population of the area is located at one centre.

Because of the possibility of converting continuous data into a discrete data and *vice versa*, we are able to show the distribution of various phenomena either by density patterns or by point symbols. In the former case, we use density figures whereas in the latter, we use absolute figures. We may yet find a set of data which show neither of these characteristics. We saw that in continuous series the densities are calculated with respect to the area unit selected for the purpose. In one unit the density may be high and in the other quite low. The boundary between the two units will mark a sharp break in the density patterns. But we know that many of the phenomena do not respect man-made boundaries. To solve this problem we can take one of the following two steps:

1. We can generalize the area densities to do away with administrative boundaries, or
2. We can treat the given data as continuous and further generalize them to give step—like isarithmic lines.

If we follow the administrative boundaries to show densities by line or colour patterns, we get a choropleth map. When we generalize this choropleth map, we get dasymetric map. But when we further generalize the data and use isarithmic lines, we get an isarithmic map (Fig. 19.24).

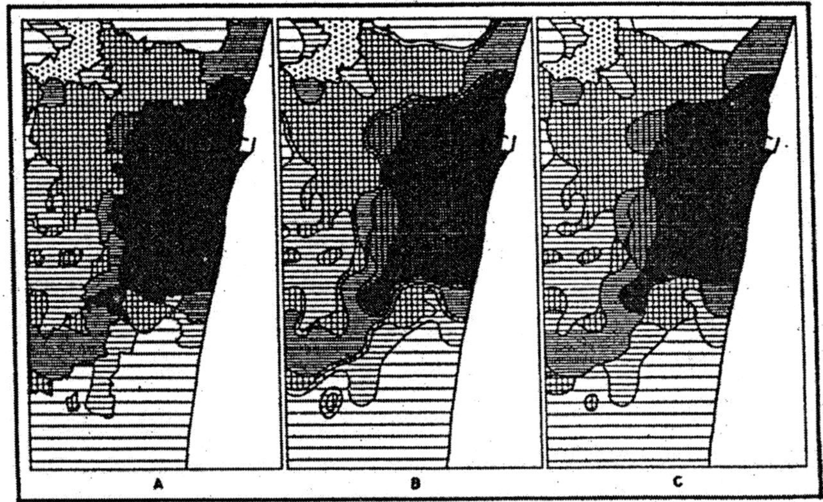

Fig. 19.24: Choropleth, Dasymetric and Isarithmic Maps

Among the three choropleth is the most commonly used technique. The simplest of the choropleth maps is the one which shows the density per square unit of area at a given level of administrative division like district, state, etc. If the total area of the selected administrative division is 100 sq km and its total and population is 10,000, its density is 100 persons per square km. This way we can calculate the density for each of the administrative divisions to get a choropleth map.

Choropleth maps can also be constructed to represent deviations from its average. For example, if we take the density of a country as the mean, and then calculate the deviations of regional densities from the mean we can prepare a choropleth map to give an idea of the density anomalies. Such maps can be prepared to show spatial variations in production of various commodities, age group distribution and so on.

Choropleth maps can also be drawn to represent certain changes. This can be done in two ways. Suppose we want to show changes in the density of population between 2001 and 2011. We can draw two choropleth maps showing distribution in 2001 and 2011. The other way of doing it is to calculate the index of change. This can be done by dividing the area total for 2011 by the area total for 2001.

PICTORIAL MAPS

Statistical data can also be represented by such pictorial symbols like sacks, bales, coins, human or animal figures, etc. (Fig. 19.25). For example, to show the distribution of human being in the world, we can show human figures. The number of figures will depend upon the data to be represented and the scale of the map. Such maps are very good illustrative materials for classroom purposes.

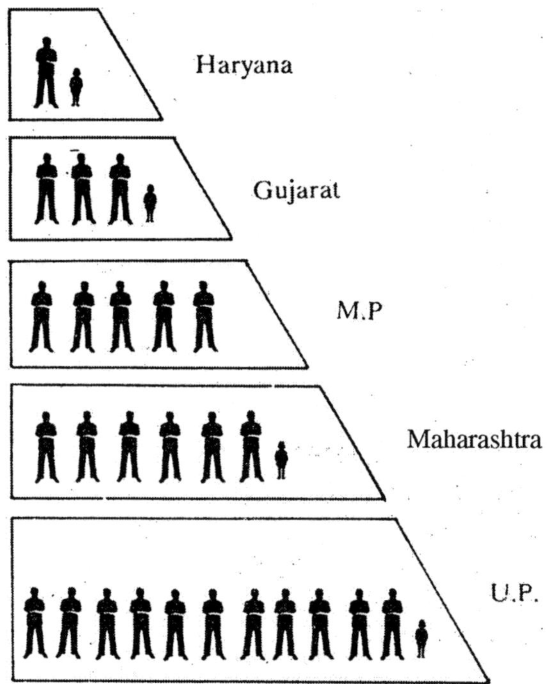

Fig. 19.25: Pictorial Symbols showing population of selected states

20
Thematic and Complex Mapping

THEMATIC MAPPING

If we rely on semantics, any map which has a theme is thematic. Such a broad view, however, leads us nowhere because there is no map which is without a theme. What we really mean by thematic map is a map which focuses on portraying certain specific information often at the cost of other information often more important but irrelevant to basic theme. The purpose of such a map is to meet the requirements of those and only those who are specialists and looking for theme specific information.

To make a distinction between thematic and general purpose maps clearer, let us look at a soil map of India drawn on a 1:1,000,000 map produced by the Survey of India. This map has all those topographical details, which can be incorporated in a map of that scale. If some features are not represented there, it is not because they were deliberately suppressed but because they cannot be represented at that scale of the map. These maps are made to cater to needs of specialist as well as common man. But the soil map is not a general map; it is made to serve the interests of a specific section of the society. It tends to give in permissible details everything about the distribution of various types of soils without caring much for the details printed on 1:1 million sheet of the Survey of India. Such a map is useful only to those who are interested in soils such as pedologists, land use planners and farmers. The soil map is a thematic map whereas the 1:1 million sheet of the Survey of India is a general purpose map.

There is another distinction that becomes obvious as soon as we see the two maps referred to above. The million sheets of the Survey of India are constructed on a much larger scale than the soil map of India. Most thematic maps are drawn on smaller scales but this should not be treated as a rule. In fact, many of the thematic maps of the *National Atlas of India* (English Edition) are drawn on 1: 1 m scale. But that must be the largest scale, possible for a country like India. Small countries with smaller territories can afford to make thematic maps on still larger scales. Nevertheless, most thematic maps are drawn on smaller scales. Sometimes, thematic maps are called statistical maps and *vice versa*. These two terms are not interchangeable, although most thematic maps do use data derived from statistical analysis of the raw data.

There are two types of thematic maps:

1. Simple thematic maps are those which concentrate on the representation of a single feature; and
3. Complex thematic maps are those, which represent combinations, correlations, associations, variations or interconnections between several aspects of the same phenomenon or between two or more phenomena.

An example of the first type is the soil map giving the distribution of various types of soils in a given area (Fig. 20.1). The example of the second type is a map showing the association or correlation between rainfall and rainfall variability (Fig. 20.2).

Simple Thematic Maps

Simple thematic maps can be classified into three types:

1. Qualitative in construction as well as in appearance.
2. Quantitative in construction but qualitative in appearance.
3. Quantitative in construction as well as in appearance.

Qualitative Thematic Maps

These maps show such natural and cultural features, whose distribution is usually not measured quantitatively. Among such features are lithology, rock types, soils, vegetation, religions, languages, etc. It is not to suggest that they are not quantifiable. It only means that the non-quantitative aspect of such phenomenon is often sought after and hence quantitative aspects do not get prominence.

One of the difficulties in constructing qualitative thematic maps is the problem of mixed or transitional areas. If we are showing religion-wise distribution of population in India, the Hindus are in majority in almost every administrative unit. How can we show the distribution of minority groups like Christians, Muslims and Sikhs, unless we base our maps on certain quantitative values? There are, of course, methods to deal with such problems, but they give only a schematic picture.

Most of the qualitative maps use colour patch or choro-chromatic methods of representation. Colours are chosen carefully so that they are distinct from each other. Similar features have similar colours. For example, in a geologic map, all tertiary formations are shown in one colour. In a historical map, different kingdoms or countries are shown by different colours.

Qualitative thematic maps may also show the distribution of such features which do not distributed spatially. For example, a railway map uses line symbols to show railway tracks. If these lines have different gauges or zones, each gauge or zone is shown by a different colour or a line symbol. At times a qualitative map gives only the

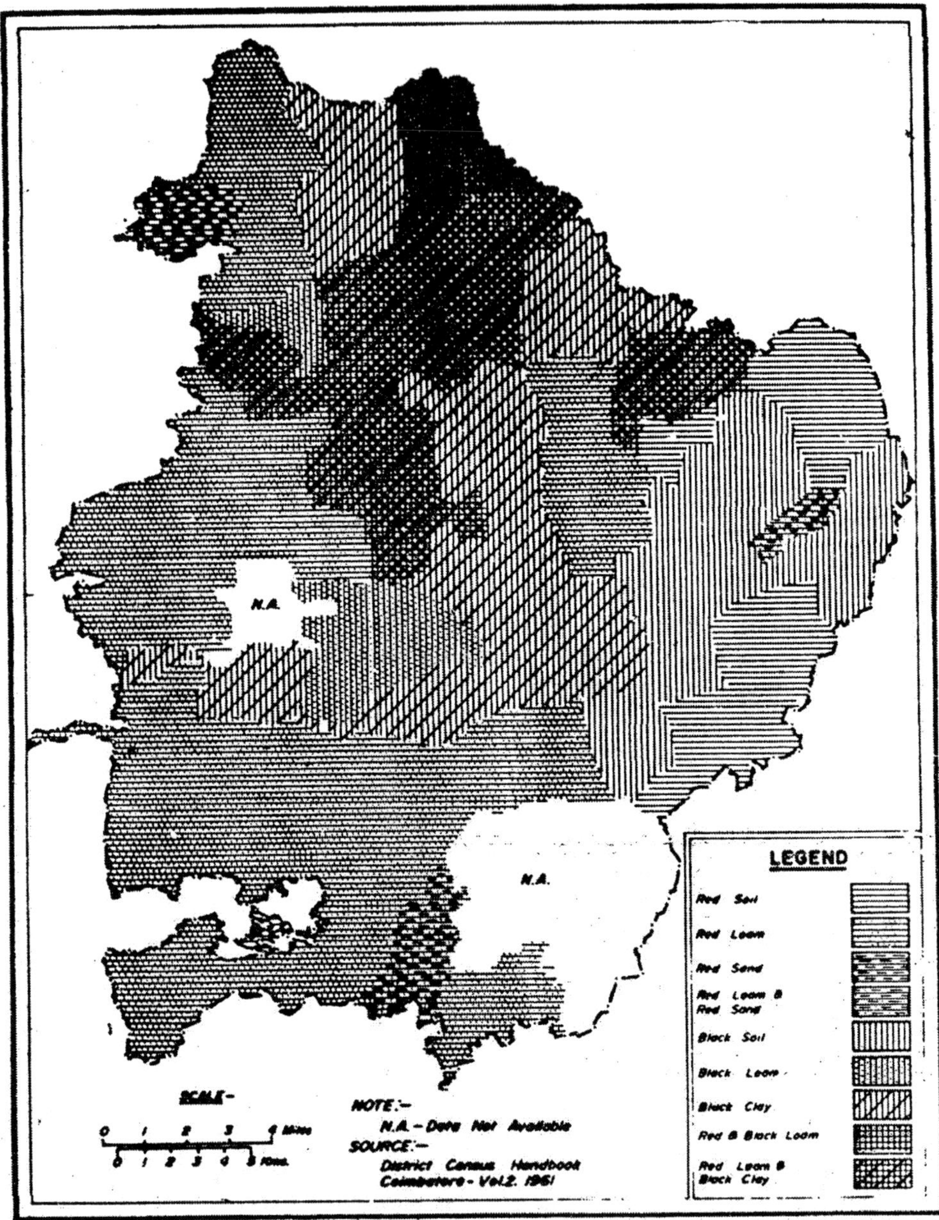

Fig. 20.1: A Simple Thematic Map Showing the Soils of Coimbatore Metropolitan Area

location. For example, a mineral map may give the location or spread of mines of various energy and mineral resources without giving anything about their size or production. All such maps are thematic but non-quantitative. The reader should, therefore, be careful in not equating a thematic map with a statistical map. The two are not always the same.

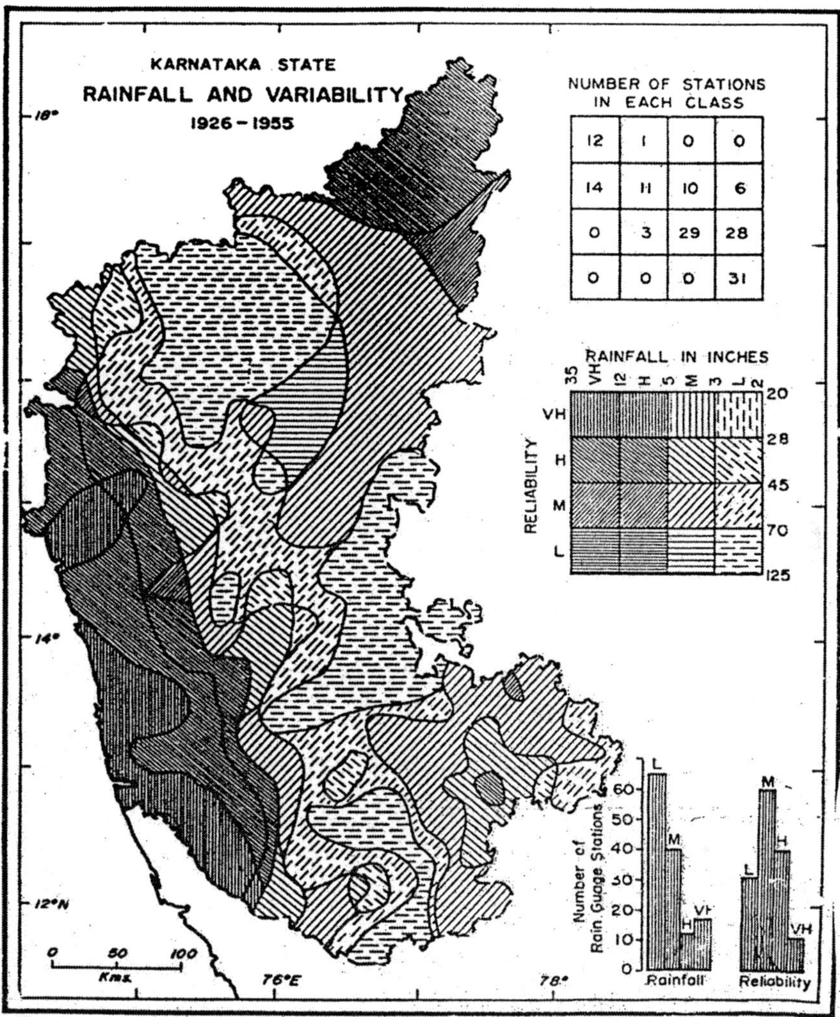

Fig. 20.2: A Complex Thematic Map showing Rainfall and Its variability in Karnataka State (after A.T.A. Learmonth)

Semi-quantitative Maps

Some cartographic techniques intermix qualitative and quantitative data. The most glaring example of this is the dot map showing the distribution of population or slope. In a dot map, each dot may represent a quantitative value but we rarely use it for quantitative interpretation. It gives only a, visual picture of population distribution; we never try to get the values represented by the dots by multiplying the number of dots by the values assigned to each dot. A dot map is different from a choropleth map. By looking into the legend of a choropleth map we can determine the density of the element represented, but a dot map does not help us to do that. Each dot stands individually and at least theoretically stands for the point at which it is placed. It represents no area and hence gives no quantitative measurement.

Quantitative Maps

Finally we come to the simple thematic maps of quantitative nature. Among such maps are the density maps which give the frequency of occurrence of a phenomenon by civic divisions such as Talukas, Districts or States. The simple way to prepare a distribution map is to colour or tint the civil divisions darker or lighter in proportion, to the density of distribution.

We can have choropleth or isopleth maps also to represent distribution. Such maps give the numerical values for different areas. We use isopleths for continuous area distribution and choropleth for discrete area distribution.

COMPLEX THEMATIC MAPS

When more than one set of equally important data are portrayed on a single map to show combination, correlation, association, or interaction, we get complex thematic maps. Almost all complex thematic maps can be called 'quantitative maps'. Such maps may show, for example, the rural and urban population by circle graphs. In a medical geography map of disease morbidity, death rate, number of hospital-beds and so on, can be shown on a single map to give a comparative picture. Most of these maps are analytical and give very specialized type of information.

Thematic Mapping involves a series of problems. Some of them are discussed below.

Choice of Map Projections

All map projections, as we have already noted, do not suit the requirements of all types of maps and data representation. Maps showing spatial distribution of a phenomenon must be drawn on equal area projection. But there is no equal area projection which shows every part of the world satisfactorily. On the margins the shapes are so distorted that the representation touches the stage of meaninglessness. With the increasing tempo of international cooperation, necessity of thematic maps with common scale, projection and symbols has not only increased but also become indispensable. Further, the variety of thematic maps that we see today is so bewildering that limited number of simple projections that we have at our disposal will not suffice. In any case, each thematic map being a unique piece in many respects, it needs a unique projection suited to it and, may be, it alone. We have, therefore, to devise new ways of deriving projections and to elaborate and develop further the already known projections to suit modern needs. Increasing use of computer in cartography helps in developing new and problem specific projections, but the possibilities of innovations are limited as there is no projection which can keep the measurements of the globe intact.

Waldo Tobler, in the USA has done considerable work in this field. He has not only given lead in the derivation of projections by feeding data in computers, but also in newer uses of projections. Projections, he says, are nothing but the means to transform a given surface into some other form. This concept, although inherent in map projection had not caught the attention of geographers. If transformation of a surface is the real purpose behind map projections, the conventions regarding the location of poles, equator and parallels and meridians have no meaning. The approach has opened up new avenues for research in this area.

Choice of Base Map

The success of a thematic map depends to a considerable extent on the quality of the base map. As the demand for thematic maps increases, the need for devising more efficient ways of preparing base maps increase. Even at present, thematic cartographers are handicapped by the non-standardized base maps available to them. Especially for the purposes of International Series of thematic maps, the quality and uniformity of base maps need further improvement.

Data and Their Representation

Whether simple or complex, a thematic map is designed to give limited information. It caters to the specialized needs of the people. The idea is to represent something which appears more prominently on the map than other things. For example, if the purpose is to show the distribution of cattle, nothing should appear on the map as prominently as the distribution of various species of cattle.

The question is: can a given phenomenon stand alone in a thematic map? How prominently should we show thing that are related to the main theme of the map. While showing the density of population, should we also show the physical built of the country, river courses, mountain chains, rainfall, etc. also? If we decide to do so, then the question arises as to whether we should have a single map or a series of maps. If we give the distribution of population by dot method and show the pertinent physical features as background, we get a simple thematic map. The population sheets of the *National Atlas of India* fall in this category. But if we venture to show the relationship between, say, rainfall and density of population then we get a complex thematic map. In the latter case the objective is to show the relationship between two phenomena and not to represent the distribution of a single phenomenon.

If we are making a simple thematic map, we have to keep the less important but not irrelevant data in the background. We know that the distribution of cattle will depend on a variety of factors: physical and cultural especially. Among the physical factors are the landforms, climate and vegetation. In a hot and wet climate cattle do not thrive well. In areas of scant vegetation and sparse settlement, one should not expect a dense cattle population. In many mountainous areas too cattle are not important. In arid areas, one finds sheep instead of cattle. How to show these facts on a map showing the distribution of cattle calls for the ingenuity and innovativeness of the cartographer.

If the purpose of the map is to show the distribution of cattle, no other data should be shown as prominently as this. We can show the landforms by light plastic shading, rivers in light blue and temperature by light colour isotherms so that a map reader interested in knowing the reasons why cattle population is thin or dense in certain areas, can speculate the possible reasons by seeing the background information.

Generalization of Data

Thematic mapping being essentially a small scale mapping, the problem of generalization is more pertinent to this aspect of cartography than to others. At present we have no standard method of generalization. In fact each cartographer uses his own intuitive judgment in this regard rendering the comparison of maps produced by different organizations difficult.

The problem of generalization acquires new complexities in case of thematic mapping, especially because thematic maps are drawn on different scales, and each scale can take only a limited amount of generalization. The time has, therefore, come to elaborate and further refine the theory of generalization so that the intuitive approach is replaced by 'standard measures'.

Standardization of Symbols

So long as each country considered itself to be an island for cartographic purposes, the problem of standardization of symbols did not arise. With increasing contacts among the nations of the world, many of the topographic symbols have acquired international standards. But not much has been done to standardize the symbols used in thematic mapping. For efficient estimation, comparison and wise use of our resources, it will help if thematic maps portraying a given information pertaining to different areas of the world use the same symbols. This is more so with respect to national maps. The International Geographical Union (IGU) has recommended a set of symbols to be used by all the member countries engaged in land use mapping. If land use maps of all countries of the world use these symbols, it will be possible for us to determine and compare the land resources of different parts of the world more precisely. Unfortunately not all the countries have accepted or implemented the recommendations of the IGU.

Compilation of Data

As the number of map users increase, and the opportunities to further refine and analyze map information improve, demand for more specialized maps will increase. The existing process of compilation is slow and cumbersome and more suited to the days when maps were used only by a privileged group. Methods which give quick and also cost effective results will have to be adopted. The mechanical process of compilation involving the use of computer and its accessories appears to show the direction in which cartographer should move for this purpose.

DESIGN OF MAPS

The data which thematic cartography now uses are highly complex. But we do not have cartographers who can use this data and make the maps simple and legible. It is the question of developing new designs. Digital cartography and computers have added new tools in the hands of cartography as a profession is no longer a popular pursuit. Earlier every geography department in India used to have a couple of cartographers. Now all of them have been promoted as assistant professors. Cartography is considered to be a non-academic discipline. Even those who are trained in the subject do not want to be called cartographers.

The down grading of cartography among the geographic disciplines has hampered the development newer designs of maps which are attractive, informative and simple at the same time.

Map Reproduction

Mass production of thematic maps is a major problem in terms of cost and time. As mentioned earlier, maps become old the day they are produced. As such newer edition have to come as often as possible. That means increase in cost. It is especially so with regard to colour maps. Now it is possible to bring out updated multi-colour maps faster than ever before. Even then, the cost factor is still a major constraint.

Inter-disciplinary Coordination

The discipline of cartography is dualistic in nature. At its core are its methods and techniques but its application is spread over a huge canvas consisting of disciplines ranging from physics and music; and activities ranging from sports to nuclear physics. Thus its use is not limited to geography; it extents to geology, pedology, biological sciences, demography, management, engineering, and even nuclear science. Cartography portrays spatial distributions, combinations, and interdependencies of various natural and social phenomena and the temporal changes therein by means of symbols it devises and innovates. It can be divided into as many branches as the disciplines it deals with. As new disciplines emerge, cartography has to evolve new techniques and symbols to handle their problems. And if thematic cartography has to continue to develop and cater to emerging needs of the newer disciplines, ways and means must be found to bring cartographers and other experts together to exchange views and learn from each other.

None should expect a cartographer to be acquainted with each discipline using illustrations and maps. His role is similar to a designer or engineer. In order that a cartographer can help the specialist, the latter must explain his needs to the cartographer. Cartographer's role is to suggest the design that caters to the needs of the specialist most.

ATLAS CARTOGRAPHY

Atlas cartography like thematic cartography, presents unique problems of cartographic representation of various natural and human phenomena. An atlas is similar to a book. Each map forms an integral and essential part of the atlas. If it is not there, the atlas would not be complete in the same way as a chapter or a section of a chapter. As the success of a book depends on how well its chapters express its theme/title, so the success of an atlas depends on the maps included and organized, and the cartographic skills used to portray the relevant data.

Some of the atlases give specialized information; but a great majority of them are of general nature. Thus we have climate, land forms, soils and vegetation atlases. Then there are atlases which are meant for schools, and general readers; they contain maps of various kinds ranging from physical geography to population. As the popularity of maps increases, the number of specialized atlases is also on the increase. An example, of such a specialized atlas is the Atlas of Disease Mortality in U.K. The general purpose atlases are only too common to need further elaboration.

The specialized atlases generally contain thematic maps of both qualitative and quantitative nature. But the general purpose maps contain locational and other information required by the general map user and some simple thematic maps. Such maps give a variety of information such as location of towns and villages, transportation and communication lines, political and administrative boundaries and physical geographic information such as types of landforms-mountains, plateaus and plains-rivers, vegetation, trade routes, etc. All this information has sometimes to be shown on the same map and sometimes on separate maps.

Depending on the nature and amount of the data to be represented, atlas maps can be drawn on a variety of scales. If the objective is to give very detailed information, it is desirable to divide the area to be represented—world, a country, or a part of a country—into convenient number of sheets. And each sheet will form a page of the Atlas. But in case the data are not so detailed, and the atlas is to give only generalized information, the whole area can be represented in a single sheet, and each sheet in turn may represent part of the total information to be represented. Thus some maps may show physical data while others may represent cultural data.

The atlas maps which try to give several pieces of information on a single map are not easy to make. Mapping this type of data is considered to be an aspect of complex cartography. Complex cartography presents problems which are in some respects different from those presented by thematic cartography. In general purpose complex cartography each piece of information has to be so balanced with others that it is legible and easily distinguishable yet it appears as an integral part of the whole map. There is nothing which is suppressed in order to highlight something which is considered to constitute the theme of the map. In complex cartography more than one information has to be brought together to give a simple map.

Atlas mapping comes within the purview of small scale cartography. The data shown in atlas maps are invariably generalized. Generalization and symbolization are therefore two of the most crucial processes in atlas mapping.

The design, contents and the size of the maps vary with shape of the area and the amount of data to be represented and the market it is to cater. The design of the atlas for the children as well as the maps included in it will be different from those meant for adults. Atlas for the children is marked with a simple design. More specialized an atlas, more complex is likely to be the design. Atlas mapping is another aspect of cartography which needs further research.

21
Constructing Maps for Children, Visually Challenged and Neo-literates

As perceptual abilities of different persons differ according to age, physical abilities, and educational background, a cartographer has to apply different designs and techniques to satisfy the specific needs of each group. Most of the map users are, however, educated adults whose perceptual abilities are considered to be 'normal'; a great majority of the maps are, therefore, made to satisfy the needs of 'normal' persons. But there are also persons among us who perceptual abilities are either not yet developed or they are impaired because of certain physical disabilities. Important among them are children, visually impaired and neo-literates. Children not only form a sizable number, they are the makers of the future. They find the maps made for educated adults too complex to understand. They need maps which are simple, clear, informative and beautiful to look at.

The visually disabled people cannot see. They lack the faculty of visual perception. They lack the faculty for visual perception. So they have to have maps which are very simple and generalized. They can feel an image, if drawn in relief, with the help of fingers but they cannot see it. This being the case, a cartographer not only has to design a map which can be easily understood by touch method. Even the lettering has to be in raised relief (Braille) to enable them to read by touch.

Similarly, being illiterate during the formative periods of their life, the neo-literates are aware of things of only local importance. They know their trade in minute details, although the methods used may be traditional. They are, however, better equipped to understand things beyond their experience if presented in a familiar form.

We can think of several other groups which require special attention from cartographers, but in this chapter we will confine ourselves to only children, neo-literates, and the blinds.

MAPS FOR CHILDREN

Children are the future map users. If they become interested in maps in their childhood, they will continue to use maps in their adulthood also. There are many educated adults who do not use maps. It is not difficult to find a book written about country and its people without a single map. Books with crude and repulsive maps are only too many to need emphasis here. Some of these maps give wrong information and mislead people. In order that we get successive generations of people interested in map as a source of information and as a visual means of communication we must expose our children to good maps from an early age. In other words, we should pay special attention to designing, and producing maps for the children.

Preparation of maps for the children involves the understanding of child psychology and the process of learning. It also involves the knowledge of the needs of the heterogeneous group called children. A four years old child surely needs different maps than an eight year old child. To write about the types of maps, which will suit the children in different age-groups is beyond the scope of this chapter. What is attempted here is the setting out of a few general principles which can be applied to develop maps for various age-groups of children.

A child has only a limited experience with the help of which he interprets the world he sees around him. The day he is born, he starts sensing the world through various organs. Mother is the first teacher of a child. Gradually he develops motives towards the satisfaction of which he comes in contact with the things and events. The motivated organism senses its world, interprets it, responds to it, and then responds to the consequences of its own responses. Once the organism has passed through this cycle, it is never the same again. The first behaviour of the child has consequences in the form of resultant comfort, relief or satisfaction. These consequences of its own behaviour are registered and stored and becomes a part of the organism. It has gained new experience. It is never the same again. It continues to register the consequences of its own behaviour and thereby embarks on the spiraling and dynamic process of living, a process in which growth and change, over a period of time and with experience, play a central role.

The map is a means to provide a child with the familiar as well as unfamiliar experiences about the world in which he lives and grows. It is also the means to change his attitudes about the world around and to perceive it in its true perspective i.e., interrelatedness. It can also be a means to change motivations and thinking. As a matter of principle, a child should learn to read and draw maps as early as possible. It cannot, however be done before he reaches the age of six or seven. It must, however, be kept in mind that a child of six or seven cannot realize the extent to which the representations on maps are only symbolic. And, therefore, he cannot be expected to form an adequate conception of the relations which they bear to reality. If he is given a complex and 'meaningless' map, his mind will tend to fix and relate impressions which are grotesque and untrue and which by their very clearness in his mind, as well as because they are being fixed so early in life, will be difficult, if not impossible to correct in later stages. It is unfortunate indeed that, in the absence of suitable maps for the children, we make them use the maps made for adults and thus force them to develop wrong ideas about maps and so also about geography.

Up to seven years of age a child learns though his personal observations alone. He is curious to know everything he sees. He does not, however, concentrate on anyone object; he likes to change his interests frequently. One thing is clear that by this time the character of the child is formed. Whether, he should lie or be honest is ingrained in his and what is ingrained in his mind is received from the parents, family members and neighbours. By the time he is nine years old, the power of observation and muscular skills increase, and he develops a better mechanical memory and an equally good receptive mind. He uses his newly acquired skills and intelligence to sort and analyze the impressions he gathered earlier indiscriminately. He is able to understand some simple explanations and relationships. Maps for a child of seven or below should represent only the home and school surroundings to which he is familiar. These maps should be attractive, colourful and simple in design.

Mapping the class room is a time-honoured method that still has much relevance. Class room is chosen because the whole area to be mapped is in sight, and is equally familiar to all. Problems in making such a plan are those connected with the representation of space relationship of one object to another. Relative size (horizontality), distance and direction of all important pieces of furniture have to be considered. Regional distinction showing a distributional pattern can also be made when part of the room is filled with desks is differentiated from that which is clear. The class room plan is indeed a map. It must be oriented with north at the top.

A nine or ten year old child should be initiated to the plan of his neighbourhood. His first step in the use of a neighbourhood map is to go out, like a geographer, with the map in his hand, and relate it to the area it represents. This procedure will help him discover how truly and by what means the maps show facts that he has come to know and perhaps reveal some of those he did not know. The procedure can be reversed also. The children may be taken out with a sheet of paper and stopped at important points to draw the plan of the routes and features visited to them.

Maps of local areas and neighbourhood on 1:2500 to 1:10,000 provide a child with a fascinating occupation. He enjoys walking over the area and trying to identify what he sees with symbols on the map. Even 8 year old gets excited if something on the landscape, a new house or a pond, does not appear on the map. Gradually such a map can be reconstructed to depict simple differences in the distribution of things such as agricultural area, school area, and so on. Emphasis should be laid on the representation of the phenomena which are familiar to the boys and girls of this age with familiar symbols. For example, the distribution of animals can be shown by locating recognizable figures of animals on the map (Fig. 21.1). Children should also be introduced to the maps of the world.

A child of 11 or 12 years has a better critical sense to sense the conflicts between the realities of the world and the 'theories', 'concepts' and experiences he acquired earlier. The period between 12 and 15 years is the period of adolescence when a child tries to discard what is appropriate to childhood but is unable to be a part of the world of adults which is too complex to understand. This is the age of correlation and classification. Maps for the children of this age-group should show the interrelationships underlying the various geographic phenomena-natural as well as

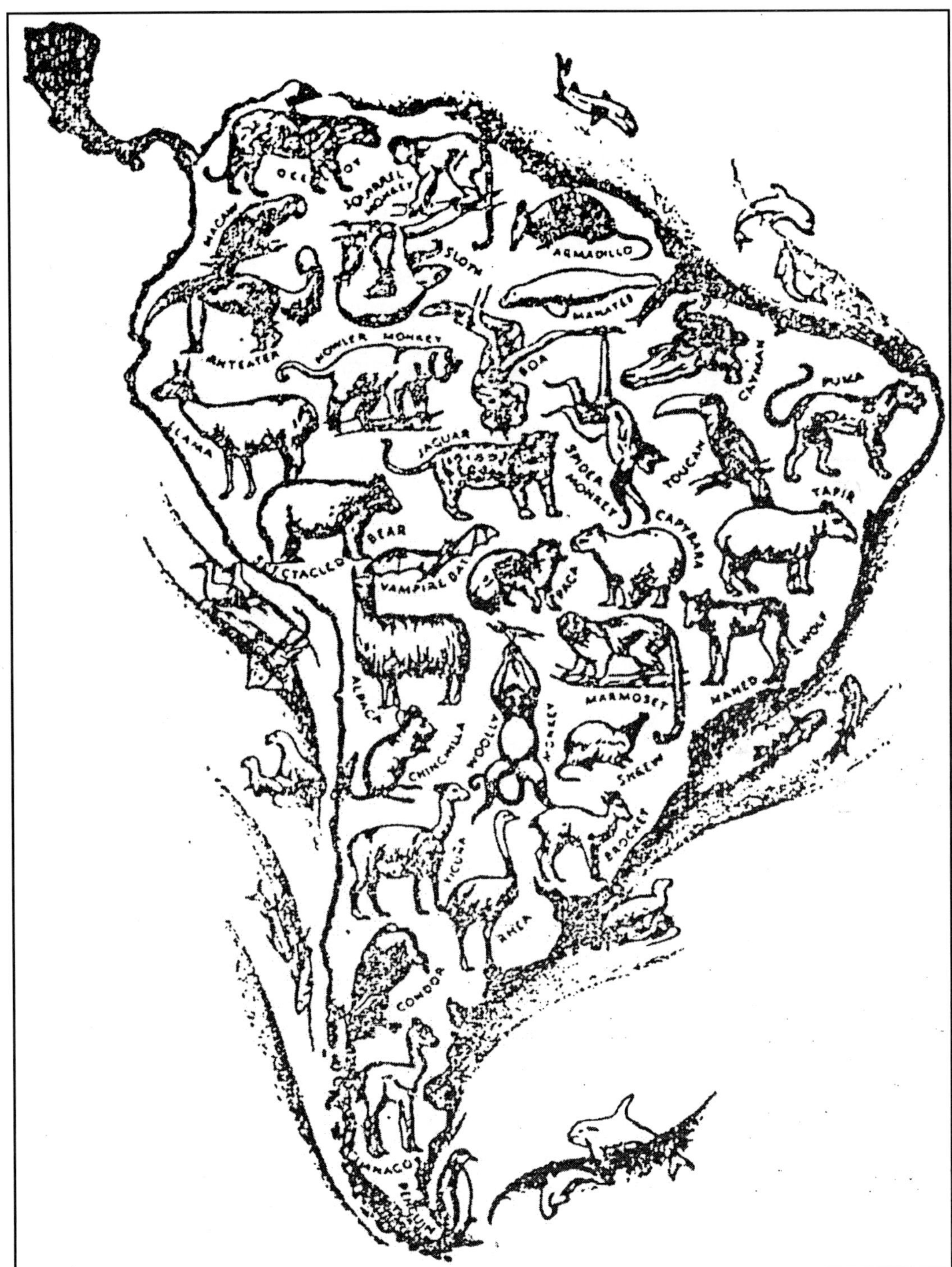

Fig. 21.1: A Pictorial Map for the Children (South America)

cultural. This can be done by relating the home area to the district, state and the country.

The use of the globes and world maps should go together. Only broad details should be shown on the world maps to be used by the children of thirteen years or less. For children above thirteen years, the symbols used in the maps need not have direct relationship with the shape of the phenomena represented. Various abstract symbols like dot, circles and lines can be used to depict the data.

After the age of 15, children use reason more often to understand the world and their own place in it. This is the period of generalization. Knowledge is classified and discarded or synthesized into workable instruments according to its usefulness in every day life. It is at this stage that complex maps made for educated adults should be given to them.

Details on the maps for the children should be generalized and the portrayal should not be devoid of aesthetic beauty, colour and attraction. The first map used by a child should be one which he has himself drawn and the teacher should continue to provide opportunities for the child to develop his skills in the making and reading of maps.

In conclusion, we can say that we are not yet sure about the contents and design of maps for the children. We adults appear to underestimate the capabilities of a child to comprehend and understand the world. Considerable research is needed before we can arrive at any satisfactory conclusion.

MAPS FOR THE VISUALLY HANDICAPPED

For the purpose of this chapter we will not make any distinction between a blind and a partially blind person. This is a group of people which lacks one of the most important senses of perception. Our eyes help us recognize the shape, size and colour of various objects. They help us know the position of one object in relation to the other. They also help us visualize the facts of the earth through the medium of maps.

When this important sense that we call 'sight' is not functioning properly, we lack the faculty to know the relative shape, size, colour and position of different things visually. What we, then, need is some other mechanism to perform this vital function. In other words we need to train some other senses to do the job that our eyes do, at least partially.

The part of our body that does this job best is the hand. Our hands with five fingers in each, can feel the shape and size of various objects and they, with the help of feet, can feel their relative position also if the objects are located within a walking distance. This naturally limits the extent and number of things a blind can know to expand his knowledge. We have to devise a method by which he will also be able to comprehend the shape, size and relative position of objects which cannot be reached by hand or feet. In other words, we have to prepare maps for the blind. Such maps must show the symbols in raised relief so that the blind can feel them with his fingers.

As our maps also show symbols by letters we should know something about the letters devised for the blind. The system of these letters is called the 'Braille system'.

Braille was developed by Barbeir, an officer in Nepoleon's army who called his system 'Night Writing'. He was seeking a means of sending messages which could be read by touch at night without light by the men in forward positions. Braille has since been modified and greatly improved. It is based on the positioning of dots within a cell in raised relief. There are six dots in all for each Braille cell. When reproduced in various combinations they form all the letters of the alphabet, Indian numerals, and 63 different symbols including some dipthongs and other combinations of letters, punctuation marks, capitals and commonly used abbreviations.

Maps for the blind are almost as old as the Braille but the amount of effort expanded to improve and standardize the Braille has been far greater than that devoted to graphics. The demand has been and is still very great. The latest estimate is that there are as many as 6 million legally blind persons in India alone. Of these only a few attend schools for the blind. The highly developed and standardized Braille fills the requirements for embossed verbal communication but the means available for embossed graphic communication are still relatively crude.

The construction of maps for the blind involves two basic problems: (1) the selection of symbols which the blind can easily read and associate with reality, and (2) the development of a cheap and efficient reproduction process. To devise suitable symbols for the blind is full of difficulties. In the case of the maps for the sighted people, we develop symbols which can be easily distinguished. We use standard psychological tests to judge the suitability of these symbols. But these tests, in the case of symbols selected for maps for the blind, present unique problems. More often than not, the persons tested have never seen a map before and if the person administering the test has no training or experience in graphics, the test results are likely to be highly suspicions.

Symbolization is a very complex process. The drawings and their designs must be kept simple and the symbols should be distinct from each other so that, while being read by touch, they do not get mixed up. Pictorial symbols which are so useful for the sighted people are of no use for the blind because they convey meaning only if one has seen what has been depicted. The tactual symbols are easily confused by the blind. The symbol X, for example, is confused as a square though visually the two are quite distinct. Every symbol used must be explained and if the symbol is rotated this must also be carefully spelled out. Tests conducted at the Virginia State School for the Deaf and Blind show that the symbols indicated by asterisk (in Fig. 21.2) are tactually the most distinct ones. In particular, symbols number 11 and 23 proved to be clear, distinct and readily understandable.

The problems involved in symbolization are great indeed, but the problem of reproducing multiple copies of raised relief maps is still greater. Several experiments point to the use of a Thermocraft process which can give a raised relief of 0.007 inch. With research and experimentation, it can further be raised which would facilitate quick reading. The maps must be printed on glossy paper because rough paper interferes with the reading of fine symbols. The more practical method at present seems to be the Thermoform process which is slower than the above mentioned Thermocraft process but is more uniform in quality and more readily available. The Thermoform copy is also a plastic material, hence more durable.

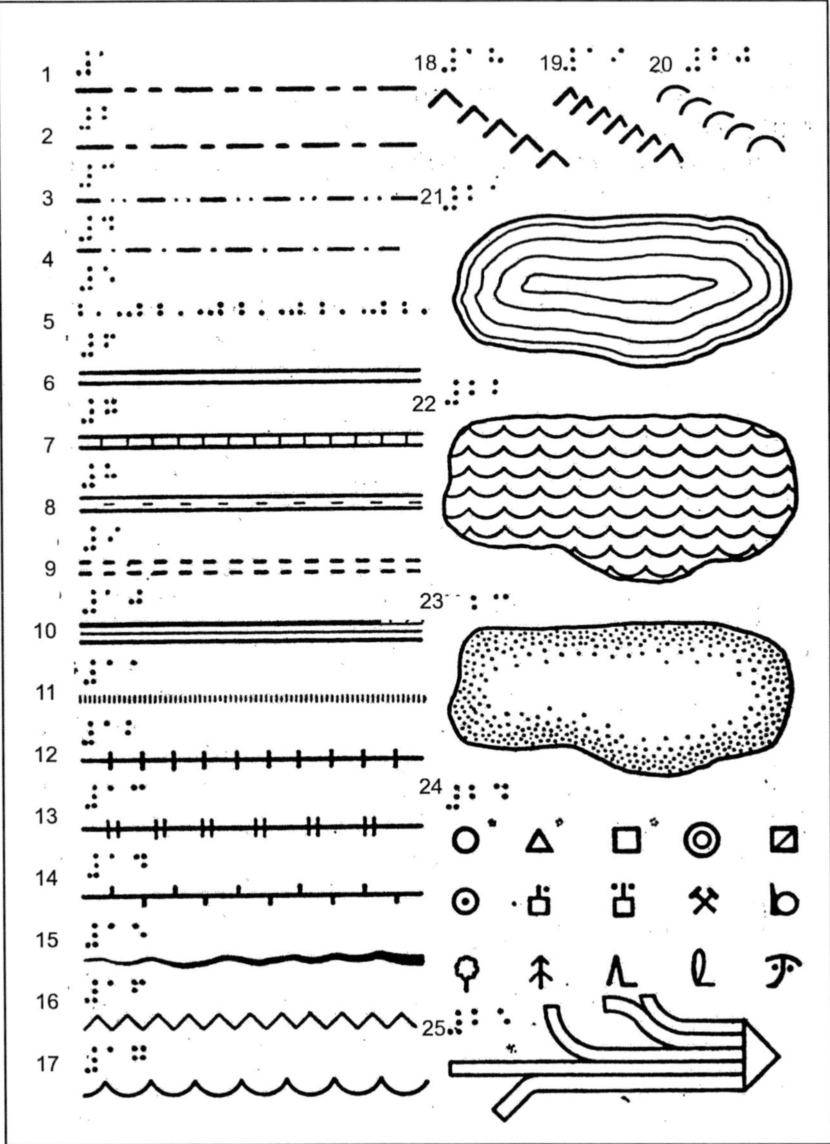

Fig. 21.2: Symbols for Maps for the Blind

Perhaps the foremost difficulty in designing maps for the blind lies in lettering. The size of the Braille letters cannot be reduced since it is set as a standard by the International Braille Authority. The existing size of Braille letters takes too much space, making lettering on maps very difficult. One must, therefore, design various components of the map around the lettering rather than the usual method of selecting lettering to fit a given map design. The space between the Braille cells and the line-work is critical and though it does not bother some tactual readers to turn the page if the lettering is placed vertically or diagonally, for others it causes confusion. The second grade Braille is more convenient in the design of graphics since it uses a number of contractions and is known to most Braille readers. Fig. 21.3 gives a map for the blind in which Braille letters are reduced by 50 per cent.

It should also be kept in mind that nothing more than what is absolutely necessary should be included in a map for the blind. There is no need of borders and neat lines; the title should be at the top and be followed by necessary explanations and keys. The symbols should be kept to a minimum, clearly explained in the key and constructed with as much contrast as possible. Second grade Braille should be used horizontally as far as possible. If the maps are to be used by partially sighted persons also, large type can be used in conjunction with the Braille.

Lastly it must be mentioned that considerable research is needed to develop suitable designs for maps for the blind. At present there are less than half-a-dozen cartographers engaged in this type of work. Perhaps no other field of cartography needs creative and welfare minded cartographers more than this.

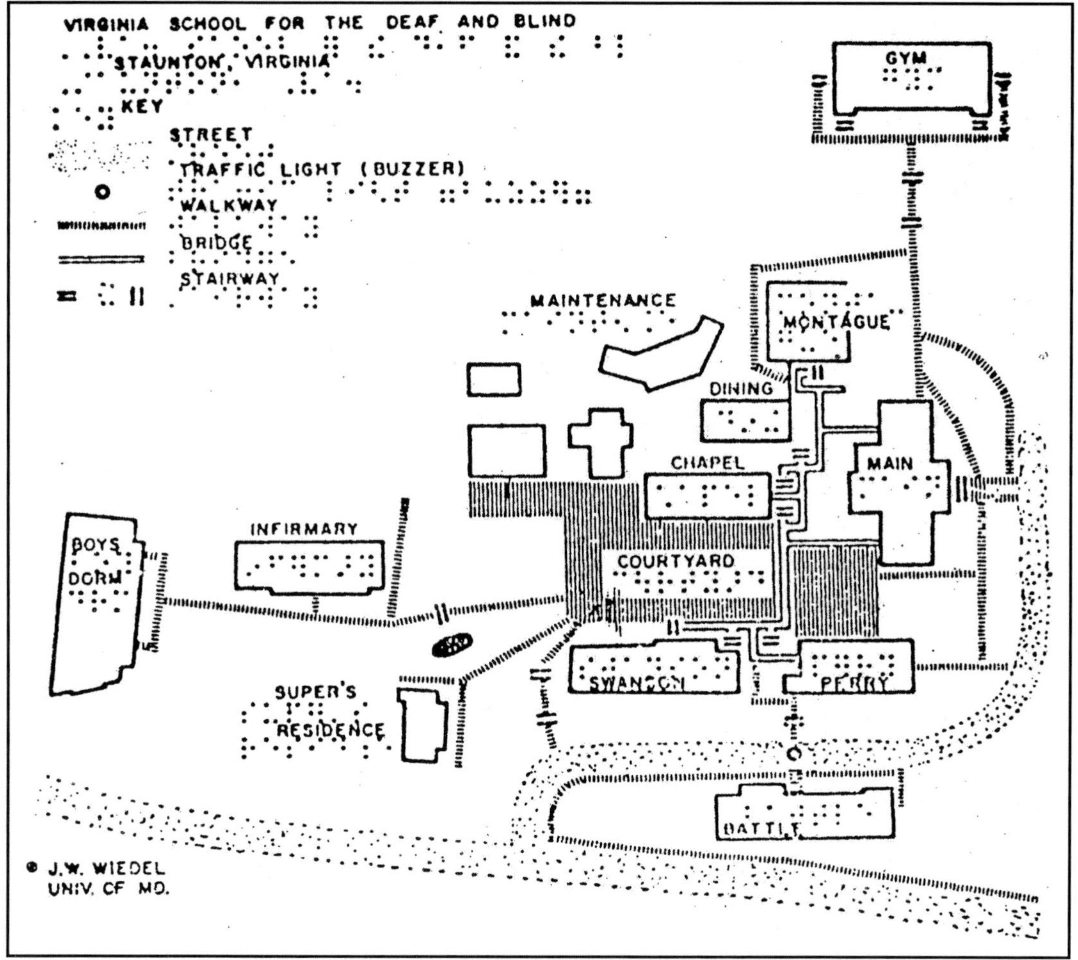

Fig. 21.3: A Map for the Blind

The American Telephone and Telegraph Company (AT&T) has introduced an easy to read Braille AT&T Calling Card that offers several new features and access to a number of services it offers through its Worldwide Intelligent Network. The users are given a permanent AT&T Braille Calling Card numbers which they can use even if they are moving to place local, long-distance, and international calls from any phone, without using change. There is yet another facility available to the visually challenged. They may secure the Braille AT&T Custom Calling (TM) Card, which enables them to call up to ten telephone numbers and thus they can reach ten most important people in their lives anytime, from almost anywhere, without the need for change.

In the USA, the National Library Service for the Blind and Physically Handicapped (NLS), Library of Congress, Washington D.C. 20542 distributes free copies of "Assistive Technology: A Selective Bibliography." It cites books, articles, and pamphlets, published since 1985. This bibliography, as well as other bibliographies and Reference Circulars, are available free on request from: Reference Section of the NLS.

Among other services that the NLS offers to the visually handicapped include New Tactile Maps. My colleague Joseph Wiedel, at the Department of Geography, University of Maryland designed the embossed maps for NLS patrons on two regions. When the Gulf War started, the blind too wanted to know where the Suez Canal was and what was the location of the countries at war? Joe Wiedel prepared a tactile map the Middle East within no time. The map of the emergent independent states that once formed the Soviet Union was similarly produced to meet the requirements of patrons of the NLS. NLS is trying to offer tactile maps of other parts of the world too.

MAPS FOR THE NEO-LITERATES

Neo-literates are those which acquire literacy in their adulthood. For all practical purposes they are 'normal' adults except that they do not have as much experience with the world outside as others have. An illiterate man might have heard of the world being round but he has no faith in the idea for he fails to understand the logic behind it. He neither has the opportunity to go round the world, nor is he able to read the accounts of such voyages. Similarly, he might have heard of wheal being produced in certain parts of India but he does not know the location of these areas. Maps for the neo-literates, then, should concentrate more on locational and distributional details.

Unlike a child, a neo-literate is more interested in locating the known facts about his neighbourhood, the country and the world. He is interested in knowing about the location of things he deals with in his day-to-day life. He is concerned about the administrative divisions of the country and state; he is interested in knowing the international boundaries; and he is curious to know about the distribution of human beings, animals, crops, towns, sea and land routes, etc.

As a neo-literate is new to maps, he is not able to interlink various details given in it if they are given in an unfamiliar way. For example, the density of population shown by patterns is less comprehensible than one shown by pictograms or bar diagrams. Similarly, the topography shown by contours is confusing but one shown by layer tinting is relatively more easily perceptible. The map in which a neo-literate is most interested is that which deals with his occupation or hobby. The representation of details in such maps should be of moderate complexity. Before he can make the best use of a map he must be given proper concept of map distance. Illiterate persons tend to equate actual distance with the distance given on maps. So if two known points 100 miles apart are separated by 2 inches on a small scale map and by 4 inches on a large scale map, he will tend to value the latter to be 200 miles because he does not know that maps are drawn on different scales. Letterings on maps for neo-literates should be big and unstylish. Pictures also play an important role in getting ideas across the neo-literates when given along with maps.

22 Map Reproduction, Cataloguing and Marketing

Printing has undergone many changes over time but the need for people living at distant places to communicate has not changed. It is a process to produce many identical copies from one master image which may range from an inscription engraved in stone to an image cut into a wood block or a text stored as digital information in a computer. The agent that enables the master copy to transfer its own image is the printing press and the medium that imprints the images is the ink. Even today, block printers print images cut in wood on bed sheets. The principle is the same; only the mechanics has changed. We have now advanced from manual printing using ink to office copiers which reproduce master images using electro-statically charged graphite toner.

The first known attempt at copying images appears to be the Phoenician stone tablets. The Chinese in 8th century AD started the first rudimentary reproduction of textual matter, when they experimented with the printing of relief, or raised, images cut in wood blocks. In India textual materials were engraved on wall of temples, etc. but not attempt was made to reproduce them in multiple copies. By the 11th century movable types made from clay and wood and, later, from bronze and iron were manufactured by the Chinese and the Koreans. Progress was slow because the Chinese characters are not alphabets; they are transformations of various images, ideas, and actions. Yet, the notion of movable type was a major advance in printing.

Four hundred years later in fifteenth century, Johanan Gutenberg, a German used the same Chinese principles and worked out a method of casting type and printing so successful that its fundamental principles were not improved until well into the 19th century. Gutenberg's invention consisted of lead-alloy type cast in an adjustable mold fitted in a wooden printing press which moved up or down against paper bed over which were laid the type. The ink was oil-based so that it could stick and also sock the paper. Gutenberg's first book, a *Latin Bible*, was completed about the year 1455.

It was around 1450 A.D. that either Johanan Gutenburg (1397-1468) or Laurens

Janszoon Cozter (1370-?) invented movable types which could be used again and again. Printing technology, however, continued to be based on letterpress principle i.e. the image to be printed was projected above non-printing areas. There were several important consequences of this development, chief among which was the explosive cultural effects of the wide dissemination of textual and graphic material. In fact, the new printing influenced the characteristics of lettering and map delineation too.

In 1796, Alois Senefelder invented flat-plane or planographic printing process later called 'lithography'. He found that even a detailed image drawn with a greasy substance on the face of a water-absorbent stone and then inked could be printed onto paper without any distortion. It was ideally suited for illustration and hence became highly popular for colour printing, which required a separate stone to print each colour. Later on, the image on the stone was transferred to a metal plate that was used for the actual printing.

Subsequent improvements in Gutenberg's screw press were aimed at improving the clarity of the printed image by increasing impression power, and developing mechanisms for the press handle to return smoothly and regularly. Until then, in India copies of books, etc. were hand written and a large number of people were employed to do this work. The next major development took place in 1800, when Charles, 3rd Earl Stanhope, developed an all-metal press. In 1813, George Clymer replaced the screw with power-multiplying levers. During 19th-century further improvements in iron hand press continued but no major innovation took place. The fundamentals remained the same. As such small job printers continued to use the platen press, invented in the early 19th century, in which the flatbed was vertically positioned and power was supplied by a foot treadle or by steam.

The first flatted cylinder press that used a revolving cylinder instead of a flat plate to press sheets of paper against a flatbed of letters (types) was patented by Friedrich Koenig in 1811. As the bed moved under inking rollers, it took fresh ink before the cylinder rolled over to cast the next impression. The *Times of London* installed a steam-powered Koenig press which could print over 1,000 sheets per hour. The speed increased when in 1844, Richard Hoe an American invented a rotary press. Hoe replaced the flatbed by metal type to the surface of a cylinder. In 1835, Rowland Hill of England patented a rotary press that printed a continuous reel of paper. It was followed in 1837 by the first operating web-fed rotary press in the United States.

The next major development in printing took place in 1870s when curved relief printing plates used in the rotary press were replaced by curved stereotype plates. Since then until the middle of the 20th century, the automatic rotary cylinder press which printed on both sides of a continuous web of paper run by steam power and then by electric power became the press of the choice. Te restriction on artists, however, continued to be many, although less stringent than before until the middle of the last century. No matter which style and form of lettering was deemed desirable and what lines, symbols or colours might best convey the concept of a map, only those available to the printer and engraver could be used. Because of the overwhelming role of the technicians in the printing process their impact on the design of maps was widely felt. The cartographic products of this period were artistic but portrayed repeatedly the same design and layout.

For colour printing, lithographic metal plates had only rarely been used for commercial printing, because the image on the plate was often worn out by the printing paper an American printer Ira S. Rubel. In 1904 accidentally discovered that the lithographic image could be transferred, or offset, to a rubber cylinder that could then print as perfectly as the plate and would last indefinitely. Rubel's three-cylinder offset press was the first in the field of offset lithography, which became the most popular printing process because of its economy, long plate life, and ability to print on many different textures.

The currently used halftone colour printing to reproduce full colour, was introduced in the 1890s. Although colour reproduction theory was fairly well understood, the lack of colour film restricted colour work to studios where the necessary separation negatives had to be made directly from the subject, under the most exacting conditions. Reliable colour film became available in the 1930s and 1940s, and colour reproduction grew both more common and more accurate.

As the printing process improved in 19th century, serious attempts were made to mechanize the type-making (casting) and composition (typesetting). The linotype machine invented by Ottmar Mergenthaler in 1887 and the monotype invented by Tolbert Lanston the same year, proved to be excellent mechanical systems for letterpress composition. The Linotype was a keyboard-operated machine that composed and was particularly suitable for newspapers. The Monotype's keyboard produced a punched tape that instructed a separate type-caster to produce individual characters in complete lines. It was used largely for book printing.

In 1954 the Photon machine became the first commercially successful electronic photo-composition system. Its key elements, which were used by later machines as well, were a stroboscopic light source and a spinning film matrix disk through which photographic film was exposed with images of type previously composed on a keyboard.

The technological developments which took place during the late nineteenth and during this century totally eliminated earlier limitations. To day there is theoretically no reproduction constraint in the demonstration of cartographic genius. Almost any design, detail and colour can be reproduced. Among the several reproduction possibilities, there are a few which are very inexpensive but do not suit the requirements of artistic details. There are others which are expensive but give copies which not only match the original but also improve it. Some of them give quick results and are within the reach of individual cartographers. Some of them can produce only line work; some others can produce line works as well as continuous tones. Which of these processes a cartographer should use will depend upon his resources, drawing, and demand on time? In this chapter we plan to provide a brief summary of the processes with which a cartographer is likely to be concerned in his capacity of print buyer.

The most recent development in printing technology is the use of computers. They now play a vital role in nearly every area of printing, from typesetting to on-press control of the many variables subject to change during a print run. When a matter is ready for printing, a central production facility electronically transmits the entire contents to printing plant, speeding up the whole process of publication and distribution. It is called Computer to Plate Technology (CPT). In this technology, an image created in a Desktop Publishing application is output directly to a printing plate. This compares well with the older technology, computer to film where the computer file is output onto a photographic film. This film is then used to make a printing plate.

The advantages of Computer to plate are the elimination of the need for film and related developer chemicals, thus increasing sharpness and detail. Plates are produced in less time, are more consistent and at a lower cost, and can significantly increase plate production outputs. High-volume litho print is used today to produce just about any smooth, mass-produced item with print and graphics on it. In offset litho printing depends on photographic processes, flexible aluminum, polyester, mylar or paper printing plates are used.

This powerful system can now provide the vast storage required for very high-resolution graphics, as well as providing methods for sophisticated image manipulation. The operator of a typical system can scan a colour photograph into the computer, and then call the image up to a display screen where a number of editing processes can be employed: rotation of the image, increased shading, colour correction or colour changing, the moving of parts of the image or its entire deletion. The final, edited image is sent to an output laser scanner, which produces a set of colour film separations that are used to make the printing plates.

CATALOGUING OF MAPS

Map construction and publication is a costly affair. It is specially so with regard to colour maps. That is why maps are costly and not within the reach of every one. Individuals cannot afford to have many of them. It is only the libraries which can collect a large number of maps. It is, however, easy to purchase a large number of maps and store them in drawers of libraries, but quite different to so classify them that people wanting to have a look at them can lay their hands on them easily and promptly. Some of the relevant aspects of map classification are discussed here.

Maps are more diversified in nature than books. The system of cataloguing and classifying them is, therefore, different from that of books. We can classify maps on the basis of scale, subject matter, use and so on. So long as the number of maps is limited, it is easy to locate them. But as the number increases, the map librarian is unable to remember each map individually. He has to classify them so that he can easily lay his hands on the shelf where a particular map is kept by looking into the map catalogue.

Catalogues should incorporate the following details:

1. Person or organization who owns it;
2. Cartographer, who constructed it;
3. Name of the publisher;
4. Name of the copyright holder;
5. Title of the map;

6. Scale of the map;
7. Edition of the map;
8. Size of the map; and
9. Place and date of publication.

According to D.M. Norris, all the above details under nine heads are not needed for map catalogues. Only four details are needed. These are:

1. Place delineated,
2. Subject, that is what the map is intended to show, as canals, railways, physical features, coal fields, industrial location, etc.,
3. Date, and
4. Author

> "Working on this principle of the place being of the first importance. The British Museum makes its main entry under place, and provides only supplementary entries under the names of cartographers, surveyors and engravers. The Library of Congress follows the same method giving only cross references under the names of cartographers, etc." —*D.M. Norris*

The question then arises as to which is the best method of classifying maps. For an answer to this we have to see whether the system used for classification of books can also be applied to maps. In the classification of books there are two prevalent systems:

1. Decimal system, and
2. Colon system.

In both the systems books are classified according to the subject they pertain to. So that geography books are numbered from 910 to 913 in the decimal system and under U in the colon system. Small libraries classify maps under geography, and this classification serves the purpose because, at least in India, the number of maps the libraries possess is small. Even the National Library at Kolkata has only 86,000 maps. This compares very unfavourably with Library of Congress collection of 5.5 million maps. The Geography and Map Division (G&M) of Library of Congress has custody of the largest and most comprehensive cartographic collection in the world with collections numbering over 5.5 million maps, 80,000 atlases, 6,000 reference works, over 500 globes and globe gores, 3,000 raised relief models, and a large number of cartographic materials in other formats, including over 19,000 CDs/DVDs. The online Map Collections represents only a small fraction that has been converted into digital form. These images were created from maps and atlases and, in general, are restricted to items that are in public domain, meaning those which are not covered by copyright.

But we know that people in India are now becoming more map minded than before and that libraries in universities and other institutions will have to increase their map collections to cater to the growing demand from the readers. That being the case, it is desirable to follow a rational system of classification. It certainly is not the purpose of this chapter to give a blue print in this regard. It is only intended to suggest a method which might suit the Indian conditions.

The first classification should be based on the size and use of maps.

1. Atlases.
2. Wall maps.
3. Topographical sheets.
4. Weather, aeronautical and navigational charts.
5. Other maps.

Each of these can be further sub-divided on the basis of area or themes. For example, the atlases could be classified as:

1. General (World)
   (a) Cultural Landscapes
   (b) Military Battles and Campaigns, and
   (c) Transportation and Communication
   (d) Cities and Towns
   (e) Conservation and Environment
   (f) Discovery and Exploration

2. Thematic

Similarly, the wall maps and topographical sheets should also be sub-divided. In case of the latter, however, the scale will also have to be taken into account. For example, the Survey of India sheets are on the scales of 1" = 1 mile, 2 miles, 4 miles, and 6 miles (1:1000, 000). It will be desirable: to keep these sheets separately scale wise. For each country there should be 4 or 5 shelves set vertically so that this separation is done systematically.

As the libraries in India will have more of the Indian topographical sheets, as they are in great demand, it will be desirable not only to classify them scale wise but also according to numbers the sheets bear.

Among the 'other maps' are a large variety of specialized maps. Some of these are multi-purpose maps but a great majority of them are thematic. It would be desirable to classify the thematic maps on the basis of their information content. There should be physical, economic, political, cultural and other classes and each of them must have the sub-classes. For example, the economic maps may have sub-classes like land use, crop distribution, mineral resources, industrial, transport, trade etc., as sub-classes.

STORING OF MAPS

Classification is only a part of the problem of map collection. Before acquiring maps from various sources, the librarian must give thought to their upkeep. Storing maps is surely more difficult than storing books. As sizes of maps vary from 2" 2" (slide) to 10"×10" special arrangements have to be made for the storage of each size of map. It must also be kept in view that except for the atlases all other maps are in loose sheets. Some of these, like the ones we use as wall maps, are cloth mounted but a vast majority of them are loose sheets. The paper on which they are printed is usually of good quality, yet within it becomes brittle after a few years. Moreover, the map users cannot handle large sheets as carefully as they handle books. As a result, life of a map is relatively short. Cloth mounting them would help but the number of maps is so large that it would not be a practical proposition. Only very costly and frequently used maps can therefore be cloth mounted.

Map mounting is a very simple operation. First step in this process is to wax the surface of mounting table with a good grade of paste wax. Put the required size of muslin impregnated with a paraffin type of stiffener called chartex on the waxed table. And then put the map sheet on this muslin and press the map with warm iron. This is called dry map mounting. Now there are mechanical ways of mounting maps. Libraries which have large collection of maps can get their maps mounted mechanically.

Once a map is mounted, it should be put on racks or in cabinets. The type of racks or cabinets which can be used for storing maps will depend upon the size of the maps. If possible such racks should be fitted with wheels so that they can be moved from place to place without noise or difficulty. For topographical sheets, which come in large numbers, we need filing cabinets. On the face of each drawer should be written the details of the maps or their numbers. For keeping original drawings or similar cartographic materials we need draw-index cabinets.

MARKETING OF MAPS

From the cartographic point of view the marketing of maps has much to do with the types of maps which are produced in a country. Maps being the vehicles of conveying certain information, they must cater to the felt needs of the people. It means that the type and quality of maps published will largely depend upon the inter-communication between the map maker and the map users.

Although it is true that the finished maps are the results of the mutual interests of map makers and map users more often than not the two do not see eye to eye. There are several reasons for this, a few of which are given below:

1. A map maker wants to give information in as clear and aesthetic way as possible. The map user also wants to get information clearly and aesthetically but he wants everything he needs in one map. This is a cartographic impossibility.
2. A map maker's mind works toward the attainment of excellence irrespective of the cost, but a map user wants inexpensive maps.
3. Even a simple but well designed map takes a lot of time to construct, but from a map user's point of view, a line is just a line. He does not realize the effort that goes in the compilation and analysis of the data and in designing a good map.

As a result of the above conflicts, map makers are rarely able to satisfy the market demand. This situation keeps cartography ever alive yet at the same time less popular than what it can potentially be. There is need of greater interaction between the map maker and the map user. The market can bring the two together. The map user must understand the limitations of the map maker and the latter, of the former. The book sellers and salesmen can play a very catalytic role in this regard. The job of a salesman is not only to inform people about the maps but also about how a given map can satisfy their felt needs. At times the felt needs of the customers remain unexpressed because they do not ask for maps which do not exist. A customer may like to see a map of the mars, but he may not demand it because he knows that such a map does not exist. It is also the job of a salesman to create awareness about new cartographic products and to inform the publishers and cartographers the nature of maps the users want. Awareness about the cartographic products generates demands, and demands create new cartographic products. This is how cartographer serves the people and the people support cartography.

REFERENCES

Akers, Robert C. (1980), 'Cleaning and restoration of maps,' in *Map Collector,* 10 (March, 1980): 19-22.

Association of Canadian Map Libraries, (1979), *Map Collections in Canada and Conservation: a Report Based on Responses to a Questionnaire Distributed by the Conservation Committee,* 1976, Ottawa: The Association, 50, 63 p.

Banks, Paul N. (1976), 'The Conservation of Maps and Atlases,' in *The 1976 Bookman's Yearbook,* p. 53-62, Clifton, N.J.: AB Bookman Publications.

Bartlett, James and Douglas Marshall (1979), *Maps in the Small Historical Society: Care and Cataloging,* Nashville, Tenn.: American Association for State and Local History, 12 p.

Brown, Lloyd Arnold (1941), *Notes on the Care and Cataloguing of Old Maps,* Windham, Conn.: Hawthorn house, 110 p.

Clapp, Anne F. (1987), *Curatorial Care of Works of Art on Paper: Basic Procedures for Paper Preservation,* New York: Nick Lyons Books, 191 p.

Ehrenberg, Ralph E. (1982), *Archives and Manuscripts: Maps and Architectural Drawings,* SAA basic manual series. Chicago: Society of American Archivists. 64p.

Farrell, Barbara and Aileen Desbarats (1984), *Guide for a Small Map Collection.* 2nd ed., Ottawa: Association of Canadian Map Libraries, 101 p.

History News 34 (January 1979): Technical Leaflet, 111.

Larsgaard, Mary Lynette (1987), *Map Librarianship: an Introduction.* 2nd ed. Littleton, Colo.: Libraries Unlimited, 382 p.

Journals

Baseline: A Newsletter of the Map and Geography Round Table. v.1 (since 1981), Chicago, Ill.: The Round Table.

Bulletin. v.1 (since 1951), New York: Special Libraries Association, Geography and Map Division, 1951-date.

Bulletin. v.1 (since 1968), Ottawa: Association of Canadian Map Libraries, 1968-date.

Information Bulletin. v.1 (since 1969), Santa Cruz, Calif.: Western Association of Map Libraries, 1969-date.

Kidd, Betty H. (December 1980), 'Preventive conservation for map collections', in *Special libraries* 71, pp. 529-538.

Larsgaard, Mary Lynette (1987), 'Storage, Care, and Repair', in *Map Librarianship: An Introduction,* 2nd ed., p. 163-197, Littleton, Colo.: Libraries Unlimited.

Library of Congress, Preservation Office, Polyester Film Encapsulation, Washington: Library of Congress, 1980. 23 p.

Meridian: a Journal of the Map and Geography Round Table of the American Library Association. v.1 (since 1989), Chicago: The Round Table, 1989-date.

Interpretation and Cataloguing of Maps

Andrews, John Harwood (1974), *History in the Ordnance Map: An Introduction for Irish Readers,* Dublin: Ordnance Survey Office, 63 p.

Anglo-American Cataloguing Rules, 2nd, revised, ed. Chicago: American Library Association, 1988. 677p. Chapter 3 on Cartographic Materials (p. 92-121) edited by Michael Gorman and Paul W. Winkler has been the model for much of the map cataloging done in the English-speaking world for over a decade.

Boggs, Samuel Whittemore and Dorothy Cornwell Lewis (1945), *The Classification and Cataloging of Maps and Atlases.* New York: Special Libraries Association, 175 p.

Buisseret, David (1990), *From Sea Charts to Satellite Images: Interpreting North American History Through Maps.* Chicago: University of Chicago Press, 1990, 304 p.

Cook, Andrew S. (1989), *Edition, Printing, Issue, and State as Terms in Cartobibliography,* 6 p. Paper read at the 13th International Conference on the History of Cartography, Amsterdam.

Harley, J.B. (1964), *The Historian's Guide to Ordnance Survey Maps,* London: For the Standing Conference for Local History by the National Council of Social Service, 51 p.

Harley, J.B. (1972), *Maps for the Local Historian: a Guide to the British Sources Standing Conference for Local History.* London: For the Standing Conference for Local History by the National Council of Social Service, 86 p.

International Federation of Library Associations and Institutions (1977), *International Standard Bibliographic Description for Cartographic Materials (ISBD (CM),* London: IFLA International Office for UBC, 58 p.

Karrow, Robert W. (1985), 'The Role of cartobibliography in the history of cartography', *Association of Canadian Map Libraries,* Bulletin 57 (December): 1-13.

Karrow, Robert W. (1976), 'Cartobibliography', in *The 1976 Bookman's Yearbook,* p. 43-30, Clifton, N.J.: AB Bookman Publications.

Krummel, Donald William (1984), *Bibliographies: Their Aims and Methods,* New York: Mansell, 192 p.

Long, John H. (1977), 'Studying George Rogers Clark's Illinois Campaign With Maps', in *The French, the Indians, and George Rogers Clark in the Illinois Country: Proceedings of an Indiana American Revolution Bicentennial Symposium,* p. 67-91, Indianapolis: Indiana Historical Society.

Parry, J.H. (1976), 'Old maps are slippery witnesses', in *Harvard Magazine* (Alumni Ed.) 78, no. 8 (April), pp 32-41.

Schlereth, Thomas J. (1980), *Past Cityscapes: Uses of Cartography in Urban History,* Newberry papers in family and community history, paper 80-1, Chicago, Ill.: The Newberry Library, 41 p.

Skelton, R.A. (1965), *Looking at an Early Map*, Lawrence: University of Kansas Libraries, 29, p 3.

Stibbe, Hugo L. (1982), *Cartographic Materials: a Manual of Interpretation for AACR2*, Chicago: Ottawa: American Library Association; Canadian Library Association, 258p.

Tanselle, G. Thomas (1982), "The Description of non-letterpress material in books', *Studies in bibliography* 35, pp 1-42.

Verner, Coolie (1974), 'Carto-bibliographical description: the analysis of variants in maps printed from copperplates', *The American Cartographer*, 1 (1974), pp 77-87. 21.

Wiberley, Stephen E. (1980), 'Editing maps: a method for historical cartography', in *The Journal of Interdisciplinary History* 10, pp 499-510.

Part IV

Impact of Information Technology on Cartography

23
Computer Aided Cartography

A NEW APPROACH TO MAPPING

With the introduction of desktop computers, not only the method of making maps but also the very concept of map has changed. Instead of something printed on paper, people having access to computers like to see maps displayed on computer screen. The stage is already set to fit a screen in front of the automobile driver's seat which gives the route map and other relevant information to him as he drives to his destination. The number of people who use map-based services available in the Internet is ever growing.

It is not only the cartographers who are using computers to create maps; professionals of other fields too are now getting directly involved in map making to satisfy their specific needs. While cartographers produce maps that they think people need, the map users themselves now produce maps which they think they need. True, the time is not yet ripe to replace the cartographers, but it has certainly arrived to use maps more than it was ever done. Maps are the tools of visual thinking. They are the means to communicate ideas and information visually (see chapter 1). Earlier, maps were made to communicate ground realities to common people; now they help analyze a given situation, in solving problems, and in making-decisions.

To meet the above requirements, cartography has to shift its focus from the traditional approaches and methods of making maps to using computers and related technologies to produce visual products that meet the emerging demands of various sections of the society, from professionals to common people. It has to:

1. Examine the data and phenomena that can be mapped with the help of the new expressive opportunities provided by the computers;
2. Evaluate the potentials of new technologies for mapping; look at their strong and weak points;
3. Develop a typology of 'mappable' things and phenomena and lay down new principles of cartographic representation;

4. Examine the purposes for which computer maps could be used to meet the analytical and communicative needs of the people and then see how maps can meet these needs.
5. Find out what technique is suited for what purpose;
6. Work out new ways of cartographic expressions using new technologies to meet the emerging needs of and demands from various sectors of the society;
7. Make cartographic knowledge and products available to the broad community of map-users, especially those who use existing maps as raw materials to produce their own maps.

WHAT IS NEW THAT THE NEW TECHNOLOGIES OFFER?

We are flooded with information on what is going on in different parts of the globe. To convert this information into knowledge which can help solve the emerging problems, it has to be analyzed speedily. Computers have proved to be a great help in this respect for they can store and process voluminous amount of data and give the desired results. Computers accept data and instructions and then process the data according to the given instruction and give the results in the form of numbers and words or 'graphic patterns' including maps.

Computer technology offers unparalleled opportunities for cartographic expressions. Most of these are available even on standard personal computers. Among the new opportunities available to are:

1. Unlimited number of colours and patterns to represent finer details of reality;
2. High resolution of graphical displays to distinguish one phenomenon from the other and also to present even a point or line to the scale;
3. Generation of realistic 3 or 4-dimensional scenes to take the map nearer the real world;
4. Making the scales and projections dynamic;
5. Making cartographic products dynamic by introducing animation;
6. More opportunities for user interaction with the displays;
7. Making displays dynamic;
8. Interlinking of multiple perspectives and views of the phenomena displayed;
9. Increasing possibilities of inter-conversion of digital maps and mathematical equations; and
10. Making maps more user-friendly.

So far maps were conceived by map makers and the users had to fit their requirements to the existing maps. Now the table is turned in favour of the users. The user not only uses maps but also creates maps on the computer screen to suit his requirements. The symbols used in the paper maps encoded a given set of data which the map user had to decode using the legend printed along with the map. It is now possible to prepare maps whose symbols once clicked by the computer mouse would present to the map user all the detailed data which underlies them. The map user no longer needs to take the trouble of decoding the values. The map maker need no longer use his time in symbolizing data; he can now use his time to design and implement various interactive techniques which enhance the role of maps as tools of 'visual thinking'.

While preparing the choropleth and isopleth maps, cartographers had to work very hard to select patterns and colours. And once a map is done, it is there giving certain information which may change after a month. It is now possible to prepare a choropleth map giving the distribution of a phenomenon in two colours, one below the mid-point and the other above the mid-point. If the mid-point is changed, the distribution pattern would also change. It can be of great help in the investigation of spatial patterns and trends on the one hand and in getting a comparative picture on the other. One can also use interactive pie charts and several other graphical symbols to compare the structure of different objects. While comparing the age structure of population of different states of India, the pie chart representing the average national figure or figure of given state can be used as a norm to find out which of the states are above or below the national average. For doing all this, one need not prepare a paper map. It can be done on the computer. By clicking the mouse on the pie chart of any state, one can get full details in tabular or whatever form one desire.

It is also possible to modify the map symbols to achieve specific objectives. To remove congestion from general maps, one can now take off certain elements. To show the railway tracks alone in a general transport map, it is now possible to lift the roads and waterways leaving the railways there. One can enlarge maps to accurately measure distances between any two points located close to each other. This would not be possible on a small scale map.

While preparing interactive maps one should keep in view that the properties of traditional, well-known cartographic representation methods are no longer applied. In traditional cartography, choropleth maps are not considered suitable for comparative analysis. However, the use of a diverging colour scale and the change of the midpoint of this scale allow one not only to see which of the two objects has higher value but also to compare an object to all other objects present on the map. For this one only needs to set the value associated with this object as the mid-point of the colour scale and to click on the object in the map. Computer technologies thus greatly enhance the capabilities of traditional cartographic techniques.

Further, the computer technologies allow us to do many things that we could not do with traditional cartographic methods. For example, representation of phenomena changing in time was difficult to map. Now it is possible to do it by animated maps. Software packages for it are now in the market. It is also possible to a add third dimension (time) in cartographic products. In order to use the full potentials of computer technologies, the theoretical cartography must investigate which phenomena to represent in a particular way; how expressive they can be; and what cartographic symbols should be used in different cases? Further, cartography must prepare itself to map even those things that were hardly ever considered as mappable. For example, mapping alternative ideas, models, arguments, and techniques being discussed in a conference/brainstorming session. The 3-dimensional maps are already too common to emphasize here.

Maps are becoming indispensable tools of 'visual thinking'. The ICA Commission on Visualization and Virtual Environments is working hard to develop typology of analytical tasks that can be supported by maps. It is, therefore, important to develop a typology of communicative tasks and relate these tasks to appropriate cartographic tools/products. For example, how maps can be used to convince people to do or not to do something or to refute a widely spread opinion, is one of the tasks of modern cartography. How to mix maps with text, images, diagrams or other communication media, needs further exploration. Further, maps can be combined with sound, and used in multimedia presentations.

Since more and more of the people make not only the use of maps but also involve in its creation, cartographers must ready themselves to make cartographic knowledge accessible to the masses. This can be done by preparing appropriate books in a language general readers can understand, and by participating in activities that involve the use of computer maps. In Germany, the Naturdetektive project (*www.naturdetektive.de*) helps German school children to register what they see on the map and then trace routes of migratory birds on animated map displays. It is also possible now to prepare softwares that tell the users how to represent their data on a map.

One of the major problems underlying the maps prepared manually was the time involved in collecting data, analyzing it and then converting it into maps. By the time the maps were in the hands of the users, they were outdated. This lacuna is on the way out. Computer helps in preparing maps with great speed and accuracy because the collection, processing, storage and graphic output of cartographic information such as charts, graphs, maps and photographs can now be accomplished simultaneously. All that we have to do is to spatially order the data in terms of their x, y vocational coordinates for use in the computers. This can be done through manually operated or automatic electronic digitizers.

The data so coded, can be directly fed into the computer along with the necessary mapping instructions to get the final cartographic outputs through line printers, coordinate plotters and Cathode Ray Tube (CRT) displays.

A number of computer mapping programmers are being used today. More commonly used among them are:

- SYMAP system for mapping both qualitative and quantitative spatial data;
- GRID displays data values by zones formed by grids;
- CALFORM produces shaded maps based on data zones like census tracts;
- SYMVU converts girded spatial data into 3-dimensional surfaces;
- DOT MAP produces contour, shaded and dot maps based on girded data;
- MAPIT is a package that gives flow maps, dot maps and graduated symbols;
- COLMAP is a colour map system 'used to get maps printed separately in plates for different colours on the basis of instructions.'

Note: This chapter draws heavily from 'Computer Cartography and Cartographic Knowledge' by Gennady Andrienko and Natalia Andrienko of Fraunhofer Institute for Autonomous Intelligent Systems (AIS), Schloss Birlinghoven, Sankt-Augustin, D-

24
Geographic Information Science (GIS)

With the development of reconnaissance technologies using aircrafts and artificial satellites, collection, retrieval and transformation of geographic data into maps has gained speed and precision. Until the advent of computers, maps were drawn using symbols like points, lines and patterns. The explanatory notes accompanying them constituted the database and the analysis of the data, the maps contained was essentially qualitative and visual.

When the size, range, complexity and quality of data increased, the visual observation proved inadequate. The gap was bridged by the computer based Geographic Information Science (GIS) otherwise known as Geographic Information System, which could analyze large volumes of complex geographic data speedily. Today, GIS has become an important field of study all over the world; it is one of the fastest growing sectors of the computer industry and an essential component of the Information Technology (IT) infrastructure.

Use of GIS is no longer limited to Geography. It has become a common intellectual property of all disciplines and departments of life. As such it defined differently. Given below are a few of the commonly known definitions:

1. 'GIS is a system of hardware, software and procedure designed to support the capture, management, manipulation, analysis, modelling and display of spatially referenced data for solving complex planning and management problems.' (Rhind, 1989).
2. 'GIS is a computer system capable of assembling, storing, manipulating and displaying geographically referenced information i.e. data identified according to their locations' (USGS, 1973).
3. GIS is a powerful set of tools for collecting, storing, retrieving as well as transforming and displaying spatial data from the real world. (Burrough, 1986).
4. GIS is a database of a particular kind containing spatial information from micro to macro levels. It can also be considered as a tool to analyse resource and environmental information in a spatial frame of reference.

5. GIS is a computer-based processing and analysis of geo-referenced multivariate data. It provides easy access to stored geographical information, customised maps and selective database queries. It is possible to get land cover, administrative boundaries and transport and communication network, and location attributes of various natural and human phenomena. Among the various kinds of GIS, are Land Information System (LIS) and Environmental Information System (EIS).

Figure 24.1 graphically presents the linkages of the GIS to various disciplines and technological fields. Some understanding of all these fields is essential for acquiring GIS expertise.

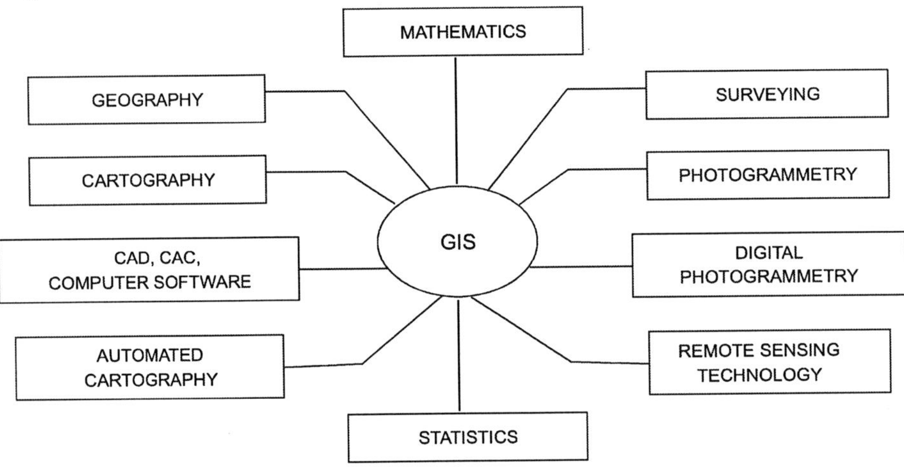

Fig. 24.1: Linkages of the GIS with Other Fields

Being a remotely sense information system, GIS provides an opportunity to interpret various topographic and socio-economic data for spatial analysis and modelling. The emphasis is on integration of conventional land based and remotely sensed data so as to develop databases comprehensive enough for multi-level spatial analysis and planning.

GIS thus helps in:

1. The creation of precise and adequate structured resource inventory or data base,
2. Monitoring of geographical, hydrological and socio-economic activities,
3. Analysis of current and time variant information,
4. Testing of scientific hypothesis about the interaction and functioning of environment and predicting resource potentials, and
5. Preparation of development plans and resource management strategies.

The GIS technology has revolutionized the mapping process. Softcopy images have replaced printed maps. The time lag between data collection and mapping has been reduced drastically as the high-resolution images of designated sites can be obtained in minutes by moving a nearby satellite into a new orbit. It is particularly useful in monitoring the ever-changing ground realities. Assessment and monitoring of natural calamities, analysis of resources, gathering of military intelligence, land use planning and spot location of fast moving air masses, storms, vehicles, animals and even human beings. Satellite remote sensing and computer-assisted cartography have given new impetus to spatial data storage and exchange. The combined use of Geographic Information System (GIS), and Digital Image Processing (DIP) offers excellent opportunities for geographical monitoring and forecasting of both natural and human phenomena over wide areas within limited time-span. It is now easy to develop predictive models for planning, development and policy-making.

Definitions of Terms Used

As is the case with other fields, GIS too uses terms having very specific and contextual meaning. In order to make the full and best use of the system, one must be conversant with them. Given below are some of the frequently used terms in GIS related literature.

Data

'Data' means facts or figures pertaining to places, people, things, events and concepts. These facts and figures are generally represented by numerical values, alpha-numeric characters, symbols or signals. Geographic data record the location and characteristics of natural features or human activities that occur on or near the earth surface.

Information

When the data collected by various sources and means are transformed into meaningful and user-friendly forms, it is referred to as information. The difference between data and information is the same as between knowledge and intelligence.

Information System

The word 'system' is used in different contexts to mean different things. It can be used to describe physical entities, such as the 'solar system' and the "immune system of the human body". A system consists of several sub-systems. Our body is a system; so is a factory, a tree, a municipal authority, a government, etc. When several previously separate systems come together and enter into a close relationship, they form a higher-level system and the systems forming the next level of system become its sub-systems.

A system is not, however, a random conglomeration of sub-systems. All parts are interrelated and subject to a given order. Thus a heap of sand is not a system because its elements, the grains of sand, are not linked to each other. Any addition or subtraction makes no difference. It remains a heap of sand. A tree is a system, and so is factory, a city, a nation; a human body.

The 'Information System' is a special class of systems set up to achieve the specific objectives of collecting, storing, analyzing and presenting information in a manner most conducive to the goals and objectives behind the system. It is made up of inter-related components that include data, information, technical tools to handle them and the human resources involved in processing the data/information according to a well-defined set of operational procedures (Fig. 24.2).

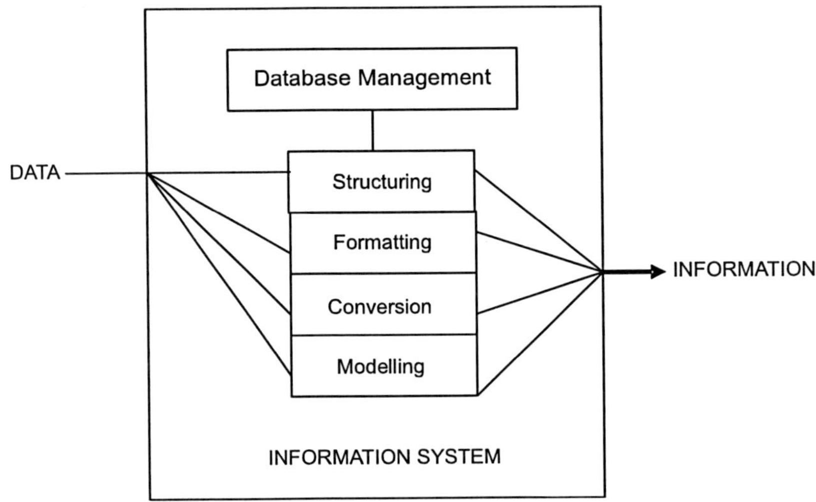

Fig. 24.2: Transformation of data into information using information systems

GIS is a special class of information system. The word 'geographic' in GIS carries two meanings: 'Earth' and 'geographic space'. It implies that all data in the system are related to Earth's features and resources including human activities based on or associated with these features and resources. It implies that the data and the problems that the system is developed to solve is geographic, i.e. location, distribution, patterns and relationship. The focus on geographic data and their use in solving spatial problem makes GIS unique among the information systems.

The relationship between GIS and different types of information systems can be seen from the topology of information systems given in Fig. 24.3.

We can see from Fig. 24.3 that there are two types of information systems: spatial and non-spatial. The non-spatial information systems are those, which are designed to process data that are not referenced to any position in geographic space. Systems for accounting, banking, human resource management and goods inventory are typical examples of non-spatial information systems. It is important to note that not every spatial information systems can be regarded as GIS. Computer Assisted Draping (CAD) and Computer Assisted Manufacturing (CAM) systems are typical examples of spatial information systems that are not parts of GIS. These systems are spatial, but they are different from GIS in purpose and data processing procedures and requirements.

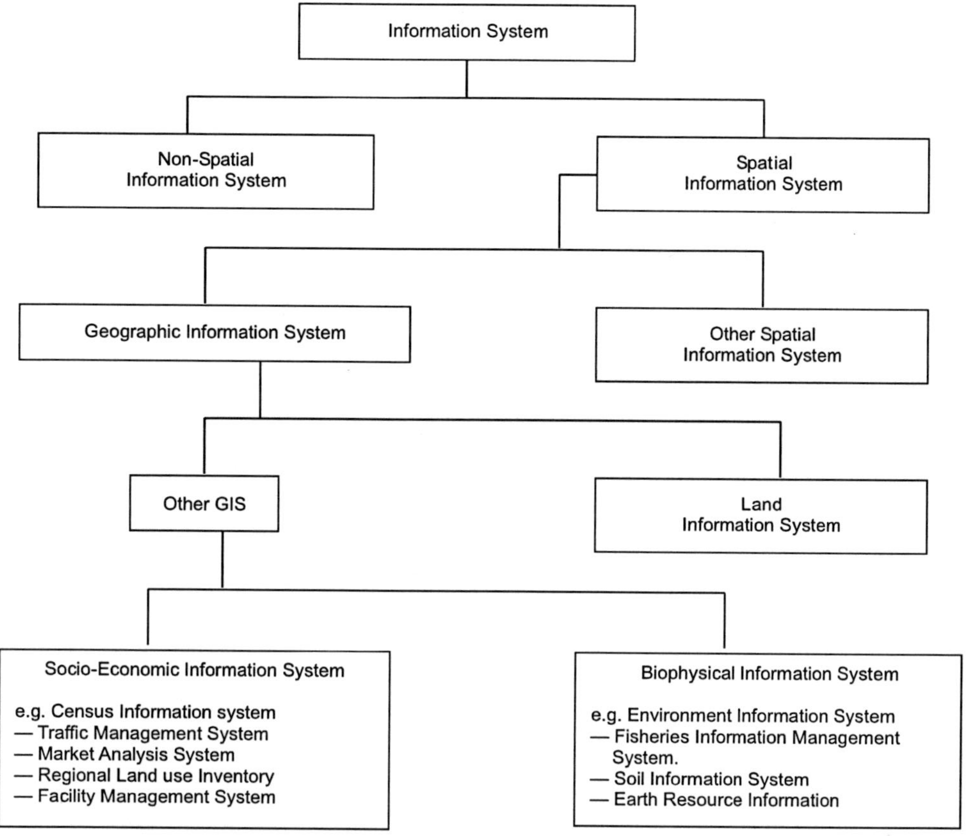

Fig. 24.3: Topology of Information System

GIS Infrastructure

The term 'GIS infrastructure' implies that geographic information must be treated in the same way as other infrastructures, such as transportation, power and communication networks, education system, etc. The geographic information infrastructure is global in coverage. There are international and regional organizations to deal with issues that transcend national boundaries, such as international navigation, air and maritime transportation, weather forecasting, monitoring of natural of hazards and global changes in climate, vegetation, etc.

TYPES OF GIS

GIS analyses the spatial attributes and the inter-relations among them within a region/ space. There are two different types of GIS:

1. Vector or polygon structure; and
2. Raster or grid structure.

Vector structure denotes spatial units of the features to be represented. Such units are of uniform size and shape and relative location within an area. In the raster (image) data, spatial objects are made of contiguous cells sharing the same attribute within the regions. Attribute dimension is described as set of grids in which the value assigned to each cell expresses its overall characteristics. The merits and demerits of the two GIS modes are given in Table 24.1.

Table 24.1: Properties of the GIS Modes

Properties	*Object Mode (Vector Structure)*	*Image Mode (Raster Structure)*
Spatial Sampling	Variable	Constant
Units of Observation	Phenomenon Dependent	Phenomenon Independent
Spatial Measure (X, Y)	Precise	Imprecise
Continuous Spatial Distribution	Complex	Simple
Data Structure	Complex	Simple
Storage Compactness	High	Low
Spatial Analysis	Complex	Simple

APPLICATION OF GIS

GIS has many practical applications. Important among them include:

1. Mapping (Cartography),
2. Database Management,
3. Data Queries based on spatial location, attribute or features
4. Spatial modelling such as capability assessment, suitability analysis, sensitivity analyses, simulation, etc.

GIS can answer question such as: where is a particular phenomenon located; what is the speed of the flow of say a river at a given place; where is all the land within 200 meters of a river in an area; what has changed through time; what changes take place in water when the water level is lowest; which is the best way to go; which is the quickest way to reach a particular place; what is the pattern of a given phenomenon; is there some pattern in the distribution of a phenomenon; what if climate changes continue in future; which areas of the earth will be affected by a 20 CM rise in sea level?

Thus, GIS can be used with great advantage to resolve problems related to:

1. Network-based services management for:
   - Address matching
   - Vehicle routing and scheduling
   - Location analysis, and site selection
   - Development of evaluation plans

2. Natural resources management for:
    - Conservation of Forests
    - Wildlife habitat, migration routes
    - Wild and scenic rivers preservation
    - Recreational resources planning
    - Floodplains and wetlands preservation
    - Agricultural lands
    - Groundwater modelling and contamination tracking
    - Environmental Impact Analysis (EIA)
    - View shed analysis
3. Land parcelling:
    - Zoning, subdivision plan review
    - Land acquisition
    - Environmental impact statements
    - Water quality management
    - Maintenance of ownership
4. Facilities management:
    - Locating underground pipes, cables
    - Balancing loads in electrical networks
    - Planning facility maintenance
    - Tracking energy use

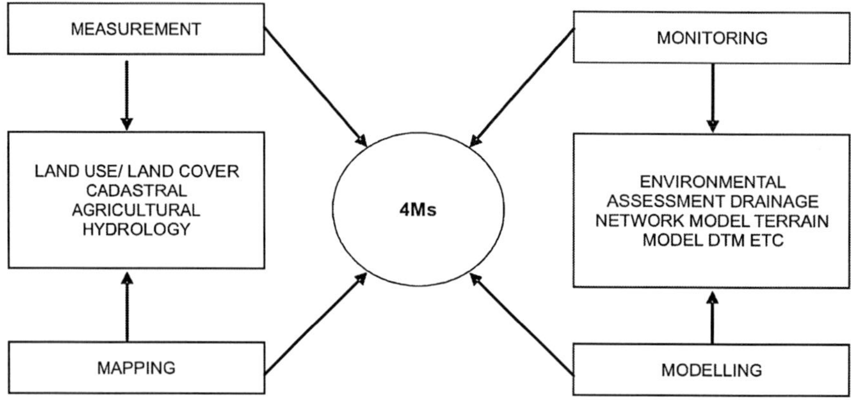

Fig 24.4: The 4-Ms of GIS: Measurement, Mapping, Monitoring and Modelling

HISTORY OF GIS

The origin of GIS can be traced to R & D in electronic data processing dating back to the 1940s and 1950s. Research and Development efforts in those years finally led to the successful implementation of computer assisted geographical data processing at the Massachusetts Institute of Technology (MIT) in 1955 and Database Management Systems (DBMS) by General Electric in 1965. The idea of processing large amount of complex data electronically was quickly picked up by many government agencies to build

computer system for geographic data handling. In this connection the following four applications could be considered as the pioneering works in the development of GIS.

1. Canadian Geographic Information System (CGIS). Conceptualized in early 1960s and made operational in 1971, it was built to address the needs of land and resource information management of the federal government of Canada. (Tomlinson *et al.* (1976).
2. Geographical Information Retrieval and Analysis System (GIRAS) of the USGS. It was evolved in 1973, to analyze land-use/land cover data pertaining to the USA.
3. Swedish Land Data Bank (SLDB). It was developed in 1970s to automatize land and property registration in Sweden.
4. Local Authority Management Information System (LAMIS) and Joint Information System (JIS). Both these were used in Britain by the local governments for land-use control and monitoring.

The 1960s and 1970s represent the formative decades of the GIS. It was the period of main-frame class of computers. Running in the batch mode as stand alone software applications, hundreds of software packages for handling and analyzing geographically referenced information were produced. These early systems were application driven and constructed on the relatively simple concept of raster cells.

In 1979 when Corbelt published his technical paper on the concept of topology as applied to statistical data. Perhaps the most important factor in the rapid development of GIS technology was the technical paper of Corbelt (1979). By topology Corbelt meant spatial relationships of adjacency, connectivity and containment among topographic features. Using this concept, geographic data could be stored in a simple structure that is capable of representing their attributes as well as their relationships. This concept solved the data representation problem that hindered the development of GIS in early years. It significantly reduced the complexity of applying geographic data for spatial analysis, thus making GIS much easier to use. In 1982 ESRI Launched Arc Info was based on minicomputers. It was one of the first vector based GIS to use geo-relational data model that employed a hybrid approach to geographic data. Once the theoretical complexity of data structure were resolved by the mid-1980s, the focus of GIS development gradually shifted to methods of data collection, quality and standards as well as data analysis and database organization. The new focus fundamentally changed the way the GIS technologies developed. It led to the integration of raster and vector geographical data with other types of business data in what is called the enterprise-computing environment. It also nurtured the concept of open GIS that aimed to develop systems capable of sharing both technical and data resources with others. By mid 1990s, GIS became mature in terms of both technology and application and specialized sectors to handle specific problems in the traditional computer industry came into being.

By mid-1990s, GIS entered a new era that can be called the Age of Geographic Information Infrastructure (Table 24.2). In early 1990s the United States Government

launched the National Information Infrastructure (NII). The objective of this initiative was to provide all US citizens access to information affecting their lives such as healthcare, education and community development. Being the primary vehicle for delivering geographic information services, GIS is a critical component of Geographic Information Infrastructure.

A typical GIS today has the following technical and operational characteristics:

1. A part of computer network rather than a stand-alone but computer configured using advanced processors and memory systems, and sharing input/output connectivity.
2. It is made up of multiplier, distributed software components that allow geographic information processing function to be performed across multiple processors.
3. It is connected to Internet for access to data and technical resources available on the World Wide Web (WWW) and for distributing the geographical information, it contains to any interested person or organization anywhere in the world.
4. It is based on standards applicable to computer systems development methodology, software development tools, database connectivity and communication protocol as well as user interfaces.

The GIS in 2005 is very different from what it was in 1960s when the Canada Geographic Information System (CGIS) was first developed. Table 24.2 gives the stages through which it has passed before it developed to the level we find it today.

Table 24.2: Evolution of GIS

Stages of Development	*1960-80 Formative Years*	*1980-1990s Technological Developments*	*1990s-2005 Development of GIS Infrastructure*
Technical environment	* Main frame computers * Proprietary software * Proprietary data structure * Mainly vaster based.	* Main frame and mini-computers * Georelational data structure * New data acquisition (GPS, redefinition of datum, remote sensing).	* Workstation and PCs. * Network/Internet * Open system design * Multimedia * Data Integration * Enterprise Computing * Object relational data model.
Major Users	* Government * Universities * Ministry	* Govt. * Universities * Business * Utilities * Military	* Govt. * Universities and schools * Business * Military * Utilities * The general public
Major Application Area.	* Land and Resource Management * Census * Surveying and Mapping	* Land and resource management * Census * Surveying and Mapping * Facilities Management * Market analysis	* Land and resource management. * Census * Surveying and Mapping * Facilities Management * Market Analysis * Utilities * Geographical data browsing.

COMPONENTS OF GIS

Conventionally, the construction and function of GIS were explained by dividing them into subsystem of input, processing, analysis and output. However, an information system can be conceived of as having the following components: data, technology, application and beneficiaries.

DATA/INFORMATION

Depending on their nature and use, geographic data can be categorized into three distinct types:

1. Geocoded Control Network.
2. The Topographic Base.
3. Graphical Overlays.

The *Geocoded Control Network* is the foundation of all geographic data. It provides a framework for different sets of geographic data to be spatially cross-referenced with one another. Geocoded Control Networks are established by high precision surveying methods and vigorous computation at the national or, as in the case of North America, continental level. The *Topographical Base* is created as a result of a base mapping programme by national, state/provincial and local government agencies. The contents of topographical database can be obtained by land surveying, but photogrammetry is now more commonly used. *Graphical overlays* constitute the thematic data pertaining to physical and cultural features often derived directly from the topographic base. Graphical overlays pertaining to such socio-economic activities as population distribution, land boundaries, natural resource values and land use status are obtained only by sight investigation, field surveying, remote sensing and other forms of data collection. Within a digital geographic database, there are three basic forms: raster, vector and surface.

TECHNOLOGY

GIS was developed as a distinct class of information systems that could use only one of the above mentioned data types. However, as the result of recent advances in computer technology and data processing methodology, limitations associated with different data structures and segregation of vector and raster systems have been largely eliminated. In GIS data component is more concerned with data sharing, quality and standards rather than with data creation, structure and types. The CPU performs all the data processing and also controls the input/output connectivity with data acquisition, storage and display systems. Depending on the data processing power of CPU, computers are classified as supercomputers, mainframes, minicomputers, work stations, and micro computers or PCs. All these classes of computers can be used as the hardware platforms for GIS.

APPLICATION

Today's GIS are mostly implemented in a network environment using the Client/server modes of computing.

On the software side, GIS was conventionally developed using a hybrid approach that handled geographical and descriptive geographic data separately. In such an approach, the geographical data engine, which was usually proprietary to a GIS software vendor, handled the graphical data while a commercial DBMS took care of the associated descriptive data. The connection between graphic data engine and the DBMS was provided by the software vendors in the form of a proprietary interface. This form of software configuration, which is known as the geo-relational data model, was the norm for GIS implementation until the late 1990s.

Following the thrust of computer industry towards the use of object oriented technology GIS software vendors have adopted new concepts and techniques in design and development of their software products. The geo-relational data model has been rapidly replaced by what is known as object-relational model. GIS constructed on this new database model allow both graphical and descriptive data to be stored in a single database. They also allow user to identify facts about the real world and transform these facts into geographic objects useful for geographic data processing and analysis. The use of object-oriented technologies has effectively transformed GIS from automated filling cabinets of maps into smart machines for geographic knowledge.

Areas of Application

The application of GIS can be explained from three perspectives: areas of applications, nature of applications and approaches of implementation.

Research and Training

GIS is now an indispensable tool for research in humanities, sciences, engineering and technology. Most universities and college offer courses in GIS leading to certificates, diplomas and degrees both at graduate and post-graduate levels.

Planning and Mapping of Public Services

GIS are used for national topographic mapping, resource and environmental management, weather services, public land management, population census, election and voting. Public safety and law enforcement, land surveys and mapping, resource and environmental management, highways planning and management, social and community development, land registration and property assessment are the other areas of application.

Business and Industry

GIS is now extensively used in banking and insurance, real estate—development, project planning and management, sales and renting services, building management, retail and marketing analysis. The engineers use it for surveying and mapping site and landscape development, pavement management. The transportation sector uses it for route selection for goods delivery, public transit vehicle tracking. It is widely used in the management of utilities and communications: electricity and gas distribution pipelines, telecommunications networks. Among the other important uses are: forest resource inventory, harvest planning, wildlife management and conservation, mining and mineral exploration and systems consulting and integration.

Military Intelligence

The armed forces are the major users of GIS. They are able to locate the enemy positions and track down the terrorist camps. GIS constitutes an important component of training for command control and intelligence gathering.

THE USERS OF THE GIS

To begin with the users of the GIS were few and were seldom formally recognized. Many users applied GIS to their own limited area of interest. Others used turnkey systems that were commercially obtained from the software vendors. Thus, the human factor was taken largely for granted. Early in 1990s, GIS technology and use grew fast that a GIS community has come into being. It is large and highly sophisticated in requirement. This has in turn led to the growing awareness of the interest in the human factor in GIS. According to the information they need and the way they interact with the system, GIS users can be classified into three categories—viewer, general users and GIS specialist.

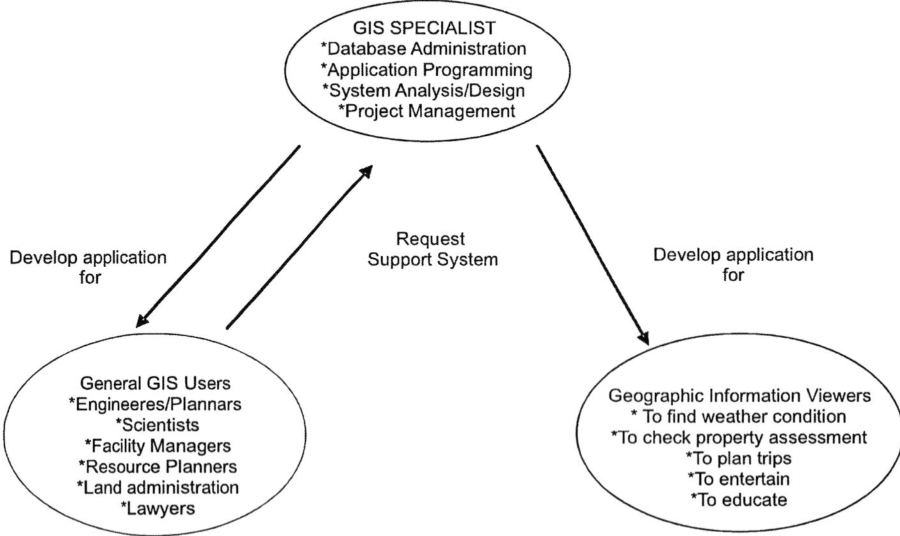

Fig. 24.5: GIS Users and their relationships

General users are people who use GIS for conducting business, performing professional services and making decisions. This group of users includes facility managers, resource planners, land administrators, lawyers, business entrepreneurs and politicians unlike viewer general users are active users because GIS is implemented to support their informational needs.

GIS specialists are people who actually make GIS work. They include GIS Managers, database administrators, application specialists, system analysts and programmers. They are responsible for the maintenance of geographic database and the provision of technical support to other two classes of users.

POTENTIAL USES OF GIS

We have now become acquainted with the potential use of GIS in solving our problems such as land and forest cover change, water and air quality assessment, soil erosion risk analysis, natural hazard monitoring, selection of afforestation areas, land capability and suitability analysis, and data directory study, etc. Worldwide experience has shown that remote sensing and GIS are very effective tools for resource management. The new age of microcomputers and increased potential for information exchange aids in its use in India too.

Geographical Information System (GIS) offers such capabilities as they integrate multi-sector, multi-level and multi period database. Subsequently, GIS can be used as decision support system identifying and integrating monitoring and predicting natural hazards, which are the major environment risks in India. Integrated Watershed Management Plans that consequently result could help reduce the disastrous impacts of disasters that continuously threaten us.

Land-use Monitoring

Land use is an important input in the geographical monitoring of rural as well as urban resources of a region. The general land cover classification can be derived from past changes in land use in general and forest cover in particular. Deforestation for example is a serious ecological problem in the Himalaya. Forest maps of two time periods can provide information of forest cover changes (Fig. 20.4). Satellite remote sensing can be effectively used for land use mapping and for monitoring forest degradation process in the country. Due to rapid change, the information collected by conventional methods becomes outdated.

Land Suitability Analysis

Land use suitability analysis helps in taking decision on different land related/based developments. The GIS analysis integrates various natural, anthropogenic and interactive factors to produce three types of maps. (1)maps showing the land use most conducive to ecological balance and environmental quality; (2) maps showing qualitative predictions of environmental impacts of proposed development activities and remedial steps to be taken to forestall any damage; and (3) maps showing the best and least suitable locations for those actions.

Land use planning in marginal regions like Himalayan region, Rajasthan desert requires such mapping. The analysis of land suitability for agriculture in highlands will produce information and maps related to soil erosion, slope, altitude, water availability and nutrient availability, etc. The U.S. Department of Agriculture Soil Conservation Service has developed the agricultural Land Evaluation and Site Assessment (LESA) system to determine the quality of land for agricultural uses and to assess agricultural land areas for their economic viability. India should also take up such a project.

Many countries and state governments have attempted to computerise land and other geographical information for municipal operations. Planning departments have made attempts to automate parcel maps. Environmental planners have attempted to automate environmental overlays for land suitability/capability. Such information includes base map, environmental, engineering, plan/profile drainage, parcel/street address tabular data, area tabular data, street network file and area boundary maps etc. The GIS software tools allow different data sets in mapped or tabular form to be corrected, interrelated and interpreted in a variety of ways.

Spatial Analysis and Modelling

Spatial data modelling in GIS is basically map data manipulation such as polygon over lay and buffering. It may also refer simply to spatial summarization of the data. Spatial data are retrieved selectively for a region of interest. Basic summary statistics are then computed and mapped. Spatial analysis is the manipulation of the spatial data into different forms in order to extract additional meaning. The major concerns are to investigate the patterns in spatial and to discover possible relationship between such patterns and other attributes within the study region. The ultimate goal is to model such relationships for the purpose of understanding and prediction.

Four main areas have been identified where statistical spatial data analysis techniques may be strengthening current GIS practices:

1. Sampling objects from the database and the choice of an adequate spatial scale of analysis;
2. Data rectification to computer variables for a different and incompatible set of zones;
3. Exploratory spatial data analysis aiming to explore and exploit the GIS database to arrive at new insights and an understanding of the data characteristics; and
4. Confirmatory or explanatory spatial data analysis and hypothesis testing based on some specified assumptions.

The linking of the spatial analytical functions to standard GIS may be achieved by three major approaches:

1. Full integration of the spatial analysis procedures within the GIS
2. Close coupling between the spatial analysis software and GIS by sharing the same data structures, and
3. Loose coupling in which an independent spatial data analysis module relies on a GIS for its input data and for functions such as geographic display via the import and export of the data in common format

The first approach is not yet realized although IDRISI for Windows developed by Professor Ran Eastman of Clark University has integrated a number of useful spatial statistical procedures in the programme, which are further enhanced in IDRISI 32. The second approach is more realistic but it requires a good user interface in the linkage. The S+ SPATIAL STATS is added on module to the S-PLUS system for data analysis and graphics. It develops spatially written commands to transfer spatial data back and forth between Arc INFO and S-PLUS. The third approach is used most often and has the advantage of using fully the advantages of the GIS and the spatial packages. It is common to use statistical packages such as SAS or SPSS or Minitab with the Arc INFO programme for this purpose.

DESCRIPTIVE STATISTICS

As an exploratory tool, spatial analysis facilitates the visualization of geographic data one variable at a time as maps, histograms bar harts, box plots or pie charts. The method of visualization chosen is a function of both the data and the objective of a particular analysis. The purpose is to discover outliers in the spatial data. Obviously a map is the most effective way to reveal any spatial clustering of similar and dissimilar values or a pattern of randomness. The descriptive statistics include measures of central tendency and dispersion. All these descriptive statistical techniques are fully integrated into GIS.

SPATIAL AUTOCORRELATION

The descriptive statistics shed light on the nature of the statistical distribution of the geographic data; but spatial autocorrelation deals with both the attribute and location of spatial features. In other words spatial autocorrelation makes it possible to measure interdependence in the spatial distribution and to use formal statistical method and to test hypothesis about spatial interdependence. It is important to note the scale of spatial pattern affects that spatial autocorrelation. The same spatial pattern at different scales may produce different spatial autocorrelation results.

The general spatial autocorrelation is present when similar values cluster together on a map. Spatial autocorrelation allows use of formal statistical procedures to measure the dependence among nearby values in the spatial distribution, test hypothesis about geographically distributed variables and develop statistical methods of spatial patterns. There are two commonly used measures of spatial autocorrelation Geary's Index and Moran's Index.

QUARDANT COUNTS AND NEAREST NEIGHBOURHOOD ANALYSIS

In a spatial distribution made up of points it is necessary to describe its pattern. In real life situations these points could be birds, nests, persons, trees, cities, and galaxies. The spatial pattern is the result of spatial processes, which reveal both physical and socio-economic forces at work. A spatial pattern of points may be described as regular or irregular. Under each type one can further subdivide into clustered, random, etc. The best-known ones are the quadrate counts and nearest neighbour analysis. In this type of analysis the main concern is with the location characteristics of the points rather than their attributes.

The quadrant counts require covering the spatial pattern with a quadrant, or a sampling area with any consistent size and shape, which can be circular, rectangular, square or both. Usually square quadrants are used. The number of points counted in each quadrant is counted and the observed frequency of occurrence of points is tabulated. A distribution model is used to determine weather the spatial pattern observed is random or not.

The quadrant count method is used in IDRISI for windows for raster data employed in GIS. However, it computes the mean number of points per cell (quadrant), the variance in cell totals, the variance in mean ratio, and the degree of randomness. The quadrant analysis has found applications in both social and physical areas. There are two type of point's pattern that the quadrant analysis deals with: the contagious processes in human populations and the structural features of the cultural and physical landscape in homogeneous regions. The study of spread of disease, diffusion of innovations and impact of social contacts on voting behaviour, involve the contagious processes in human populations and the structural features of the cultural and physical landscapes in homogeneous regions.

To avoid the problem of quadrant size, determination in quadrant counts, nearest neighbour analysis is the alternative approach for mapped point pattern analysis. This method may be easily integrated in both vector and raster GIS because it makes use of distance measurement between points to generate the statistics. Nearest neighbour analysis measures distance between two sample points and their nearest neighbouring points. The mean of the nearest neighbour distance (observed) for the study area is then compared with the expected mean distance.

Trend surface analysis is very useful data exploratory technique that will shed light on the spatial trend of the two point data. Basically, this is the application of the general linear model to spatial data with the intension of discovering the regional and local trend. Geologists for oil prospecting originally developed the technique. The trend revealed may be the underlying geological structure, topographic structure, and broad trends of rainfall belts on the pattern of housing and income surface; according to weather you are a geologist, a geomorphologist, a climatologist or a population geographer respectively.

CONCLUSIONS

To make the best use of the GIS, effective networking is essential. It bridge the gap between the users, end-users and data providers. It is important that sharing digitised GIS data as well as socio-economic and statistical data is a key issue to promote spatial technology for sustainable development. Therefore, GIS technology has become a key science for decision support of sustainable development and management of environmental resources.

It is apparent that building a career in GIS requires knowledge of geography, GIS fundamentals and GIS software. For those interested in exploring the full potentials of GIS, knowledge of geography, cartography, GPS, aerial photography and remote sensing is a must.

REFERENCES

Duckham, Matt and Michael Worboys (2004), *GIS: A Computing Perspective*, 2nd edition, London: CRC Press.

Herzog,, David (2003), *Mapping the News: Case Studies in GIS and Journalism*, Esri Press.

Lovett, Andrew A., Katy Appleton, editors (2007), *GIS for Environment Decision Making*, London: CRC Press.

Lunetta, Ross S. and John G. Lyon, editors, (2004) *Remote Sensing and GIS Accuracy Assessment: A Research Agenda for Geographic Information Science*, London: CRC Press.

McMaster, Robert B. and E. Lynn Usery, editors (2004), London: CRC Press.

Murlikrishna, 'Introduction to GIS' in *GIS Theory and Practice*, R. Ram Mohan Roi and Afral Sharieff, eds.

Peterson, Gretchen N. (2009), *GIS Cartography: A Guide to Effective Map Design*, New York: CRC Press.

Roger Tomlinson (2007), Thinking about GIS: Geographic Information System Planning for Managers.

van Sickle, Jan (2010), *Basic GIS Coordinates*, 2nd Edition, Denver, Colorado, USA.

Wilson, John P. (2008), *The Handbook of Geographic Information Science*, Blackwell Pub./OUP.

25
Land Information System (LIS)

There can be no life without nature and Land is the very base of the nature. Everything else is seen in relation to land. Atmosphere, hydrosphere and even ionosphere would have no meaning and purpose for man without lithosphere. Land is the base and air and water are the conditioning factors for life to emerge and survive. While other life forms live on nature, Man lives with nature. His relationship with land is dynamic; he is part of it with capacity to change it within a given threshold limit. Thus the land we have today is the product of interaction among lithosphere, hydrosphere, atmosphere and biosphere.

History is witness to the fact that early civilizations flourished wherever fertile land was in plenty to sustain human needs. It also shows that anti-nature activities of man led to the decadence of many civilizations of the past. No wonder then that the Indian thinkers and philosophers from earliest times called for a symbiotic relationship between man and land. *Mata Bhumih Putroham Prithvyaha* (earth is the Mother and we are its children), said they. Here Earth includes lithosphere, hydrosphere and atmosphere. They thus tried to define the complex and interactive role played by nature in the sustenance of life on planet earth.

Land has played a pivotal role in shaping the life and living of human beings. In the pre-industrial era, when agriculture was the principal source of income, ownership of land determined the status of a person. In British days in India, a big *Zamindar* (land owner) was something like a mini-king. Similar was the situation in other parts of the world too. The dawn of the industrial age changed the man-land relationship. Land became a commodity instead of a life support system, to meet the needs of manufacturing industries. This attitudinal change was further accelerated by unprecedented pressure of population on land especially in 19th and 20th centuries. The result was over exploitation of natural resources in general and land resources in particular and consequent damage to the life supporting eco-systems. Alarmed by the threat to the very survival of human life the United Nations convened a Conference of its members (Earth Summit) in Rio de Janeiro, Brazil on 3-14 June 1992 to discuss the emergent problems and take appropriate steps to keep the environment as natural as possible to ensure human survival and development.

The Earth Summit enunciated twenty-seven principles to guide its members on environmental protection and development. Its first principle reads, "Human beings are at the centre of concerns for sustainable development. They are entitled to a healthy and productive life in harmony with nature". It recognized the sovereign rights of the member states to use its resources without detriment to the environment. Principles 3, 4 and 5 read as follows:

Principle 3: "The right to development must be fulfilled so as to equitably meet the development and environmental needs of present and future generations".

Principle 4: "In order to achieve sustainable development, environmental protection shall constitute an integral part of the development process and cannot be considered in isolation from it".

Principle 5: "All States and all people shall cooperate in the essential task of eradicating poverty as an indispensable requirement for sustainable development in order to decrease the disparities in standards of living and better meet the needs of the majority of the people of the world".

Apart from enunciating 21 principles, the Earth Summit put forth work plans for sustainable development called Agenda-21 which is not a legally binding document but it became a moral commitment for all the nations to work towards achieving the objectives of the Earth Summit. It has four sections covering: Social and Economic issues including the need to eradicate poverty and hunger and to accelerate sustainable development in Developing Countries; Conservation and Management of Resources for Development including Integrated Planning and Management of Land Resources and high lights the need to allocate land to uses that provide the greatest sustainable benefits; the role and importance of major groups in implementing the sustainable development; and issues relating to means of implementing Agenda-21. Agenda-21 also requires creation of efficient and accessible land markets to meet community requirements, streamlining land transaction procedures by improving land registry systems and to establish appropriate land tenure to provide security to all land users.

IMPORTANCE OF LIS IN INDIA

More than 70 per cent of the people of India live in villages and over 60 per cent of the people depend on agriculture for their livelihood. It is clear then that land is the most important resource of the country. It has to be conserved and developed in such a way that it helps the country overcome the age old problem of poverty and ensures food security for the 1.20 billion people of the country.

Despite impressive economic performance of the country in terms of economic growth, the stark reality of the situation is that nearly half of the people of the country living in villages and urban slums live below the poverty line. One third of the people are illiterate; many villages even today lack proper approach roads and easy access to safe drinking water. Health facilities are poor at best. Unless the country ushers in another Green Revolution, it would not be able to feed its growing population by 2030.

The land resources of the country are being misused and diverted to uses other than agricultural on a massive scale. This calls for a better management of our land resources. The first step in this direction is the development of a land information system that enables the government, the planners, and the people to know what is happening to the land resources of the country, to what use they are being put, who owns them, and in what physical condition they are?

Fortunately, we have now access to modern information technology which can help us keep tap on our land resources. It is the Geographic Information System (GIS) which includes collection of data, and their storage and analysis with relation to their spatial distribution at various stages of information handling. When the data in the GIS is related to various quantitative and qualitative aspects of land resource, it is termed as Land Information System (LIS).

An LIS may be briefly defined as an on-line repository of information that enables the most efficient use of land and a system that relates macro level development programmes and land use practices to the individual landholdings and their owners. Application of LIS implies that the spatially referenced land data stored in the database is related to the land parcel available in cadastral records discussed in chapter 11. The LIS will comprise of graphic elements derived from cadastral maps and non-graphic attribute data obtained from cadastral records suitably supplemented with additional information useful for planning purposes.

But the task is not easy. The cadastral system is outdated in most the states of the country. It was supposed to be updated every 30 years, but in most of the states no new survey was carried out since independence in 1947. The original cadastral surveys were based on technologies available 100 years back. Further, lot of tempering has been done since then at the local levels in the numbering of parcels and their boundaries. The legal cover for the land owners is provided through a sale deed on stamp paper. Although it is a legal document but it does not ensure that the piece of land purchased is necessarily the one which exists on the ground.

In view of the deficiencies in the cadastral system, the Planning Commission called for proper maintenance of land records in First, Second and Third Five Year Plans. A target time of five years was laid by the Sixth Plan from 1980-1985 for compilation and updating of land records. The Seventh Plan went further and advocated scientific survey of unmeasured lands. Computerization of land records was started under a centrally sponsored scheme during this plan period and was continued during the Eighth, Ninth, and Tenth Plans.

Unfortunately, a computer based LIS is not yet operational in the country. If we consider the traditional cadastral system as the base for LIS, then India has a village level LIS comprising of two distinctive components: one, cadastral map showing the boundaries and numbers of the parcels of land and two, descriptive land register in which the title and other land related information is recorded. However, both the entries are defective and hence of limited value for the planners. Further the access to information was limited to only those who held the records.

NEED FOR A MODERN LIS

A country like India with its rural based economy largely dependent on agriculture and its planning process is multi-level (Nation, State, District, Block and Village) the most suitable model for an LIS is one which is multi-level wherein data pertaining to each parcel of land is fully described in the cadastral records, and generalized picture becomes available at higher territorial levels.

The village-LIS will be useful for micro-level planning as it can contain all information, both graphic and non-graphic pertinent to the village planning requirements. The graphic database will store all data relating to graphic elements which describe the limits of parcels, its adjacency and other feature information. The non-graphic database, apart from containing information available in the land registry records can also include other attribute data such as house hold, cultural, socio-economic data etc. which would be relevant for planners at village level.

At higher territorial levels, the issues addressed by the planners are different. It is not the individual land holdings which are important. They, instead look at the needs of a group of contiguous villages or a larger area of land with identified common requirements such as all lands falling in different watersheds, land under forests, land under water bodies, etc. An LIS for Block planners could be generated from village LIS using the village cadastral maps and generalizing and aggregating the non-graphic attribute data. Here it is not the individual parcel of land that is important; it is the grouping of the parcels that matter. The non-graphic land data will also cover the common land attributes of the grouping and in addition any other land data which will be relevant for inter-village planning at this level.

The next level of hierarchical plan formulation could be a District. At this level the issues to be addressed will be far more general and will focus on infrastructure facilities like communications network, power, hospitals, colleges, etc. which constitute elements of meso level planning. The data requirements at this level will therefore be of a vastly different nature. Here, the data in the database needs to be referenced to the villages and towns, e.g. development of road networks, infrastructure, marketing facilities etc. will be with reference to various sizes of human settlements. The Survey of India topographic maps on 1:25,000 in which the village boundaries and municipal boundaries are depicted, could serve as ideal base maps for generating the LIS at District Level. While the Block or Tehsil level LIS's will contain the village boundaries, it will be possible to aggregate the non-graphic attribute data village wise in the LIS and map it into the District LIS. Any other information necessary for the district level of planning can also be incorporated in the databases.

At the state and nation levels too, we need LIS. First, the role of the state and national governments is very crucial in improving the District, Block/Tehsil, and Village LIS. Fund allocation for this purpose is made by the state governments. Further, there need for accurate land records for land use planning at the state level. Lack of land use planning leads to a great number of problems such misallocation of land for various activities, land valuation leading to confrontation with farmers, adverse environmental impact, etc.

We need LIS at the national level too that would be far more generalized. The national government would have to initiate and fund the multi-level LIS to encourage the state governments to take up this work seriously.

Such a multi-level LIS will ensure that plans generated at all levels are compatible as they will be based on information derived from a common source. It will also reduce redundancy in storage and eliminate duplication of efforts at data collecting and recording stage.

NEED TO MODERNIZING CADASTRAL SURVEYS

An accurate map is a basic requisite of an LIS. It is also desirable that these maps are on some standard scales and with accuracy standards to meet the contemporary needs. The projection system for these maps should be such that can facilitates seamless integration of map data when generating maps for higher level LIS. The present day cadastral maps, when viewed in the context of these requirements, are wholly inadequate and unsuitable because they are not based on a common scale. Scales differed from state to state and range from 1:500 to 1:5000. The accuracy level also varies from state to state.

The village maps are generally very old and either need heavy updating or have gone through various stages of manual settlement resulting in degradation of quality. The cadastral maps are drawn is Cassini projection wherein the scale is true along the central Meridian and in a direction at right angles to it. But the scale along the other meridians is exaggerated as we go away from the central meridian. Hence, the projection is suitable only for areas having small extent in East-West directions. This therefore, renders it unsuitable for larger areas and for seamless integration of maps. Yet another drawback of the cadastral maps is the lack of information about the elevation of the land.

Thus there is an urgent need to modernize the cadastral surveying techniques and to update the cadastral records if we want to develop a modern LIS. The first step in this direction would be to determine a common scale, common projection, and a common surveying technique for the Base Map.

SCALE OF BASE MAP

Factors determining the scale of surveying are desired accuracy, size of parcels and land value. In rural and sparsely populated areas the parcel sites are generally larger and land value lower, and hence a relatively smaller scale survey is adopted. Urban and semi-urban areas, where the land value is high and plot/parcel sizes are small, require higher measurement accuracy and larger scale mapping. Hence, it is suggested that the new cadastral surveys be carried out using three scales:

- 1:2000 for rural areas where the parcel sizes are bigger and land value is lower
- 1:1000 for rural areas where the parcel sizes are small and land value is high
- 1:500 for urban/semi-urban areas with high land value

The Tehsil/Block—LIS should be on 1:10,000 scale and the district LIS on 1: 50,000 scale. Both could be based on original surveys or on village level cadastral maps.

PROJECTION FOR BASE MAP

A common projection system will not only facilitate ease of merger of adjacent sheets but also upward integration of graphic elements from Village level to Tehsil level and District level—an essential pre-requisite for the hierarchical LIS to evolve to GIS. An ideal projection would be one which is conformal, equidistant, and equivalent i.e. represents shape, direction, and area correctly. Since, there's no projection which can simultaneously meet all the three criteria, we have to select a projection wherein the distortion in shape, distance and area are the least.

The Cassini projection on which Cadastral mapping is done is not suitable because it gives a highly distorted picture. It would be better to use Polyconic, Lambert orthomorphic conical or Transverse Mercator projection for the distortions of shape, direction and area are the least on these projections. Moreover, the existing cadastral survey practices are too outdated to be used now. The pilot project carried out by the Survey of India in Orissa for Angul Nalco Project used Analogue photogrammetric instruments fitted with encoders to capture 3-D data digitally. The results achieved were consistent with the accuracy requirements of the cadasters. The other options are: use of GPS and Total Station; use of Soft Copy Photogrammetric Work Station with scanned aerial photography; and plotting on analogue photogrammetric instrument and digitizing the photogrammetric plotted section.

These new techniques can give accuracy that modern cadastral surveys require and the output thereof comes in the digital form which could be fed directly in the LIS. The first two methods automatically record the third dimension (height) too. The third method produces a 2-D plot, but by tagging heights to contours during digitization, and interpolation, height too can be assessed.

The task of carrying out fresh surveys for cadastral mapping in a country like India is a huge one and can be done only if the national government takes it up as Mission. It is time consuming, needs lot of money, man power and sophisticated equipment. It cannot be done by the state or local governments. The Survey of India alone would not be able to accomplish this task. Fortunately, a number of private agencies are in a position to participate in this venture. And ideal approach would be PPP with Central, and State Governments together with relevant agencies in the private sector join hands to take up the cadastral re-survey in a time bound manner to produce appropriate base maps for LIS.

Introduction of LIS in India on the lines suggested above would be immensely useful in implementing land reforms and land consolidation, and would bring down the number of land related disputes in the country. The Planning Commission of India has been trying to impress on the state governments to distribute the surplus land to the landless in rural areas. LIS would point out the vacant land. It will also help in the consolidated of the fragmented holdings of the farmers into economically viable parcels. It would also help in identifying land that can be allocated for common community uses through proportional deduction from each individual holder.

The networking of the LISs and GISs at the state level will further enhance the utility of the information system especially because any modifications in the land use or land ownership carried out at the grassroots level would be instantaneously reflected at the district, state and national levels. The common man would benefit most from this development. All the relevant information about land would be easily accessible to farmers, and other stakeholders.

GEOTEC'S CADASTRAL MAPPING

Gujarat State

Geotech has undertaken a series of cadastral mapping applications and evaluated the feasibility of an automated LIS development in Gujarat State. The work has been done at the initiative of the Settlement Commissioner and Director of Land Records in the state. It involved the computerization of the existing land survey records, called Tippans and generation of Cadastral level village maps for the Government of Gujarat. The agency has developed software to computerize the land records from the old records some of which are 200 hundred years old and in vulnerable physical condition and to generate an updated revenue land record with MS unit measurements. The LanRec software has the facilities to automatically generate individual survey plots (TIPPANS), Unit conversion, field measurement tables (FMT), plotting/reproduction of the TIPPANS and finally mosaic them together to create a village cadastral map. In the absence of individual plot ownership drawings, LanRec software can generate individual survey plots from the village cadastral map. The same can be printed automatically with a specific map layout.

Rajasthan State

Geotech had earlier developed village level digital database for Rajasthan Government. It digitized individual village maps with administrative boundaries, cadastral data, transportation networks, various amenities and natural resources etc. the number of information layers were nearly 10 in number. For a quick query and retrieval under Arc/Info GIS environment more than 120 attributes were hot-linked with spatial map data.

And finally, all these village maps were mosaiced to make a seamless district/ state level digital database. Further, these digitized spatial layers were integrated with the existing tabular data facilitating options for performing queries and analyses. Thereby the system could generate various socio-economic parameters and spatially depict them in the form of maps for better understanding of the regions in the state.

Overseas Experience

Geotech has executed such projects for several Cities/Counties of US Federal Government departments. Every parcel of land has been allocated a unique code, called Parcel

Identification Number (PIN). The source information for the above code is acquired from paper drawn measurement sheets, ammonia prints and Ortho-photo maps prepared on varied scales like 1/100, 1/200 or 1/400. Computerized COGO based CAD tools have been developed in-house and were employed in these projects to deliver accurate databases. Automated the process for calculating the acreage of each parcel and linked with other attributes. The same data has been further integrated in a GIS environment and used for creation of city map guides and address geo-coding applications.

COMPUTERIZATION OF LAND RECORDS IN INDIA

The importance of computerization of land records in rural areas cannot be over emphasized, particularly in this IT age and where India is an IT leader. But unfortunately rural India is lagging far behind in e-governance.

Land has been the main source of livelihood from ancient times. Maintenance of land records in India is therefore as old as Indian civilization. The system has however undergone changes in response to administrative and socio-economic changes. The present system of preparing and maintaining land records originated from the Moghal period and reached its scientific form during the British rule. Since Independence in 1947, only minor changes have been effected without altering the overall system. With growing population and demand for land from other sectors of the economy, land records have acquired new significance and hence their revision and updating have become necessary.

The cadastral maps depicting land parcels are supposed to be updated every 30 years through the survey and settlement operations. But most of the States of India have not done any survey or settlement operations during the last 60 years. As such the land records available from revenue officials at the local level cannot be relied on while processing the claims of ownership and possession. The problem has been taken note of at various forums. The First Plan (1950-54) took note of this fact that in a primarily agrarian economy, it has serious implication for socio-economic development. In December 1988, the Conference of Revenue Secretaries of States recommended immediate updating of land records. According to the Sixth Plan document, "Systematic programmes would be taken up for compilation/updating of land records for completion within a period of five years, i.e. 1980-1985. In states, where the backlog is heavy, aerial survey techniques may be employed for expeditious survey operations. Each cultivator would be given a passbook indicating his status/title to description of the land viz. area, cess, etc. along with a copy of khasra/map and other details that are considered necessary. Appropriate provision will be made in revenue laws to confer legal status on these documents as proof of title and rights in land". Similarly, the Seventh Plan document also emphasized the need for updated and accurate land records. According to the Seventh Plan document, "Land records form the base for all land reform measures and, therefore, regular periodic updating of land records is essential in all states. This will necessarily have to include a scientific survey of unmeasured land and recoding of rights of tenants and share-croppers which have remained unrecorded up till now".

The Eighth Plan (1992-1997) and the Ninth Plan (1997-2002), Tenth Plan (2002-07) and Eleventh Plan (2007-12) also envisaged the updating of land records. Land being the main asset providing the primary and secondary needs of the people, maintaining land records is a pre-requisite for rural development. Fortunately, we have modern information technology at our service. It can help us store, process and retrieve data pertaining to land. Thus, computerization of land records is a must for effective development from below as envisaged in the 73rd Amendment to the Constitution of India.

A Conference of Revenue Ministers of states/UTs in 1985 suggested that the Central Government take up the computerization of land and crop-based data on a pilot basis in one Tehsil/Revenue Circle of each state/UT. The same year (1985) a Study Group representing the Ministry of Agriculture, the Central Statistical Organization and State Governments of Karnataka, Madhya Pradesh, Maharashtra, Tamil Nadu and Uttar Pradesh recommended computerization of Core Data in land records. The Planning Commission, however, did not pursue the proposal at that point of time.

Two years later in 1987, a national workshop reviewed the experience of different states in computerization of land records made at their own initiative and recommended to the Government of India to fund this programme on a pilot project basis. In pursuance of this recommendation, the Department of Rural Development selected 8 districts one each in Andhra Pradesh (Rangareddy), Assam (Sonitpur), Bihar (Singhbhum), Gujarat (Gandhinagar), Orissa (Mayurbhanj), Madhya Pradesh (Morena), Maharashtra (Wardha), and Rajasthan (Dungarpur). The Government approved the proposal in 1988 on the condition that the project should be completed in 6-8 months; the state governments should commit to computerization of the land records once the pilot project is over and the benefits thereof are brought about; and an officer with knowledge, training and experience in handling computers should be made incharge of the project and should be posted in the district chosen for the purpose.

The objectives of the scheme as stated in the Memorandum for Expenditure Finance Committee (EFC Memo), submitted by the Ministry of Rural Development in 1993 were:

1. To facilitate easy maintenance and updating of changes which occur in land database such as changes due to availability of irrigation/natural calamities/ consolidation/or on account of legal changes like transfer of ownership, partition, land acquisition, lease etc.
2. To provide for comprehensive scrutiny to make land records tamper-proof and to reduce the menace of litigation and social conflicts associated with land disputes.
3. To provide the required support for implementation of development programmes for which data about distribution of land holdings is vital.
4. To facilitate detailed planning for infrastructural as well as environment development.
5. To facilitate preparation of an annual set of records in the mechanized process and thereby producing accurate documents for recording details such as collection of land revenue, cropping pattern etc.

6. To facilitate a variety of standard and ad-hoc queries on land data.
7. To provide database for agricultural census.

By 1991-92, the scheme was extended to 24 districts in different states viz., Haryana, H.P., J&K, Karnataka, Kerala, Manipur, Punjab, Tamil Nadu, Tripura, Sikkim, Uttar Pradesh, West Bengal and Delhi UT. The Eighth Plan (1992-97) approved the scheme as a separate centrally-sponsored scheme on computerization of land records and allocated Rs. 594.2 million covering 299 new districts. Separate funds (Rs. 50.2 million) were allocated for the ongoing pilot projects in 24 districts.

Since 1994-95, the National Informatics Centre (NIC) has taken the responsibility of the supply, installation and maintenance of hardware, software and other peripherals needed to expand the scheme further to cover all the districts of the country. Apart from the technical support for proper implementation of the scheme, the NIC is also responsible for providing training to the revenue officials. All expenses related to the implementation of the scheme such as site preparation, data entry work, furniture, etc. are met by the Ministry of Rural Development, Government of India.

During the first year of the Ninth Plan i.e. 1997-02, 177 new districts were included in the scheme bringing 475 taluks/tehsils under the scheme during 1997-98. A year later in 1998-99, the scheme covered 625 more tehsils/talukas. During 1990-2000, the scheme was operational in 544 districts of the country leaving only those districts where there are no land records.

As in 2009, all the districts of the country where land records exist have been covered by the scheme. But lot more needs to be done. Whatever has been done so far is not available to the landowners even after two decades of pumping in funds and time. To encompass the computerization of land records in its entirety instead of just trying to deliver the text database, three parallel activities need to be done:

1. Digitization of the cadastral maps.
2. Transfer of existing database of parcel information into ORACLE or some other such appropriate RDBMS and defining which year's data is required to be kept while the rest of it may be either erased or left on CDs/floppies in DBFs for a certain defined period of time.
3. Building up of cadastral maps (in areas where these are not available) by, either—

   (i) Resurveying with GPS so that digital maps with geo-coordinates are produced, or
   (ii) Converting the data from FMB (Field Measurement Books) into maps, if a survey has already been done.

The LIS system would take off from GIS system which consists of digital cadastral maps linked with parcel information of landowners in ORACLE or some other appropriate RDBMS. It would be a "Query and Modify Software" working on an extensive network initially built across each state from the districts and subsequently tied up at the office of the Commissioner Land Records and/or Settlement Commissioner/Director Land Records at the state level. It would be done by NIC on lines of the work they have done in Kerala where they have connected 152 blocks in 14 districts. It is possible to cover the entire country.

The government scheme of computerization of land records since 1988-89, has not proceeded with the desired pace. The main objective of the scheme was to provide landowners with computerized copies of ownership, crop and tenancy and updated copies of Records of Rights (RoRs) on demand. According to the recent records maintained by the Union ministry for rural development only 13 states out of 35 states and Union territories are in a position to provide RoRs on demand. These States are Andhra Pradesh, Chhattisgarh, Goa, Gujarat, Karnataka, Madhya Pradesh, Maharashtra, Rajasthan, Tamil Nadu, Sikkim, Uttar Pradesh, Uttarakhand and West Bengal.

The scheme of Computerization of Land Records (CLR) was extended to cover 4,423 talukas or tehsils (revenue circles) in 1021 sub-division in 582 districts of the country. Computer centres have been set up in 3,356 tehsils and in only 2902 tehsils computerized copies of RoRs are being issued to landowners on demand. Till January 31, 2008 the Central Government released Rs. 5613 million to state governments. Only 62 per cent of this amount was utilized.

There is another centrally sponsored scheme called Strengthening of Revenue Administration and Updating of Land Records (SRA&ULR), launched in 1987-88 on 50:50 contribution basis between the Centre and the states. The scheme aims at strengthening of the survey and settlement organizations of the states. Funds are provided for purchase of modern survey equipment like GPS, EDM, total stations, theodolites, work stations, aerial survey office, photocopiers, and laminating machines. Since the inception of SRA&ULR, Centre has provided Rs. 4172 million to state governments, out of which only Rs. 2831.3 million was utilized, implying 68 per cent utilization of funds.

It has been proposed to merge the above two scheme (CLR and SRA&ULR) into a new scheme called National Land Records Modernization Programme (NLRMP). The new scheme would generate three layers of data namely spatial data from satellite imagery or aerial photography, topographic maps and other data from the Survey of India and Forest Survey of India and land record data and maps.

26
Global Positioning System

Global Positioning System (GPS) is a navigational and positioning system developed by the US Department of Defense by which the location of a position on or above the earth can be determined by a special receiver placed at that point. The receiver interprets signals received simultaneously from several of a constellation of satellites especially commissioned for the purpose. Most common use of the GPS is the location-based service for mobile devices.

GPS, earlier known as NAVSTAR (Navigation System with Time and Ranging) or Navigation Star, is an Intermediate Circular Orbit (ICO) satellite navigation system used for determining one's geographical location with respect to latitude, longitude and altitude and providing accurate time reference almost any where in the world.

The G of GPS stands for Global but it cannot be used in places, which are not in direct view of sky, such as:

1. Interior of buildings
2. Underground shelters, caves or mines
3. Heavy downpour of rain or snow
4. Under heavy tree canopy
5. In "Urban Canyons" connecting tall buildings
6. Near powerful radio transmission antennas.

The P answers questions like: Where are you? How fast and in which direction are you moving? In what direction should you go to reach some other location and how long would it take at your speed to get there? And above all: Where have you been? And the S of the GPS stands for System, which consists of a number of components, which are inextricably linked with each other both vertically and horizontally.

Designed and controlled by the United State Department of Defense, the GPS is now accessible to all free of charge for geo-referencing, positioning, navigation, and time and frequency control.

SEGMENTS OF GIS

One can gain a better understanding of GPS by studying it in three segments;

1. Space segment;
2. Control segment;
3. User segment.

The Space Segment

The system consists of constellation of 24 satellites in 6 orbit planes. Each satellite circles the earth twice every day at an altitude of 20,200 km. Number of operational satellites at a particular time is, however, more than 24 as new ones are launched to replace older satellites still in service. The satellite orbit tracks (as the earth turns i.e. rotates beneath them) once each day. The orbit altitude is such that the satellites repeat the same track and configuration over any point approximately every 24 hours (4 miles a minute earlier each day). There are six orbital planes, equally spaced (6.0 degree apart) and inclined at about 55° with respect to equatorial plane (Fig. 26.1).

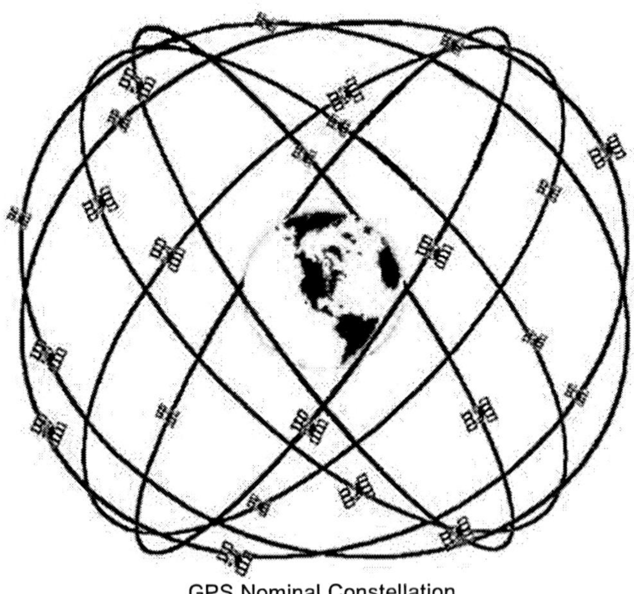

GPS Nominal Constellation
24 Satellites in 6 Orbital Planes
4 Satellites in each Planes
20,200 km Altitudes, 55 Degree Inclination

Source: www.colorado.edu

Fig. 26.1: GPS Nominal Constellation

Control Segment

The control segment consists of a system of tracking stations located around the world. The tracks of the GPS satellites are influenced by the gravitational effect of moon and sun. Further, they are crammed with electronics making it necessary to monitor their track and innards. This is accomplished by four ground base stations near the equator, one on Ascension Islands in the South Atlantic at Diego Garcia in the Indian ocean, the second on Kwajalein Atoll in the Pacific, the third in Hawaii, in North Pacific, and the fourth, the master control station, at Sehriever (formerly Galcon) Air Base near Colorado Springs, Colorado, USA.

Each satellite passes over at least one monitoring station twice a day. Information developed by monitoring station is transmitted back to the satellite, which in turn rebroadcasts it to GPS receivers. Subjects of a satellite's broadcast are the health of the satellite's electronic, how the track of the satellite varies from what is expected, the current almancre for all the satellites and other, more exoteric subjects, which need not concern us at this point. Other ground-based stations exist primarily for uploading information to the satellites (Fig. 26.2 and 26.3).

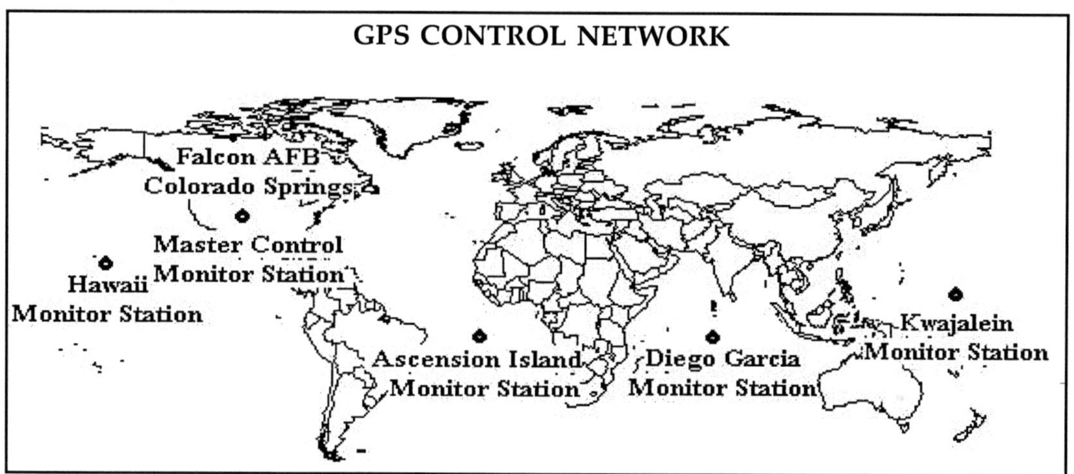

Global Positioning System (GPS) Master Control and Monitor Station Network

Source: www.colorado.edu

Fig. 26.2: GPS Master Control and Master Station Network

User Segment

The user segment consists of GPS receivers and the user community. The GPS receiver converts space vehicle signal into positions, velocity and time estimates. The receivers receive signals from four satellites in order to find out its own longitude, latitude, elevation and precise time. The receivers compute the distance to four satellites by the difference between local time and the time satellites signals were sent and then it decodes the satellites location from their own radio signals and an internal database.

CONTROL AND NAVIGATION SYSTEMS OF GPS

The receiver should be located at the interaction of four spheres, one around each satellite, with a radius equal to time delay between satellite and the receiver multiplied by the speed of radio signals. The intersection gives the precise location of the receivers, if elevation information is not required, and only signals form three satellites are needed (Fig. 26.3).

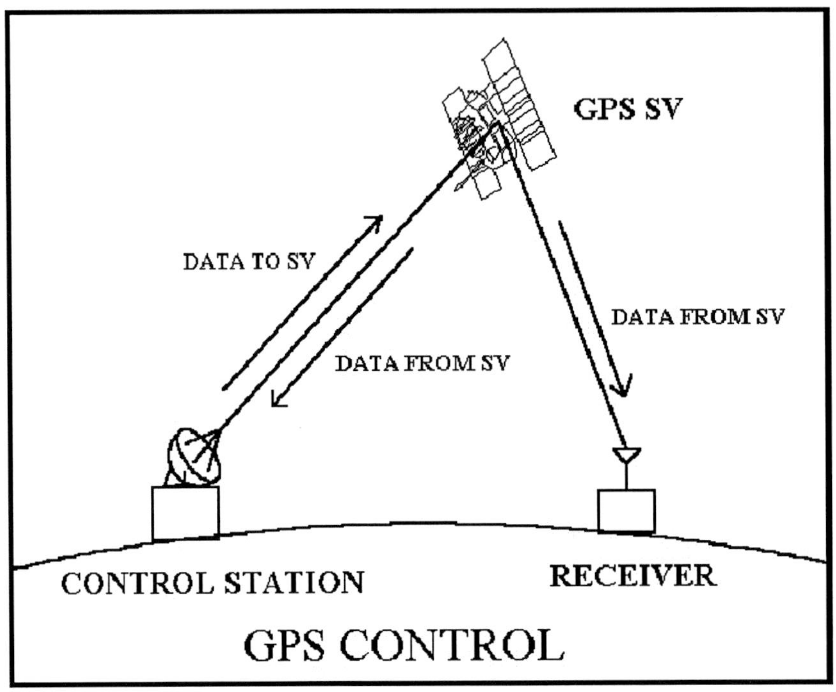

Fig. 26.3: GPS Control System

In reality four satellites rarely intersects. There are several reasons for it: the initial local time may be a guess: the relatively imprecise clock of receivers, the radio signals more slowly as they pass through ionosphere, the receiver may be moving. The receiver then applies an offset to local time, and, therefore, to the sphere radii, so that the spheres finally do intersect in one point.

The GPS receivers are used for navigation, positioning, time dissemination, and other research purposes. Navigation in three dimensions is the primary function of GPS navigation. Receivers are made for aircraft, ships, ground vehicles, and for hand carrying by individuals. Precise positioning is possible using GPS receivers at reference locations providing corrections and relative positioning data for remote receivers. Surveying, geodetic control, and plate tectonic studies are some examples.

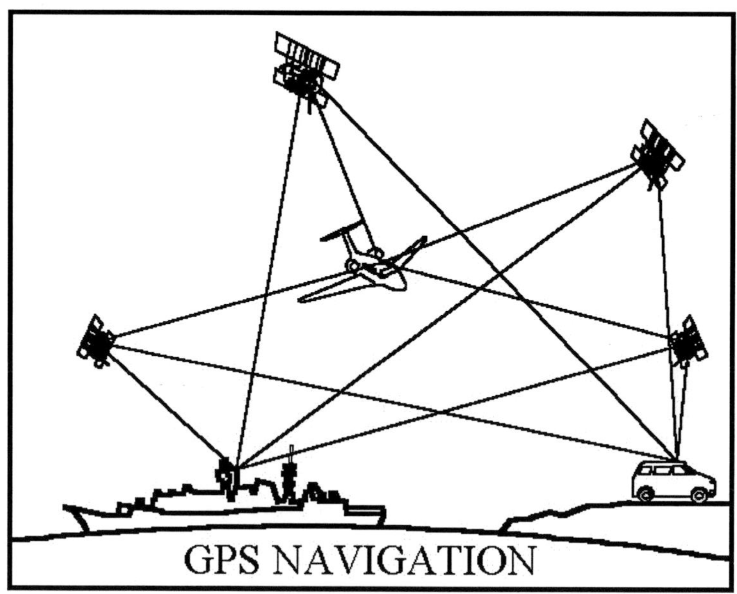

Reference: www.colorado.edu

Fig. 26.4: GPS Navigation

Time and frequency dissemination based on precise clocks on board the space vehicles are controlled by the monitor station, is another use of GPS astronomical observatories, telecommunication facilities, and laboratory standards can be set to precise time signals or controlled to accurate frequencies by special purpose GPS receivers.

USERS OF GPS

There are two kinds of users:

1. Authorized Users, and
2. Civil Users

The Authorized Users use what is known as Precise Positioning Service (PPS). The Authorized users have cryptographic equipment, key and special receivers. US and Allied armed forces, certain US department, government agencies, and selected civil authorities approved by US government use the PPS.

PPS Predictable Accuracy of the PPS is as follows:

- Horizontal accuracy: 22 meters
- Vertical accuracy: 27.7 meters
- Time accuracy: 200 nanoseconds

The Civil Users also use Standard Positioning Service (SPS) without charge or restriction. The SPS accuracy is internationally degraded by the US Department of Defense. The SPS accuracy is as follows:

- Horizontal accuracy: 100 meter
- Vertical accuracy: 156 meter
- Time accuracy: 340 nanoseconds

DIFFERENCE BETWEEN GPS AND GPRS

The space vehicles (satellites) transmit two microwaves carrier signals. The L1 frequency (1575.42 MHZ) carries the navigation massage and the SPS code signals. The L2 frequency (1227.60 MHZ) is used to measure the ionospheric delay by PPS equipped receivers. Three binary codes shift the L1 and L2 carrier phase.

The C/A code (coarse acquisition) modulates the L1 carrier phase. The C/A code is repeating 1 MHz pseudo random noise (PRN) code. This noise like code modulates the L1 carrier signals, spreading the spectrum over a 1 MHz bandwidth. The C/A code repeats every 1023 bits. There is a different C/A code PRN for each SV. The PRN number identifies GPS satellites. The unique number identifies for each pseudo-random-noise code. The C/A code then modulates the L1 that is the basis for the civil SPS.

The P-code (precise) modulates both L1 and L2 carrier phase. The P-code is very long (seven days) 10 MHz PRN code. In the anti-spoofing (AS) mode of operation the Y-code. The encrypted Y-code requires a classified AS module for each receiver channel and for use only by authorized by cryptographic keys. The P (Y) code is basis for PPS.

The navigation message also modulates the L1-C/A code signal. The navigation message is a 50 HZ signal consisting of data traits that describes the GPS satellite orbits, clock corrections and other system perimeter (Fig. 26.5).

SOURCES OF GPS ERRORS

GPS errors are a combination of noise, bias and blunders.

Noise errors are the combined effects of PRN code noise within the receiver noise (around, meter). Bias errors result from selective availability SA) and other factors

Selective Availability (SA)

1. SA is intentional degradation of SPS signals by a time varying bias. SA is controlled by the department of defense to limit accuracy for non-US military and government users. The potential accuracy of CA code of around 30 meter is reduced to meters.
2. The SA bias on each satellite signal is different, and so the resulting position solution is a function of the combine SA bias from each SV used in the aviation

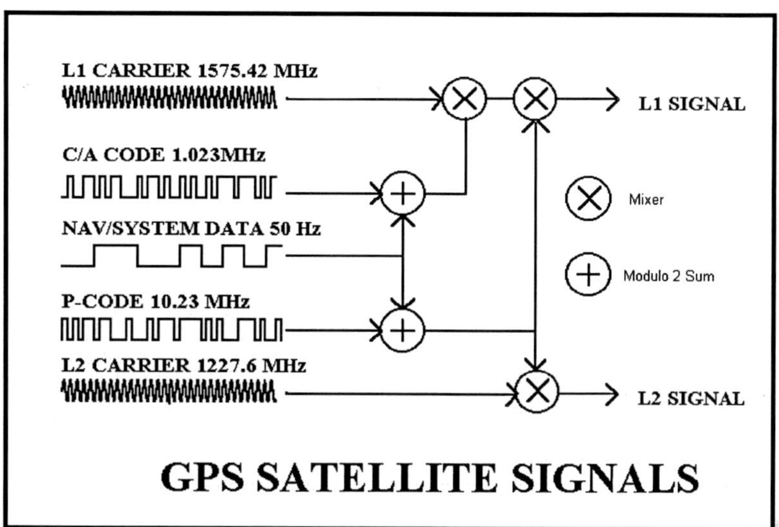

Reference: www.colorado.edu

Fig. 26.5: GPS Satellite Signals

solution. Because SA is a changing bias with low frequency terms in excess of a few hours, position solution or individual SV. Pseudo-range cannot be effectively averaged over periods shorter than a few hours.

Other Sources of Bias Errors

1. SV clock errors uncorrected by control segment can result in 1-meter error.
2. Ephemeris data error: q meter.
3. Troposphere delay: 1 meter. The troposphere is lower part of atmosphere. The experiences of changes in temperature, pressure, and humidity associated with weather change. Complex modes of topospheric delay require estimates or measurement of these parameters.
4. Unmodeled ionosphere delay: 10 meter. The ionosphere is the layer of atmosphere form 50 to 500 km that consists of ionized air. The transmitted model can only remove about half of the possible 70 ns of delay leaving a ten meter un-modeled residual.
5. Multipath: 0.5 meter. Multipath is caused by reflections form surface, means the receiver that can either interface with or be mistaken for the signal that follows the straight linepath from the satellite. Multipath is difficult to detect and some time hard to avoid.

Blunder Can Result in Error of 100 kilometres

1. Control segment mistake due to computer or human error can cause error from one meter to hundred of kilometre.

2. User mistake, including incorrect geodetic datum solution, can cause error from 1 to 100 metres.
3. Receiver error from software of hardware Faviusers can cause blunder error of any size.
4. Noise and bias error combined, resulting in typical ranging errors of around fifteen meters for each satellites used in position solution.

USES OF GPS

There are many uses of GPS, important among them are:

1. For mapping: the most basic use of GPS is for mapping. It can be used to update the reference map or to map object whose locations were previously unknown or inaccurate.
2. For spatial analysis: the next logical step for mapped point data location is to perform spatial analysis data. The simplest form of spatial analysis is point pattern analysis, in which the researcher uses computer algorithm to determine if the point location are clustered, evenly distributed or randomly distributed.
3. Environmental scientist are using GPS to locate fields plots and analyze the characteristics of the vegetation, find animal roots and examine spatial determinants of roost locations based on surrounding landscape, and identify locations of fishing debris in the ocean.

A GPS was used to locate temporary grazing shelter and pasture in addition to collect ground over types with in the study area. Pasture location were mapped and GIS modelling was used to predict the most heavily grazed land. Land cover GPS data were used to classify a Landsat Thematic mapper sȧtellite image and map the land cover in the study area. Apart from this GPS technology is being heavily used in the management of transport, health care researches, and defense.

REFERENCES

Crampton, Jeremy W. (2001) 'Maps as Social Constructions: Power, Communication and Visualization', *Progress in Human Geography* 25, no. 2 : 235-52.

Dixon, Conrad (1994), *Using GPS*, paperback, 90 pages, Published by Sheridan House.

Dye, Steve and Frank Baylin (1997), *The GPS Manual: Principles and Applications*, published by Baylin Systems.

Hoffmann-Wellenhof, B., Lichtenegger, H., and Collins, J. (1993), *GPS Theory and Practice*, 2nd ed., New York: Springer-Verlag.

Kaplan, Elliott D. (1996), *Understanding GPS: Principles and Applications*, Artech House.

Kennedy Michael (2002), *The Global Positioning System and GIS: An Introduction*, London: Taylor and Francis.

Kleusberg, Alfred and Peter J.G. Teunissen, Eds. (1996) *GPS for Geodesy (Lecture Notes in Earth Sciences*, Vol. 60, New York: Springer Verlag.

Larijani, L. Casey (1998), *GPS for Everyone: How the Global Positioning System Can Work for You,* paperback, 384 pages, published by American Interface Corporation.

Letham, Lawrence (1995), *GPS Made Easy: Using Global Positioning Systems in the Outdoors,* paperback, published by Mountaineers Books.

Pickles, John, ed. (1995), *Ground Truth: The Social Implications of Geographic Information Systems,* New York: Guilford.

Spencer, J. *et al.,* (2000), Global Positioning System: A Field Guide from the Social Sciences, Blackwell Publishing, UK.

Trimble Navigation (1989), GPS: A Guide to the Next Utility Paperback, Published by Trimble Navigation.

Wells, D, (1987), *Guide to GPS Positioning,* 2nd ed, Canadian GPS, Associates, Frederiction, New Brunswick, Canada.

27
International Cooperation in Cartography

Founded in 1945 after the Second World War, the United Nations (UN) has a long history of working to maintain peace and security around the world. The international organization, composed of 192 member states, strives to improve relations among nations and protect human rights.

In the UN Secretariat alone, approximately 40,000 staff members work each day to implement UN programmes and policies. From Darfur in western Sudan, Africa, to Lebanon, Haiti, and beyond, GIS is supporting those who work in departments and offices within the UN Secretariat so they can provide essential security for millions of people, as well as fragile institutions, emerging from conflict and recovering from humanitarian emergencies Providing Geographic Knowledge

With increasing role of cartography, geospatial analysis and cartographic products in various aspects of human life, international cooperation through the United Nations in has become indispensable. This cooperation has been promoted by the cartographers themselves but earliest initiative came from the United Nations. Three front organizations have spearheaded this cooperation. We would discuss each of these in some detail.

Progress of cartography depends as much on cooperation between cartographers and subject specialists, as among cartographers of different parts of large countries like India and different countries of the world. United Nations has been taking keen interest in the development of cartography ever since its formation in 1948.

India has Indian Cartographic Association which meets annually and provides a forum for discussion and exchange of views and experiences. Then there is International Cartographic Regional Conferences (UNRCC) which meet once in two years in different regions of the world. It was established under ECSOC Resolution 131 (VI), adopted on 19 February, 1948. Text of the Resolution is as follows:

> "WHEREAS accurate maps are a prerequisite to the proper development of the world resources which in many cases lies in relatively unexplored regions;

WHEREAS such maps facilitate international trade, promote safety of navigation, both in the air and in the sea, and provide information required for the study of measures of peaceful adjustment under Chapter VI of the Charter, and for application of security measures contemplated in Chapter VII of the Charter;

WHEREAS several Member Governments have already indicated their interest in a coordinated programme of international cartography;

The Economic and Social Council Therefore Recommends:

1. That the Member governments stimulate the accurate survey and mapping of their national territories;
2. That the Secretary-General take appropriate action within the limits of budgetary availabilities:
   (a) to further such efforts by promoting the exchange of technical information and other means, including the preparation of a study on modern cartography, methods and development of uniform international standards.
   (b) to coordinate the plans and programmes of the United nations and specialized agencies in the field of cartography, taking into account the work of the various inter-governmental and non-governmental organizations and to report on the subject to a subsequent session of Council;
   (c) to develop close cooperation with cartographic services of interested Member Countries."

On 27 July, 1954, the ECSOC resolved vide its Resolution No. 556 (XVIII), to hold regional cartographic conferences for Asia and the Far East, Africa and Americas. The first Asia and Far East Conference was held at Dehradun, India 1955; the first such conference for Africa was held in 1963, and the one for Americas in 1976. Since 1994, the Asia and Far East Conference has been renamed as Asia and Pacific Conference.

UN GEOGRAPHIC INFORMATION WORKING GROUP (UNGIWG)

The United Nations Geographical Information Working Group (UNGIWG) is a network of professionals working in the fields of cartography and geographic information science to building the UN Spatial Data Infrastructure needed to achieve sustainable development.

UNGIWG was formed in 2000 to address common geospatial issues—maps, boundaries, data exchange, standards—that affect the work of UN Organizations and Member States. UNGIWG also works directly with non-governmental organizations, research institutions and industry to develop and maintain common geographic databases and geospatial technologies to enhance normative and operational capabilities. UNGIWG reports periodically to the UN Chief Executive Board (CEB) on progress made and priority issues.

UNGIWG aims at:

1. Improving the efficient use of geographic information for better decision-making;
2. Promoting standards and norms for maps and other geospatial information;
3. Developing core maps to avoid duplication;
4. Building mechanisms for sharing, maintaining and assuring the quality of geographic information; and
5. Providing a forum for discussing common issues and emerging technological changes.

UNGIWG consists of 30 members. Fig. 27.1 illustrates its present structure.

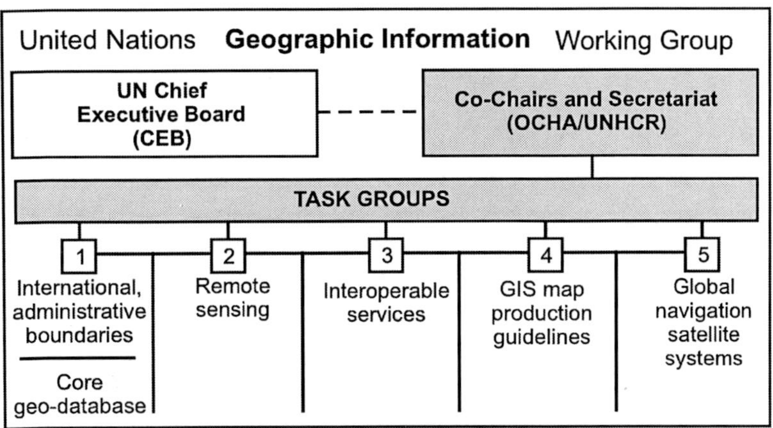

Fig. 27.1: Organizational Structure of UN GIS Working Group

Background and Recommendations

The UNGIWG's two main areas of concern are: Data and Services. Both of these areas have a common ground dealing with Standards. There are five Task Groups (see Fig. 27.1) to take care of its activities:

1. Core Geo-Database
2. Remote Sensing
3. Interoperable Services
4. GIS Map Production Guidelines
5. Global Navigation Satellite Systems

The Task Groups consist of experts from relevant organizations. Each Group is headed by a coordinator.

Task Group 1: Core Geo-Database

This Task Group has two main objectives:

(a) to identify and prepare a global framework core data layers for the UN systems;
(b) to give a working platform for the international community to collect, manage, analyze, visualize and share the sub-national data down to the second sub-national level.

It looks at the following set of fundamental layers:

(a) International and administrative Boundaries;
(b) Human Health: boundaries and facilities;
(c) Human Population: centres and distribution;
(d) Infrastructure: roads, railways, airports, harbors, and navigation;
(e) Bathymetry, Topography and Digital Elevation Model;
(f) Hydrology: drainage, water bodies, watersheds

The task group is looking at the issues that would insure that the initiatives being taken by the UN agencies with respect to these layers create a homogeneous and seamless UN Core Geo-Database. Two specific projects directly attached to this task group are: The International Boundaries project which offers 1:1 and 1:15 million-scale dataset for International Boundaries. The Second Administrative Level Boundaries (SALB) project was launched in 2001. It provides to the UN Member States, the following information and data validated by National Mapping Agencies of respective countries:

(i) The contact information of the National Mapping Agencies;
(ii) The changes observed since 1990 at the first sub-national level and since 2000 at the second sub-national level together with the relevant maps;
(iii) The digital maps (ESRI format) for each second administrative level through time since January 2000. Second administrative level

One of the major outcomes of this project is a coding scheme to manage sub-national data through time and space. For details about the project, see at http://www.who.int/whosis/database/gis/salb/salb_home.htm.

Task Group 2: Remote Sensing

The Task Group on Remote Sensing has been set to:

(a) Inform people about the very high-resolution imageries of different areas by various organizations;
(b) Ensure easy ordering, rapid access and reduced cost to data by arranging satellite imagery purchasing agreements; and

(c) Expand the range and objectives for the Space Charter to better comply with UN requirements.

Satellite imagery and other remotely sensed data are now widely used in geospatial analysis. This Task Group helps in finding and accessing very-high resolution imageries. It promotes sharing and purchasing data through the UN Systems Contract and other agreements in order to improve the use of same data among all UN organizations.

(a) The UNOSAT website provides a database of images available whole UN system. Currently populated with UNOPS Landsat and ASTER images, UN organizations are invited to upload metadata about images they wish to share with others, thus avoiding duplication of purchase.
(b) Developing a UN Systems Contract for purchasing satellite data. This contract ensures significant discounts on a wider range of data and allows for multi-agency use of the data.
(c) Developing an overview of which UN organization has expertise in specific fields of specialization within Remote Sensing.
(d) Global Landsat imagery from 1980, 1990 and 2000 has been provided free of charge to the UN in order to analyze changes during the past three decades.

Task Group 3: Interoperable Services

This Task Group has been constituted to improve access to and interactive use of spatial data to facilitate data sharing and support decision-making through international standards and specifications. It helps members to further develop their 'respective geospatial capacities in a compliant and interoperable manner.' It is now possible to exchange a substantial amount of geospatial data between agencies, and eliminate duplication. "UNGIWG has adopted the ISO/TC211 standard number 19115 on Geospatial Metadata as its baseline. From this standard a draft minimum field subset has been developed, tested and is in the process of being recognized as a UN standard. Details of the fields and tools developed for creating and editing metadata are provided in 'word format' or as an 'XML template' containing sample data."

The UNGIWG members are also developing a number of data portals with a view to aid data discovery. These portals allow GIS users to effectively document, manage, and publish their data. It is now possible to overlay datasets from disparate sources.

Task Group 4: GIS Map Production Guidelines

This Task Group brings together a related set of practice on mapping. Its activities include:

(a) Collection of documents covering GIS and mapping standards from various GIS and mapping professionals such as:
   (i) style guides,
   (ii) map templates,

(iii) symbology, and
(iv) production guidelines.

One can share one's document by submitting them to the task group. One can register to download files. For guidelines **click:** *http:www.ungiwg.org/map/guidelines*

(b) Offer monthly tips and tricks on mapping. **Click:** *http:www.ungiwg.org/map/tips_tricks*
(c) Provide the subscriber list which may reflect GIS/mapping activities around the world. **Click:** *http:www.ungiwg.org/map/contact.php*
(d) Build the community by spreading the word to other GIS/mapping colleagues to subscribe. All subscribers would receive monthly tips and tricks and be able to download files. **Click:** *http:www.ungiwg.org/map/map/subscription*

Task Group 5: Global Navigation Satellite Systems

The objective of this Task Force is to harmonize and facilitate field data collection activities undertaken by the different UN agencies. It has two main activities. These are:

(a) harmonized set of training tools; and
(b) survey the use of global positioning devices within each agency at country level to prioritize training activities.

SDI Initiatives

The United Nations has evolved its own Spatial Data Infrastructure (UNSDI) that delivers to users integrated spatial information services. It can address both technical and non-technical issues, ranging from technical standards and protocols, institutional and organizational issues, including data access policy and the creation and maintenance of geospatial information for a wide range of themes.

Members of the UNGIWG

More than 30 UN Departments, Funds, Programmes and Specialized Agencies are currently (2010) members of UNGIWG:

1. UN Department of Economic and Social Affairs
2. UN Department of Management
3. UN Department of Political Affairs
4. UN Department of Public Information
5. UN Department of Peacekeeping Operations
6. Office for Coordination of Humanitarian Affairs
7. United Nations Office on Drugs and Crimes
8. United Nations Office on Drugs and Crimes

9. UN Office of Legal Affairs
10. Nations Office for Outer Space Affairs
11. United Nations Office at Geneva
12. United Nations Economic Commission for Africa
13. Economic Commission for Europe
14. Economic Commission for Latin America and the Caribbean
15. Economic and Social Commission for Asia and the Pacific
16. Comprehensive Nuclear-Test-Ban Treaty Organization
17. Food and Agriculture Organization of the United Nations
18. International Atomic Energy Agency
19. International Civil Aviation Organization
20. Joint United Nations Programme on HIV/AIDS
21. United Nations Development Programme
22. United Nations Environment Programme
23. United Nations Educational, Scientific and Cultural Organization
24. United Nations Population Fund
25. United Nations Human Settlements Programme
26. United Nations High Commissioner for Refugees
27. United Nations Children's Fund
28. United Nations Institute for Training and Research
29. United Nations Office for Project Services
30. United Nations University World Bank
31. United Nations World Food Programme World Health
32. Organization World Meteorological Organization

UNGIWG and the International Organization for Standardization (ISO) cooperate to align UN technical specifications for geographic information with agreed International Standards. It also cooperates with the Open Geospatial Consortium (OGC) to create open and extensible applications and interfaces for geographic information systems technologies.

In addition to the member, a number of organizations and commercial Companies are recognized as observers. They are:

1. CGIAR CSI-CGIAR Consortium for Spatial Information (*http://csi.cgiar.org*)
2. ESRI (*http://www.esri.com*)
3. EUROGI—European Umbrella Organisation for Geographic Information (*http://www.eurogi.org*)
4. LEICA—Leica Geosystems Geospatial Imaging (*http://gis.leica-geosystems.com*)
5. OSCE—Organization for Security and Co-operation in Europe (*http://www.osce.org*)
6. SPOT IMAGE (*http://www.spotimage.fr*)

Data Portals

Those interested in securing relevant information through the internet could look into the following websites:

- FAO GeoNetwork *http://www.fao.org/geonetwork*
- WHO GeoNetwork *http://www.who.int/geonetwork*
- UNEP.net *http://unep.net*
- United Nations Entry Point to Geospatial Data Clearinghouse *http:// gateway.ungiwg. org*
- UNOSAT Remote Sensing Data Archive (http://www.unosat.org/ungiwg http://www.unosat.org
- Global Administrative Unit Layers (GAUL) *http://www.fao.org/geonetwork/srv/ en/metadata.show?id=12691*
- Geographical Names (*http://unstats.un.org/unsd/geoinfo* International and administrative boundaries *http://boundaries.ungiwg.org*
- International Charter "Space and Major Disasters" *http://www.disasters charter.org* WFP VAM-SIE *http://vam.wfp.org/geonetwork*

INTERNATIONAL CARTOGRAPHIC ASSOCIATION

International Cartographic Association (ICA) plays a very significant role in promoting international cooperation in cartographic fields and advancing the professions related to globally. It is the global voice of individual cartographer and all those who are directly or indirectly related to map making, map reproduction and marketing of maps, atlases and other products.

Cartography has expanded so much and so fast, that it would not be an exaggeration if we give it the status of industry. With the arrival of internet and World Wide Web our access to information has changed vastly. The importance of cartographic education and the role of both national and International cartographic entities and associations have changed. Reproduced below is a paper 'Changing Role of Geospatial Associations' Prof. William Cartwright, President, International Cartographic Association advising associations like the ICA to review how they support education in cartography, internationally, and how to best deliver educational programmes.

ICA has a number of aims

1. To contribute to the understanding and solution of problems worldwide through the use of cartography in decision-making processes;
2. To foster the international dissemination of environmental, economic, social and spatial information through mapping;
3. To provide a global forum for discussion of the role and status of cartography;
4. To facilitate the transfer of new cartographic technology and knowledge between nations, especially to the developing nations;

5. To carry out or to promote multi-national cartographic research in order to solve scientific and applied problems;
6. To enhance cartographic education in the broadest sense through publications, seminars and conferences; and
7. To promote the use of professional and technical standards in cartography.

Changes in Industry

The cartographic profession has changed to one that is now supported by contemporary digital production, storage and distribution devices and communication resources. What have also changed are the organizations that conduct mapping programmes and 'build' repositories of geographical knowledge, digital and material. Today is the workplace of small government agencies; contractors (large and small); regional, national and global publishing collaborations; and map producers/users.

In addition, there are now new players in geospatial information provision whereby consumer electronics companies like Nokia have purchased mapping companies. With Google wanting to enhance its advertising potential and Microsoft its profile in computer software through Web presence, other players are also now part of the geospatial industry.

Another change for cartography is the fact that an amateur map publisher is now a map provider as well as traditional publisher. Collaboration via Web 2.0 (the use of social software for personal publishing) allows users to produce maps as mash-ups, which combine already available maps and satellite imagery and overlay their own data. Base maps or satellite imagery is 'mashed' together with pre-existing geospatial artifacts. There are a number of social, collaborative organizations, for example, Open Street Map, that make their data and maps available for free. Through a network of Web-connected amateurs and professionals, geographical data is collected, scripts marked-up with data attributes, data placed in repositories and maps published.

Everything has changed, but the underlying need for useful (and usable), accurate and timely geospatial products remains the essential underpinnings for what we do.

In order to ensure that both the professional and amateur map producers/users are attuned to the issues related to contemporary cartography, it is essential that appropriate educational programmes be provided, face-to-face and on-line. ICA has strategies in place that addresses this. These strategies are outlined as below.

Education: Actions proposed in the ICA Strategic Plan

The ICA Strategic Plan proposes the following actions:

1. To analyze existing university curricula in cartography and GI science, and encourage promotion of the goals contained in the ICA mission;
2. To help widen the cartographic/GI science knowledge base and skills into new segments of society;

3. To increase efforts directed to capacity-building, especially in developing countries, especially with reference to human resource development;
4. In co-operation with commercial suppliers, to develop virtual academy courses on cartography and GI science to support and complement what is currently available;
5. To provide geospatial data for educational use;
6. To establish a network of university/school teachers to provide a forum for discussion and the possibility for support; and
7. To extend travel awards for young scientists; lower conference participation fees for students.

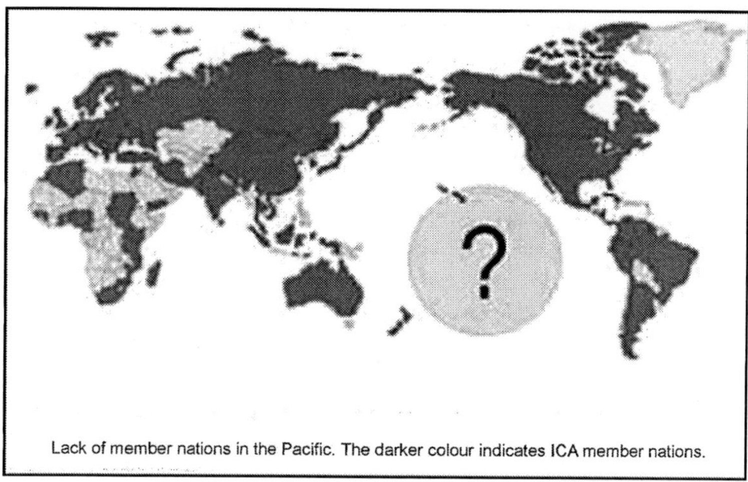

Lack of member nations in the Pacific. The darker colour indicates ICA member nations.

Fig. 27.2: Membership of the ICA

Following these goals, ICA can assist in the delivery of appropriate courses by assessing the availability of programmes; delivering education courses directly; providing support materials and publications; and providing outreach support for developing nations.

Availability of Programmes

International Cartographic Association (ICA) has investigated the possibility of offering courses and programmes online, via discrete media (CD-ROM) and conventional paper products. This investigation was undertaken by the ICA Education Commission, in association with other commissions and working groups.

Direct Delivery

Also through its education and training commission, ICA has developed a virtual course on cartography and GI science in collaboration with contributions from universities and individual academics. Users of this Web-delivered resource can assemble a hybrid course. In future, this concept could be expanded to include various implementation levels, which could be:

1. ICA designed and delivered.
2. ICA designed and commission-delivered.
3. Commission designed and delivered.
4. Working group designed and delivered.
5. Delivered by universities with ICA endorsement and branding.
6. Delivered by organizations or companies with ICA endorsement and branding.

Publications

To achieve its aims, ICA operates through a number of commissions and working groups. Commissions and working groups carry out the general operations of the ICA. These organizations have addressed the full range of scientific, technical and social research, which is the mark of ICA activity.

One important outcome of this work is the generation of extensive publications. This activity provides a focus for commissions and working groups and allows for knowledge about advances in contemporary thinking and research to be disseminated. The publications include books, journals and the ICA newsletter.

Outreach

International Cartographic Association (ICA) has always championed outreach programmes. In particular, the Association has a working group called Mapping Africa for Africa. However, we need to extend our role in Australasia and Oceania to extend the activities to nations who are not yet members. When we look at a map of member nations of the ICA, Pacific nations, in particular, are missing. For Australasia, there are national cartographic organizations in Australia and New Zealand that are members of ICA. These organizations have as their core activities the progress of the professional, academic and scientific status of all working in the geospatial professions. We need to ensure that Pacific nations benefit from belonging to ICA. We need to champion extended ICA representation in the Oceania region and promote activities and events in cartography and geographic information science that occur in the region.

We can do this by:

1. Forming a regional consortia to champion cartographic activities (these can be both physical and virtual).
2. Championing activities and alliances that advance cartographic knowledge in the region.
3. Encouraging nations to participate in ICA activities, both at a regional and international level.
4. Disseminating information at conferences, through the provision of books, journals, educational programmes, outreach programmes like workshops and perhaps a cartographic 'roadshow'.

5. Organizing a regional cartographic conference/event (perhaps in conjunction with the UN Asia-Pacific Regional Cartographic conference), linked to either Australia or New Zealand, or a trans-Tasman initiative to 'kick-start' such an event, but 'owned' by the region and offered/operated by contributing nations.
6. Eliciting journal papers that provide regionally-focused information (on topics that range from historical, to practical/production, to cultural to research).

In 2011, France will host the 25th Conference of the International Cartographic Association (ICA), marking the fiftieth anniversary of the first General Assembly of the Organization. During these fifty years, the ICA has skillfully remained true to its two fundamental objectives, namely, to maintain its position as the benchmark international conference for cartography and geomatics, thanks to its solid and progressively updated scientific content, whilst remaining a conference which is open to a variety of participating groups: researchers, decision-makers in the field of geomatics, teachers and lovers of cartography.

To this end, the conference is structured in four segments:

1. The conference proper, comprising some 500 oral presentations over five days, organized in themed sessions which reflect the twenty-eight commissions and working groups of the ICA;
2. Scientific workshops organized by the commissions of the association.
3. An exhibition for the presentation of the players involved in geomatics: suppliers of softwares, hardware, data bases and maps, consultancy firms, geomatic educational institutions or publishers; and
4. Two exhibitions of maps, one showing the most recent products from some fifty countries, and the other of children's drawings.

In its organization of the ICA in July 2011, on the theme of an enlightened view on Cartography and GIS, the French Committee of Cartography is seeking to combine thematic diversity, high-quality contributions and an up-to-date content:

1. Diversity will be ensured by the thematic variety of the twenty-eight commissions and working groups, ranging from the most theoretical (theories of cartography and spatio-temporal modelling) to the most practical (production lines, new softwares, participative data bases and Neo-Geography or data law).
2. The quality of contributions will be ensured by a process of selection which will encompass both a leading-edge scientific audience and a broader public, thanks to the delivery of three types of presentations:
   (a) reference presentations, selected from full-length papers by the scientific committee, will be published in a scientific volume for distribution at the conference,
   (b) selected presentations from abstracts. These papers will be published in digital form, and

(c) posters will also be produced in digital format.

3. The latest scientific and technical developments will be extensively represented, with an emphasis on new theories and practices associated with cartography, geomatics and information technology. Consideration of the links between Web 2.0 and geographic information will have its place, as will the links between art and digital cartography.

By inviting the major scientific and industrial players in the fields of Cartography and GIS to Paris in 2011, we hope to provide all parties with a better insight into both the history and the current state of our field, together with an opportunity to participate in the future development of a sector which is essential to public security, the sustainable development of the entire planet and the fulfillment by each individual of their role as a global citizen.

CONCLUSION

In conclusion, international organizations like the International Cartographic Association must be realistic about their contributions to international professional and amateur communities. They can't contribute to all efforts. They need to work with their strengths and promote cartographic events and educational programmes internationally and regionally, especially considering the regional needs."

—*William Cartwright, President ICA*

Email: *w.cartwright@rmit.edu.au*

APPENDICES

Appendix A

ICA COMMISSIONS
Their Objectives and Activities
by José Jesús Reyes Nuñez and Temenoujka Bandrova

Commission on Cartography and Children

1. Disseminate the outcomes of research into map use by children and young people.
   Deliverable: an enhanced set of online bibliographic tools and Web links on cartography and children.
2. Develop stronger links between the ICA and relevant international and regional organizations.
   Deliverable: a joint international conference within the EU, other regional organization or worldwide.
3. Stimulate a debate on all aspects of cartography for children with the aim of collecting and publishing the research results (in digital and/or printed format).
   Deliverable: a collection of papers on case studies in educational cartography.
4. Promote the use of digital cartography and GIS in schools. *Deliverable:* an international (and/or regional) conference and associated publication.
5. Provide consultative support in relation to the rules and judging of the Barbara Petchenik Children's World Map Competition.
   Deliverable: appropriate support as required.

Commission on Digital Technologies in Cartographic Heritage

Aim of the commission is the study and research on the possibilities of applying new Information and Communication Technologies (ICT) on issues related to the cartographic heritage, focused on the following Terms of Reference:

1. Support transformation of old maps, globes and cartographic documents into digital form.
2. Apply digital techniques in the study of old maps.
3. Develop digital tools for assisting the work of map curators and the functionality of map libraries networking and access to cartographic heritage.

4. Apply digital support for the preservation and restoration of old maps, atlases and globes.
5. Foster ICTs in teaching and diffusing history of cartography and maps to the public.

Commission on Education and Training

1. Work for reaching the general aims conceived in the ICA Statutes.
2. Develop the existing ICA-sponsored Internet cartography teaching programme (modules), with regional workshops for training the teachers.
3. Produce an ICA-sponsored Internet programme for continuing professional development, with regional workshops for training the teachers.
4. Support cartography and cartographic education in developing nations by holding seminars in these nations.
5. Promote integration by establishing linkage with the committees on education of sister surveying and mapping disciplines, by maintaining the list of institutions providing cartography course programmes, and interact as much as possible with other ICA commissions and working groups (organize joint meetings).
6. Produce appropriate publications for dissemination of results of the above efforts and publish it in proceedings or on the CET website according to the IOF publication policy.
7. Promote joint project cooperation between participating nations represented on the commission.
8. Contribute to international forums where applicable.
9. Develop mapping awareness competitions.
10. Investigate the role of emerging technologies for use in education and training.

Commission on Generalization and Multiple Representation

1. Foster research and practical experience in multi-scale issues and automated map generalization and to form a network and focal point for researchers and practitioners in this domain.
2. Foster and maintain a liaison with related commissions and working groups within the ICA as well as in related national and international organizations (e.g. ISPRS, IGU, EuroSDR), including:
   (a) identification of contact persons
   (b) timely mutual updating on planned activities
   (c) organization of joint conference sessions and/or workshops
3. Maintain a World Wide Web site for electronic dissemination of research on multi-scale issues and map generalization, including:
   (a) a membership database
   (b) a bibliographic database
   (c) examples of research projects and test data

   (d) links to related web sites
4. Organize sessions and meetings for exploring the issues named above, including:
   (a) workshops held every two years in conjunction with international cartographic conferences
   (b) special sessions and/or panel sessions at international cartographic conferences, in collaboration with local organizers
5. Produce publications for dissemination of the above efforts, including one special issue of a peer-reviewed scientific journal or a book every two years.

Commission on Geospatial Analysis and Modelling

1. Network cartographers, geographic information scientists, and other researchers involved in geospatial analysis and modelling.
2. Facilitate the interaction and communication between the computational community and cartographic community for creation of geographic knowledge.
3. Foster a new research community centreed on visual geospatial analysis and modelling.
4. Organize ICC sessions and ICA workshops on geospatial analysis and modelling.
5. Edit and publish geospatial analysis and modelling related reports, books, and special issues with some scientific journals.

Commission on Geospatial Data Standards

1. Continue the work on developing conceptual models of the Spatial Data Infrastructure (SDI) using various modelling techniques, working in the areas of science, technology and standards, at the Global, Regional and National levels.
2. Explore research in the area of spatial semantics, ontologies and standards, in collaboration with other ICA Commissions.
3. Participate at the scientific level with other organizations active in standards and SDIs. This work will be closely coordinated with the ICA Executive Committee.
4. Develop reports, conference presentations and/or journal articles on our work, or hold specialist meetings on these topics.
5. Help to arrange workshops on standards and present tutorials at such workshops.
6. Organize reporting sessions on the Commission's activities at the 2009 and 2011 ICA meetings.
7. Organize and hold full Commission meetings in the summers of 2008, 2009, 2010 and 2011.
8. Organize and hold mini-meetings of the Commission at suitable conferences

and meetings being attended by members of the Commission.

9. Continue to serve as a Category A Scientific Liaison with ISO/TC211, Geographic Information/Geomatics, and establish liaisons with various SDI organizations.
10. Exchange information and reports by the ICA member countries and relevant bodies concerning the development standards for the transfer of digital spatial data, metadata and SDIs.
11. Maintain the Commission website.
12. Identify research and education needs that arise from the standards and SDI processes.

Commission on Revisualization

Objectives

1. Promote, develop and report upon the use of cartography in its widest sense in the exploration and analysis of spatial information through interactive visual interfaces.
2. Define short and medium term research goals that address key issues associated with this work and its application.
3. Encourage a multi-disciplinary and international approach to this work that draws upon and contributes to the efforts of relevant stakeholders (such as international and national organizations promoting and coordinating research), cognate disciplines and ICA Commissions and working groups.

Activities

1. Organize expert workshops for defining key issues, promising research directions, and opportune goals and, on this basis, planning further activities of the Commission.
2. Convene annual meetings focusing on particular discipline/theme/objective hosted in different countries in conjunction with international research conferences.
3. Develop and publish research presented at these meetings through peer-reviewed papers where appropriate.
4. Maintain a web-log of activities, research undertaken and related activities.

Commission on the History of Cartography

1. Investigate the impact of technological innovations (printing methods, topographical surveying, aerial photography, satellite imagery and geographical information systems) on cartography during the last two centuries.
2. Document the recent history of GI science in general and computer-aided

mapping and Geographical Information Systems in particular.

3. Examine the changing role of maps worldwide as cartographical (as opposed to historical) documents in history during the 19th and 20th centuries.
4. Examine the significance of the cartographical work undertaken during the 19th and 20th centuries by the traditional colonial powers and all countries which extended their sovereignty over territory beyond their borders.
5. Compile a list of useful sources on the history of cartography since 1800, which do not as yet appear in published catalogues or electronic databases.
6. Maintain a dedicated website to act as a database for interested researchers.
7. Hold annual meetings/conferences to discuss and monitor research work undertaken.
8. Publish research findings in accredited journals and on the Commission website.

Commission on Management and Economics of Map Production

1. Extend Production Management knowledge in economics, planning and organizing, cost calculations, guidelines for production, quality assurance and quality control.
2. Complete publications: Compendium of Cartographic Techniques, Flow Charting—Tools for Managing and Education, Surveys of software and hardware.
3. Investigate (new) channels for publication with emphasis on technical, managerial and financial aspects. This includes also map publishing. Amongst all matters like Printing-on-demand-Webpublishing and Archiving should be researched.

Commission on Mapping from Satellite Imagery

1. Maintain and improve the digital repository of the Commission on topics related to:
   (a) analysis of existing and forthcoming satellite imagery in view of their potential inputs for producing and updating topographic and thematic maps and databases.
   (b) cartographic capabilities of different remote sensing systems.
2. Produce a special edition the ICC theme "*Cartography and satellite imagery for the management of natural resources and the environment, early warning and natural disasters mitigation*" with papers to be presented at ICC2009 or ICC 2011, in a scholarly journal of remote sensing, or a book on Mapping from Satellite Imagery.
3. Develop closer links with international organizations concerned with the use of satellite imagery (e.g. UNOOSA, UNEP, PAIGH), in view of strengthening ICA's presence, and to produce joint courses or seminars on cartographic applications of satellite imagery.

4. Promote knowledge transfer at fundamental and advanced levels on the use of satellite imagery for cartographic applications related to natural and built environments, early warning and natural disaster mitigation.
5. Prepare promotional materials reflecting the Commission's terms of reference and activities that can be used to support ICA presence at international forums (e.g. World Map, Map Middle East, meetings of Joint Societies, UN meetings, other regional conferences). Outcomes: posters, fact sheets, booklets.
6. Network with ISPRS and other ICA commissions with similar interests in applications of satellite imagery. Such networking would be in view of producing joint workshops or seminars within the period 2007-2011.

Commission on Map Projections

1. Establish a repository of knowledge on map projections.
2. Promote map projections within all disciplines.
3. Provide a platform to promote research and a common language in the field of map projections.
4. Encourage collaboration between researchers within the field of mapping sciences as well as across disciplines.

Commission on Maps and Graphics for Blind and Partially Sighted People

1. Exchange and disseminate information on the design of and production technologies for maps and graphics for blind and partially sighted people.
2. Study and evaluate developing and appropriate technologies and resources for the improvement of maps and graphics for blind and visually-impaired people.
3. Explore the potential value of maps and graphics designed for blind and visually-impaired people to other user groups (e.g. children and those with learning difficulties).
4. Develop closer links with the World Blind Union and to extend this co-operation to the International Mobility Conference and other appropriate organizations to further world-wide co-operation, communication and education in the use of maps and graphics by blind and visually-impaired people.
5. Integrate the workings of this commission more closely with other ICA commissions.
6. Disseminate, as appropriate, results of the above efforts.

Commission on Maps and Society

This Commission has parallel academic and practical rationales to facilitate and encourage the use of social scientific and critical approaches to mapping and to encourage a more active engagement between cartographers and other groups in society engaged in mapping activities. In 2007-2011 it would:

1. Develop a Maps and Society web site to act as a clearing house and one stop shop for researchers and practitioners in this area.
2. Sponsor and organize sessions at future ICA conferences.
3. Organize two specialist workshops on topics related to Maps and Society; these would be *local meetings* to generate increased ICA participation from groups in different global contexts.
4. Encourage publication in this area, specifically publication of an issue of *The Cartographic Journal* in 2008-2009 arising from this theme and publication of a research monograph.
5. Collaborate and participate with representatives of other disciplines in meetings and seminars.
6. Hold *joint meetings* with specific mapping communities focusing upon different mapping encounters. In the first four years these might focus upon, artistic practice, participatory community mapping and graphic designer practice.

Commission on Maps and the Internet

1. Focus attention on disseminating maps and spatial data through distributed electronic networks including wireless networks. *Product:* Books, journal articles and websites.
2. Examine Internet and mobile Internet map usage and project future areas of growth. *Product:* Oral or published report.
3. Examine Internet and mobile Internet map user issues to better serve user needs. *Product:* Oral or published report.
4. Promote the exchange of information about effective Internet mapping for an international audience. *Product:* Workshop(s) and websites.
5. Promote instruction on Internet mapping through collaboration/coordination with other ICA commissions and other cartographic institutions.
6. Produce materials to be used in workshops.

Commission on Marine Cartography

1. Promotion of marine cartography.
   (a) Facilitate the networking of those member nations representatives on the commission as a resource through which marine cartography can be promoted.
   (b) Establish a website for marine cartography.
   (c) Establish and maintain routine and higher profile liaison with kindred scientific bodies such as the International Hydrographic Organization (IHO), the Hydrographic Society, Fédération Internationale des Géomètres (FIG), the International Association of Geodesy (IAG) and the International Geographical Union (IGU) and the International Society for Potogrammetry and Remote Sensing (ISPRS).

   (d) Encourage authors writing in the general area of marine cartography and provide contacts and assistance in the accessing of resources and material.
   (e) Represent and promote the work of ICA at appropriate fora.
   (f) Respond to issues referred by the ICA Executive Committee for advice and information.
2. Promotion of ICA and cartography as major contributors in the field of coastal zone management.
   (a) Continue to conduct, in conjunction with the IGU Commission on Coastal Systems, the Coast GIS series of symposia.
   (b) Explore opportunities for training and workshops in GIS applications to the Coastal Zone within developing nations.
3. Encouragement of publications in the areas of marine cartography and, in particular, publications dealing with the coastal zone.
   (a) Publish proceedings for all symposia/conferences/workshops organized by the commission. Encourage and where needed, seek out the preparation of articles in the area of coastal zone cartography.
   (b) Target leading experts publishing an authoritative volume dealing with cartography and GIS within the coastal zone (planned in conjunction with IGU).
4. Ongoing co-operative activity.
   (a) Provide a list of special features, objects and attributes relating to the coastal zone to IHO and to the Digital Geographic Information Working Group (DGIWG) for consideration.
   (b) Provide any proposed change to IHO Publication S-23 (Limits of Oceans and Seas) and to the Undersea feature names gazetteers (following the guidelines stated in IHO Publication B6).

Commission on Mountain Cartography

1. Further define the topics of Mountain Cartography and promote the methods and knowledge of mountain cartography among scientists and professionals in cartography and related fields.
2. Provide an updated, attractive web-site with information about Commission activities, links to other events and theme-specific knowledge.
3. Provide an updated web-portal with links to related web-sites and bibliographic information.
4. Emphasize cartographic design issues and map related representations in large scale topographic mapping.
5. Continue the well-established workshop series.
6. Promote publication activities (proceedings, web-proceedings, journal articles and special issues) and common research activities.

Commission on National and Regional Atlases

1. Produce a cookbook for the production of (digital) national and regional atlases.
2. Promote and support the design, production and use of national, regional, city and thematic atlases and atlas information systems.
3. Conduct workshops and/or seminars that present recent and potential atlas developments and innovations.
4. Document guidelines and procedures for the design, production and use of national, regional, city and thematic atlases.
5. Elaborate and maintain an electronic inventory of national, regional, city and thematic atlases containing the main characteristics of these map works as well as relevant addresses in order to support information exchange between atlas editors as well as to disseminate information on atlases to a wider public.
6. Maintain a commission website and update it regularly with the major aims of providing for information exchange between commission members and disseminating information on activities in this field to a wider public.
7. Encourage efforts for integrating data for the development of electronic atlases and national and regional atlas information systems by:
   (a) evaluating metadata standards as they apply to atlas design, creation and dissemination.
   (b) acknowledging requirements for diverse geographic and temporal data themes.
   (c) identifying and exploring various alternatives to data integration issues.

Commission on Planetary Cartography

1. Continue the documentation and analysis of the planetary cartography status within the international community.
2. Prepare and print the multilingual Glossary of planetary cartographic terms after necessary correction and reduction (in 5 languages).
3. Continue the series of multilingual maps in two parallel directions; i.e. to organize and print existing maps for new groups of languages, and to compile and print maps of other Solar System Bodies. In the next series the moons of the giant-planets will be included. The Io map is in progress.
4. Conduct a new project on construction of a Specialized Planetary Cartography GIS.
5. Discuss the role of planetary cartography in the development of the Extraterrestrial Territory Geography, especially in connection with the global education process.
6. Publish an Education Programme and a Textbook on this field.

Commission on Theoretical Cartography by Alexander Wolodtschenko

1. Study selected topics of theoretical interest such as:

(a) general and applied cartosemiotics
(b) map language and cartographic conceptions
(c) mapping as a cognitive process
(d) cartographic ontology
(e) terminology and toponymy

2. Compare and contrast theoretical structures in cartography.
3. Publish a cartosemiotic glossary.
4. Conduct workshops or symposia about selected aspects of theoretical cartography.
5. Compile and publish a report and collected papers about the work of the commission.
6. Continue to develop the Commission website.

Commission on Ubiquitous Mapping

1. Organize regional workshop including site observation to comprehend contemporary situation of ubiquitous and pervasive mapping.
2. Clarify similarity and difference in comparing variant systems to establish an evaluation scheme.
3. Place the notion of Ubiquitous Mapping in the domain of Theoretical.

Commission on Under-represented Groups and Cartography

1. Encourage representatives of under-represented groups to participate in all ICA activities.
2. Serve as a resource on issues related to gender and other under-represented groups problems and cartography by:
   (a) preparing papers for presentation at conferences or for publication in cartographic journals.
   (b) preparing and disseminating on the Web a basic, working bibliography on publications (professional papers, atlases, maps) related to gender and under-represented groups problems that are of interest to cartography.

(3) Contribute to the understanding and solution of under-represented groups problems through the use of cartography in decision-making processes by holding the debate and distribute the professional knowledge concerning the making and use of maps presenting spatial distribution of different problems relating the under-represented groups.

4. Prepare a guide for the production and use of maps presenting the different problems of under-represented groups in the most effective way.
5. Organize a conference on *using popular computer software for visualizing problems connected to under-represented groups.*
6. Continue the cooperation with other ICA Commissions and Working Groups, and with corresponding commissions or working groups of ICA sister societies.

Commission on Use and User Issues

1. Maintain an on-line bibliographical database with a bibliography on the user, usability (including User Centered Design and methods and techniques of user research) and improvement of user abilities.
2. Maintain a database of individuals working on, or with expertise in, the various use and user issues and denote it with keywords (similar to the category listings used in the bibliography) to stimulate exchange of information.
3. Produce a major publication on use and user issues in cartography and geo-information processing and dissemination (either a special issue or a refereed journal or a text book).
4. Foster and stimulate other publications on use and user issues in cartography and geo-information processing and dissemination.
5. Organize sessions on use and user issues at future ICA conferences.
6. Organize a specialist workshop in 2008 on a focused topic related to use and user issues in cartography and geo-information processing and dissemination.
7. Participate in/contribute to other workshops/seminars organized by representatives of other disciplines (e.g. computer science) or by other ICA Commissions/Working Groups.

Appendix B

UNITED NATIONS REGIONAL CARTOGRAPHIC CONFERENCES

In February 1948, the Economic and Social Council in resolution 131(VI) recommended that Governments of Member States stimulate surveying and mapping of their national territories and that the Secretary-General of the United Nations take appropriate action to further such efforts. A Committee of Experts on Cartography was appointed by the Secretary General to study the problem and advise upon the means of its implementation. The convening of regional cartographic conferences was recommended by the Committee, which met in March-April 1949, as an effective means to attain the objectives set forth in the resolution 131(VI). At its ninth session the Economic and Social Council noted the report of the Committee and requested the Secretary-General to consult with Governments concerning the early calling of such meetings. The results of these consultations indicated a wide measure of support and a new impetus was given to the project when the Government of India offered to act as host at the first conference for Asia and the Far East, to provide a meeting place, and to be responsible for all the physical arrangements for the conference. In accordance with the Economic and Social Council resolution 476(XV), further consultations were proceeded with. Many Governments expressed their desire to participate in the conference for Asia and the Far East, and proposed agenda items for the conference.

The Economic and Social Council, after consideration of this response adopted resolution 556(XVIII) on 27 July 1954. And thus the First UN Regional Cartographic Conference was born in 1955. Since 1957, it is being held at an interval of three years. Each Conference brings out two volumes: Vol. I-Report of the Conference, and vol. II-Proceedings of the Conference and Technical Papers. The dates and venue of the Conferences held so far are given below:

I. UN Regional Cartographic Conference for Asia and the Far East, *Mussoorie, India, 15-25 February 1955;*
II. UN Regional Cartographic Conference for Asia and the Far East, *Tokyo, Japan, 20 October-1 November 1958;*
III. UN Regional Cartographic Conference for Asia and the Far East, *Bangkok, Thailand, 27 October-10 November 1961,*
IV. UN Regional Cartographic Conference for Asia and the Far East, *Manila, The Philippines, 21 November-5 December 1964.*
V. UN Regional Cartographic Conference for Asia and the Far East, *Canberra, Australia, 8-22 March 1967,*

VI. UN Regional Cartographic Conference for Asia and the Far East, *Tehran, Iran, 24 October-7 November 1970*

VII. UN Regional Cartographic Conference for Asia and the Far East, *Tokyo, Japan, 15-27 October 1973*

VIII. UN Regional Cartographic Conference for Asia and the Far East, *Bangkok, 17-28 January 1977,*

IX. UN Regional Cartographic Conference for Asia and the Pacific, *Wellington, New Zealand, 11-22 February 1980,*

X. UN Regional Cartographic Conference for Asia and the Pacific, *Bangkok, 17-28 January 1983,*

XI. UN Regional Cartographic Conference for Asia and Pacific, *Bangkok 5-16 January 1987,*

XII. UN Regional Cartographic Conference for Asia and the Pacific, *Bangkok, 20-28 February 1991,*

XIII. UN Regional Cartographic Conference for Asia and the Pacific, *Beijing, 9-18 May 1994,*

XIV. UN Regional Cartographic Conference for Asia and the Pacific, Report of the Conference, *Bangkok, 3-7 February 1997,*

XV. UN Regional Cartographic Conference for Asia and the Pacific, Report of the Conference, *Kuala Lumpur, Malaysia, 11-14 April 2000,*

XVI. UN Regional Cartographic Conference for Asia and the Pacific, Report of the Conference, *Okinawa, Japan, 14-18 July 2003,*

XVII. UN Regional Cartographic Conference for Asia and the Pacific, Report of the Conference, *Bangkok, Thailand, 18-22 September 2006* (E/CONF.97),

XVIII. UN Regional Cartographic Conference for Asia and the Pacific, Report of the Conference, *Bangkok, Thailand, 26-29 October 2009.*

Appendix C

UNITED NATIONS CARTOGRAPHIC SECTION

The UN Cartographic Section (UNCS) is the core of the UN Secretariat's geospatial initiatives. It provides geographic visualization to the Security Council, offers the analyses of the current affairs to the Departments of Peacekeeping Operations and Political Affairs, and prepares cartographic illustrations for the report of the Secretary-General. It is in charge of the GIS programme of peace operations, managing GIS offices in 13 peacekeeping and political missions. Practically all UN agencies rely on UNCS for geographic information products. GIS data and analysis are used for boundary delimitation and demarcation, field mission planning and operations, humanitarian intervention, logistics, resource allocation, and critical analysis and visualization for situational awareness and security. GIS officers are placed in most UN field missions to assist the UN in building peace and stabilizing operations.

UNCS has launched a UN International Boundary Information System (UNIBIS) project to create a worldwide geodatabase on international boundaries. It will depict, as accurately as possible, their geometry, status, claims, dates of treaty agreements, and reference materials. UNIBIS highlights potential conflict areas to support conflict prevention activities and enhance readiness for assistance on boundary demarcation issues.

United Nations has so far undertaken 63 field missions for peace which have disarmed nearly half a million combatants and organized free and fair elections in over a dozen countries. In view of critical role GIS plays, the Panel on United Nations Peace Operations has made it now part of the UN peacekeeping infrastructure and is used throughout the range of activities that peacekeeping missions must perform. In Lebanon, after fighting erupted in 2006, the United Nations Interim Force in Lebanon (UNIFIL) created the GIS-based Security Warden Information System (SWIS) to support UN security operations in the area. The SWIS provides security and military response teams with accurate visual information related to staff locations, along with other important locations, such as UN positions, evacuation or gathering locations, and incident locations. It supports the safety and security of the UN staff in position there.

As mentioned earlier, the UN uses GIS to provide humanitarian assistance to people in need. The Office for the Coordination of Humanitarian Affairs (OCHA), for example, uses GIS as a communication tool to evaluate the humanitarian situation on the ground in the event of a natural disaster or any other humanitarian emergency.

OCHA coordinates the activities of all other humanitarian organizations that are grouped into nine clusters: camp management, early recovery and reconstruction, education, emergency shelter, health, IT and communications, nutrition, protection, and water and sanitation. Most UN agencies, such as United Nations Children's Fund (UNICEF) and World Food Programme, also use GIS in their humanitarian assistance work.

Appendix D

MUNICIPAL INFORMATION SYSTEM ASSOCIATION (MISA)

MSIA holds National Conference from time to time to deliberate on various issues and problems faced by municipal administrators. The theme of 2011 Conference was '100 Years of National Topographical Mapping: 1910-2010'. MSIA also publishes a Newsletter called e-CARTO. One can get further information about it from *www.mappingsciences.org.au/ecarto.htm.*

Bibliography

Akerman, James R., ed. *Cartographies of Travel and Navigation*. Chicago: University of Chicago Press, 2006.

Andrews, J.H., "What Was a Map? The Lexicographers Reply," *Cartographica* 33, no. 4 (1996): 1-11.

Andrews, J.H., *Meaning, Knowledge and Power in the Map Philosophy of J.B. Harley, Trinity Papers in Geography*, 6, Dublin: Department of Geography, Trinity College Dublin, 1994.

Avery, Bruce, "The Subject of Imperial Geography." In *Prosthetic Territories: Politics and Hypertechnologies*, ed. Gabriel Brahm, Jr. and Mark Driscoll, 55-70. Boulder, Colo.: Westview Press, 1995.

Axelsen, Bjørn, and Michael Jones. "Are all Maps Mental Maps?" *GeoJournal* 14 (1987): 447-64.

Banerjee, R.K. and Bireswar Banerjee, *Remote Sensing Techniques for Regional Development*, Concept Publishng Company, (2000).

Barbara Bartz Petchenik, "A Map Maker's Perspective on Map Design Research, 1950-1980," in *Graphic Communication and Design in Contemporary 10. Cartography*, ed. D.R.F. Taylor [New York: John Wiley & Sons, 1983], 37-68),

Belyea, Barbara. "Images of Power: Derrida, Foucault, Harley." *Cartographica* 29, no. 2 (1992): 1-9.

Biggs, Michael, "Putting the State on the Map: Cartography, Territory, and European State Formation," *Comparative Studies in Society and History* 41, no. 2 (1999): 374-405.

Black, Jeremy, *Maps and Politics*. London: Reaktion Books, 1997, Reprinted Chicago: University of Chicago Press, 1998.

Blakemore, Michael J., and J.B. Harley, "Concepts in the History of Cartography: A Review and Perspective." Ed. Edward H. Dahl, *Cartographica* 17, no. 4 (1980): Monograph 26.

Blaut, J.M., D. Stea, C. Spencer, and M. Blades, "Mapping as a Cultural and Cognitive Universal," *Annals of the Association of American Geographers* 93 (2003): 165-85.

Brealey, Kenneth G., "Networks of Power: Cartography as Ideology," *Western Geography* 3 (1993): 15-50.

Brückner, Martin, *The Geographic Revolution in Early America: Maps, Literacy and National Identity*, Chapel Hill: University of North Carolina Press, 2006.

Buisseret, David, ed. *From Sea Charts to Satellite Images: Interpreting North American History through Maps*, Chicago: University of Chicago Press, 1990.

Buisseret, David, ed. *Monarchs, Ministers, and Maps: The Emergence of Cartography as a Tool of Government in Early Modern Europe*, Chicago: University of Chicago Press, 1992.

Buisseret, David, *The Mapmaker's Quest: Depicting New Worlds in Renaissance Europe*, Oxford: Oxford University Press, 2003.

Burnett, D. Graham, *Masters of All They Surveyed: Exploration, Geography, and a British El Dorado*, Chicago: University of Chicago Press, 2000.

Casey, Edward S., *Representing Place: Landscape Painting and Maps*, Minneapolis: University of Minnesota Press, 2002.

Conley, Tom, *The Self-Made Map: Cartographic Writing in Early Modern France*, Minneapolis: University of Minnesota Press, 1996.

Cosgrove, Denis, *Apollo's Eye: A Cartographic Genealogy of the Earth in the Western Imagination*, Baltimore: Johns Hopkins University Press, 2001.

Cosgrove, Denis, ed. *Mappings*. London: Reaktion Books, 1999.

Craib, Raymond B., *Cartographic Mexico: A History of State Fixations and Fugitive Landscapes*, Durham, N.C.: Duke University Press, 2004.

Cumming, William P., *The Southeast in Early Maps*, Ed. Louis De Vorsey, Jr. 3rd ed. Chapel Hill: University of North Carolina Press, 1998.

De Vorsey, Louis, Jr., "American Indians and the Early Mapping of the Southeast." In William P. Cumming, *The Southeast in Early Maps*, 65-98, 3rd Ed. Chapel Hill: University of North Carolina Press, 1998.

Del Casino, Vincent J., Jr., and Stephen P. Hanna, "Beyond the 'Binaries': A Methodological Intervention for Interrogating Maps as Representational Practices," *ACME: An International E-Journal for Critical Geographies* <http://*www.acme-journal.org/*> 4, no. 1 (2006): 34-56.

Del Casino, Vincent J., Jr., and Stephen P. Hanna. "Representations and Identities in Tourism Map Spaces," *Progress in Human Geography*, 24, no. 1 (2000): 23-46.

Delano Smith, Catherine, "Map Ownership in Sixteenth-Century Cambridge: The Evidence of Probate Inventories," *Imago Mundi* 47 (1995): 67-93.

Edney, Matthew H., "Cartography without 'Progress': Reinterpreting the Nature and Historical Development of Mapmaking," *Cartographica* 30, nos. 2 and 3 (1993): 54-68.

Edney, Matthew H., "Cartography: Disciplinary History," in *Sciences of the Earth: An Encyclopedia of Events, People and Phenomena*, ed. Gregory A. Good, 2 vols. New York: Garland, 1998, 1:81-85.

Edney, Matthew H., *Mapping an Empire: The Geographic Construction of British India*, 1765-1843. Chicago: University of Chicago Press, 1997.

Edward H. Dahl, ed., "Concepts in the History of Cartography: Some Responses, with the Authors' Reply, Especially to Questions of Definition," *Cartographica* 19, no. 1 (1982): 77-96.

Edward H. Dahl, ed., "Responses to J.B. Harley's Article, 'Deconstructing the Map,' Published in the Last Issue of *Cartographica* (Volume 26, Number 2, Summer 1989, pp. 1-20)," *Cartographica,* 26, nos. 3 and (Begin Page 22) 4 (1989): 89-127.

Edwards, Jess. *Writing, Geometry and Space in Seventeenth-Century England and America: Circles in the Sand,* Routledge Studies in Renaissance Literature and Culture, 5. London: Routledge, 2006.

Escolar, Marcelo, "Exploration, Cartography and the Modernization of State Power," *International Social Science Journal* 49, no. 151 (1997): 55-75.

Fremlin, Gerald, and Arthur H. Robinson, "Maps as Mediated Seeing," *Cartographica* 35, nos. 1 and 2 (1998): *Monograph* 51.

Gilbert, Pamela K., *Mapping the Victorian Social Body,* Albany: State University of New York Press, 2004.

Groot, Erlend de, *The World of a Seventeenth-Century Collector: The Atlas Blaeu-Van der Hem,* translated by Andrew McCormick being Vol. 7 of Peter van der Krogt and Erlend de Groot, The Atlas Blaeu-Van der Hem of the Austrian National Library: Descriptive Catalogue, 2006.

Hannah, Matthew G., *Governmentality and the Mastery of Territory in Nineteenth-Century America,* Cambridge Studies in Historical Geography, 32. Cambridge: Cambridge University Press, 2000.

Harley, J.B., "Deconstructing the Map," *Cartographica* 26, no. 2 (1989): 1-20. Reprinted in *New Nature of Maps,* 149-68, and with slight modifications in *Writing Worlds: Discourse, Text, and Metaphor in the Representation of Landscape,* ed. Trevor J. Barnes and James S. Duncan (London: Routledge, 1992), 231-47.

Harley, J.B., "Historical Geography and the Cartographic Illusion," *Journal of Historical Geography* 15 (1989): 80-91.

Harley, J.B., "Maps, Knowledge, and Power," in *The Iconography of Landscape: Essays on the Symbolic Representation, Design and Use of Past Environments,* ed. Denis Cosgrove and Stephen Daniels, 277-312, Cambridge Studies in Historical Geography, 9. Cambridge: Cambridge University Press, 1988; Reprinted in *New Nature of Maps,* 51-81.

Harley, J.B., "New England Cartography and the Native Americans," in *American Beginnings: Exploration, Culture, and Cartography in the Land of Norumbega,* ed. Emerson W. Baker *et al.,* 287-313, Lincoln: University of Nebraska Press, 1994; Reprinted in *The New Nature of Maps,* 169-95.

Harley, J.B., "Power and Legitimation in the English Geographical Atlases of the Eighteenth Century." In, *Images of the World: The Atlas Through History,* ed. John A. Wolter and Ron E. Grim, 161-204. New York: McGraw-Hill for the Library of Congress, 1997; Reprinted in *New Nature of Maps,* 109-47.

Harley, J.B., "Rereading the Maps of the Columbian Encounter," Ed. Karl W. Butzer and William M. Denevan, *Annals of the Association of American Geographers,* 82 (1992): 522-36.

Harley, J.B., "Silences and Secrecy: The Hidden Agenda of Cartography in Early Modern Europe," *Imago Mundi* 40 (1988): 57-76, Reprinted in *New Nature of Maps*, 83-107.

Harley, J.B., "The Map User in the Revolution," In J.B. Harley, Barbara Bartz Petchenik, and Lawrence W. Towner, *Mapping the American Revolutionary War*, 79-110, Chicago: University of Chicago Press, 1978.

Harley, J.B., *The New Nature of Maps: Essays in the History of Cartography, ed.* Paul Laxton (Baltimore: Johns Hopkins University Press, 2001.

Harvey, David, (1989), "The Time and Space of the Enlightenment Project," In *The Condition of Postmodernity: An Enquiry into the Origins of Cultural Change*, 241-59, Oxford: Basil Blackwell.

Harvey, David, "Cartographic Identities: Geographical Knowledge under Globalization," in *Spaces of Capital: Towards a Critical Geography,* London: Routledge, 2001, 208-33.

Harvey, P.D.A., *The History of Topographical Maps: Symbols, Pictures and Surveys,* London: Thames and Hudson, 1980.

Helgerson, Richard (1992), *Forms of Nationhood: The Elizabethan Writing of England,* Chicago: University of Chicago Press.

Herb, Guntram Henrik (1997), *Under the Map of Germany: Nationalism and Propaganda, 1918-1945,* London: Routledge.

Hotstetler, Laura, *Qing Colonial Enterprise: Ethnography and Cartography in Early Modern China,* Chicago: University of Chicago Press, 2001.

Janet, Abrams and Peter Hall, eds., (2006), *Else/Where: Mapping. New Cartographies of Networks and Territories,* Minneapolis: University of Minnesota Press for the University of Minnesota Design Institute.

John F. Kasson, *Rudeness and Civility: Manners in Nineteenth-Century Urban America,* New York: Hill and Wang, 1990.

Kagan, Richard L. "*Urbs* and *Civitas* in Sixteenth- and Seventeenth-Century Spain," In *Envisioning the City: Six Studies in Urban Cartography*, ed. David Buisseret, 75-108, Chicago: University of Chicago Press, 1998.

Kagan, Richard L. *Urban Images of the Hispanic World, 1493-1793,* New Haven, CT: Yale University Press, 2000.

Kain, Roger J.P., and Elizabeth Baigent, *The Cadastral Map in the Service of the State: A History of Property Mapping,* Chicago: University of Chicago Press, 1992.

Karrow, Robert W. *Mapmakers of the Sixteenth Century and Their Maps: Bio-Bibliographies of the Cartographers of Abraham Ortelius, 1570.* Chicago: Speculum, Orbis Press for The Newberry Library, 1993.

Keates, J.S., *Understanding Maps,* New York: John Wiley and Sons, 1982. London: Addison Wesley Longman, 1996, 2nd. ed.

King, Geoff, *Mapping Reality: An Exploration of Cultural Cartographies.* New York: St. Martin's Press, 1996.

Kivelson, Valerie A., *Cartographies of Tsardom: The Land and its Meanings in Seventeenth-Century Russia,* Ithaca, N.Y.: Cornell University Press, 2006.

Konvitz, Josef W., *Cartography in France, 1660-1848: Science, Engineering, and Statecraft,* Chicago: University of Chicago Press, 1987.

Krogt, Peter van der, *Koeman's Atlantes Neerlandici,* 10 vols. 't Goy-Houten, Netherlands: HES Publishers, 1997.

Krygier, John B., "Cartography as an Art and a Science?" *The Cartographic Journal* 32 (1995): 3-10.

Lane, K. Maria D., "Mapping the Mars Canal Mania: Cartographic Projection and the Creation of a Popular Icon," *Imago Mundi* 58, no. 2 (2006): 198-211.

Lewis, G. Malcolm (ed.), *Cartographic Encounters: Perspectives on Native American Mapmaking and Map Use,* Chicago: University of Chicago Press, 1998.

MacEachren, Alan M., *How Maps Work: Representation, Visualization and Design,* New York: Guilford Press, 1995.

Mapline: A Quarterly Newsletter Published by the Hermon Dunlap Smith Centre for the History of Cartography, 1, March 1976, Chicago, Ill.

Mignolo, Walter D., *The Darker Side of the Renaissance: Literacy, Territoriality, and Colonization,* Ann Arbor: University of Michigan Press, 1995.

Monmonier, Mark, *How to Lie with Maps,* Chicago: University of Chicago Press, 1991; 2nd ed., 1996.

Monmonier, Mark, *Technological Transition in Cartography,* Madison: University of Wisconsin Press, 1985.

Mukerji, Chandra, "A New World-Picture: Maps as Capital Goods for the Modern World System," in *From Graven Images: Patterns of Modern Materialism,* 79-130. New York: Columbia University Press, 1983.

Mundy, Barbara E., *The Mapping of New Spain: Indigenous Cartography and the Maps of the Relaciones Geográficas,* Chicago: University of Chicago Press, 1996.

Nag, P. and M. Kudrat, *Digital Remote Sensing,* Concept Publishing Company, (1998).

Nag Prithvish (Ed.), *Thematic Cartography and Remote Sensing,* Concept Publishing Company, (2008).

Nag, P. & Smita Sengupta, *Introduction to Geographical Information System,* Concept Publishing Company, (2008).

Nag, Prithvish and Smita Sengupta, *Geographical Information System,* Concept and Business Opportunities, Concept Publishing Company, (2006).

Orlove, Benjamin, "Mapping Reeds and Reading Maps: The Politics of Representation in Lake Titicaca," *American Ethnologist,* 18, no. 1 (1991): 3-38.

Orlove, Benjamin, "The Ethnography of Maps: The Cultural and Social Contexts of Cartographic Representation in Peru," *Cartographica* 30, No. 1 (1993): 29-46.

Padrón, Ricardo, *The Spacious Word: Cartography, Literature, and Empire in Early Modern Spain,* Chicago: University of Chicago Press, 2004.

Pal, Saroj K., *Statistics for Geoscientists: Techniques and Application,* Concept Publishing Company, (1998).

Paulston, Rolland G., ed. *Social Cartography: Mapping Ways of Seeing Social and Educational Change,* Garland Reference Library of Social Science, 1,024; Reference Books in International Education, 36. New York: Garland, 1996.

Pedley, Mary, *The Commerce of Cartography: Making and Marketing Maps in Eighteenth-Century France and England,* Chicago: University of Chicago Press, 2005.

Petchenik, Barbara Bartz, "Maps, Markets and Money: A Look at the Economic Underpinnings of Cartography," *Cartographica* 22, no. 3 (1985): 7-19.

Pickles, John, "Texts, Hermeneutics and Propaganda Maps," In *Writing Worlds: Discourse, Text and Metaphor in the Representation of Landscape,* ed. Trevor J. Barnes and James S. Duncan, 193-230. London: Routledge, 1992.

Pickles, John, *A History of Spaces: Cartographic Reason, Mapping and the Geo-Coded World,* London: Routledge, 2004

Piper, Karen, *Cartographic Fictions: Maps, Race, and Identity,* New Brunswick, NJ: Rutgers University Press, 2002.

Ramaswamy, Sumathi, "Catastrophic Cartographies: Mapping the Lost Continent of Lemuria," *Representations,* no. 67 (1999): 92-129, and "History at Land's End: Lemuria in Tamil Spatial Fables," *Journal of Asian Studies* 59, no. 3 (2000): 575-602.

Ramaswamy, Sumathi, "Enshrining the Map of India: Cartography, Nationalism, and the Politics of Deity in Varanasi," In *Visualizing Space in Banaras: Images, Maps, and the Practice of Representation,* ed. Martin Gaenszle and Jörg Gengnagel, 165-88.

Ramaswamy, Sumathi, "Maps and Mother Goddesses in Modern India." *Imago Mundi* 53 (2001): 97-114.

Ramaswamy, Sumathi, "Visualizing India's Geo-Body: Globes, Maps, Bodyscapes," *Contributions to Indian Sociology,* 36, no. 1 and 2 (2002): 151-189.

Ramaswamy, Sumathi, *The Lost Land of Lemuria: Fabulous Geographies, Catastrophic Histories,* Berkeley: University of California Press, 2004.

Rampal, K.K., *Handbook of Aerial Photography and Interpretation,* Concept Publishing Company, (1999).

Rampal K.K., *Mapping and Compilation: Methods and Techniques,* Concept Publishing Company, (2008).

Ristow, Walter W., ed. *Guide to the History of Cartography: An Annotated List of References on the History of Maps and Mapmaking.* Washington, D.C.: Geography and Map Division, Library of Congress, 1973.

Robert A., "Mapping, Postmodernism, Indigenous People, and the Changing Direction of North American Cartography," *Cartographica* 28, no. 2 (1991): 1-12.

Robinson, Arthur H., and Barbara Bartz Petchenik, *The Look of Maps: An Examination of Cartographic Design,* Madison: University of Wisconsin Press, 1952. See also his, *Early Thematic Mapping in the History of Cartography,* Chicago: University of Chicago Press, 1982.

Robinson, Arthur H., and Barbara Bartz Petchenik. *The Nature of Maps: Essays Toward Understanding Maps and Mapping,* Chicago: University of Chicago Press, 1976.

Schein, Richard H., "Representing Urban America: 19th-Century Views of Landscape, Space, and Power," *Environment and Planning D: Society and Space* 11 (1993): 7-21.

Schulten, Susan, *The Geographical Imagination in America, 1880-1950,* Chicago: University of Chicago Press, 2001.

Schulz, Juergen, "Jacopo de' Barbari's View of Venice: Map Making, City Views, and Moralized Cartography before the Year 1500," *Art Bulletin* 60 (1978): 425-74.

Stone, Jeffrey C., "Imperialism, Colonialism and Cartography," *Transactions of the Institute of British Geographers* no 13 (1988): 57-64.

Turnbull, David, *Masons, Tricksters and Cartographers: Comparative Studies in the Sociology of Scientific and Indigenous Knowledge*, Amsterdam: Harwood Academic Publishers, 2000.

Turnbull, David. *Maps are Territories: Science is an Atlas: A Portfolio of Exhibits*. Geelong, Victoria: Deakin University, 1989. Reprinted, Chicago: University of Chicago Press, 1993.

Winichakul, Thongchai, "The Quest for 'Siwilai': A Geographical Discourse of Civilizational Thinking in the Late Nineteenth and Early Twentieth-Century Siam," *Journal of Asian Studies*, 59, no. 3 (2000): 528-49.

Winichakul, Thongchai, *Siam Mapped: A History of the Geo-Body of a Nation*, Honolulu: University of Hawaii Press, 1994.

Withers, Charles W.J., *Geography, Science and National Identity: Scotland since 1520*, Cambridge Studies in Historical Geography, 33. Cambridge: Cambridge University Press, 2001.

Wolff, Larry. *Inventing Eastern Europe: The Map of Civilization on the Mind of the Enlightenment*, Stanford: Stanford University Press, 1994.

Wood, Denis, "Pleasure in the Idea: The Atlas as Narrative Form," *Cartographica* 24, no. 1 (1987): *Monograph* 36, 24-45.

Wood, Denis, "The Fine Line between Mapping and Mapmaking," *Cartographica* 30, no. 4 (1993): 50-60.

Wood, Denis, "The Map as a Kind of Talk: Brian Harley and the Confabulation of the Inner and Outer Voice," *Visual Communication*, 1, no. 2 (2002): 139-61.

Wood, Denis, and John B. Krygier, eds. *Art and Cartography*, Special Issue of *Cartographic Perspectives*, no. 53. n.p.: North American Cartographic Information Society, 2006.

Wood, Denis, with John Fels. *The Power of Maps*, New York: Guilford, 1992. Wallis, Helen, and Arthur H. Robinson, eds. *Cartographical Innovations: An International Handbook of Mapping Terms to 1900*. Tring, Hertfordshire: Map Collector Publications for the International Cartographic Association, 1987.

Wood, Denis. "Cultured Symbols: Thoughts on the Cultural Context of Cartographic Symbols." *Cartographica* 21, no. 4 (1984): 9-37.

Wood, Denis. "Introducing the Cartography of Reality." In *Humanistic Geography: Prospects and Problems*, ed. David Ley and Marwyn S. Samuels, 207-19. Chicago: Maaroufa Press, 1978.

Wood, Denis. "Now and Then: Comparisons of Ordinary Americans' Symbol Conventions with Those of Past Cartographers." *Prologue* 9 (1977): 151-61.

Woodward, David, ed. *Art and Cartography: Six Historical Essays*, Chicago: University of Chicago Press, 1987.

Woodward, David, ed. *Five Centuries of Map Printing*, Chicago: University of Chicago Press, 1975.

Woodward, David, *Maps as Prints in the Italian Renaissance: Makers, Distributors and Consumers*, Panizzi Lectures, 1995. London: The British Library, 1996.

Woodward, David. "The Study of the History of Cartography: A Suggested Framework," *The American Cartographer* 1 (1974): 101-15.

Yonemoto, Marcia, *Mapping Early Modern Japan: Space, Place, and Culture in the Tokugawa Period (1603-1868)*, Berkeley: University of California Press, 2003. See also Yonemoto's "Silence without Secrecy? What is Left Unsaid in Early Modern Japanese Maps," *Early Modern Japan: An Interdisciplinary Journal*, 14 (2006): 27-39.

Index